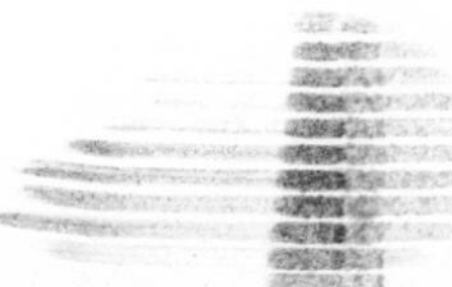

# $C_3$, $C_4$: Mechanisms, and Cellular and Environmental Regulation, of Photosynthesis

Myself when young did eagerly frequent
Doctor and Saint, and heard great Argument
About it and about; but evermore
Came out by that same Door as in I went.

# $C_3$, $C_4$: mechanisms, and cellular and environmental regulation, of photosynthesis

GERRY EDWARDS Professor, Department of Botany,
Washington State University

DAVID WALKER Professor of Biology,
Department of Botany University of Sheffield

University of California Press
BERKELEY AND LOS ANGELES

University of California Press
Berkeley and Los Angeles, California



**Library of Congress Cataloging in Publication Data**

Main entry under title:

$C_3$, $C_4$: mechanisms, and cellular and environmental regulation, of photosynthesis
1. Photosynthesis
I. Edwards, G. II Walker D.

QK882.1′3342 82-049-298

Printed in Great Britain

A Packard Publishing Limited book

Other titles in the biology series are:

Practical Invertebrate Zoology
edited by R. P. Dales

Mechanisms of Insect Behaviour
by P. E. Howse and P. L. Miller

Distributed in the United States of America and Canada by
University of California Press
Berkeley, California

# Contents

# Preface

As the title implies, this book seeks to set out the principal features of photosynthetic carbon assimilation in $C_3$ and $C_4$ species and in those which exhibit Crassulacean Acid Metabolism (CAM). It is aimed at university students and graduates at all levels and, for completeness, includes a concise and elementary description of the photochemical apparatus and the photochemical events which lead to the generation of assimilatory power (ATP + NADPH) [Part A, Chapters 1–5].

Section B goes on to consider the most important aspects of photosynthetic carbon assimilation including a number of topics which have only been elucidated in the last decade and are therefore inadequately covered in many contemporary texts. For example the chapters devoted to $C_3$ photosynthesis not only consider the essential features of the reductive pentose phosphate pathway but also aspects of its regulation, and its relationship to the movement of metabolites between the chloroplast and its cellular environment (Chapters 6–9). In addition photorespiration (Chapter 13) is covered in some detail. Similarly the treatment of $C_4$ photosynthesis, which is, in any case, a relatively 'new' topic includes sections on the formulation and function of the pathway, the compartmentation and regulation of enzymes, metabolite transport, photochemical requirements, taxonomic diversification and physiological implications (Chapters 10–12).

The discussion of Crassulacean Acid Metabolism (Chapter 15) covers its discovery, and its biochemical and ecological aspects. Chapter 16 covers photosynthesis in reproductive tissue and the diversity of malate metabolism in plants. General information on techniques of chloroplast isolation is provided (Appendix A) and enzyme nomenclature is given for enzymes in $C_3$, $C_4$ and Crassulacean Acid Metabolism (Appendix B).

The textbook is written from our personal point of view, describing photosynthetic carbon assimilation as it now stands. It is not written as a review since to do so would make the coverage of references and controversial issues unmanagable. References are often listed at the end of a subsection and specific citations are generally avoided within the text. Reviews are cited which will allow the student to go readily into greater depth in specific areas.

Throughout, we have aimed at the simplest possible approach and have not assumed that our readers will necessarily have more than an elementary knowledge of biology and the physical sciences. Although the book is primarily concerned with $CO_2$ assimilation, which is covered in some depth, many other aspects of photosynthesis are considered in sufficient detail for most non-specialist courses, of a broadly biochemical nature.

The reference numbers in the text for Chapters 6 to 16 are in ordinary type for general references and in bold type for specific citations.

# Acknowledgements

Alice Herold and Joyce Foster contributed extensively through discussions and reading of the text. We are grateful to a number of people for reading various chapters, for their criticisms and their suggestions. These include Randy Alberte, Mordhay Avron, Jim Barber, Doug Doehlert, Christine Foyer, Gary Harris, Ulrich Heber, Hans Heldt, Robert Hill, Peter Horton, Steve Huber, Maurice Ku, Richard Leegood, Joseph Neumann, Barry Osmond, Simon Robinson, Mark Schmitt, Marty Spalding, Hideaki Usuda, Klaus Winter. We would also like to thank Mark Schmitt for arranging some of the micrographs for publication and Kathryn Oldman for preparing much of the final typescript and, together with Shirley Walker, helping sanity prevail. We are grateful to Richard Walker for his original illustrations in Section A.

# List of Abbreviations

| | |
|---|---|
| ADP | Adenosine diphosphate |
| AMP | Adenosine monophosphate |
| APS | Apparent rate of photosynthesis |
| ATP | Adenosine triphosphate |
| $C_4$ pathway | $C_4$-dicarboxylic acid pathway |
| $C_4$ plants | Plants having the $C_4$ dicarboxylic acid pathway |
| $C_3$ plants | Plants fixing atmospheric $CO_2$ directly through the RPP-pathway |
| CAM | Crassulacean Acid Metabolism |
| CE | Carboxylation efficiency |
| $CF_0$ | Proton channel of coupling factor |
| $CF_1$ | ATP synthesizing component of coupling factor |
| Chl | Chlorophyll |
| DBMIB | Dibromothymoquinone-2,5dibromo-3-methyl-6-isopropyl-p-benzoquinone |
| DCMU | 3-(3′,4′-Dichlorophenyl)-1,1-dimethylurea |
| DCPIP(DPIP) | Dichlorophenolindophenol |
| DHAP | Dihydroxyacetone phosphate |
| DPGA | 1,3-diphosphoglycerate |
| DTT | Dithiothreitol |
| Fd | Ferredoxin |
| FDP | Fructose 1,6-bisphosphate |
| G3P | Glyceraldehyde 3-phosphate |
| G6P | Glucose 6-phosphate |
| hv | light (Planck's constant × velocity of radiation) |
| $NAD^+$ | Nicotinamide-adenine dinucleotide oxidized |
| NADH or $NADH_2$* | Nicotinamide-adenine dinucleotide reduced |
| NAD-ME species | NAD-malic enzyme species |
| $NADP^+$ | Nicotinamide-adenine dinucleotide phosphate oxidized |
| NADPH or $NADPH_2$* | Nicotinamide-adenine dinucleotide phosphate reduced |
| NADP-ME species | NADP-malic enzyme species |
| nm | nanometre, $10^{-9}$ metres |

* In the reduced state the nicotinamide dinucleotide coenzymes acquire two electrons and one proton (Section 5.15) and cannot, therefore, be easily abbreviated in an accurate manner (except, possibly, by the clumsy $NADH+H^+$). Accordingly, the simpler forms (NADH, NADPH) are preferred but $NADH_2$ and $NADPH_2$ have occasionally been used where it seemed desirably to remind the reader that, in principle these coenzymes participate in the transfer of hydrogen e.g., $H_2O \rightarrow NADP \rightarrow CH_2O$

| | |
|---|---|
| OAA | Oxaloacetate |
| *P680* | Reaction centre chlorophyll of PSII |
| *P700* | Reaction centre chlorophyll of PSI |
| $P/e_2$ | Ratio of ATP to 2 electrons |
| PEP | Phosphoenolpyruvate |
| PEP-CK species | Phosphoenolpyruvate carboxykinase species |
| 2-PGA | 2-phosphoglycerate |
| PGA, 3-PGA | 3-phosphosphoglycerate |
| Pi | Orthophosphate, inorganic phosphate |
| PMF | Proton motive force |
| PMS | N-methylphenazonium methosulfate |
| PSI, PSII | Photosystem I and II |
| PPi | Inorganic pyrophosphate |
| Q | Primary electron acceptor of PSII, Quencher. |
| QY | Quantum Yield |
| $r_a$ | Boundary layer resistance to diffusion of $CO_2$ |
| $r_a'$ | Boundary layer resistance to diffusion of water |
| RBP | Ribulose 1,5-bisphosphate |
| $r_m$ | mesophyll resistance to $CO_2$ fixation |
| RPP-pathway | Reductive pentose phosphate pathway |
| $r_s$ | Stomatal resistance to diffusion of $CO_2$ |
| $r_s'$ | Stomatal resistance to diffusion of water |
| TCA cycle | Tricarboxylic acid cycle |
| TPS | True rate of photosynthesis |
| X | The primary electron acceptor of PI |
| $\delta^{13}C$ | A measure of carbon isotope composition |
| $\Gamma$ | $CO_2$ compensation point |

# PART A

# Chapter 1
# Introduction



## 1.1 What is it?

Photosynthesis is the process in which light energy is used to drive the conversion of carbon dioxide to sugars. It is often represented by Equation 1.1 or some similar equation which summarizes the same essential features

$$CO_2 + H_2O \xrightarrow{nh\nu} CH_2O + O_2 \qquad \text{Eqn. 1.1}$$

In this equation two symbols are used. Light energy is represented by the letters $nh\nu$, for reasons which will be described later (Section 3.5) and '$CH_2O$' is put into italics to emphasize that it is not a real compound but something with the same general structure as a carbohydrate. Sometimes the whole equation is multiplied by 6 so that the non-existent $CH_2O$ becomes $C_6H_{12}O_6$. This has the advantage that it makes $CH_2O$ into a real sugar (glucose, for example, is one of several sugars which has the empirical formula $C_6H_{12}O_6$) but since glucose is not an important photosynthetic product it is important to bear in mind that $C_6H_{12}O_6$ is, in this context, just as much a symbol or term of convenience as $CH_2O$ and that other organic carbon compounds are formed in secondary reactions (Section 12.9). The important point is that $CO_2$ has been converted into some sort of carbohydrate by a process of reduction (Section 1.3) in which $H_2O$ serves as a hydrogen donor and oxygen is evolved. This, in turn, brings us immediately to a consideration of two fundamental aspects of physical chemistry.

## 1.2 Breaking bonds

In order to convert $CO_2$ to $CH_2O$ in Equation 1.1, it is necessary to remove $H_2$ from $H_2O$ and this involves breaking H – O bonds. Energy is always required to break

[H—O] → [H] + [O]

Light is used to break H—O bonds.

chemical bonds and always released when bonds are formed. In photosynthesis light energy is used to bring about the photolysis (light-splitting) of water and the hydrogen is passed to the $CO_2$ which is accordingly 'reduced' to the level of carbohydrate.

## 1.3 Oxidation and reduction

Reduction is the opposite of oxidation. In every day usage an oxidation is regarded as a process in which oxygen from the atmosphere is combined with something. When fuels such as oil, gas or coal are burnt they are oxidized. Most natural fuels contain various impurities but charcoal is mostly carbon and its oxidation may be represented by Equation 1.2.

$$C + O_2 \rightarrow CO_2 \qquad \text{Eqn. 1.2}$$

Chemists also apply the term oxidation to a reaction such as that in Equation 1.3 in which a compound $XH_2$ becomes oxidized by transferring $H_2$ to an acceptor (A) which becomes reduced.

| $XH_2$ | + | A | → | $AH_2$ | + | X | Eqn. 1.3 |
|---|---|---|---|---|---|---|---|
| (hydrogen donor) | | (hydrogen acceptor) | | (reduced acceptor) | | (oxidized donor) | |

Some compounds do not have hydrogens which they can donate to acceptors but only electrons. Equation 1.4 describes the oxidation of the ferrous ion which becomes oxidized and gains a positive charge by donating an electron (which carries a negative charge) to an acceptor.

| $Fe^{2+}$ | + | A | → | $Fe^{3+}$ | + | $A^-$ | Eqn. 1.4 |
|---|---|---|---|---|---|---|---|
| (ferrous ion) | | (electron acceptor) | | (ferric ion) | | (reduced acceptor) | |

In some circumstances a molecule may also undergo an excitation process (e.g. the excitation of chlorophyll by light)

$$Chl \xrightarrow{h\nu} Chl^* \rightarrow Chl^+ + e^-$$

and this also involves a transient and short-lived movement of an electron which can be represented as follows

$$\oplus \rightarrow e^- \qquad \text{Eqn. 1.5}$$

**Table 1.1** Oxidation and reduction summarized

| Oxidation | Reduction |
|---|---|
| Addition of oxygen | Removal of oxygen |
| Removal of hydrogen | Addition of hydrogen |
| Removal of electrons | Addition of electrons |
| Displacement of electrons (away from) | Displacement of electrons (towards) |

(Displacement of electrons is a common feature)

as though the removal of the negatively charged electron leaves behind a positively charged hole. All of these examples (and all other oxidation/reductions) share a common feature (Table 1.1). In each case an electron is (or electrons are) moved from the thing which is being oxidized towards the thing which is being reduced. A hydrogen atom is comprised of a proton plus an electron (Eqn. 1.6)

$$H = H^{+} + e^{-} \qquad \text{Eqn. 1.6}$$

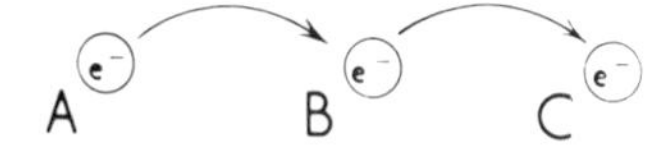

Oxidation involves electron transport.

so it is evident that hydrogen transfer cannot occur without electron transfer. The movement of electrons in Equation 1.2 is less obvious. It occurs because oxygen is an element which attracts electrons more strongly than carbon and accordingly the electrons in $CO_2$ are displaced, to some extent, towards the oxygen (which is reduced) and away from the carbon (which is oxidized).

$$C + O_2 \rightarrow O \; C \; O \qquad \text{Eqn. 1.7}$$

In all of these examples it can be seen that the oxidation (electron loss) is accompanied by a reduction (electron gain), even if this occurs within different parts of one molecule.

## 1.4 Photosynthesis as an oxidation/reduction

The basic equation for photosynthesis (Eqn. 1.1) is very useful as a simple summary but it is misleading in one important respect. It seems to imply that some of the $O_2$

which is evolved is derived from the $CO_2$. This is because it has been over-simplified. In fact, all of the $O_2$ comes from the water and the photolytic process (Eqn. 1.8) involves the transfer of H to an acceptor (or oxidant) which then donates this H to the $CO_2$ (Eqn. 1.9)

$$2H_2O + 2A \xrightarrow{nh\nu} 2AH_2 + O_2 \qquad \text{Eqn. 1.8}$$

$$CO_2 + 2AH_2 \rightarrow CH_2O + H_2O + 2A \qquad \text{Eqn. 1.9}$$

Together Equations 1.8 plus 1.9 become Equation 1.10

$$CO_2 + 2H_2O \xrightarrow{nh\nu} CH_2O + O_2 + H_2O \qquad \text{Eqn. 1.10}$$

which constitutes a more meaningful representation of photosynthesis but the extra molecule of $H_2O$ is usually cancelled out on both sides of the equation in order to arrive at the briefest possible equation (Eqn. 1.1). As these equations show, photosynthesis is an oxidation/reduction reaction. Water is oxidized by the removal of hydrogen and oxygen is released. Carbon dioxide is reduced to the level of carbohydrate.

## 1.5 Winding the biological mainspring

In order to define photosynthesis more fully than in Equation 1.1 we have been obliged to consider the concept of oxidation/reduction and the fact that energy is required to split chemical bonds. This also enables us to see that photosynthesis is much more than the source of organic carbon (although, as we shall see, this is hardly an aspect which can be easily dismissed). Thus photosynthesis uses light energy to wind the biological mainspring by splitting H—O bonds. Almost all other biological processes lead ultimately to the restoration of these bonds and depend on the energy

Light winds the biological mainspring.

which is released by their reformation. In this sense photosynthesis is concerned with the splitting of H—O bonds and respiration is concerned with their re-establishment (Fig. 1.1). In this regard man also occupies a special ecological niche because unlike most other species he consumes fossil fuels in order to meet his energy requirements. In so doing he re-establishes H—O bonds which were originally broken by primordial photosynthesis.

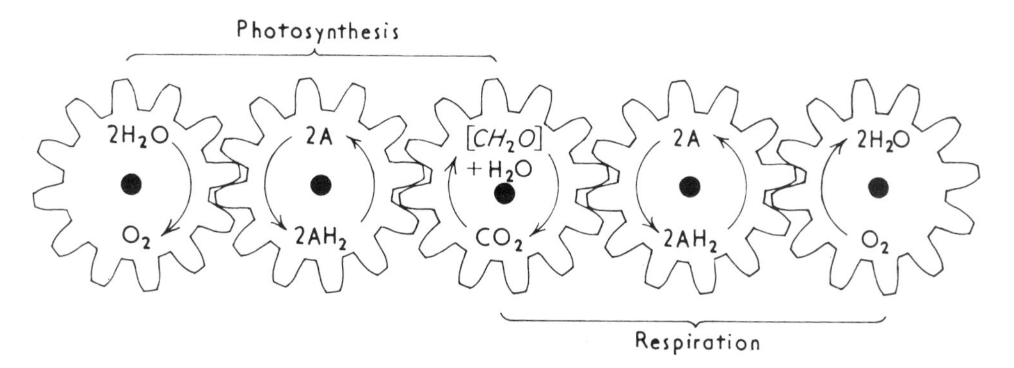


**Fig. 1.1.** In photosynthesis H—O bonds are broken, $O_2$ is evolved, and $CO_2$ is reduced to $CH_2O$. In respiration $CH_2O$ is oxidized, $CO_2$ is released and H—O bonds are re-established.

## 1.6 Photosynthesis as the source of organic carbon

Some informed sources think that we are wrong to burn oil because it is too valuable a source of organic carbon to squander in this fashion. The point which is made is a valid one and we tend to forget that if we could solve our fuel energy requirements by harnessing the power of sea waves or creating nuclear fusion on earth we would still be left with a need for organic carbon. Practically everything that we see about us has involved photosynthesis at some stage or other. The gardener often talks about 'feeding' plants when he applies fertilizers and the notion that plants derive their nourishment from the soil is one that is commonly held. They do not. Plants take up minerals from the soil, they derive their nourishment from the air. That plants are not made of soil was shown in the seventeenth century by van Helmont who planted a slip of willow in carefully weighed soil and watered it with rain water (Fig. 1.2). After 5 years the plant had gained 164 lb and the soil had lost 2 oz. Being an honest man van Helmont ascribed the latter to experimental error and concluded that the willow had grown entirely at the expense of the water. Today we can improve on this interpretation. Most of the fresh weight of a plant is water from the soil. Much of its dry weight is carbon and oxygen, derived from atmospheric $CO_2$ (Table 1.2). Only a

**Fig. 1.2.** Van Helmont waters the tree.

**Table 1.2** Chemical composition of leaves (percentage dry weight)

| | |
|---|---|
| C | about 45% |
| O | about 45% |
| H | about 5% |
| N | often 1–3% |
| P, S, K, Na, Ca, Mg and trace elements (together with N) | usually 1–5% |

little, usually about 1% or less, is mineral. The mineral constituent is, of course, essential but it does not contribute much to the mass of the plant.

When we look at living organisms we are looking at the products of photosynthesis. Virtually all of the things that man surrounds himself with are also either products of photosynthesis or have involved fossil fuels in their manufacture. This is true of natural and artificial fibres, plastics, steel, furnace-fired pots and brick and so on. At present, the world's oil reserves are thought to be sufficient for about 50 years and its coal reserves for 200–300 years. During this period it seems inevitable that man must move towards a society which relies on contemporary photosynthesis rather than past photosynthesis, in the shape of fossil fuels, for its elaborate carbon. Modern agricultural practice alone would scarcely be sufficient for this purpose because of its heavy reliance on fossil fuels, i.e. there would be little point, in most circumstances, in expending more energy to produce a crop than the crop actually yielded in terms of its chemical energy content.

## 1.7 Analogy and epitaph

Some of the essential features of photosynthesis can be illustrated by analogy. Almost everyone who has had lessons in elementary physical science will have seen a demonstration of electrolysis, in which water is split into hydrogen and oxygen. This is achieved by taking advantage of the fact that water dissociates, to a very small extent into hydrogen ions ($H^+$) and hydroxyl ions ($OH^-$). More complex ions may also be formed but in principle this reaction may be represented by Equation 1.11 and it proceeds until there is one $H^+$ and one $OH^-$ for every 554 million molecules of $H_2O$ [This is the basis of pH as a measure of acidity. The symbol 'p' means 'the negative logarithm of' and, accordingly, (since there are $10^{-7}$ moles of $H^+$ per litre of water) the negative logarithm of the hydrogen ion concentration is 7. In acid solutions there are more hydrogen ions so that pH is less than 7].

$$H_2O \rightleftharpoons H^+ + OH^- \qquad \text{Eqn. 1.11}$$

Water will continue to dissociate according to Equation 1.11 if the products of the reaction are removed. In practice this is done by introducing two platinum electrodes (Fig. 1.3) which are joined by wires to a chemical cell or battery. Within

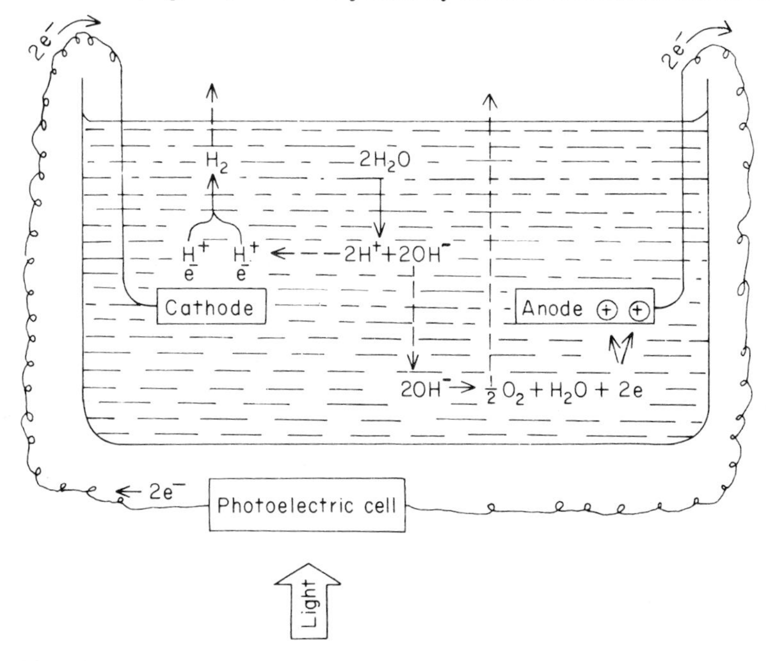


**Fig. 1.3.** Photolysis of water in a non-biological model. In conventional electrolysis the photoelectric cell is replaced by a chemical cell or battery. As illustrated, the photoelectric (solar) cell plays a role which is similar to that of the chloroplast in photosynthesis.

this cell, oxidation/reduction reactions result in electron transport. The electrons flow through the wires to the cathode which becomes negatively charged and from the anode (vacating positively charged holes) into the battery. (Corresponding changes occur within the battery itself). Because 'unlike' charges attract, hydrogen and hydroxyl ions migrate to the cathode and anode respectively. At the cathode the protons ($H^+$) accept electrons and $H_2$ is evolved.

$$2H^+ + 2e^- \rightarrow H_2 \qquad \text{Eqn. 1.12}$$

At the anode, hydroxyl ions donate electrons into the positively charged holes and $O_2$ is formed.

$$2OH^- \rightarrow H_2O + 1/2O_2 + 2e^- \qquad \text{Eqn. 1.13}$$

In this way electrical energy is used to provide 'sinks' for $H^+$ and $OH^-$ so that an unfavorable equilibrium is displaced and $H-O$ bonds are broken.

If both gases are collected in one vessel and a spark is passed, the energy which has been trickled into the system, probably over a period of several hours, can be recovered instantaneously as hydrogen oxygen bonds are reformed with explosive violence.

$$H_2 + 1/2O_2 \rightarrow H_2O + \text{Energy} \qquad \text{Eqn. 1.14}$$

Hydrogen-oxygen bonds are re-established.

In recent years, space exploration has furthered the development of solar (photoelectric) cells, which can deliver electrons in sufficient number and at a sufficiently high potential to bring about electrolysis of water (Fig. 1.3). The basic principles are the same as before but this time the primary act is excitation by light. This causes electrons to flow from the solar cell and into the vacant holes thus created. The full sequence of events now involves the conversion of light energy to electrical energy to chemical energy. As in photosynthesis, the light energy initiates electron transport and this brings about the dissociation of H—O bonds.

'Photoelectrolysis' (Fig. 1.3), as it could be called, is evidently very similar to photosynthesis in many ways. In both processes light excitation leads to electron transport and this, in turn, to the dissociation of H—O bonds. In most circumstances photosynthesis does not bring about the simultaneous evolution of $O_2$ and $H_2$ (although chloroplasts can be made to evolve either $O_2$ or $H_2$ in the presence of suitable electron acceptors or donors). Instead 'hydrogen' is passed, through a series of carriers to $CO_2$ which becomes reduced (Section 5.9). The product $CH_2O$ is both a chemical energy store and the fabric from which living organisms are made. Respiration releases energy from these reserves by allowing hydrogen to reunite with oxygen but it does it piecemeal so that opportunities for conserving energy (as ATP) are allowed and the violence of direct re-combination is avoided. Similarly, in respiration, $CO_2$ is returned to the atmosphere so that the entire cycle may be repeated. Virtually all of the carbon in our bodies has already passed through the process of photosynthetic carbon assimilation on many previous occasions and, in the fullness of time, it will inevitably start upon its journey once again.

'In the fullness of time'.

[All references for Part A are given at the end of Chapter 5].

# Chapter 2
# Energy and Laws



## 2.1 The laws of thermodynamics

In Chapter 1 it is claimed that photosynthesis winds the biological mainspring, that energy originates in the sun and that photosynthesis is the means of making this energy available to living organisms. If photosynthesis is to do with energy on the one hand and carbon on the other we need to consider both more fully. This requires a passing acquaintance with the laws of thermodynamics and since C. P. Snow held that anyone with any claim to being a reasonably educated person should be familiar with these laws, it seems that they might also be worth considering in their own right. Happily, natural laws of this sort are simply expressions of human experience and therefore we can phrase them more or less as we wish. For biological purposes they have been stated as follows (more formal definitions are given in Scheme 2.1):

First Law: you can't win
Second Law: you can't break even
Third Law: you can't stay out of the game.

Expressed in these terms they have a nice philosophical ring and could be said to summarize not only the world of energy but much of the human condition. In more mundane terms they provide the basis for a lot of useful arithmetic and predictions. (Scheme 2.1) The first law states that energy (like mass) can be neither created nor

**Scheme 2.1**

| | |
|---|---|
| First Law | Energy can be neither created nor destroyed.<br>The energy of the universe is constant. |
| Second Law | The entropy of the universe always increases.<br>There is a universal tendency to chaos and disorder. |
| Third Law | The entropy of a perfect crystal of any element or compound is zero at absolute zero temperature. |

destroyed. It can pass between the system (Section 2.2) and its surroundings and the different forms of energy (heat, mechanical, electrical etc.) may be interconverted but there can be no net increase or decrease. When temperature, pressure and volume are constant the second law can be summarized by Equation 2.1.

$$\Delta H = \Delta E = \Delta F + T\Delta S \qquad \text{Eqn. 2.1}$$

In this equation $\Delta$ is a symbol which means 'the change in', $E$ is the total energy of the system, $F$ is the free energy, $T$ is the absolute temperature ($K$ or $^\circ C + 273$) and $S$ is the entropy (Section 2.3).

## 2.2 The system

The 'system' referred to above (Section 2.1) may be anything the investigator chooses (such as a chloroplast, a cell, an organism or a society) but, whatever it is, it has to be set in the framework of its surroundings and ultimately these surroundings extend to the limits of the known universe.

[Strictly, thermodynamic principles apply to *closed* systems (i.e. those which do not exchange matter with their surroundings) and quite clearly a chloroplast, a cell or an intact plant is an *open* system which may enter a steady-state in which synthesis (anabolism) is exactly balanced by consumption (catabolism) but is unlikely ever to attain true thermodynamic equilibrium. Nevertheless classic equilibrium thermodynamics offer a simplified, idealized basis for the analysis of biological processes which will often suffice until 'irreversible' or 'non-equilibrium' thermodynamics can be developed and applied.]

## 2.3 Entropy ($S$)

This is an elastic and somewhat romantic concept. Entropy (designated '$S$') describes the state or conditions of matter in regard to the random motion of the atoms and molecules of which it is composed. For example, as the temperature of water increases, or as it changes state from ice to liquid or from liquid to gas, this motion becomes more vigorous and less ordered. Entropy is therefore a measure of disorder and the second law supposes a universal tendency towards disorder. Herein lies the notion of cosmic futility or entropic doom. Only within a crystal at absolute zero does order reign supreme in the absence of thermal movement. Fleetingly, living organisms can reverse the general trend towards the random state by becoming more organized at the expense of their surroundings. In the end however there is death and decay, ashes to ashes and dust to dust.

In Eqn. 2.1 (which assumes constant temperature, volume and pressure) entropy change ($\Delta S$) is always positive in any real process (Section 2.4) and $T\Delta S$ is that fraction of the total energy change which is unavailable for useful chemical work.

## 2.4 Free energy

Biological reactions are often carried out at constant temperature and pressure and it is here that the fraction $\Delta F^\circ$ (the standard free energy change) in Eqn. 2.1 is of prime importance. In this context $F^\circ$ (sometimes written as $G$ after Gibbs who first introduced it) represents the fraction of the total energy $E$ which is available for useful chemical work. The *standard* free energy change ($\Delta F^\circ$ or $\Delta G^\circ$) is the gain or loss of free energy (at 25° C and pH 7.0) as 1 mole of reactant is converted to 1 mole of product. It should not be confused with the actual change ($\Delta F$ or $\Delta G$) which varies according to the conditions under which the reaction occurs. [It should be noted that the biological standard (usually written as $\Delta F'$ or $\Delta G'$) is different from the physical standard ($\Delta F^\circ$ or $\Delta G^\circ$). The physicists place everything, including hydrogen ion concentration, on a molar basis but the biologist prefers a neutral pH (i.e. $H^+$ at $10^{-7}$ g ions/l) as a more appropriate reference point. Biological standards are sometimes based on atmospheric $CO_2$ (0.03 %) for the same reason. It should also be emphasized that the values used and quoted are approximate, based as they are on a great many assumptions and on data which can not always be derived with great accuracy.]

The term $\Delta F$ in Equation 2.1 represents that fraction of the total energy change ($\Delta E$) which can perform useful chemical work. In real processes (as opposed to partial reactions of an endergonic or energy-requiring nature which can not proceed in isolation but only when linked to other partial reactions of an exergonic or energy-yielding nature) the free energy always decreases as the system moves towards equilibrium, whereas the entropy always increases. The advantage of $\Delta F$ over $\Delta S$ in predicting the feasibility of a chemical reaction is due to the fact that for the former it is sufficient to know the changes in the *system*, whereas for the latter changes in the *system* and the *environment* should be known.

## 2.5 Free energy and equilibria

The extent and sign of the free energy change allows us to predict the extent and direction of chemical reaction. These matters are governed by the law of mass action which states that if A and B react to give C + D the initial rate of the forward reaction in any given circumstances is maximal at the outset and related to concentration (depicted by [A], [B] etc.) as follows.

**Scheme 2.2**

---

For a reaction: $A + B \underset{k_2}{\overset{k_1}{\rightleftharpoons}} C + D$
Velocity of forward reaction $= V_1 = [A] \times [B] \times k_1$
Velocity of back reaction $= V_2 = [C] \times [D] \times k_2$
(where $k_1$ and $k_2$ are rate constants)
Similarly the rate of the back reaction is initially zero but as C and D are formed and grow in concentration its rate increases (as the rate of the forward reaction falls) until at equilibrium the two rates are equal and opposite. Then

$[A] \times [B] \times k_1 = [C] \times [D] \times k_2$

or $\frac{k_1}{k_2} = K = \frac{[C] \times [D]}{[A] \times [B]}$

(where $K$ is called the Equilibrium Constant)

---

From Scheme 2.2 it will be seen that if the reaction goes virtually to completion in the forward direction (so that [A] and [B] fall to very low values) $K$ will be very large. On the other hand if [A] and [B] are equal to [C] and [D] at equilibrium the value of $K$ would be 1 and if the equilibrium position favours the conversion of C and D to A and B the value of $K$ will be less than 1. In this regard the $\Delta F'$ value is a disguised equilibrium constant and is indeed related to $K$ by Equation 2.2.

$$\Delta F' = -RT \ln K \qquad \text{Eqn. 2.2}$$

[where $R$ is a factor called the gas constant, $T$ is the temperature in degrees absolute ($^\circ C + 273$) and $\ln K$ is the natural logarithm log to the base $e$ of $K$]. For a temperature of 25°C, if the values of $R$ and $T$ are inserted in this equation and, at the same time, log to the base $e$ is converted to log to the base 10, it becomes

$$\Delta F' = -1.36 \log K \qquad \text{Eqn. 2.3}$$

The standard free energy change is a measure of the difference between the standard conditions and the equilibrium state (where $\Delta F$ is zero). It is usually expressed in kcal or J (2.6). If, for example, $\Delta F'$ happened to be $-1.36$ kcal it would follow from Equation 2.3 that $\log K$ would equal $\frac{-1.36}{-1.36}$ (i.e. 1) and since the antilog of 1 is 10, the value of $K$ would be 10. Similarly, if $\Delta F'$ were $-2.72$, $\log K$ would be 2 and $K$ would be 100. From Table 2.1 it can therefore be seen that a large negative $\Delta F'$ value is equivalent to a large positive $K$ value and that both indicate that a reaction will go a long way in the forward direction before equilibrium is reached. Neither say anything about the likelihood of the reaction occurring. This is governed by other factors (including the presence or absence of a catalyst). For example, the energy

**Table 2.1.** Relationship between $K$ and $\Delta F'$ (From $\Delta F' = -RT \ln K$)

| $\Delta F'$ (kcal) | $K$ | Type of reaction |
|---|---|---|
| +6.81 | 0.00001 | very difficult |
| +5.45 | 0.0001 | difficult, unfavourable |
| +4.09 | 0.001 | can be pushed fairly |
| +2.73 | 0.01 | easily |
| +1.36 | 0.01 | |
| 0.00 | 1.0 | freely reversible |
| −1.36 | 10 | |
| −2.73 | 100 | reasonably favourable |
| −4.09 | 1000 | to very favourable |
| −5.45 | 10 000 | |
| −6.81 | 100 000 | goes virtually to completion |

Biological reactions (other than photochemical reactions) with $\Delta F'$ values much greater than 7–8 kcal are virtually irreversible

released when hydrogen combines with oxygen is about −56 kcal (−235 kJ)

$$H_2 + 1/2O_2 \rightarrow H_2O \qquad \text{Eqn. 2.4}$$

but the appropriate mixture of gases can be stored indefinitely without detectable change. If a spark is passed however there is a violent explosion as energy is released.

In theory, all reactions, even ones like the combination of hydrogen and oxygen, are reversible. If this were not so, photosynthesis (dependent on the photolysis of water) would not be possible. Similarly an unfavourable reaction can sometimes be pulled in a forward direction by an appropriate sink which removes one or more products (e.g. by physical translocation or by consumption in a subsequent reaction). In practice the limit of biological feasibility lies at about 8 kcal (33 kJ) for most reactions and the achievements of photochemistry are the exception rather

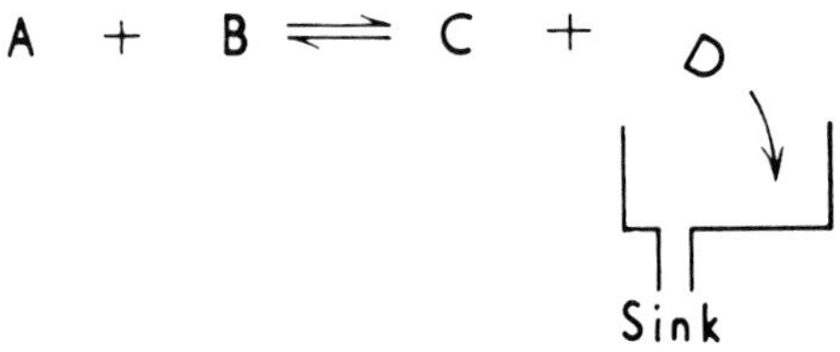

$$A + B \rightleftharpoons C + D$$

Sink

A reaction may be pulled in a forward direction by a sink.

than the rule. It should also be noted that reactions with large positive $\Delta F$ values (*end*ergonic or energy-requiring reactions as opposed to *ex*ergonic or energy-yielding reactions) do not 'go' at all, in a practical sense, unless they are linked to a reaction with a sufficiently large negative $\Delta F$ value to ensure that the net change in free energy is also negative (Section 2.4).

## 2.6 Energy units

In recent times there has been a move towards the general acceptance of S.I. units (Système d'Unités International) in the measurement of energy, light, distance etc. This has some attractive features particularly in regard to its use of multiples of 1000 with common prefixes (e.g. $k = 10^3$, $m = 10^{-3}$, $\mu = 10^{-6}$, $n = 10^{-9}$ etc) so that we can, for example, express metres, grams and moles in the same basic terminology. However, the S.I. system is not used exclusively in this account for two reasons. The first is that all the old texts and literature employ the old terminology and the contemporary reader will, therefore, need to be familiar with the old and the new for many years to come. The second is that the old system contained some useful reference points. It may be old-fashioned to measure distance in feet but we carry two around with us. This permits an immediate appreciation of the size of the unit. Much the same argument can be applied to our concept of energy units. The S.I. unit is the joule. This, in turn, is based on the dyne and the erg. [The dyne is the force

Joules are a girl's best friend.

which, when exerted on a mass of 1 g, produces an acceleration of 1 cm/s and the erg is the amount of energy delivered by 1 dyne acting through 1 cm. Ten million ergs equal 1 joule (or 1 erg = $10^{-7}$ J).] The joule is a very useful unit, not least because it is so readily converted to watts (1 watt = 1 J/s) but in some respects it is less easily referred to something tangible like the calorie. Moreover, the large calorie (when dieticians talk about 'calories' they really mean 'large calories' or kcal) is still the most common term in many areas of biological energetics. Accordingly, the calorie will usually be preferred. [One calorie (or gram calorie) is the amount of heat needed to raise the temperature of one gram of water 1°C (from 14.5–15.5°C). There are approximately 4.2 J (4.1855) to the calorie.]

## 2.7 The energy required for the formation of carbohydrate

If 180 g (one mole) of glucose is burned in a calorimeter the heat output ($\Delta H$) is 673 kcal. At constant temperature the standard free energy change $\Delta F'$ for the oxidation of glucose has been calculated to be $-686$ kcal (and since under these conditions $\Delta H = \Delta E$ it follows from Equation 2.1 that the fraction $T\Delta S = +13$ kcal).

In this and many similar reactions (at constant temperature) the difference ($T\Delta S$) between the accurately measured $\Delta H$ and the calculated $\Delta F'$ is a convenient approximation of $+112$ kcal ($673 \div 6$) for the formation of the hypothetical carbohydrate $CH_2O$ from $CO_2$ and $H_2O$ according to Equation 1.1. Strictly speaking, however, the higher value of 114–115 (based on 686) should be used and this may rise to about 120 in circumstances in which it is necessary to take into account the actual rather than the standard free energy changes. Reluctance to switch from the lower value stems from the fact that a single calorimetric measurement is accurate whereas the higher value is less precise (based as it is on a number of measurements and assumptions) and, for most purposes, the difference between the two is relatively unimportant.

## 2.8 Bond energies

Similar values to those mentioned above can be derived from empirical bond energies. These are based on the heat of formation of various compounds but again difficulties arise from the fact that bond energies are not absolute but vary in relation to the molecular environment in which they exist. For example, less energy is required to break the first H—O bond in water than the second. Similarly a double bond, C=O, in an acid is 'stronger' than two single C—O bonds (Table 2.2) and the C = O bonds in $CO_2$ are stronger than in an aldehyde $-C{<}^{H}_{O}$. One consequence of this variation is that a value derived from the heat of formation of one compound cannot be applied to another with a very high degree of accuracy and, within limits, it is possible to select from a range of values those which best suit a particular case.

**Table 2.2**

| Empirical bond energies | (kcal/mol) |
|---|---|
| C–H | 99 |
| C–C | 83 |
| C–O (in alcohol) | 77 |
| C=O (in acid) | 180 |
| C=O (in $CO_2$) | 190 |
| H–O | 110 |
| O=O | 118 |
| H–H | 104 |

*Carbohydrate Formation (Eqn. 1.1)*

| Bonds broken | Bonds formed |
|---|---|
| +190 | −83 |
| +190 | −99 |
| +110 | −77 |
| +110 | −110 |
| | −118 |
| +600 | |
| | −487 |
| Difference = +113 kcal | |

Granting the resulting loss of objectivity it is nevertheless comforting to arrive at approximately the same answer whether this is based on the oxidation of glucose (above) or bond energies derived from quite different measurements. Thus, applying the values shown in Table 2.2, and assuming that the hypothetical $CH_2O$ compound would, in reality, be one of a number of such units joined by C–C bonds (as in glucose, sucrose or starch) the energy requirement (Eqn. 2.5) is again approximately 112 kcal (470 kJ)

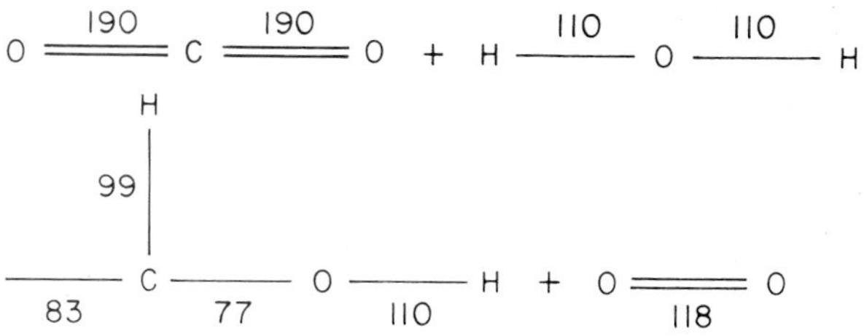

$$O \overset{190}{=\!=} C \overset{190}{=\!=} O + H \overset{110}{—} O \overset{110}{—} H$$

$$\underset{83}{—} C(\overset{99}{|}H) \underset{77}{—} O \underset{110}{—} H + O \underset{118}{=\!=} O \qquad \text{Eqn. 2.5}$$

## 2.9 Splitting water

The same sort of exercise allows an estimate of the total energy associated with the formation and dissociation of water. As already shown H–O bonds in water are broken in photosynthesis and reformed in respiration so that this also permits an estimate of the total energy changes relating to hydrogen transfer (electron transport) in these processes (Equation 2.6).

$$\mathrm{H} \text{---} \mathrm{H} + \tfrac{1}{2}(\mathrm{O} \overset{118}{=\!=} \mathrm{O}) \longrightarrow \mathrm{H} \overset{110}{\text{---}} \mathrm{O} \overset{110}{\text{---}} \mathrm{H} \qquad \text{Eqn. 2.6}$$

| *Bonds broken* | | *Bonds formed* |
|---|---|---|
| 1/2 (118) = | 59 | 2(110) = −220 |
| | 104 | |
| | +163 | |

Difference = −57 kcal (−240 kJ)

In the reverse direction +57 kcal are needed (Fig. 4.13) to split $H_2O$, as a vapour, into $H_2$ and $O_2$. In most circumstances, however, the photosynthetic process does not lead to the formation of hydrogen gas as well as the formation of oxygen. Instead hydrogen is transferred from water to the coenzyme $NADP^+$ which functions as the terminal oxidant in the photosynthetic electron transport system (Section 4.9).

Similarly, in respiration, hydrogen is transferred from NADPH (or the related coenzyme NADH) to oxygen so that water is reformed. The free energy change associated with the reduction of NADP (or the reoxidation of NADPH) is about 52 kcal.

## 2.10 Resonance

Reference has already been made to the fact that bond energies may be different in different circumstances; that is they may change in relation to their molecular environment. This is partly because of resonance. For example, in Equation 1.7, $CO_2$ was represented thus to emphasize the electron attracting quality of oxygen. This sort of representation also shows how C and O share electrons in such a way that each acquires 8 within its sphere of influence.

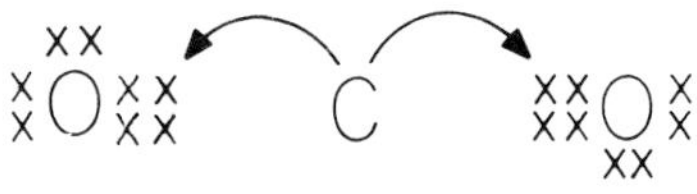

```
  XX                        
X     XX              XX    X
X  O  XX      C       XX  O X
  
                           XX
```

[Oxygen needs 2 electrons to complete its full set of 8 in this way and is therefore divalent, whereas carbon needs 4 and is tetravalent.] This sharing of two pairs of electrons between C and each O constitutes a double bond so that $CO_2$ is usually written as O=C=O if it is felt necessary to indicate its bond structure in a simple

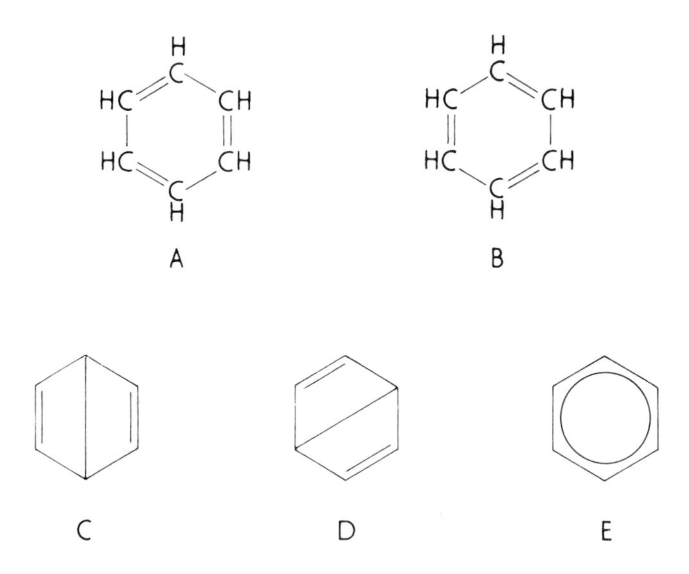


**Fig. 2.1.** Different ways of representing the benzene ring.

fashion. In the same way the benzene ring which contains six carbons, can be written as shown in Fig 2.1.

In form A it will be seen that all the valence requirements (H is monovalent and C is tetravalent) are met if there is an arbitrary alternation of single and double bonds between the carbon atoms in the ring. However, form B (or the abbreviated forms C and D) would be equally appropriate and, in recent times, form E has been preferred because it does not specify precisely where the double bonds lie. In short, C and D represent extremes of structure whereas E is something between the two. In fact, E is a much better picture of the real situation than either C or D because measurements of the distances between the carbon atoms in the ring show that the bonds are really neither single nor double bonds but something between the two. In other words the real structure (for which E is the best representation in these rather inadequate terms) lies between the extremes defined by A and B or C and D. [In terms of wave mechanics, atomic orbitals are designated *s*, *p* etc. and further defined by quantum numbers. The C–C and C–H bonds in benzene are formed from $sp^2$ hybrid orbitals. This leaves each carbon with an unpaired electron in the $2p_z$ orbital which overlaps with the corresponding orbital for adjacent carbons. Accordingly the combination of pairs of atomic $p_z$ orbitals does not lead to $\pi$ orbitals localized between 2 carbon atoms but rather to completely delocalized $\pi$ orbitals embracing all 6 carbons.] Moreover, in a real sense, the benzene ring gains strength and stability from this resonance. It no longer has relatively strong links (double bonds) alternating with relatively weak links (single bonds) but uniformity of bonding between all the constituent atoms. Similarly $CO_2$ may be written as

```
    XX                                                         XX
   XOXX   C   XXOX   but also as   XOX      C     XOX
   XOXX       XXOX                 XOX            XOX
               XX                   XX

       XX                                 XX         XX
 as   XOX   C   XX                       XOXX   C   XOX
      XOX       XXOX   or as             XOXX       XOX
       XX       XX                        XX         XX
```

As before, these resonance possibilities shorten the distance between carbon and oxygen and make $CO_2$ a particularly stable and well adjusted molecule.

[In terms of wave mechanics, hybrid orbitals ($sp_x$) and $\pi$ bonds are also probably involved in the formation of $CO_2$ (and indeed the involvement of $\pi$ bonds is implicit in the structures above) but if the molecule is described in terms of localized orbitals it is still necessary to invoke resonance in order to account for its axial symmetry and its high degree of stability.]

## 2.11 The free energy of hydrolysis of adenosine triphosphate (ATP)

Photosynthetic electron transport normally leads to the generation of reducing power in the form of ATP and $NADPH_2$. ATP (adenosine triphosphate) is often referred to as an energy rich compound and it is relevant in this context to ask why. As we have seen, energy is always required to break chemical bonds and if a reaction is energy producing it is sometimes possible to determine how much energy is released simply by adding up all the energy expended in breaking bonds and subtracting the total from the sum of all the energy derived from bond formation. If this is attempted for the hydrolysis of ATP (Fig. 2.2) it can be seen that no net change would be predicted because O—P bonds and O—H bonds are broken and re-formed in equal numbers.

Actual measurements, however, show that there is a standard free energy change of approximately 7 kcal ($\Delta F' = -7$ kcal). This is again (Table 2.2) because the bond energies vary according to their molecular environment. Like $CO_2$, a molecule of inorganic phosphate (orthophosphate or Pi) is extremely stable because of resonance. In ATP many of these resonance possibilities are denied because there is a tendency for electrons to be pulled equally in two directions at the same time

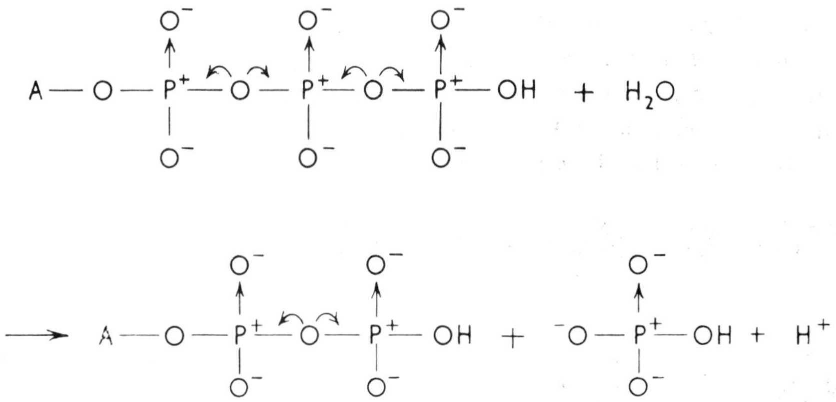

$$A—O—P^{+}(O^{-})(O^{-})—O—P^{+}(O^{-})(O^{-})—O—P^{+}(O^{-})(O^{-})—OH \; + \; H_2O$$

$$\longrightarrow A—O—P^{+}(O^{-})(O^{-})—O—P^{+}(O^{-})(O^{-})—OH \; + \; {}^{-}O—P^{+}(O^{-})(O^{-})—OH \; + \; H^{+}$$

**Fig. 2.2.** Hydrolysis of ATP to ADP and Pi

(opposing resonance). In addition, ionization and the development of fractional positive and negative charges within the polyphosphate part of the molecule lead to the juxtaposition of electrical forces which oppose each other because they are of the same sign (Fig. 2.2). The situation which arises is roughly analogous to three bar magnets tied together by two lengths of string and to a piece of wood by a third. If these were arranged on a flat surface so that their north and south poles were adjacent (and prevented from lifting upwards from the surface by a sheet of glass) they would push away from each other to the limits of the string. If the first bit of string were then cut, the first magnet would leap away (Fig. 2.3). Similarly, the second

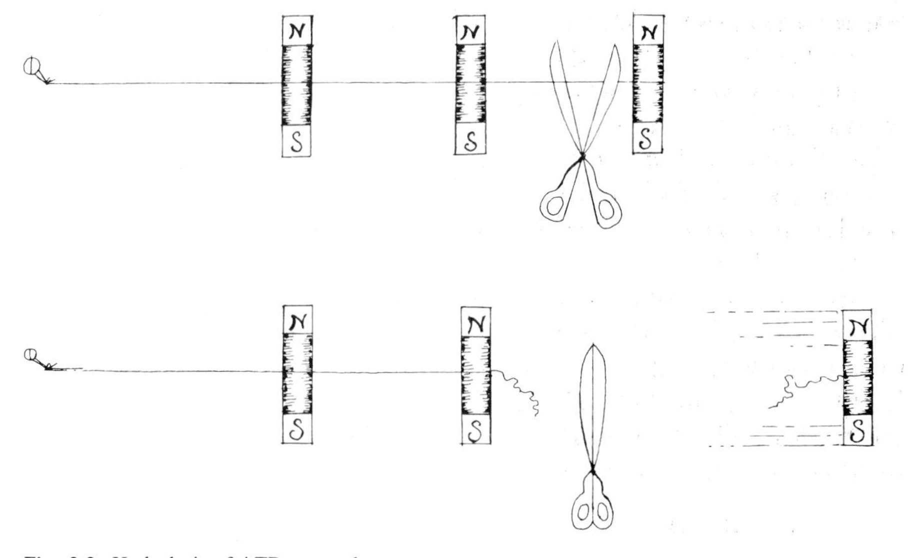


**Fig. 2.3.** Hydrolysis of ATP, an analogy.

magnet would be propelled away if the second bond were severed but when the third string was cut the third magnet would rest, unmoved, next to the wood. The nature of the bonds (string) would be the same in each case except that they would be placed under different stresses according to their magnetic environment. For analogous reasons ATP may be regarded as a marriage of partners in perpetual conflict. Divorce (hydrolysis) then generates heat (releases energy) and leads to a more stable situation (increased resonance in the products). The $\Delta F'$ for the hydrolysis of ATP to ADP is about $-7$ kcal ($-29$ kJ) and a similar amount of energy is released when ADP is hydrolysed to AMP. Conversely, further hydrolysis of AMP releases much smaller quantities of energy because although the bonds involved are the same the molecular environment has been modified by the removal of the other phosphate groups.

## 2.12 ATP as a component of 'assimilatory power'

The photochemical events in photosynthesis lead to the generation of 'assimilatory power' (ATP + NADPH) which Arnon and his colleagues showed (Fig. 2.4) to be necessary for the conversion of $CO_2$ to $CH_2O$. For purposes of arithmetic it is permissible to assume that in a reaction such as Equation 2.7, ATP contributes as much energy as it would if it were hydrolysed

$$\text{PGA} + \text{ATP} \rightarrow \text{DPGA} + \text{ADP} \qquad \text{Eqn. 2.7}$$

but this does not imply that ATP is actually hydrolyzed or that energy can exist as an independent entity which can be transferred without direct participation of the reactants (for further details of this reaction see Section 6.11 and Fig. 9.6). In the Reductive Pentose Phosphate Pathway (RPP pathway) or Benson–Calvin Cycle (Chapter 6) ATP is consumed at two points (of which Equation 2.7 is one) and there is only one reductive step (in which $NADPH_2$ serves as the hydrogen donor or reductant). Three molecules of ATP and two molecules of $NADPH_2$ are consumed for each $CO_2$ reduced. The standard free energy of hydrolysis of ATP is about $-7$ kcal (the *actual* free energy change may be as much as $-12$ due to the differences in concentration and the presence of Mg which binds with ATP and ADP to different extents, thereby altering the values). On this basis the energy contribution made by ATP ($3 \times 7 = 21$ kcal) is really much less than that made by $NADPH_2$. The latter serves as a hydrogen donor just as ATP serves as a phosphate (Pi) donor. The free energy change associated with the transfer of hydrogen from $NADPH_2$ to DPGA (the reductive step in the RPP pathway—Eqn. 6.17) is not in itself so large as that ($-52$ kcal, cf. Section 2.9) associated with the transfer of hydrogen from $NADPH_2$ to oxygen. In the sense, however, that the transfer process could continue until the

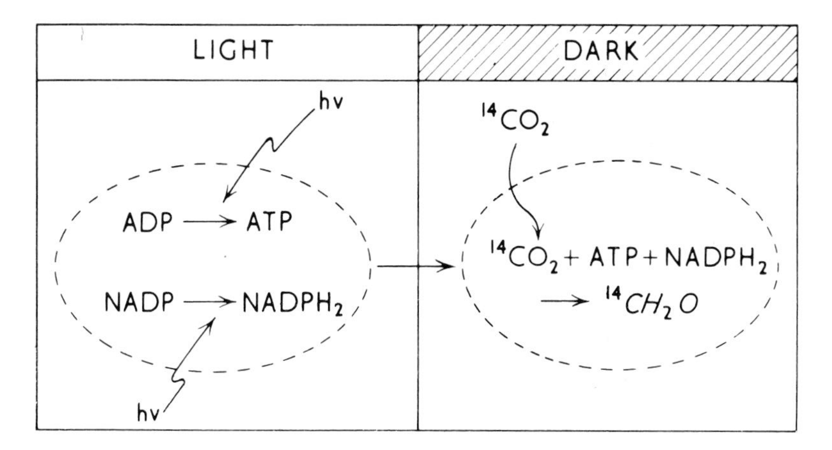


**Fig. 2.4.** Experimental separation of photosynthesis into two phases. On left illuminated thylakoids give rise to ATP and $NADPH_2$. On right enzymes of the chloroplast stroma use ATP and $NADPH_2$ to reduce (radioactive) $CO_2$ to $CH_2O$ in the dark.

hydrogen was recombined with $O_2$ (as it is in respiration) there is a potential injection of 2 × 52 kcal as 2 molecules of $NADPH_2$ are utilized in the reduction of $CO_2$. On this basis 3 ATP + 2 $NADPH_2$ are 'worth' 21 + 104 = 125 kcal and with due regard to the fact that many of these values are imprecise it is evident that the actual process of carbon assimilation (with its basic requirement for 112 kcal) proceeds with a much higher degree of efficiency than the generation of assimilatory power (Chapter 5).

## 2.13 Redox potentials

As discussed above, the extent and sign of the free energy change can be used to predict the extent and direction of a reaction. Similarly, compounds capable of oxidation/reduction may be arranged in a 'league table' on the basis of their potential ability to react with one another. Some examples of redox potentials are listed in Table 2.3. In a table of this sort those compounds at the top of the league (those with the largest negative values) are able to reduce those below (with smaller negative or positive values). Some potentials are influenced by pH so that although they are all related to the arbitrary zero of the standard hydrogen electrode (which is at pH 0) they are usually adjusted according to the relationship

$$E' = E^\circ - 0.06\,\mathrm{pH} \qquad \text{Eqn. 2.8}$$

(where $E^\circ$ is the standard state, at pH 0 and $E'$ is the value at the desired pH). Thus at

**Table 2.3.** Some Oxidation/reduction potentials

| *Couple* Reduced | Oxidized | Redox potential, (E′, in mV) |
|---|---|---|
| $Fe^{2+}-S$ | $Fe^{3+}-S$ | −600 to −700 |
| $Ferredoxin^{2+}$ | $Ferredoxin^{3+}$ | −430 |
| $H_2$ | $2H^+$ | −420 |
| $NADPH + H^+$ | $NADP^+$ | −320 |
| malate | oxaloacetate | −165 |
| Q red | Q oxid | −100 |
| cytochrome $f^{2+}$ | cytochrome $f^{3+}$ | +300 to +370 |
| $plastocyanin^+$ | $plastocyanin^{2+}$ | +390 |
| *P700* | *P700*$^+$ | +430 to +530 |
| $H_2O$ | $\frac{1}{2}O_2$ | +815 |

For further details of some of the electron carriers listed see Chapter 4.

pH 7.0 the value for $H_2/2H^+$ is decreased from zero by $7 \times 0.06$ to become −0.420 volts. Where necessary, the free energy values ($\Delta F'$) can be derived from the redox potential by the following equation

$$\Delta F' = nf\Delta E' \qquad \text{Eqn. 2.9}$$

[where n = the number of electrons transported and $f$ = the value of the Faraday (= 23)]. For example, it is evident from Table 2.3 that $H_2$ will reduce $O_2$ ($H_2 + 1/2O_2 \rightarrow H_2O$) and if the appropriate redox values (−0.420 and +0.815) are subtracted and inserted in Equation 2.9, together with the values of n and $f$, it becomes

$$\begin{aligned}\Delta F' &= -2 \times 23 \text{ kcal/volt} \times [0.815 \text{ volts} - (-0.420 \text{ volts})] \\ &= -\underline{56.8 \text{ kcal}} \qquad \text{Eqn. 2.10}\end{aligned}$$

In regard to the injection of energy into the RPP pathway (which occurs at 3 points) it should be noted that the 10 remaining reactions run 'down hill'. The cycle (Chapter 6) revolves, like a hoop, under the impact of blows delivered by a stick. Most of the blows (all of the NADPH and two thirds of the ATP) serve to bring about the reduction of PGA to triose phosphate. In effect 5/6 of this triose phosphate is reoxidized to PGA (Chapter 6) to maintain the momentum of the revolving cycle.

# Chapter 3
# Energy and Light



## 3.1 Where it all starts

The ultimate source of the energy utilized in photosynthesis is the sun where the extremely high temperatures permit nuclear fusion of hydrogen and the consequent emission of energy. That fraction of the energy which reaches the surface of this planet at wavelengths between 400 and 700 nm (see below) is called 'visible' light. The limits of visibility (Fig. 3.1) are set by the human eye which sees best in the green part of the spectrum and is unable to detect light much below 400 nm (the upper limits of

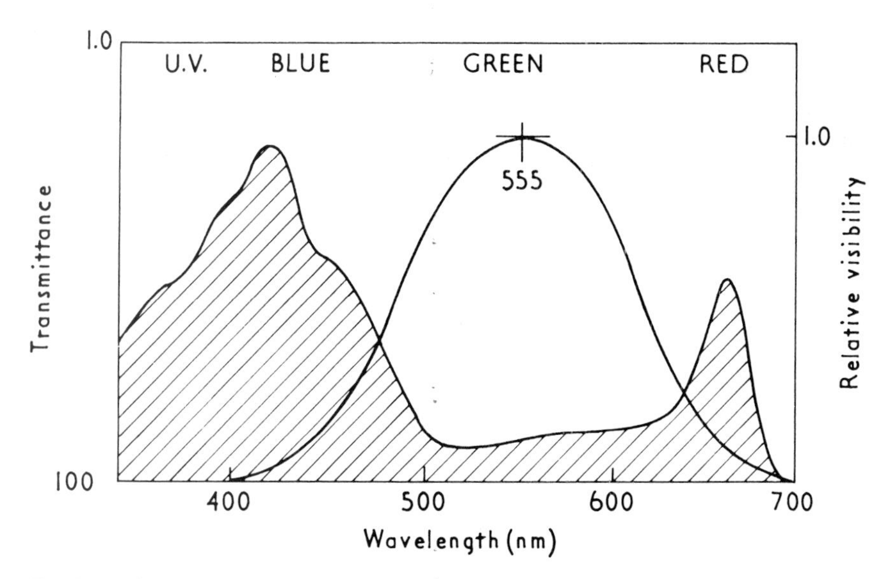


**Fig. 3.1.** What man sees and what the green plant 'sees'.
By definition 'visible' light is that perceived by the human eye (wavelengths between about 400 and 700 nm with a perception maximum at about 555 nm). The green plant also uses visible light in photosynthesis but chlorophyll absorbs best in the blue and the red.

the ultraviolet) or much above 700 nm (the lower limit of the near infra red). [1 nm = 1 nanometre = 1/1 000 000 000 metres = $1 \times 10^{-9}$ metres. This (S.I.) unit (Section 2.6) is now usually preferred to the millimicron (m$\mu$) or Angstrom (Å) which were often used to record distance or lengths of this sort of size. 1 nm = 1 m$\mu$ = 10 Å].

In a sense, the green plant is like the human eye because it is also largely incapable of utilizing energy which falls outside the above limits. Unlike the human eye, however, the green plant is less effective in its use of green light.

## 3.2 Light as a waveform

In some ways light behaves as though it is a moving force-field which has both a magnetic and electrical component. These may be regarded as waves which rise and fall together but in planes which are at right-angles to one another. The shorter the distance ($\lambda$) between the crests the greater their frequency ($\nu$) and their energy content. Other forms of electromagnetic radiation include $\gamma$-rays and X-rays, which have much shorter wavelengths (higher-frequency) than light and radiant heat and radio waves which have longer wavelengths (the latter often measured in metres rather than nanometres). (Figs. 3.2 and 3.3)

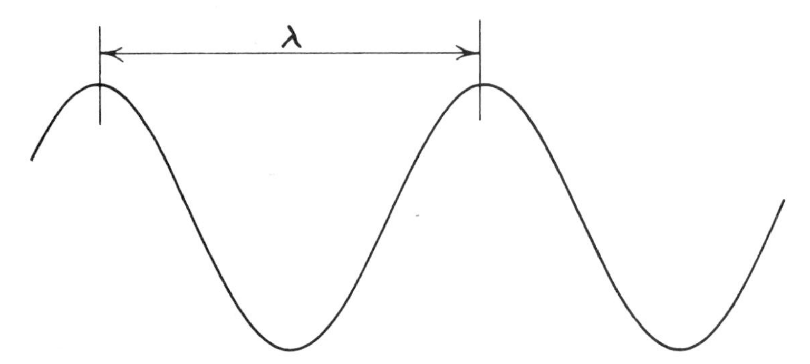


**Fig. 3.2.** Light as a waveform. The wavelength is the distance between peaks.

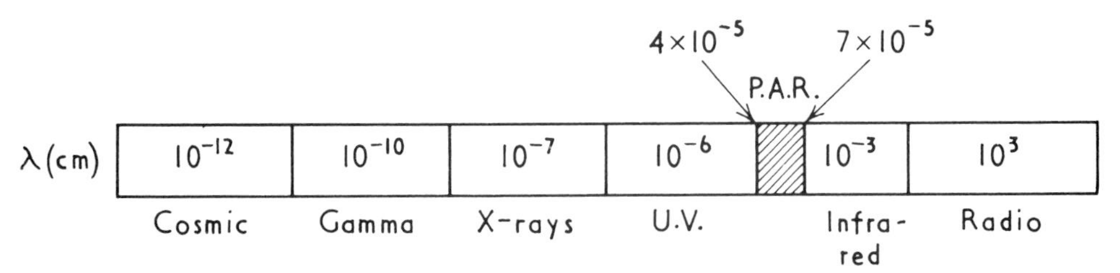


**Fig. 3.3.** Electromagnetic radiation. Photosynthetically active radiation (PAR) accounts for only a very narrow waveband.

The relationship between wavelength ($\lambda$), frequency ($v$) and the speed of light may be expressed as

$$\lambda \text{ (nm)} \times v \text{ (per second)} = c \text{ (nm per second)} \quad \text{Eqn. 3.1}$$

Alternatively, the frequency may be obtained by dividing the speed of light (which is constant in a vacuum) by the wavelength ($v = c/\lambda$).

## 3.3 Light as a stream of particles

In other ways, light also behaves as though it were a stream of particles. Since both Newton and Einstein took this view we might feel safe in describing light solely in this fashion but our acceptance of this concept need not imply rejection of the other (above). When good experimental evidence seems to point in opposite directions it often turns out that both directions are right rather than one being wrong. For the present at least we must therefore accept the fact that our understanding of the nature of light is imperfect and continue to use these different concepts where they are most useful.

## 3.4 Quanta

The particles of light are called quanta or photons. In accordance with the idea of light as a waveform their energy content is related to wavelength (the closer the waves the higher the energy). Accordingly, blue light at, say, 450 nm has more energy than red light at 650 nm.

## 3.5 The energy of light

The energy content of one photon of light at a given wavelength may be found by multiplying the frequency ($v$) by a value known as Planck's Constant ($h$).

$$E = hv \quad \text{Eqn. 3.2}$$

and it is for this reason that one photon of light energy is often represented by the letters '$hv$' and a number ($n$) of quanta by '$nhv$' (as in Equation 1.1). Since $v$ is also equal to the speed of light ($c$) divided by the wavelength ($\lambda$) this important equation can also be written

$$E = \frac{hc}{\lambda} \quad \text{Eqn. 3.3}$$

Although chlorophyll absorbs blue photons as well as red it is usual to select a wavelength near to the red absorption peak of chlorophyll (say 680 nm) when calculating the maximum amount of energy available for photosynthesis. This is because the extra energy of the blue photons cannot be usefully employed (Section 4.16). The speed of light in a vacuum is $3 \times 10^{10}$ cm/s (or $3 \times 10^{17}$ nm/s) and the value of Planck's constant is $6.6 \times 10^{-27}$ erg sec so that, for red light at 680 nm, $E = \frac{hc}{\lambda}$ becomes

$$E = 6.6 \times 10^{-27} \text{ erg sec} \times \frac{3 \times 10^{17} \text{ nm}}{\text{sec}} \times \frac{1}{680 \text{ nm}}$$

$$= 2.9 \times 10^{-12} \text{ ergs} \qquad \text{Eqn. 3.4}$$

In Section 2.6 it was seen that there are $10^7$ ergs to the joule and 4.2 joules to the calorie so that if we wish to express this value in calories we get

$$E = \frac{2.9 \times 10^{-12} \text{ ergs}}{4.2 \times 10^{7} \text{ ergs/cal}} = 6.9 \times 10^{-20} \text{ cal} \qquad \text{Eqn. 3.5}$$

which is evidently a very small fraction indeed. The mass of a single molecule of oxygen ($O_2$) is a correspondingly small fraction of a gram and to make it a more convenient size we can multiply it by Avogadro's number (the number of atoms in a gram molecule). One gram molecule or mole of $O_2$ then weighs 32 g. Similarly

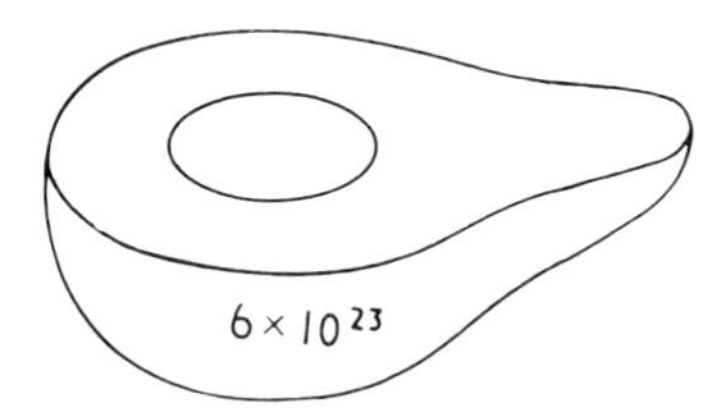


Half an Avogadro Pair.

Avogadro's number (approximatély $6 \times 10^{23}$) of red photons (at 680 nm) has an energy content of

$$6.9 \times 10^{-20} \text{ cal} \times 6 \times 10^{23} = 42 \text{ kcal (approx.)} \qquad \text{Eqn. 3.6}$$

[Avogadro's number ($N$) of photons (mole of photons) is sometimes referred to as an 'Einstein' but, as this term is also sometimes understood to mean the *energy content* of a mole of photons, we shall simply refer to the number of photons (quanta) on a molar basis. For example in Table 3.1 we refer to '$\mu$ mole quanta' rather than 'microeinsteins' in order to avoid this ambiguity].

**Table 3.1.** Full sun-light (Global Irradiance) expressed in several ways.

| Illuminance | | | Photosynthetically active component | | | | |
|---|---|---|---|---|---|---|---|
| | | | per square metre per second | | | | |
| lux | foot candles | Watts per square metre | ergs | joules | cals | photons | $\mu$ mole quanta |
| 100 000 | 10 000 | 500 | $5 \times 10^9$ | 500 | 120 | $1.3 \times 10^{21}$ | 2200 |
| Photometric | | radiometric (400–700 nm) | | | | quantum (400–700 nm) | |

This table shows the approximate equivalence between full sunlight (expressed, at left, as illuminance) and the energy content of *its photosynthetically active component*, (right) assuming a mean wavelength of 575 nm. The last line defines the type of measurement.

N.B. All of the above are measures of *flux density*. 'Flux' is the rate of flow of a substance expressed in units of quantity/time. 'Flux density' is flux through a unit surface area (i.e. quantity time$^{-1}$ unit (area$^{-1}$). A watt (1 J/s) is a measure of flux and the corresponding flux density term is *irradiance* (W/m$^2$ or J s$^{-1}$sec$^{-1}$ m$^2$). Thus photosynthetically active *radiation* (PAR) is simply radiation in the 400–700 nm waveband whereas photosynthetic *irradiance* (PI) is PAR/unit time/unit area. [Although commonly used in this way in plant studies it is not strictly correct to describe light energy arriving at a surface in terms of 'light intensity'. Intensity is a property of the radiating *source* e.g. a standard candle *emits* a flux of 1 candle power or 4 $\pi$ lumens. The corresponding measure of flux density or light *received* (Section 3.6), at the surface of a sphere of 1 foot radius with 1 standard candle at its centre is 1 lumen/ft$^2$ or 1 ft candle—Fig. 3.4].

[1 lux = 1 lumen/m$^2$, 1 ft. candle = 1 lumen/ft$^2$, 1 phot = 1 lumen/cm$^2$, therefore 1 lux = $1 \times 10^{-4}$ phot = 0.1 milliphots = 0.0929 ft candle. 1 ft candle = 1 lumen/ft$^2$ = 1.0764 milliphots = 10.764 lumens/m$^2$ = 10.764 lux. In S.I. units the standard candle is replaced by the candela which, acting as a point source, similarly emits $4\pi$ (= 12.566 lumens)].

For some purposes it is instructive to convert ergs to electron volts (by dividing by a factor of $1.6 \times 10^{-12}$) and we then find that 1 photon (at 680 nm) can accelerate 1 electron through a potential of 1.8 volts i.e.

$$\frac{2.9 \times 10^{-12}}{1.6 \times 10^{-12}} = \frac{2.9}{1.6} = 1.8 \text{ eV} \qquad \text{Eqn. 3.7}$$

## 3.6 Light intensity

The energy which enters a leaf not only depends on the quality of the illuminating light but also on how much there is of it (its intensity). If we think of light as a waveform the distance between peaks ($\lambda$) would determine the quality, whereas the height of the waves (the amplitude) would determine the intensity. If we regard it as a stream of particles the incident energy is determined by the energy content of each particle (photon or quantum) and the rate at which these particles arrive at the leaf surface. (see Section 3.2)

The human eye is very effective when given the task of comparing low light intensities but a very poor instrument for determining intensities in the absence of a standard for comparison. This is because the iris diaphragm closes in bright light to protect the retina and the brain can only make very rough allowances (based on experience) of how much light is getting through in given circumstances. For this reason we are mostly surprised when we learn for the first time that the intensity in a well lit room may be 50 times less than full daylight.

Light intensity was originally related to the output of a standard candle (Fig. 3.4). Such a candle emits a luminous flux of 1 candle power or $4\pi$ lumens so that if it is placed at the centre of sphere of radius = 1 foot the density of flux at the inner surface is 1 lumen/$ft^2$. If the radius of the sphere were increased to 1 metre the

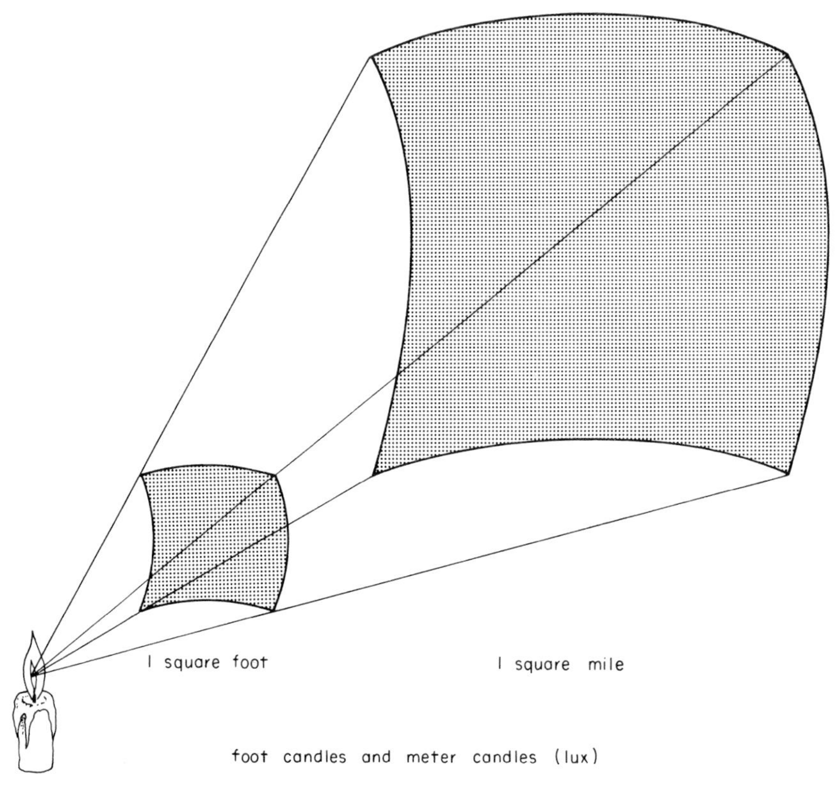


**Fig. 3.4.** Light Intensity in terms of candles. A standard candle illuminates the inner surface of a sphere of radius of 1 foot with a flux density of 1 lumen/$ft^2$ (equal to 1 foot candle). If the radius is 1 metre, the flux is 1 lumen/$m^2$ (or 1 metre candle or 1 lux or $1 \times 10^{-4}$ phots).

intensity would fall to 1 lumen/$m^2$ (1 metre candle or 1 lux). Since the decrease in flux is proportional to the square of the distance from the source (and since there are 3.2808 feet to the metre) 1 foot candle = $3.2808^2$ lux = 10.8 lux.

### 3.7 Sunlight and candles

Measuring light in foot candles or lux is useful for everyday purposes but it is inexact in the determination of energy requirements because anything other than a standard candle will emit light with a different spectral composition. Bearing these limitations in mind, we can still arrive at some rough approximations. Full summer light at sea level (direct light from the sun, at noon, plus the diffuse blue light from the sky) is approximately equivalent to 10 000 ft candles. The total energy delivered (including the infra-red) is about 100 ergs $cm^{-2}\ s^{-1}$ ft $candle^{-1}$ but since only about 50 % of the total ('global irradiance') lies in the visible region (Fig. 3.5) the true energy equivalence from photosynthetically active radiation (PAR-see Table 3.1) falls to about 50 ergs/ft candle. (Fig. 3.5). This allows us to 'rate' 1 $m^2$ (10 000 $cm^2$) of the earth's surface in full sunlight in the same way as we would an electric light or electric radiator.

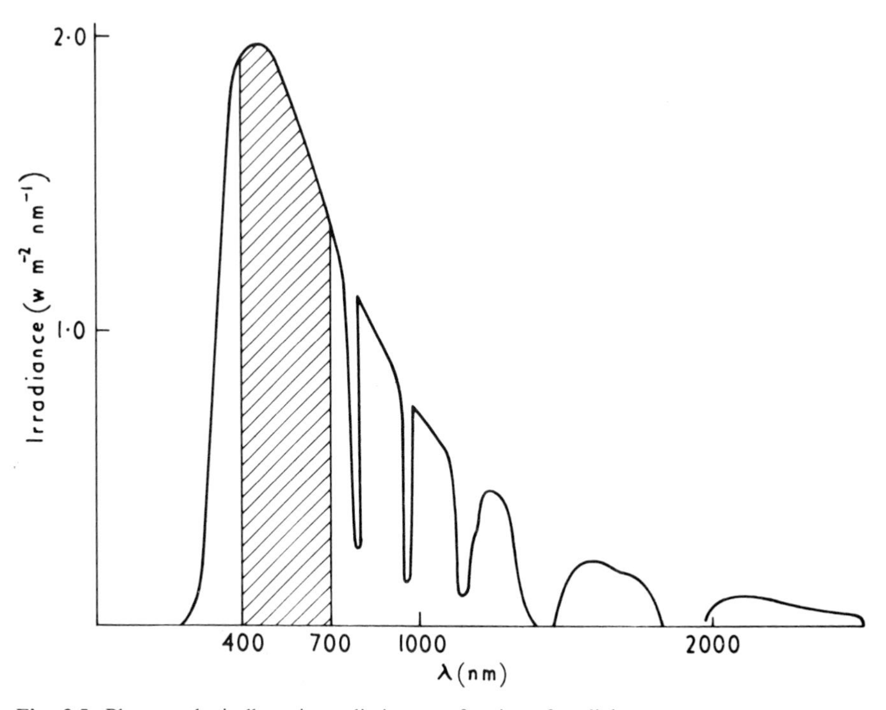


**Fig. 3.5.** Photosynthetically active radiation as a fraction of sunlight.

Thus, 1ft candle $= 50\,\text{ergs cm}^{-2}\,\text{s}^{-1}$

$= 50 \times 10^4$ ergs $\text{m}^{-2}\,\text{s}^{-1}$

and full sunlight ($1 \times 10^4$ ft candles)

$= 50 \times 10^8$ ergs $\text{m}^{-2}\,\text{s}^{-1}$

and dividing by $10^7$ to convert ergs to joules

$= 50 \times 10 = 500\,\text{J m}^{-2}\,\text{s}^{-1}$

$= 500\,\text{W/m}^2$ Eqn. 3.8

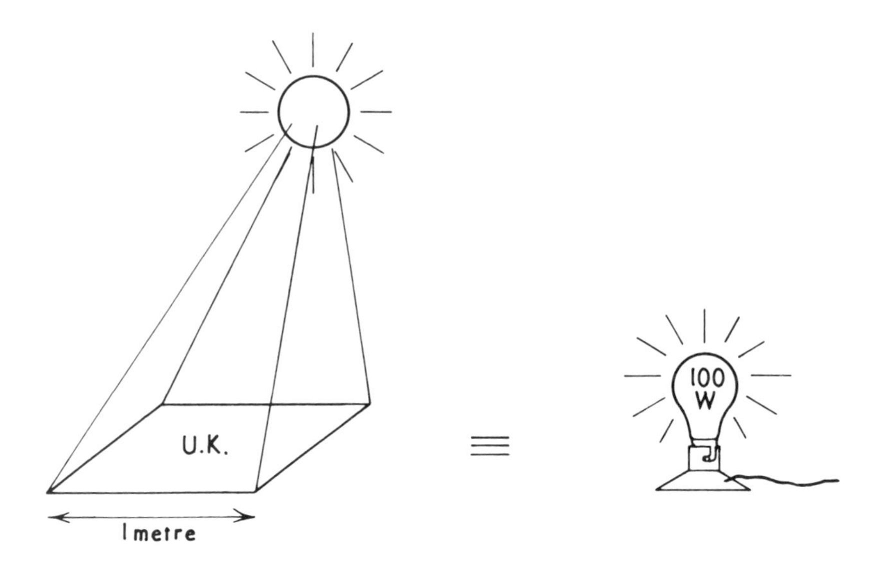


The mean annual irradiance for the U.K. is about 100 $\text{W/m}^2$.

Although sunlight, as perceived by the human eye, is a mixture of all the colours of the rainbow (i.e. light between about 400 and 700 nm) the intensity maximum is in the green at about 575 nm (close to the peak of visual perception at about 550 nm). One photon of 575 nm light has an energy content of $3.5 \times 10^{-12}$ ergs (obtained by substituting 575 for 680 in Equation 3.4) so that if 10 000 ft candles is roughly equivalent to $5 \times 10^9$ ergs $\text{m}^{-2}\,\text{s}^{-1}$ the number of photons in full sunlight is approximately

$$\frac{5 \times 10^9}{3.5 \times 10^{-12}} = 1.4 \times 10^{21} \text{ photons m}^{-2}\,\text{s}^{-1} \qquad \text{Eqn. 3.9}$$

As we have seen, a 'mole of quanta' (mole quanta or quantum mole) is often a more convenient measure than a single quantum or photon and

$$1.4 \times 10^{21} \text{ photons} = \frac{1.4 \times 10^{21}}{6.3 \times 10^{23}} \text{ mol quanta}$$

$$= 2.2 \times 10^{3}\ \mu\text{mol quanta m}^{-2}\ \text{s}^{-1} \qquad \text{Eqn. 3.10}$$

Thus full sunlight gives approximately 2 200 $\mu$mol quanta (or "microeinsteins" $m^{-2}$ $s^{-1}$—cf. Section 3.5) between 400–700 nm.

Instruments are available for making:

(a) *Photometric measurements.* These are expressed as illuminance or luminous flux, e.g. the *luminous flux* was 10 000 ft candles.

(b) *Radiometric measurements.* These are expressed as irradiance or radiant flux density e.g. the *radiant flux density* was 500 W/$m^2$.

(c) *Quantum measurements* (Lambda quantum sensor measuring between 400–700 nm range). These are expressed as quantum flux density e.g. the *quantum flux density* was 2200 $\mu$mol quanta $m^{-2}\ s^{-1}$ (400–700 nm).

Any of the above can be converted to the other on an approximate basis. For example, as shown above, the photosynthetically active component of sunlight at 10 000 ft candles is approximately equivalent to 500 W/$m^2$ or 2 200 $\mu$mol quanta $m^{-2}s^{-1}$. However, different light sources are commonly used in experiments with plants and interconversions cannot be made without knowledge of their spectral emission of light. Thus, if one worker used ft candles another could not convert this to W/$m^2$ or $\mu$mol quanta $m^{-2}s^{-1}$ without having the emission spectrum of the light source. Measurements in ft candles or total W/$m^2$ from different light sources cannot be compared quantitatively for their effectiveness as photosynthetically active radiation (PAR). Obviously, a measure of W/$m^2$ would overestimate the PAR if a large part of the emission spectra is outside of 400–700 nm (e.g. incandescent or xeon arc lamps). Likewise, a measure of ft candles will underestimate the PAR with lamps having emission primarily in the blue or red region of visible light. Rather more meaningful comparisons can be made with quantum or irradiance measurements between 400–700 nm which then more closely indicate the PAR. [Even so, some further limitations exist. For example, the sodium vapour lamp has a strong emission in the range 560–620 nm. This gives a high PAR value although plants obviously show a preference for blue and red light].

### 3.8 The green man

As a point of reference (see also Chapter 2) it may be noted that a sedentary male human has an energy requirement of something like 100 kcal/h (slightly more than the electrical energy consumed by a 100 Watt light). The mean annual irradiance over the earth's surface does not vary as much as might be imagined. In the Red Sea area it is about 300 $W/m^2$, in Australia 200, in the United States 185 and in the United Kingdom about 105. Thus, even in the United Kingdom, man could derive all his food energy from 1 $m^2$ if he could convert solar energy into metabolic energy with 100 % efficiency. The fact that the maximum efficiency of conversion by the green plant is about 5 % implies that an equally efficient photosynthesizing green man would still need to derive 95 % of his nutriment in the usual way unless he could contrive to increase his surface area by a factor of 10. The actual area of land needed to supply a normal man with his annual needs via agriculture would vary enormously, according to circumstance and diet, but would be unlikely to fall below a minimum value of 400 $m^2$.

### 3.9 Quantum efficiency

Since light energy is delivered in packets (photons) of specific energy content and the energy needed to reduce $CO_2$ to $CH_2O$ is known with some accuracy, it may be asked how many photons are required to bring about this process. At present the minimum quantum requirement is usually put at 8, and 10 is often taken as a more realistic figure. [Quantum efficiency or quantum yield is the moles of $CO_2$ fixed/mol quanta *absorbed*. The reciprocal of this will give the quantum requirement which is the mol quanta absorbed/$CO_2$ fixed. For example a quantum efficiency of 0.1 mol $CO_2$ fixed/mol quanta = a quantum requirement of $1/0.1 = 10$ mol quanta/mol $CO_2$. Quantum efficiency is not to be confused with action spectrum of photosynthesis, the latter showing the $CO_2$ fixed/quanta provided at various wavelengths (Section 4.20)]. But in the 1930's a quantum requirement of 4 was widely accepted in view of the work on the alga *Chlorella* by the German Nobel prize-winner Otto Warburg. Ironically, acceptance of the higher values followed research in Urbana, Illinois by Robert Emerson who had studied for his Ph.D. in Warburg's laboratory. One reason for these disparities is that the measurement of quantum requirement is technically very difficult. With intact organisms it is complicated by the fact that allowance has to be made for the contribution of respiration to gaseous exchange. Further problems derive from the fact that it is unlikely that respiration continues unchanged in the light. Some authorities even take the view that 'dark respiration' ceases entirely in the light although this disregards evidence that an active transfer of tracer throughout the entire Krebs cycle follows the feeding of labelled intermediates in the light, just as it does in the dark. On the other hand there can be little doubt that photosynthesis must affect some aspects of respiration and, particularly at high temperatures and high light intensities, the process of 'photorespiration' (Chapter 13) becomes increasingly important. In addition, the problems experienced in some of the earlier measurements based on manometry were compounded by a variety of transient 'bursts' and 'gulp' which occurred in passing from light to dark and vice versa.

In principle, quantum yields can be derived from curves in which the rate of photosynthesis is plotted as a function of light intensity (Fig. 3.6). This is done at a very low light with non-limiting [$CO_2$] such that a linear response is obtained. If we wish to carry out a comparable exercise for a car we would be more likely to see how far it would travel on a gallon of petrol but, given suitable instruments, the efficiency could be derived in the same way as it is for photosynthesis. Thus (for the car) a plot of miles/h against gals/h would yield a value in miles/gal (Fig. 3.6) and similarly (for the plant) $O_2$/s against quanta/s will give $O_2$/quantum (Fig. 3.6).

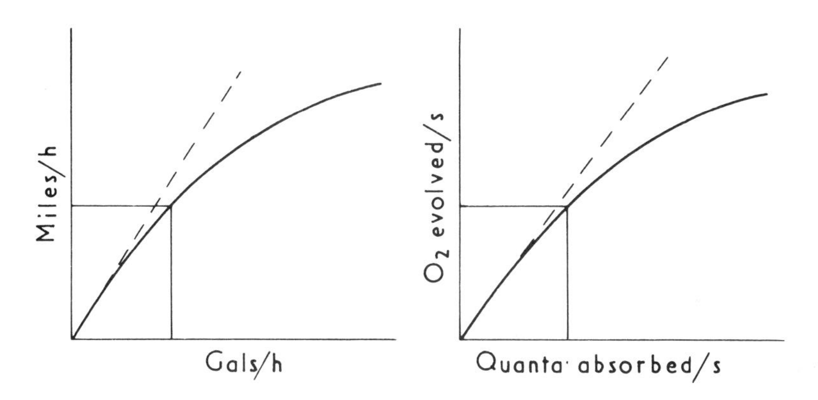


**Fig. 3.6.** Quantum requirements. The 'quantum yield' of a car (left) expressed in miles per gallon compared with the quantum yield of a photosynthetic organism (molecules of $O_2$ per quantum absorbed).

## 3.10 Photosynthetic efficiency at the molecular level

Eight 'quantum mol' photons of red light at 680 nm have an energy content of (8 × 42 = ) 336 kcal. One gram molecule of *$CH_2O$* yields (672/6 = ) 112 kcal when burned in a calorimeter. On this basis the conversion of light energy to heat energy has occurred with an efficiency of 112/336 or 33 %. This can be taken as a maximum value for photosynthetic efficiency at the molecular level. The major energy loss occurs in the utilization of electrical or excitation energy in the formation of assimilatory power (ATP + $NADPH_2$). Light energy may be assumed to excite chlorophyll with little or no loss (Section 4.16) and, as already seen (Section 2.12), a value of 125 kcal can be placed on the free energy available from assimilatory power. The actual (as opposed to the maximal) efficiency at this level is considerably lower.

While a quantum requirement of 8 is feasible, values of 10 are regarded as normal rather than high. Moreover, the quantum requirement does not vary much across the spectral range because of the rapidity with which the blue excited state of chlorophyll decays (Section 4.16). Accordingly, photosynthesis in blue light (with its higher energy content) is much less efficient. Even taking the mean value of 3.5 $\times 10^{-12}$ ergs/photon ( $\equiv$ 50 kcal) for the energy content of the visible spectrum (cf. Section 3.5) and a quantum requirement of 10, the efficiency falls from 33 % to 22 %. If photosynthetically active light (PAL) is put at 50 % of the total radiation (visible + UV + IR etc) this value again declines to 11 %.

### 3.11 Maximum photosynthetic efficiency of crops

The energy conversion factor for a crop (i.e. its photosynthetic efficiency) is considerably lower than the corresponding value at the molecular level. There are several things which combine to lower the efficiency but collectively they fall into two main categories. The first includes all those factors which limit the effective absorption of light and the second all those processes which dissipate energy which would otherwise contribute to photosynthetic gain. For want of better terms, the first can be designated absorption loss and the second, dissipation loss.

Plants look green because they reflect and transmit green light (Fig. 3.1). The extent of losses attributable to this and to various forms of inactive absorption is very difficult to assess but bearing in mind that some allowance has already been made at the molecular level by taking into account the wasteful aspect of blue excitation (Section 4.16) a further decrease of 25 % (i.e. from 11 to 8 %) would not seem unreasonable. (This assumes that the crop completely covers the illuminated surface, whereas many crops will cover only a small part of the available ground, particularly at the beginning of a growing season).

Dissipation losses are mostly brought about by respiration and photorespiration. Again, it is only possible to arrive at the crudest sort of general approximation but if something in the region of 40 % of the carbon assimilated in photosynthesis is dissipated in metabolic processes rather than accumulated as storage products or new plant material the final figure is near to 5 % (i.e. something like 5 % of the total incident light energy could be recovered as heat if the plants were burned in a calorimeter after illumination). This is in broad agreement with a number of published estimates ranging from about 4 to about 6 %. These values gain in credibility from the fact that the highest recorded figures fall into the same range. For example *Pennisetum typhoides* grown at Katherine, W. Australia, where it received a daily energy input of 5100 kcal/m$^2$ is reported to have increased in dry weight by 54 g m$^{-2}$ day$^{-1}$. In the absence of calorimetric measurements typical plant material is often accorded a calorific value of 4.25 kcal/g (rather more than carbohydrate which is about 3.7 kcal/g). On this basis the percentage conversion would be

$$\frac{54 \times 4.25}{5100} \times 100 = 4.5\% \qquad \text{Eqn. 3.11}$$

### 3.12 Maximum yield

If a crop can convert as much as 5% of the light energy it receives into chemical energy we can also derive values for maximum yield based on the mean annual irradiance (i.e. the light energy average over 24 h and 365 days) and the figure (above) of 4.25 kcal/g dry wt.

For the U.K. the annual irradiance is 105 $W/m^2$ (approximately 25 cal $m^{-2}$ $s^{-1}$ or 90 kcal $m^{-2}$ $h^{-1}$). Five per cent of this value (4.5 kcal) would give a dry weight equivalence of a little over 1 g $m^{-2}$ $h^{-1}$ (about 9 kg $m^{-2}$ $year^{-1}$, 90 t $ha^{-1}$ $year^{-1}$ or 38 tons $acre^{-1}$ $year^{-1}$). The corresponding values for Australia and the United States would be approximately twice as great.

Under normal field conditions these yields would rarely be obtained. For example, in cold climates much of the light would be received at temperatures well below those required for optimal photosynthesis, only part of the ground would be covered and so on. Conversely, light at very high intensities would be largely wasted on most $C_3$ species which saturate at 1/5 full sunlight and photorespiration could become excessive in bright light at high temperatures. Nevertheless we have seen that *Pennisetum typhoides* in Australia grew at 54 g[dry wt] $m^{-2}$ $day^{-1}$ (slightly higher, if anything, than the value based on mean annual irradiance) and even in the U.K. a crop of potato tubers of 88 t(fresh wt)/ha has been reported (a good 30% or more of the theoretical maximum if the entire 'biomass' of the plant is taken into account). The potato is particularly high yielding in temperate climates, partly because of its long growing season and the fact that it can achieve appreciable growth at temperatures which would be below the survival minimum for some $C_4$ species with potentially greater performances.

# Chapter 4
# The Photochemical Apparatus and its Function.

## THE PHOTOCHEMICAL APPARATUS



## FUNCTION



## THE PHOTOCHEMICAL APPARATUS

In a classic experiment Trebst, Tsujimoto and Arnon separated photosynthesis into light and dark phases (Fig. 2.4). They showed that, in its simplest form, the reduction of $CO_2$ to *$CH_2O$* (Chapter 1) could be driven, in the dark, by 'assimilatory power' ($ATP + NADPH_2$) generated separately in the light. The dark phase was catalysed by the soluble proteins of the chloroplast stroma and the light phase by what is sometimes called 'the photochemical apparatus'.

### 4.1 Chloroplast structure

The principle red pigment in people is restricted to the red blood corpuscle. Similarly, the major green pigment in plants (chlorophyll) occurs in small bodies or organelles called chloroplasts (Fig. 4.1) which are often not greatly dissimilar in size and shape to red corpuscles. Again, the pigments are not distributed equally throughout the chloroplast but are restricted to an internal association of thylakoid membranes (Section 4.2) which constitute the photochemical apparatus. The name 'thylakoid' comes from a Greek source which was once used to describe the baggy trousers worn by the Turks and is applied to these membranes because they are also like sacs. Since the sacs are normally flattened, the membranes appear double in higher plant chloroplasts. Often, the thylakoid membranes are stacked into piles which appear as grains or 'grana' when the chloroplast is viewed under the light microscope. The photochemical apparatus is embedded in a stroma or jelly-like mass of soluble proteins which includes all of the enzymes concerned in the assimilation of carbon (the reduction of $CO_2$). The whole is surrounded by two membrane envelopes, the inner of which plays an important role in the movement of compounds between the chloroplast and the cytosol which surrounds it (Chapter 8). [A chloroplast, like a plant cell, responds to osmotic pressure. If a strip of onion epidermis is placed in a strong solution of sucrose it becomes plasmolysed i.e. the cell contents shrink away from the cell walls as they lose water to the surrounding solution. This is because water tends to move outwards across the cell membrane (the plasmalemma) so that the external solution is diluted (exosmosis) and the concentration of water is more equal on both sides of the membrane. Similarly water will move into a semi-permeable membrane bag containing a molar solution of sucrose to such an extent that it can exert a hydrostatic pressure equivalent to that of

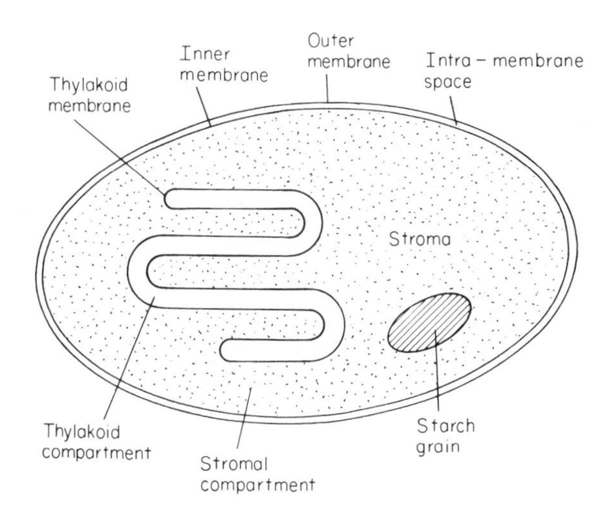


**Fig. 4.1** Chloroplast structure: diagramatic representation of main features.

22.4 atmospheres. (The osmotic pressure of a solution is governed by the number of solute particles, whether these are molecules, atoms or ions. Thus, equimolar salt solutions exert higher osmotic pressures than sugar solution because of ionization, e.g. $NaCl \rightarrow Na^+ + Cl^-$)] In order to isolate intact chloroplasts from the leaf it is therefore necessary to macerate the tissue in an isotonic solution (one which exerts the same osmotic pressure as the organelle which is to be preserved intact). Sugar solutions are often used for this purpose because sugars do not readily cross the inner-envelope of the chloroplast (Chapter 8) and the osmotic balance on either side of this membrane is maintained. Once isolated, however, chloroplasts are readily ruptured by osmotic shock, i.e. if they are re-suspended in water containing no additives they will swell rapidly as the result of endosmosis. The ruptured chloroplast will then lose its stroma but the insoluble thylakoid membranes will remain relatively intact. (These thylakoids will swell in turn if they are left in water but the outer envelopes burst within about 30 seconds and if the osmotic potential of the external solution is then restored by the addition of a sugar, the thylakoid membranes may be recovered intact). When separated from the chloroplast the illuminated photochemical apparatus retains the ability to generate assimilatory power (reducing power, $ATP + NADPH_2$) even if subjected to further fragmentation. Such preparations of isolated thylakoid membranes are sometimes referred to as broken, fragmented or envelope-free chloroplasts and sometimes as free-lamellar bodies (in the sense that a lamella is a flattened sheet of membrane).

## 4.2 The thylakoid compartment

As indicated above the thylakoid membranes occur within the chloroplast as flattened sacs or double sheets of membrane. Indeed it now seems likely that there may be one such sheet per chloroplast (Fig. 4.1) and that however this ramifies or is folded (sometimes into stacks, like piles of coins, to form grana) or is penetrated by holes (frets) it still comprises one, single, membrane-bound, compartment (the photochemical apparatus).

## 4.3 The stromal compartment

If the above interpretation is correct, the single continuous thylakoid compartment lies within the single stromal compartment which is, in turn, bounded by the innermost chloroplast envelope. (Fig. 4.1) The stroma is a protein gel of which ribulose bisphosphate carboxylase (Fraction 1 protein) is the major catalytic component. (This carboxylase is believed to be absent from the mesophyll chloroplasts of $C_4$ plants—Table 11.2). Not all higher plants produce starch, (the snowdrop, *Galanthus nivalis*, is a classic example) but where starch is produced in such photosynthetic tissues it occurs as grains (in the stroma) which do not appear to be associated with any specific membrane system. The stroma contains a variety of additional bodies including plastoglobuli (globules of lipid material) and 70S ribosomes (particles involved in protein synthesis). In most circumstances it will also contain ions (such as $Mg^{2+}$), coenzymes such as ATP and $NADP^{+}$, those soluble proteins concerned with electron transport (such as ferredoxin) and enzymes involved in carbon assimilation.

## 4.4 Membrane structure

The detailed internal structure of the thylakoid membranes remains to be certainly established. It is known that they are composed of roughly equal parts of lipid (fat) and protein and it is extremely probable that these molecules are packed into an organized semi-fluid state. Some proposals favour a unit-membrane structure in which a bi-molecular leaflet of lipid and its associated pigments is sandwiched between two layers of protein. The lipids have polar (hydrophilic or water-attracting) heads usually containing one or more molecules of the 6-carbon sugar galactose, and long, non-polar hydrocarbon tails. They are associated with smaller quantities of phosphate-containing fats (phospholipids) which are more polar in nature. More recent work suggests that the protein may exist, wholly or partly, in a globular or granular form (rather than as a continuous sheet) or that it may penetrate the entire bilayer at regular intervals (Fig. 4.2). In any event the result is a more or less continuous hydrophobic (water-repelling) structure which makes the membrane insoluble in water and houses the lipid-soluble pigments but still permits the interaction of the membrane protein (Section 5.12) with the water-soluble components of the stroma. In addition to chlorophyll, the thylakoids contain many additional components of the electron transport system (Section 4.9).

The chloroplast envelopes are also composed of a mixture of protein and lipids (of which a phospholipid, (phosphatidylcholine) is now a much more prominent component, accounting for as much as 20% of the total fat content). Douce and Benson have shown that the envelopes are yellow (Chl-free) membranes with a unique carotenoid (Section 4.7) composition which may conceivably act as a blue light filter protecting the Chl in the thylakoids from harmful photoxidation. They clearly differ in many important respects from the thylakoid membranes and, if their permeability characteristics are any guide, (Chapter 8) there are probably significant differences between the composition and/or structure of the inner and outer envelopes.

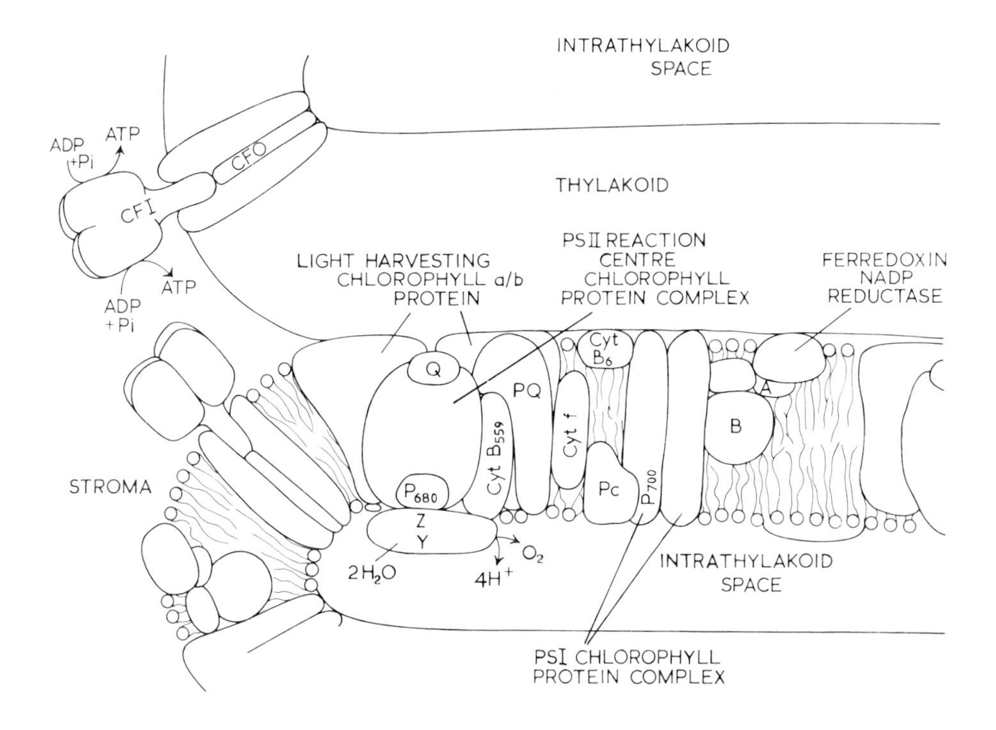


**Fig. 4.2** Schematic diagram of the thylakoid membrane showing the structural arrangement of the photosynthetic carriers as required by electron transport. The thylakoid proteins and electron carriers are embedded in a semi-solid lipid matrix. Light energy collected by the photosynthetic pigments of the light harvesting Chl *a*/*b*-protein complex is funnelled to the photochemically active Chl *a* molecule of the photosystem II (PSII) reaction centre, *P680 (P690)*. Excitation of *P680* results in the loss of an electron and the formation of a strong oxidant, $P680^+$, in the reaction centre. This promotes the oxidation of the primary donor to PSII, Z, which in turn causes the loss of an electron from the water-splitting system, Y. When four electrons are removed from Y the complete oxidation of two water molecules can be achieved. The electron from *P680* is captured by the primary acceptor PSII, Q, which is a specialized plastoquinone molecule. It is then passed to secondary acceptor, B, and thence to a pool of plastoquinone molecules (PQ). The plastoquinone pool also facilitates the transfer of protons across the membrane. An iron-sulphur protein then transfers the electron from the PQ pool to cytochrome *f*(cyt *f*). From cyt *f* the electron is passed to plastocyanine (PC) which is in contact with the photosystem I (PSI) reaction centre Chl, *P700*. Excitation by light energy causes the loss of an electron from *P700* and this is replaced by the one carried by PC. The primary acceptor to PSI which receives the electron from *P700* is probably a dimeric Chl molecule and this transfers the electron to a series of iron-sulphur centres which in turn are called X, B and A. Iron-sulphur centre A finally transfers the electron to a ferredoxin molecule.

## 4.5 Chlorophylls

Chlorophyll (Chl) occurs in two chemically different forms (*a* and *b*) in higher plants (Fig. 4.3). Like haem, which gives blood its red colour, the chlorophylls are porphyrins but the four pyrrole units which constitute the porphyrin ring are linked to Mg, whereas in haem they are linked to Fe. In addition to its tetrapyrrole head, Chl has a long phytol tail (composed of 4 isoprene units) which makes it very hydrophobic in character. Chl *a* differs from *b* only in one very minor respect (having a $-CH_3$ on one side chain rather than a $-CHO$) but this is enough to shift the absorption peaks to a significant extent so that *a* is blue-green and *b* is yellow-green (Fig. 4.4). Chl *a* is the major photosynthetic pigment and is found in all green plants. In most higher land plants Chl *a* is 2.5–3.5 times as plentiful as Chl *b*. The chlorophylls were first separated from one another on an icing-sugar column by the Russian botanist Tswett. (Fig. 4.4) In acetone, Chl *a* has an absorption maximum in the red at about 663 nm but *in vivo*, presumably as a consequence of its association with protein and other chlorophylls in the thylakoid membrane, this maximum is shifted to longer wavelengths and spectroscopic evidence shows that *a* can then exist in several forms (Section 4.11). Green plants owe their colour (in white light) to the fact that chlorophylls absorb in the red and the blue and transmit and reflect in the green.

## 4.6 Chlorophyll synthesis

Chl *a* has been synthesized in the test tube by Woodward and a great deal is also known about its formation *in vivo*. The synthesis of α-aminolevulinic acid (ALA) can be regarded as the first step leading to Chl synthesis. In animals and bacteria, a primary means for synthesizing ALA for tetrapyrrole synthesis is from succinyl CoA and glycine through ALA synthetase. However, in greening leaves there may be another pathway since glutamate or 2-oxoglutarate, and not glycine, are precursors

---

This is in contact with the ferredoxin NADP-reductase which is a flavoprotein able to utilize the electron received from ferredoxin to carry out the reduction of $NADP^+$ to produce NADPH. Cyclic electron flow may occur via the mediation of cytochrome *b6* (cyt $b_6$) which can transfer the electron back to the PQ pool. The transfer of electrons across the thylakoid membrane builds up a proton gradient which provides a suitable environment for the formation of ATP from ADP by the coupling factor which has two components $CF_1$ and $CF_0$. $CF_0$ is embedded in the thylakoid membrane and contains the proteolipid proton channel while $CF_1$ protudes into the stroma and possesses the ATPase activity.

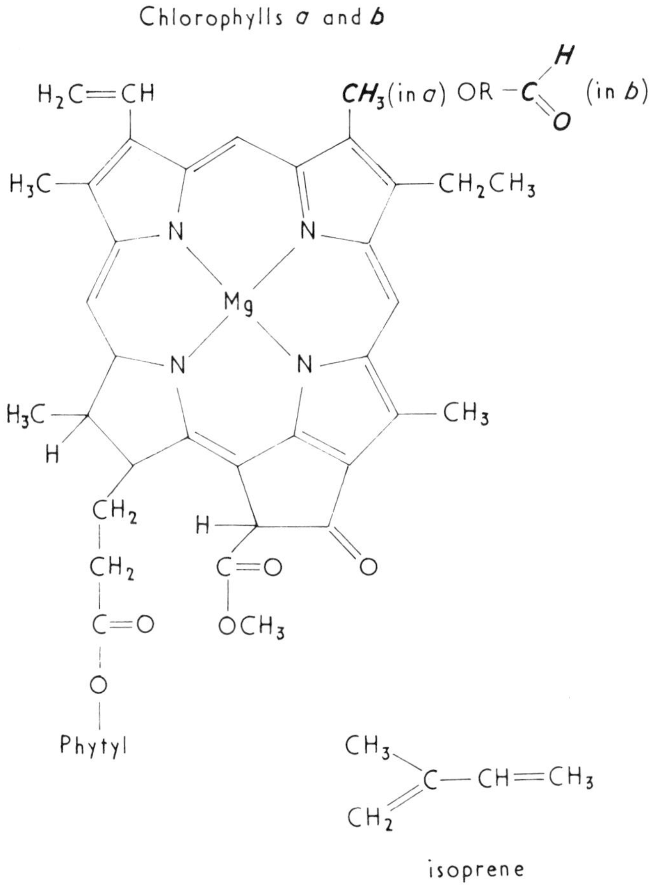

Fig. 4.3 Structure of chlorophylls *a* and *b* and isoprene.

of ALA synthesis. Since the synthesis of 2-oxoglutarate is usually considered to occur in the mitochondria, the synthesis of Chl may be partly dependent on mitochondrial metabolism. This, along with evidence that the small subunit of ribulose bisphosphate carboxylase is synthesized outside the chloroplast, does not strengthen the possibility that chloroplasts were evolved from symbiotic blue-green algae. (Fig. 4.5)

In the final stages of synthesis a yellow, protein-bound, protochlorophyllide (an oxidized Chl precursor minus the phytol tail) is photoreduced to chlorophyllide

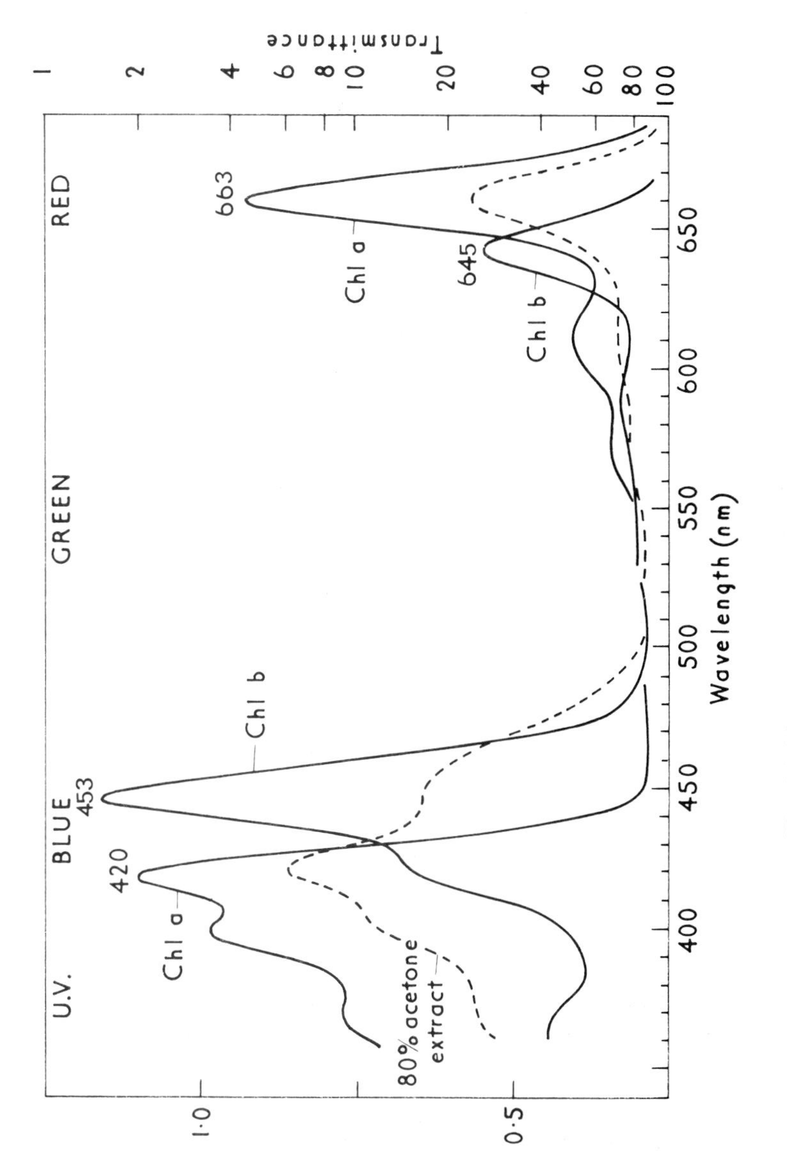


**Fig. 4.4** Absorption spectra of chlorophylls in acetone.

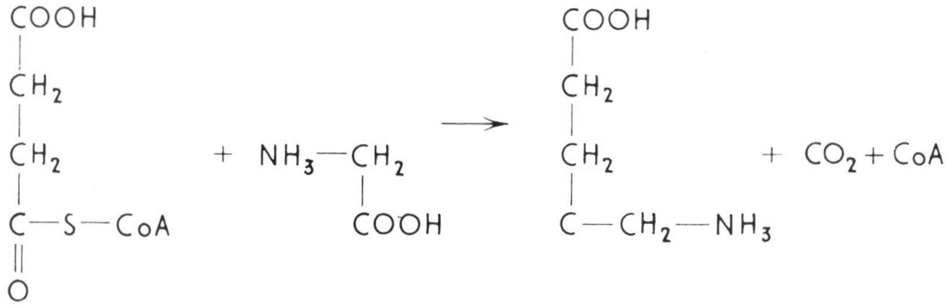

$$\underset{\text{Succinyl CoA}}{COOH-CH_2-CH_2-\underset{\underset{O}{\|}}{C}-S-CoA} + \underset{\text{Glycine}}{NH_3-CH_2-COOH} \longrightarrow \underset{\delta\text{-aminolevulinic}}{COOH-CH_2-CH_2-C-CH_2-NH_3} + CO_2 + CoA$$

**Fig. 4.5** Chlorophyll synthesis: formation of aminolevulinic acid from succinyl CoA.

which is green prior to the addition of the phytol (phytolation). Only a little of the protochlorophyllide complex is made in the dark and is very rapidly photoreduced so that dark grown seedlings can be greened by a succession of millisecond light flashes and long (15 min) dark intervals during which the protochlorophyllide is reformed. (Conifers and several green algae are able to make Chl in total darkness and many ferns are reputed to do so. Greening of citrus seeds appears to be caused by the light which penetrates the fruit).

## 4.7 The accessory pigments

In addition to Chl *a* and *b*, chloroplasts contain carotenoids; principally $\beta$-carotenoids, lutein, violaxanthin and neoxanthin (Table 4.1). These are long-chain isoprene (polyene) compounds with a cyclohexene ring at either end. Carotenoids are believed to act as supplementary light receptors (broadening the range of photons which can be used in photosynthesis) and possibly to protect Chl against photoxidation.

## 4.8 Chicken and egg

Anyone who attempts to describe the photochemical apparatus is faced at this point with the difficulty that its fine structure has been defined largely in terms of its function and yet it is not easy to understand the function without some concept of

**Table 4.1** Primary and accessory photosynthetic pigments of higher plants, and their absorption maxima (approx)* in nanometers.

| | In organic solvents | *In vivo* |
|---|---|---|
| Chlorophyll *a* | 420, 663 (in 80% acetone) | 435, 670–680 |
| Chlorophyll *b* | 453, 645 (in 80% acetone) | 480, 650 |
| β Carotene | 425, 450, 480 (in hexane) | 445, 470, 500 |
| Luteol | 425, 445, 475 (in ethanol) | 445, 465, 495 |
| Violaxanthol | 425, 450, 475 (in ethanol) | 445, 470, 495 |

* The values for specific compounds in named solvents are precise. In-vivo values are less certain, particularly for the carotenoids (β-carotene, luteol and violaxanthol) which overlap with one another and the blue bands of the chlorophylls.

structure. The reader is therefore asked, for the moment, to accept the premise that the function of the photochemical apparatus is energy transduction, i.e. the conversion of light energy via electrical energy to chemical energy (assimilatory power, Section 5.2). The contemporary view is that this is achieved by a structural and functional arrangement which allows photons to be channelled into two photosystems (Fig. 4.6). These are connected by, and (in a sense) comprised of, electron carriers which form a bridge between them. Together the two photosystems (Section 4.13) and the connecting bridge constitute the Z-scheme (Section 4.23) and electrons from water traverse the entire system en route to $NADP^+$. Further details of structure and operation are given below.

## 4.9 Components of the electron transport chain

These are electron carriers which are associated with, and which link together, the photosystems (Section 4.13) which constitute the Z-scheme (Section 4.23). A complete list would contain all the intermediates between $H_2O$ and $NADP^+$ but there is great uncertainty about location, quantity and identity. Some of the components, in the order in which they are believed to accept electrons, are listed below, together with their redox potentials (E′) where known (see Section 2.13).

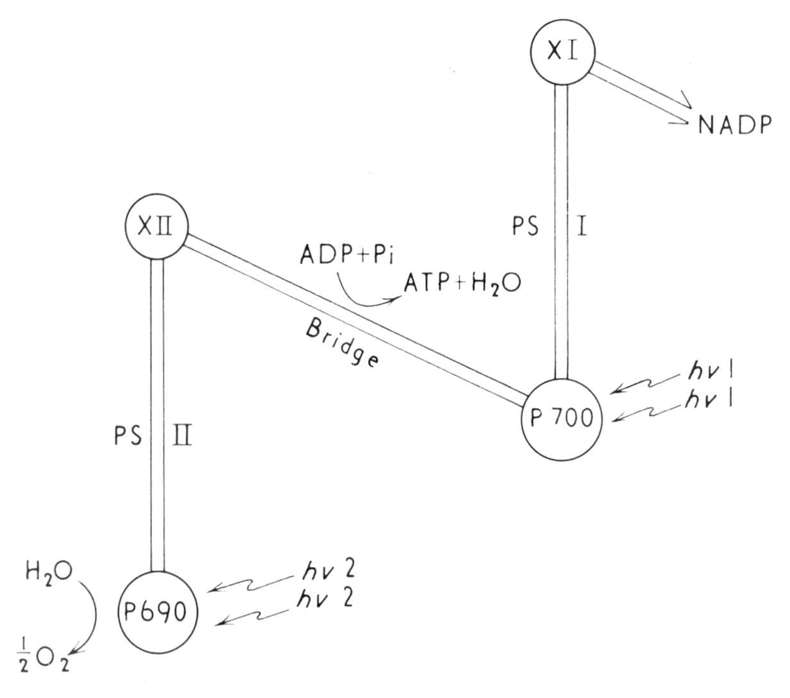


**Fig. 4.6** The Photochemical Apparatus: major aspects of function. There are two photosystems (Section 4.13) joined by a 'bridge' of linking electron carriers (Section 4.9). Two photons are required to 'lift' two electrons from $H_2O$ via the reaction centre (*P690*, Section 4.12) of PSII to its primary electron acceptor. From there they run down the thermochemical gradient of the bridge to fill 'holes' vacated by the excitation of *P700* in PSI. Finally they are passed via X to $NADP^+$ at the total energy cost of 4 photons per 2e transported from water to $NADP^+$. For more detailed pictures see Figs. 4.2 & 4.18

### Manganese protein (*Y*)

PSII is believed to contain a manganese protein (*Y*) which may catalyse the early stages of $O_2$ evolution. (For example, mild heating and exposure to hydroxylamine or Tris at high pH, which all deplete chloroplasts of manganese (Mn), inhibit $O_2$ evolution but not photosynthetic electron transport with many donors other than $H_2O$. Algae grown in Mn-deficient media lose the ability to evolve $O_2$ but not the ability to photoreduce $CO_2$ with $H_2$. The difficulty of detecting Mn in chloroplasts by Electron Spin Resonance techniques implies that it is protein bound). There are about 5–8 atoms of manganese per 400 molecules of Chl. The minimal amount of

Mn required for full $O_2$ evolution capacity seems to be 4 atoms per reaction centre (E′ for $Mn^{3+}/Mn^{2+}$ is about +1100 mV).

### Chloride ion

This is a requirement which is associated with the oxidizing side of PSII. When chloroplasts are washed free of $Cl^-$ they lose the ability to utilize $H_2O$ as an electron donor whereas electron donation by ascorbate or diphenylcarbazide is largely unaffected.

### *Z*

The primary electron donor to PSII. Detectable as an Electron Spin Resonance signal but not otherwise characterized.

### Cytochrome $b_{559}$

Bendall has described a high potential (E′ = 370 mV) form of cytochrome $b_6$ which has an absorption peak at 559 nm and may be associated with Y (above). The fact that $b_{559}$ is oxidized by PSII at low temperatures is consistent with this view whereas oxidation by PSI and reduction by PSII is not. A possible rationalization of these effects has been proposed by Butler.

### *P690*

(Also designated *P690 P680* or *P682*). The reaction centre (RCII) of photosystem II (PSII). May be regarded (Section 4.12) as a molecule (or dimer) of Chl *a* in PSII which traps photons, or photon energy, harvested by an array or antenna of chlorophyll molecules (Fig. 4.18) and, in so doing, exhibits spectral changes at 690 nm (hence the name Pigment 690 or P690). May exist as a complex (Z.RCII.Q) with Z(above) and the primary acceptor, Q (below).

### Q (quencher)

In its oxidized state, this component quenches fluorescence from the Chl *a* molecules which preferentially transfer their energy to RCII (or *P690*) and is the primary electron acceptor in PSII. A component with absorbance changes at 550 nm (C-550) is related to the redox state of Q but is not considered to be synonomous with Q. Chl fluorescence (Sections 4.13 & 4.16) is most pronounced when 'normal' channels of energy dissipation are blocked. The quenching or suppression of fluorescence can be brought about by a large number of oxidants which thereby become reduced as the Chl is returned to its oxidized state. The 'natural' channel of de-excitation would therefore be expected to act as an effective quencher. Present in the ratio of about 1/400 Chl molecules. E′ is about −35 mV. Is a bound plastoquinone which undergoes a single-electron reduction to a semiquinone, QH. (It should be noted that Krause and others have shown that fluorescence quenching is also governed by the high energy state of the thylakoid membrane).

### Plastoquinone

First characterized by Kofler and basically a dimethylbenzoquinone with a 9-isoprenoid side chain (therefore very similar to ubiquinone in mitochondria and with roughly the same sort of 'tail' (Section 4.5) as the carotenoids and Chl) but there are a number of analogues and isomers. This is the most abundant electron

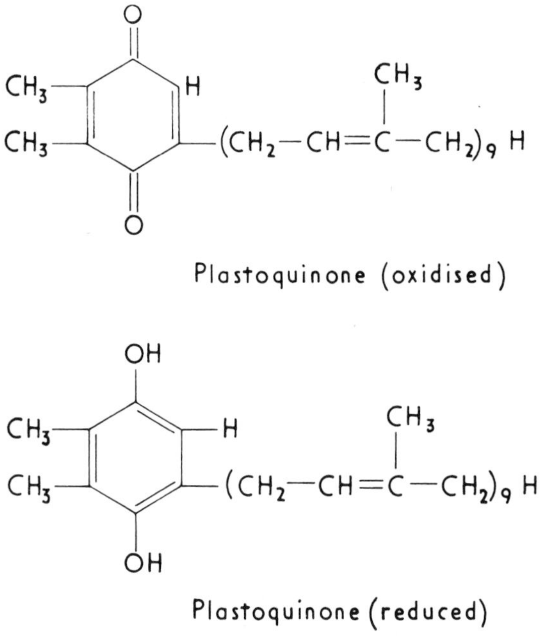

$$\text{Plastoquinone (oxidised): } (CH_3)_2C_6HO_2\text{—}(CH_2\text{—}CH{=}C(CH_3)\text{—}CH_2)_9H$$

Plastoquinone (oxidised)

$$\text{Plastoquinone (reduced): } (CH_3)_2C_6H(OH)_2\text{—}(CH_2\text{—}CH{=}C(CH_3)\text{—}CH_2)_9H$$

Plastoquinone (reduced)

**Fig. 4.7** Plastoquinone

carrier in chloroplasts (1 molecule/10 Chls) existing as a mobile pool in the membrane. Has a redox potential of zero and for this and other reasons is more closely associated with Photosystem II. Can accept two hydrogens. The redox reaction between $Q^-$ and plastoquinone requires the participation of a bound quinone, B, which can be reduced to $B^-$ by the first photon and $B^{2-}$ by a second, thence transferring both equivalents to the plastoquinone pool.

### Cytochrome *f*

A *c*-type cytochrome (from L. *frons*, a leaf) discovered by Hill and Scarisbrick has absorption maxima (depending on the source) in the range of 550–555 nm and redox potentials of +300 to +370 mV. About one molecule/400 Chl molecules. Associated with PSI. Carries a single electron.

### Iron-sulphur protein

An iron-sulphur protein ($E' = 290$ mV) is located between plastoquinone and cytochrome *f*.

### Plastocyanin

A copper protein (2 atoms Cu per molecule protein) first reported by Katoh. Blue when oxidized, colourless when reduced and a potential of +370 mV. About one molecule per 400 Chl molecules. Associated with PSI and also a carrier linking the two photosystems (Section 4.13) Carries a single electron.

### Pigment 700 (*P700*)

The reaction centre of PSI, as characterized by Bessel Kok. A modified form of Chl *a*, existing as a dimer, with a red peak at about 700 nm which is bleached upon oxidation. About 1 molecule per 400 Chl molecules. Redox potential +430 to +530 mV. Carries a single electron.

### X (*P430*)

The primary electron acceptor in PSI (Section 4.13). An unknown compound with a potential more negative than ferredoxin, and possibly (on the basis of the ability of chloroplasts to reduce low potential viologens) as low as 600–700 mV. Possibly a complex between an iron-sulphur protein (membrane bound ferredoxin?) and some other compound as detected by spectral changes at 430 nm upon illumination and by electron paramagnetic resonance. Two Fe-S compounds, designated B and A are thought to be intermediate between *X* and ferredoxin.

### Ferredoxin

This, together with the flavoprotein (below) first appeared in the literature as the methaemoglobin reducing factor of Davenport, Hill and Whatley and then as the photosynthetic pyridine nucleotide reductase of San Pietro and Lang. Following the isolation of an iron-sulphur protein from bacteria by Mortenson, a similar compound was extracted from spinach by Arnon and Tagawa which, in the presence of the flavoprotein, is able to catalyse the reduction of $NADP^+$ by illuminated chloroplasts. Redox potential $-430$ mV. Carries single electrons. Contains 2 atoms of iron per molecule. (Section 5.15)

### Flavoprotein (ferredoxin—$NADP^+$ oxidoreductase)

This enzyme transfers electrons from reduced ferredoxin to NADP. It also acts as a transhydrogenase ($NADPH_2 + NAD^+ \rightarrow NADP^+ + NADH_2$) and a diaphorase (an enzyme concerned in the transfer of electrons from $NADPH_2$ to artificial acceptors). It contains one molecule of FAD (flavin adenine dinucleotide). Weakly membrane bound with a molecular weight of 40 000 and $E' = -380$ mV.

### Nicotinamide adenine dinucleotide phosphate (NADP)

The terminal acceptor of photosynthetic electron transport in the Z-scheme (Section 4.23). In the reductive pentose phosphate pathway, $NADPH_2$ donates electrons to 1,3-diphosphoglycerate which may therefore be regarded as the last in the entire sequence of natural Hill 'oxidants' (Section 4.13). $NADP^+$ does not cross the chloroplast envelope and transport to the cytoplasm of this component of 'reducing power' occurs by shuttle mechanisms (Sections 8.20 & 8.26). On reduction NADP accepts 2 electrons and one hydrogen and $NADPH + H^+$ is a more exact representation of NADP in its reduced form than '$NADPH_2$'. On the other hand 'NADPH' is convenient, implies that two electrons are involved, and is often preferred (Fig. 4.8).

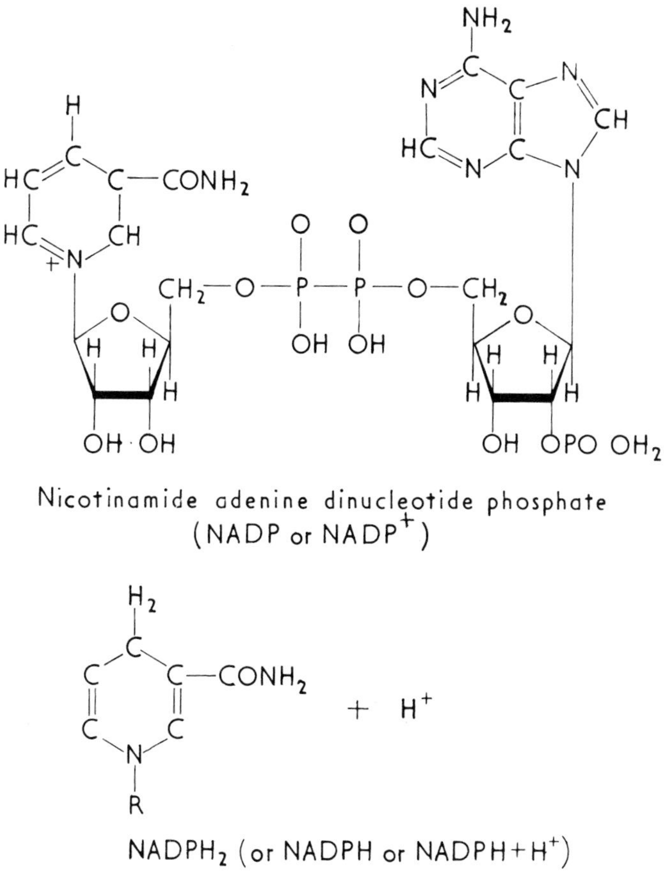

Nicotinamide adenine dinucleotide phosphate
($NADP$ or $NADP^+$)

$+ H^+$

$NADPH_2$ (or $NADPH$ or $NADPH + H^+$)

**Fig. 4.8** NADP

## 4.10 The photosynthetic unit

[Note that this was originally a totally theoretical concept but that it now has much in common with the greater physical actuality of the 'pigment system'–(Section 4.11)]

In Section 3.9 it has been seen that the quantum requirement for photosynthesis is now believed to be at least 8 (i.e. the reduction of one molecule of $CO_2$ to $CH_2O$ according to Equation 1.1 requires 8 photons of red light). These quanta are absorbed by Chl so it is also reasonable to ask how many molecules of Chl are involved in the reduction of one $CO_2$. In order to answer this question it was necessary to ensure that all (or most) of the Chl molecules present in the test system received a photon of light during the period of the experiment (otherwise the idle pigment molecules would have been counted along with the working molecules and a false answer obtained). This need to ensure photosaturation led to the use of flashing light (Fig. 4.9). In continuous light, the rate of photosynthetic carbon assimilation increases with increasing light intensity but in many plants a ceiling is approached at about 1/5 full sunlight (whereas some of the purely photochemical events in photosynthesis continue to go faster with increasing light intensity long after this point is passed). This was attributed to the fact that carbon assimilation could not keep pace with the photochemical reactions. In 1932, Emerson and Arnold had shown that if light was supplied, in flashes, to *Chlorella* the interval between the flashes allowed the limiting dark reaction (the so-called Blackman reaction) to go to completion in about 0.02 second at 25°C. Accordingly, Emerson and Arnold carried out experiments in which they established the intensity needed for photo-saturation when the light was provided as flashes of $10^{-5}$ seconds duration at intervals of about 0.09 or 0.05 seconds. Defining one 'unit' as 'the mechanism which must undergo the photochemical reaction to reduce one molecule of $CO_2$', they argued that, at photosaturation, the number of units would equal the number of molecules of $CO_2$ reduced per flash. Because of the time required (0.02 seconds) for the completion of the dark reaction there was no possibility that a unit could work twice during the duration of a flash lasting $10^{-5}$ seconds. Experiments

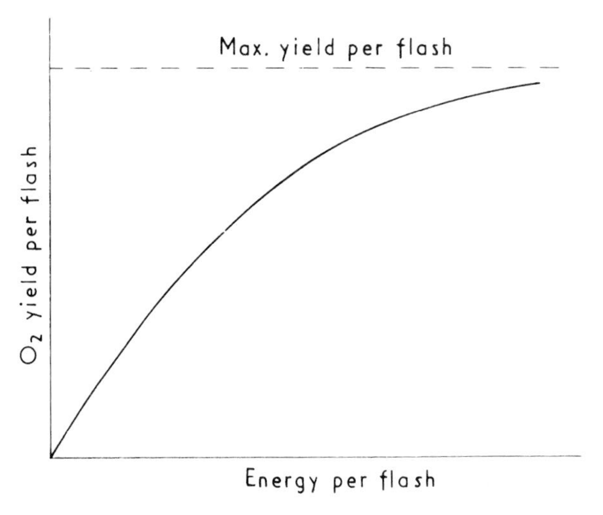


**Fig. 4.9** Light intensity required to saturate photosynthesis when applied in flashes. As the energy per flash is increased, a maximum yield is approached but the ratio of $CO_2$ fixed (or $O_2$ evolved) to Chl present does not exceed about 1 to 2500. This suggests that individual Chl molecules are organized into a photosynthetic unit.

of this sort were carried out at a number of Chl concentrations (This was achieved by adding either different quantities of *Chlorella* or samples of this alga grown in such a way that it contained more Chl). As anticipated, the yield per flash increased with increasing Chl but the ratio of $CO_2$ reduced to amount of Chl present was, on average, 1/2480 (Fig. 4.9). Emerson and Arnold felt that they were unable to offer an adequate explanation of the fact that their reduction unit contained so many molecules of Chl. Later, however, Gaffron and Wohl proposed the concept of a 'photosynthetic unit' in which a photon could migrate through densely packed molecules until it encountered a substrate which would permit a photochemical reaction leading to $O_2$ evolution. If 2500 molecules of Chl cooperate to bring about the evolution of one molecule of $O_2$ and this process proceeds with a quantum requirement of 8 then the photosynthetic unit could be comprised of 8 subunits (each containing 300 molecules of Chl and one reaction centre or any other appropriate combination in which 2500 light-harvesting pigment molecules funnel photons into 2–8 reaction centres).

An important aspect of the photosynthetic unit and its reaction centre is that in the above experiment, although virtually all the molecules were excited, the vast majority could not discharge their excitation usefully because the reaction centre was engaged (i.e. the unit cannot usefully employ more than 8 photons per flash). Maximum yield per flash can, however, only be determined in this wasteful fashion to ensure that no chlorophylls escape excitation. Conversely, maximum yield per quantum (which implies no wasted energy) results when only 8 out of the 2500 molecules are excited even though this excitation may pass through many more molecules, up to a maximum of 2500, on its way to the reaction centre.

## 4.11 The pigment systems

In contemporary usage there is little if any difference between the photosynthetic unit (Section 4.10) and the pigment system. A pigment system is regarded as a light-harvesting array (or antenna) of Chl molecules which intercepts photons and funnels their excitation energy to a reaction centre (Section 4.12). As will be seen, in Fig. 4.18 and Section 4.23, however, it is now believed that Chl is involved in the full photosynthetic process at two points and that there are in fact, two photosystems (designated I and II by Duysens, see Section 4.13) each with its own *pigment* system. The composition of the two pigment systems has been inferred, primarily from spectrophotometric measurements (rather than derived by direct analysis) and is a notional rather than a functional or precise description. Pigment System II is believed to contain the bulk of accessory pigments and most, if not all, of the total Chl *b*. In addition, it is thought to contain several membrane-bound forms of Chl *a* with absorption maxima at 650, 663, 669, 673, 677, and 680 nm. Excitation energy is channelled from these light harvesting pigments to a trap or reaction centre (*P690*).

Pigment system I contains little, if any, Chl *b*, but 4 major spectral forms of Chl *a* (*a* 663, 669, 677, 684), and another 'long wave' Chl (*a* 695) which is entirely missing from pigment system II. This array of pigments funnels energy to the trap or reaction centre called *P700*. In both systems energy is transferred (below) from the short wave pigments (i.e. those which require shorter wavelength, higher energy, photons for peak excitation) to the long wave pigments (requiring longer, wavelength, lower energy photons). Chl-protein complexes can be isolated from chloroplasts using the detergent sodium dodecyl sulfate (SDS). Recently, as many as 4 major Chl-protein complexes were obtained when subjecting SDS extracts to gel electrophoresis. One of these, *P700*-Chl *a* protein, is considered the heart of photosystem I (PSI) and accounts for about 15% of the total Chl. The other 3 complexes all contain Chl *a* and *b* and together may make up the light harvesting Chl *a*/*b* protein which accounts for up to 50% of the Chl in $C_3$ chloroplasts. [The differences between $C_3$ chloroplasts and $C_4$ chloroplasts and the plants which contain them are considered at length in Part B of this text]. The remaining Chl in the SDS extract is Chl *a* which is free from protein. This is thought to be derived from Chl-protein complexes of the antenna Chl of PSI and the reaction centre for photosystem II (PSII) which are unstable to SDS treatment. The light harvesting Chl *a*/*b* protein is thought to funnel energy primarily into PSII. (There is some evidence that the PSII reaction centre complex, containing *P690*, is a Chl-protein of approximately 32 000 mol wt). However, this complex may be so arranged in the chloroplast that 'spill-over' of energy to PSI can occur under some conditions through the Chl *a* cone surrounding the reaction centres, for example when the acceptor pool to PSII is strongly reduced or limiting. The light harvesting Chl may then exist as a continuous array with capacity to funnel energy into both photosystems rather than a distinct distribution of Chl types between PSI and PSII.

It is not known whether the total *pigment* system (I & II) comprises one unit with 2400 chlorophylls (cf. photosynthetic units above), two subunits of 1200 chlorophylls, or 4 subunits of 600 chlorophylls. Joliot, who measured the $O_2$ evolution brought about by short intense flashes of light, found that no $O_2$ was released by the first flash but then increasing amounts with maximum yield on the third, seventh, eleventh, etc. This is consistent with the present view that pigment system II is excited 4 times (Section 4.18) during the evolution of one $O_2$ and implies that its reaction centre (or 4 linked centres if there were, e.g. 4 subunits), can acquire and hold the charges and energy required for this to happen.

## 4.12 Reaction centres

The energy trap or reaction centre (*P700*) of pigment system I is better characterized than that of pigment system II (*P690*). It is thought to be a modified form of Chl *a* (a dimer which exists in a unique environment), and is called *P700* (P for pigment) because its absorption peak is close to this wavelength. Bessel Kok, who discovered *P700*, showed that it has a redox potential (Section 2.13) of about +400 mV. It constitutes about 0.3% of the total Chl *a*. It is normally oxidized in red light, becoming bleached, and regains its colour as it is reduced in the dark.

Reaction centre II (*P690*) is less well defined than *P700* and although (as the similarity in terminology suggests) it may also be a form of Chl *a*, it does not follow that it is an entity entirely comparable to *P700*. It might easily be more complex and there is no doubt that $O_2$ evolution, in which it plays a central role, is more readily inhibited by mechanical action or heating than many reactions brought about by PSI. In other respects, however, it is as stable as PSI.

## 4.13 The photosystems (PSI and PSII)

These constitute the physical basis of the Z-scheme (Section 4.23 and Chapter 5), in which two light reactions combine to bring about the transfer of electrons from water to $NADP^+$ (Fig. 4.18). Each photosystem contains a pigment system (Section 4.11) and its associated electron carriers. The nature, quantity, and precise site of these components in the electron transport chain is not known with any great certainty but Section 4.9 summarizes some contemporary views. Originally Duysens and others talked of the reduction of $NADP^+$ as the 'first' photoreaction and the evolution of $O_2$ as the 'second' because chemotrophic photosynthetic bacteria do not evolve $O_2$ and could be regarded as 'first' in an evolutionary sense. The physical entities responsible therefore became photosystems I and II respectively even though electrons passing from $H_2O$ to $NADP^+$ enter PSII before PSI. Some confusion is also associated with the terms 'oxidant' and 'reductant' when used in this context. An 'oxidant' is an electron acceptor. A great many compounds, such as ferricyanide, will act as 'Hill oxidants', i.e. they will accept electrons from water in a reaction brought about by illuminated chloroplasts.

The difficulties in terminology which arise in this way, do so because the natural electron carriers are linked together in a sequence. Thus a compound which acts as an oxidant (electron acceptor) at one moment may function as a reductant (electron donor) in the next, as electrons pass along a chain. For example, the coenzyme $NADP^+$ is often regarded as the 'natural' Hill oxidant associated with the photochemical apparatus even though it is only one of several which accept electrons at different stages in the reduction of $CO_2$. Similarly *P700* is an oxidant of plastocyanin and a reductant of $NADP^+$ just as NADPH eventually brings about the reduction of $CO_2$.]

The pigment systems (and therefore the photosystems) can be at least partly separated by mechanical disruption (i.e. French press or sonication) or treatment with detergents such as digitonin or Triton X-100. Particles enriched in PSI are composed primarily of Chl *a* and have reaction centre *P700* while PSII particles contain both Chl *a* and *b*. In $C_3$ chloroplasts, the total Chl per *P700* molecule is about 300/1 to 500/1. In PSI particles the Chl/*P700* ratio is about 200/1.

When excited with short wavelength light (435 nm) at room temperature, the fluorescence emission spectra of chloroplasts and PSII enriched particles show a peak emission at 685 nm which increases at liquid nitrogen temperature (77° K). PSI particles show little fluorescence at room temperature but a major band of fluorescence occurs at 730 nm at liquid nitrogen temperature. The fluorescence at 730 nm is thought to be from the antenna Chl *a* of PSI. The fluorescence at 685 nm is thought to be principally from the light harvesting Chl *a*/*b*-protein which is associated primarily with PSII.

There is also delayed light emission (emission of light long after cessation of illumination lasting from a few seconds up to several minutes) at room temperature specifically associated with PSII. DCMU (which inhibits electron flow between Q and plastoquinone) inhibits delayed light emission while DBMIB (a plastoquinone antagonist) stimulates delayed light emission. Delayed light emission is influenced both by the concentration of reduced acceptor and by the concentration of the oxidized donor of PSII. It is also associated with the dissipation of a protonmotive force, possibly generated at the plastoquinone site. If neither the reaction centre nor the pool of electron acceptors associated with PSII is limiting, delayed light emission will be reduced. Thus, it can be taken as a measure of PSII capacity. PSI lacks delayed light emission because, for unknown reasons, the fluorescence yield of PSI is very low.

## FUNCTION

As indicated in Section 4.8, the structure of the photochemical apparatus is not easily divorced from its function but whereas in the preceding sections the emphasis has been on structural apparatus, the remaining sections in this chapter are more concerned with its operation and also touch on some basic principles (as in Sections 4.14, 4.15 etc.).

## 4.14 Atomic absorption

Hydrogen, the simplest atom, consists of a nucleus (a proton, $H^+$) with an electron ($e^-$) in orbit around it, like a satellite orbiting the earth. Energy is needed to lift a satellite into orbit and the more distant the orbit the greater the energy. Electrons orbit atomic nuclei in definite shells and atoms are in their most stable state when the electrons occupy the shells closest to the nucleus (the state of least energy or electronic ground state). Sometimes an electron may then be promoted from the ground state to a more distant orbital (excited state) by a quantum of appropriate energy content (Fig. 4.10). Because the permitted shells are very well defined, only a quantum with precisely the right energy content is able to effect this transition. A quantum of higher or lower energy is ineffective. The energy gap between the ground state and the excited state for atoms is very large but the energy content of ultra-violet and visible light is often sufficient for this purpose and accordingly an atom may exhibit an absorption spectrum which appears as a very intense black line corresponding to the utilization of the requisite quantum of the correct wavelength. Thus the position of the line may be derived from the relationship $\lambda = hc/\Delta E$ where $\Delta E$ is the energy required to bring about the transition from the ground state to the excited state. The probability of an electron in an atom undergoing such a transition depends upon the intensity of the incident light because there will be more chance of it intercepting a quantum of precisely the right energy content.

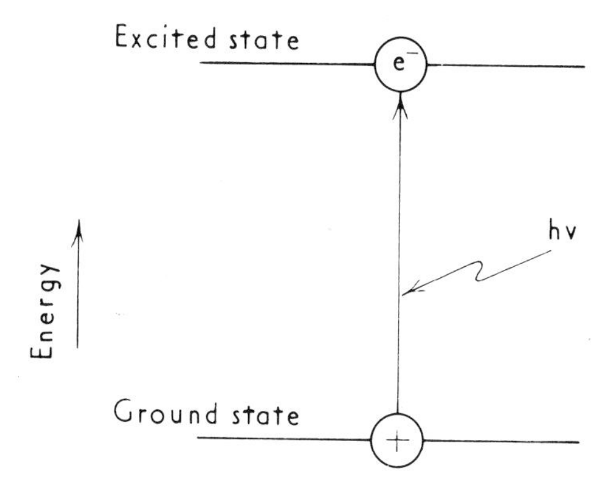


**Fig. 4.10** Atomic absorption

Photoexcitation is almost instantaneous (taking about $10^{-15}$ seconds) and involves the movement of an electron of extremely small mass within a much larger atom (or molecule). The time of oscillation is related to $\nu$, the frequency of the light absorbed.

## 4.15 Molecular absorption

In molecules there are a number of nuclei which vibrate, rotate and, in addition, electrons are shared between nuclei so that the molecular orbitals are smaller (corresponding to a lower energy content). Accordingly energy is released as molecules are formed from their constituent atoms (Chapter 2). For these reasons the electronic ground states and excited states are divided into a number of vibrational sub-states. As before, one quantum of precisely the correct energy content will lift one electron from one ground sub-state to one excited sub-state but since there are many sub-states the total range of quanta capable of effecting these transitions is wider.

The sharp lines of atomic absorption spectra are therefore replaced by a series of lines which usually coalesce to form relatively broad absorption bands. Each individual line in a band corresponds to the discharge of a number of quanta of the same wavelength and, collectively, the lines correspond to the distribution of energies about a given wavelength peak. (This picture is further complicated by the fact that the electron transition adds to the existing vibration within the molecule, causing the energy levels to shift slightly. Absorption spectra tend to be sharper at very low temperatures because the probability of electrons being found in a large number of sub-states is diminished).

## 4.16 Excitation of chlorophyll

There are two principal excitation states (1 and 2) in chlorophyll corresponding to the absorption bands in the red and the blue. The tails of these bands overlap, indicating that the excited sub-states also overlap. Excited state 1 has the lower energy content of the two, resulting as it does from the absorption of 'red' photons contributing close to 40 kcal/Einstein (Fig. 4.11). Excited state 2 has a higher energy content resulting from the absorption of 'blue' photons (at about 50 kcals/Einstein). Since the excited vibrational sub-states overlap, excited state 2 can decay very rapidly (in about $10^{-13}$ s) to excited state 1 by a process of heat dissipation (or radiationless de-excitation) involving a series of small transitions between sub-states. This occurs so rapidly that the additional energy of state 2 is not made available for useful work (photochemistry) or even re-emitted as light (fluorescence) which takes about $10^{-9}$ s. In effect, therefore, the common starting point in the photosynthetic process is the red excited state (state 1) of chlorophyll because, even if this is created indirectly by the absorption of a blue photon, the additional energy absorbed is lost before it can usefully be employed.

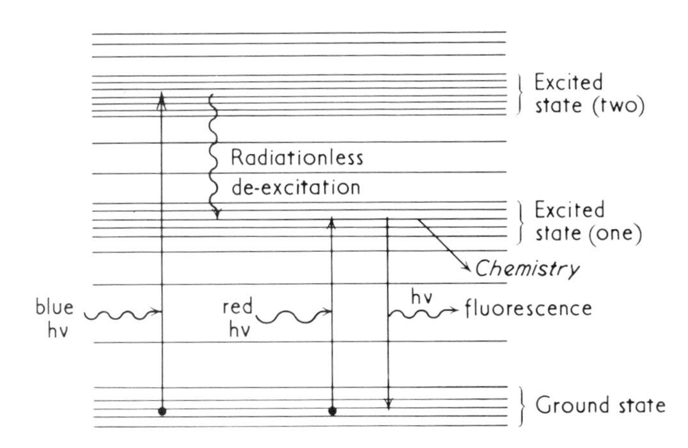


**Fig. 4.11** Excitation of Chl. The parallel lines represent energy sub-states or electronic orbitals. Thus the energy delivered by the absorption of a blue photon (left) is sufficient to raise an electron to excited state 'two' from where it rapidly returns by a process of radiationless de-excitation, 'cascading' through sub-states, to excited state 'one'. A photon of red light (centre) only has enough energy to raise an electron to excited state 'one' but this excited state is sufficiently stable to permit useful chemical work and is, in effect, the starting point of all other events in photosynthesis. Particularly in organic solvents, excited state one can also dissipate energy by re-emiting light as (deep red) fluorescence.

## 4.17 De-excitation

As indicated above excitation energy can be dissipated as heat and this can be extremely rapid if it occurs as a cascade of small consecutive transitions. Energy can also be re-emitted as visible light although in Chl, as in many other molecules, fluorescence only derives from the lower, more persistent, excited state (state 1). Even within state 1, cascade heat dissipation will decrease the energy content to give the lowest sub-states and therefore fluorescence is at longer (redder) wavelengths (the Stokes shift). When chlorophyll is extracted from leaves (e.g. in aqueous acetone) and brightly illuminated it looks green to the observer if he places the illuminated vessel directly between him and the light source. This is because the 'red' and 'blue' photons are absorbed. The green photons which are transmitted are not only much more plentiful, because of this selective absorption, but also (as it happens) more readily perceived by the human eye. In organic solvent, however, pathways of energy dissipation which exist in the chloroplast are no longer available. Excited state 2 decays rapidly to excited state 1 and a relatively high proportion of the excitation energy is re-emitted as red fluorescence. Because of the Stokes shift (above) this is at a longer wavelength. The observer at right-angles to the illuminated vessel therefore sees it as a rich blood-red solution because (in this position) his retina is not flooded with green transmitted light.

The strong fluorescence observed under these conditions is not so effective an energy dissipating mechanism that it protects the molecule from photoxidation and the clear green of the transmitted light soon gives way to yellow as decomposition proceeds.

## 4.18 Photochemistry and $O_2$ evolution

In addition to radiationless de-excitation and fluorescence, excitation energy can be dissipated in useful work (photochemistry). Chlorophyll fluorescence is much less marked *in vivo* where the possibilities of 'useful' photochemistry are much greater than in illuminated solutions. Similarly chlorophyll can normally withstand long periods of strong illumination in the intact leaf although bleaching (solarization) can occur, particularly if the principal photochemical dissipation channels are blocked (e.g. by removing the $CO_2$ from an enclosed atmosphere).

In order for a chemical reaction to occur between two stable molecules an energy barrier has to be breached. Clearly, the molecules must collide in order to react and the greater the impact the more likelihood that the barrier will be breached and that collision will be followed by reaction. Ordinary chemical reactions go faster at high temperatures because the molecules move faster and accordingly collide more often and with greater force. Conversely the reaction is facilitated in the presence of a catalyst because the barrier is weakened (i.e. the activation energy is lowered).

In some sorts of photochemistry the energy barrier is surmounted rather than breached. Chl appears to function in this way. As indicated (Section 4.16) excitation by either red or blue light eventually leads to excited state 1 from which an electron can drop over the barrier to an electron acceptor or oxidant. In the unexcited state the barrier imposed by the activation energy is too great to be crossed at biological temperatures even though it may have been lowered by the presence of a suitable catalyst.

The Hill reaction, with ferricyanide as the oxidant (electron acceptor) is an example of this process. First, light excitation raises electrons in Chl from the ground state to the excited state (Fig. 4.12, centre). In order to evolve one molecule of oxygen two molecules of water must serve as donors (Eqn. 4.1).

$$2H_2O \rightarrow O_2 + 4H^+ + 4e^- \qquad \text{Eqn. 4.1}$$

Four molecules of ferricyanide ($Fe^{3+}$) are reduced to ferrocyanide ($Fe^{2+}$), by virtue of accepting 4 electrons, and 4 hydrogen ions ($H^+$) are released into solution. (The

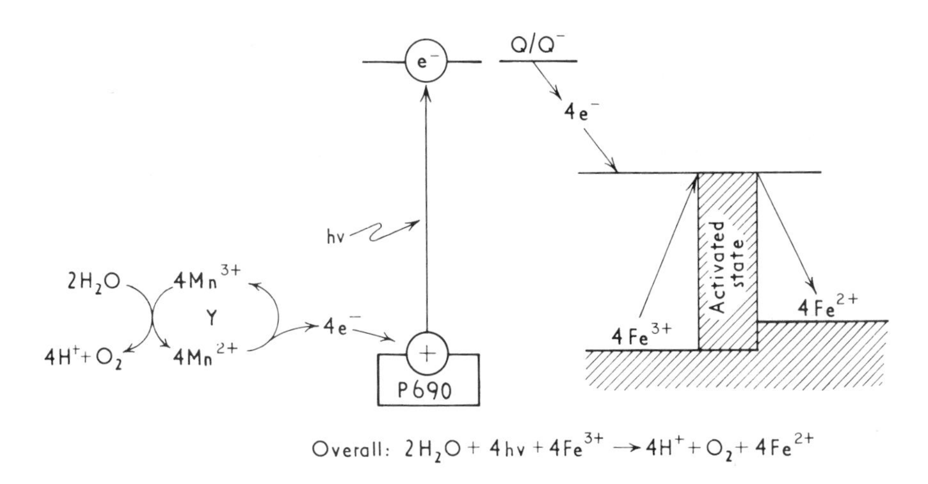


**Fig. 4.12** Photochemistry of Hill reaction with ferricyanide as the acceptor. In the centre of the diagram a molecule of Chl (*P690*) in PSII is excited by a photon, raising an electron to a sufficiently high energy level to reduce its acceptor, Q, which becomes reduced to $Q^-$. This process is repeated 4 times and Y (the manganese $H_2O$-splitting protein) accumulates 4 oxidizing equivalents. These are discharged, simultaneously, by 4 electrons as 2 molecules of water release one molecule of $O_2$ and $4H^+$. The electron acceptor, Q, transfers the electrons which it receives, in a sequential fashion, to other electron carriers in the Z-scheme. In-vivo these electrons are accepted by $NADP^+$ and eventually by $CO_2$. In the Hill reaction no $CO_2$ is reduced because an artificial oxidant, in this case ferricyanide, intercepts electrons. (The point of interception is governed by the redox potential of the oxidant and would not necessarily involve direct donation by $Q^-$ as this diagram might suggest). Although this scheme is a considerable over-simplification it attempts to illustrate the accumulation of 4 oxidizing equivalents by Y for each molecule of $O_2$ evolved (left) and the fact that photochemical excitation of this nature lifts electrons over the activation energy barrier (right) which would otherwise preclude the reduction of an artificial or natural acceptor molecule.

Hill reaction can be monitored by recording the increase in acidity associated with this process).

Each excitation of Chl in Fig. 4.12 requires one photon and each of these photochemical events produces one $Chl^+$ (or $P690^+$) and one $Q^-$ as an electron is passed to the Quencher (Section 4.9). For each $O_2$ evolved the water-oxidizing complex (Y) must accumulate four oxidizing equivalents. In Fig. 4.12 these are represented by 4 manganic ions ($Mn^{3+}$) which are formed as manganous ions in Y donate 4 electrons, in turn, to each positively charged hole in *P690* created by light excitation. Work by Joliot and Kok has shown that one molecule of $O_2$ is released on every fourth flash when dark adapted algae or chloroplasts were exposed to saturating flashes of approximately 1 $\mu$s duration. [In fact the first molecule of $O_2$ is released on the 3rd flash but thereafter on the 7th, 11th, 15th etc. In the formulation illustrated in Fig. 4.12 this would imply that, in the dark, there was already one $Mn^{3+}$ and three $Mn^{2+}$. Thereafter 4 flashes (excitations) would be required to form 4 $Mn^{3+}$ which would accept $4e^-$ from 2 molecules of $H_2O$ giving one molecule $O_2$]. This also implies that the quantum requirements of the above sequence would be 4 (per $O_2$).

In practice ferricyanide will accept electrons from PSI if intact thylakoids are used (with a quantum requirement of 8) and from both photosystems if damaged thylakoids are employed.

## 4.19 Energy transfer within the pigment systems

Before the reaction centre can donate electrons to other carriers within their respective photosystems energy must be absorbed by, and transmitted through, the antenna (light-harvesting chlorophylls) to the centres.

As already described, de-excitation of a pigment, 'A,' may occur by emission of a quantum of fluorescence (Section 4.17 and Fig. 4.11). If this light is of a wavelength which can be absorbed by a second pigment,'B,' it is clear that energy can be transferred from 'A' to 'B' by a process of excitation, de-excitation and reabsorption. The efficiency is so great that the trivial explanation of energy transfer given above may be ruled out. Instead, transfer occurs by resonance. This is an inductive process of coordinated de-excitation and excitation in which an electron falls from one excited sub-state to a ground state and the energy which would have been released as fluorescence is used more or less instantaneously to promote an electron from an adjacent ground state (Fig. 4.13). The efficiency of this process depends on the actual physical distance between the pigment molecules. For Chl to Chl, the distance must be less than 7 nm and at 2.5 nm the efficiency rises to virtually 100 %. Evidence of the operation of this process in photosynthetic organisms exists in the phenomenon of sensitized fluorescence in which, for example, Chl *a* will fluoresce as a result of excitation of Chl *b*. Energy can also be transferred with great efficiency between molecules of the same kind when an exciton (an electron-hole pair) takes a 'random walk' through densely packed molecules. In this process an electron is thought to change from one orbital to an adjacent orbital at a similar level of excitation and corresponding movement of electrons between ground state orbitals ensures that the 'hole' follows the migrating electron. Alternatively, (and perhaps more correctly) the aggregated chlorophylls in the pigment system can be regarded as a super-molecule in which the excitation is essentially delocalized. De-excitation of the entire molecular array would then occur as the energy became localized in, and transferred to, the reaction centre.

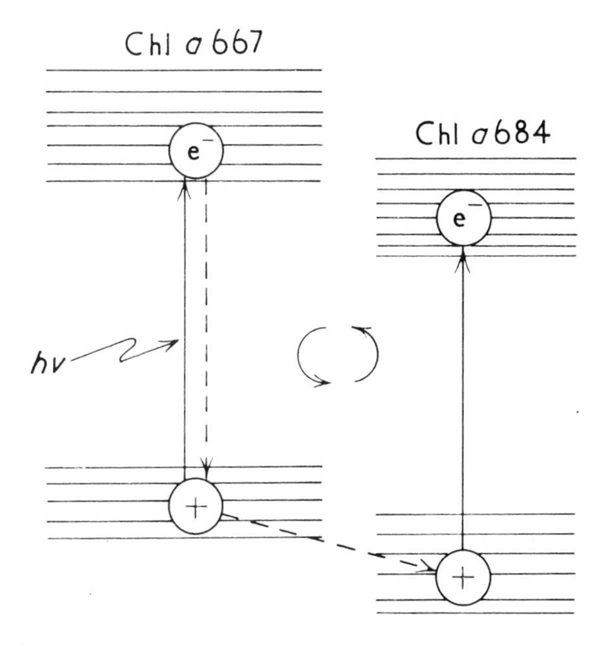


**Fig. 4.13** Energy transfer between two molecules of Chl by coordinated excitation and de-excitation.

## 4.20 Action spectra

It has been seen (Section 4.16) that chlorophyll absorbs in the blue and the red and that the red excited state of chlorophyll is the common intermediate from which all other photosynthetic events derive. With this in mind it could be predicted that illumination with blue or red light would bring about maximal photosynthesis, whereas illumination with green light would bring about a minimal result. This prediction is borne out by the simplest type of action spectrum (Fig. 4.14) in which the intensity of light required to bring about a given photosynthetic response is found to be inversely related to the effectiveness of absorption (e.g. it takes much less red light than green light to fix a given quantity of $CO_2$). In experiments of this sort the action spectrum and absorption spectrum are very similar. Although this allows the identification of Chl as the major photosynthetic pigment more useful information can be obtained by plotting the quantum yield across the visible spectrum (Fig. 4.15). From Section 4.16 it can be seen that blue photons and red photons should produce the same response because the extra energy of the blue photon cannot be harnessed to useful work. At first sight, green would be much less effective but although only a little photosynthesis occurs in green light only a few

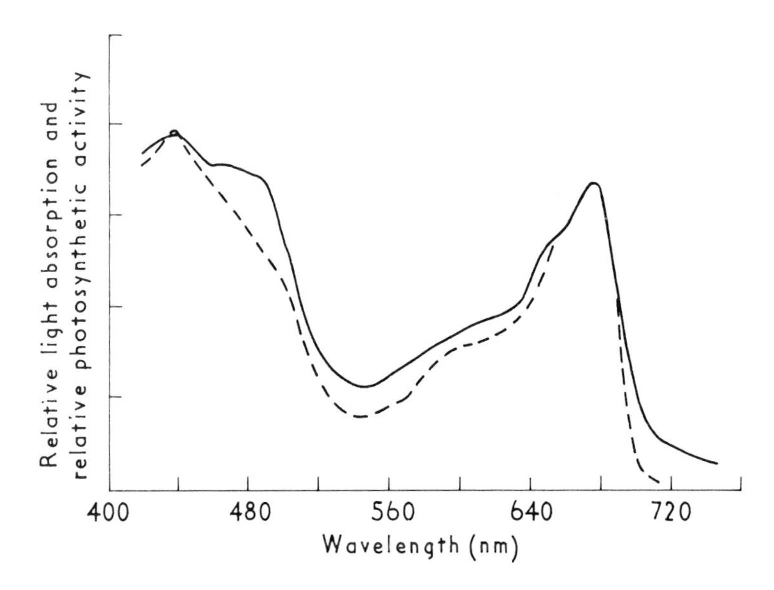


**Fig. 4.14** Absorption and action spectra compared. Solid line, absorption spectra for *Ulva*. Broken line, relative photosynthetic rates for equal numbers of *incident* quanta at different wavelengths. (After Haxo & Blinks.)

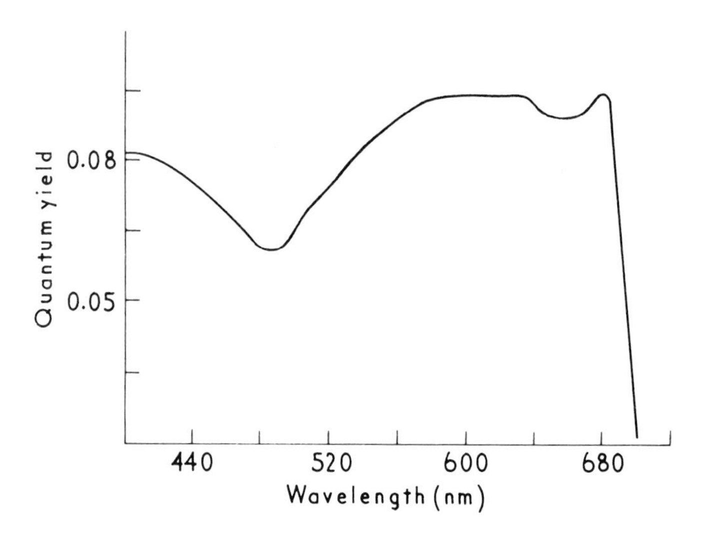


**Fig. 4.15** Action spectrum of $O_2$ evolution by *Chlorella*. Maximum quantum yield as a function of wavelength (after Emerson & Lewis) (Note the dramatic fall in yield at wavelengths above 680 nm (the 'red-drop', see Section 4.21))

photons are absorbed and the yield per quantum remains much the same in the green as it is in the red or the blue. However, action spectra have three distinctive features (Fig. 4.15). The first is a notable dip at about 480 nm. As this coincides with the main absorption range of the carotenoids it seems likely that these compounds do not transfer energy to the chlorophylls with any great efficiency (i.e. the carotenoids appear to prevent absorption by Chl rather than facilitate it). The second is the dramatic fall beyond 680 nm (the red drop) and the third is the small dip at 660 nm (see Section 4.21 below).

Action spectra for PSI and PSII (Fig. 4.16) have been obtained by Ried using intact *Chlorella*. These are based on the assumption that the $O_2$ burst in the first few seconds of illumination is derived from PSII and that the transient inhibition of $O_2$ uptake which is observed in the subsequent dark period is a function of PSI activity. It will be seen (Fig. 4.16) that PSII exhibits an additional peak in the blue because of its carotenoids and its larger content of Chl *b* and that the absorption associated with *a* type chlorophylls in PSI is shifted further into the red by approximately 10 nm (see Section 4.11 and Fig. 4.16).

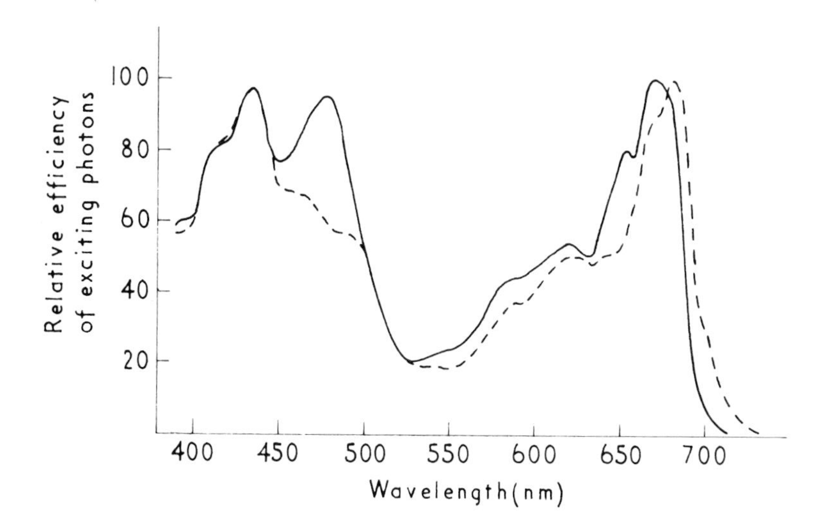


**Fig. 4.16** Action spectra of PSI (broken line) and PSII (continuous line) in *Chlorella* (after Ried).

## 4.21 The red drop

Although Chl in-vivo continues to absorb red light at wavelengths longer than 680 nm (Figs. 4.14 and 4.17) the fall in the action spectrum (Fig. 4.15) is more marked than the fall in the absorption spectrum. Emerson was the first to draw

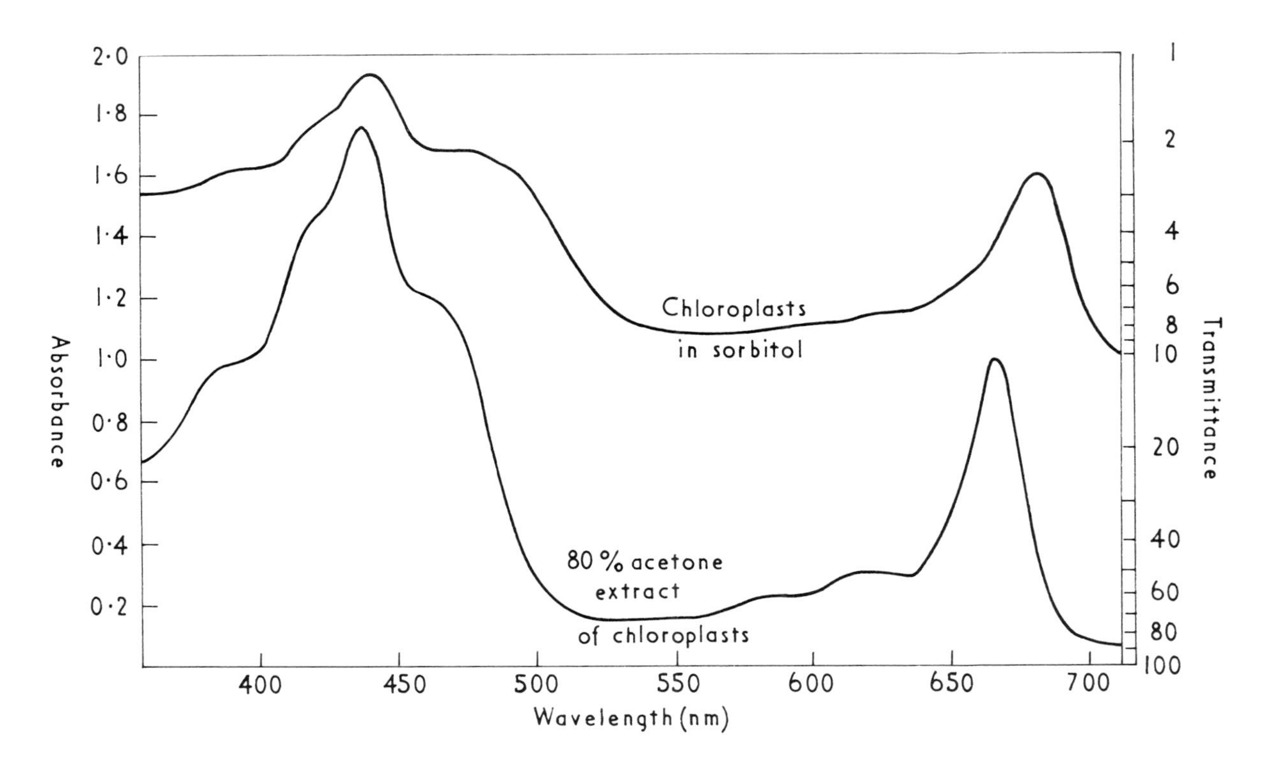


**Fig. 4.17** Absorption spectra of isolated spinach chloroplasts in aqueous media and a 99% acetone extract of the same organelles.

attention to this fact which points to the possibility (Section 4.20) that red light at long wavelengths cannot bring about photosynthesis as effectively as red light at short wavelengths. (The shorter wavelength photons have a higher energy content but the difference in effectiveness is more marked than the energy difference would suggest).

### 4.22 Enhancement

The red drop (Section 4.21) led Emerson to investigate the effect of mixing long $\lambda$ red light with relatively weak light at lower wavelengths. Such mixtures supported more rapid photosynthesis than the sum of the rates achieved by applying either wavelength singly. The improvement of the relatively inefficient long $\lambda$ red light (at wavelengths between 700–730 nm) is known as 'enhancement'. Contemporary interpretation of enhancement is based on the Z-scheme (Section 4.23) which it helped to create. For example, in 700–730 nm light PSI might easily oxidize its reaction centre (*P700*) faster than electrons from PSII could reduce it. Conversely, in shorter wavelength light, PSII could produce reduced intermediates (such as reduced plastoquinone) faster than PSI could reoxidize them. The dip in the action spectra at 660 nm (an incipient 'blue drop') might reflect this imbalance.

(As can be seen from Fig. 4.16, light at 650–680 nm will activate *both* photosystems even though it will be more effective in PSII. Conversely light at 680 nm and above, while still very effective in PSI will not activate any component of PSII at its absorption maximum. Moreover, while energy transfer from a short-wave to a long-wave Chl is acceptable, an 'up-hill' transfer from long to short is not. For these reasons any loss of efficiency at long wavelengths (red drop) is likely to be much more marked than a corresponding decrease in the blue.)

Enhancement action spectra can be determined by measuring the extent of enhancement brought about by light of variable (short) wavelength added to background light of fixed (long) wavelength. The presence of a peak at 670 nm in such spectra supports the presence of a relative excess of Chl *a* 670 in pigment system II. In experiments with spinach chloroplasts Joliot *et al.* have produced separate action spectra for the two photosystems (a) by following $O_2$ evolution with 720 nm background light plus light at varying wavelengths and (b) by following reduction of methyl viologen in constant 650 nm light + light of varying wavelengths. Using isolated chloroplasts, action spectra (Section 4.20) have been determined for $H_2O \rightarrow NADP^+$ and $H_2O \rightarrow$dichlorphenolindophenol and (in the presence of DCMU, an inhibitor of PSII) for the transport of electrons from an ascorbate/dichlorphenol-indophenol couple which donates electrons to PSI. In accordance with the Z-scheme the two former show a red drop but the latter, which only utilizes PSI, actually increases in efficiency at the longer wavelengths.

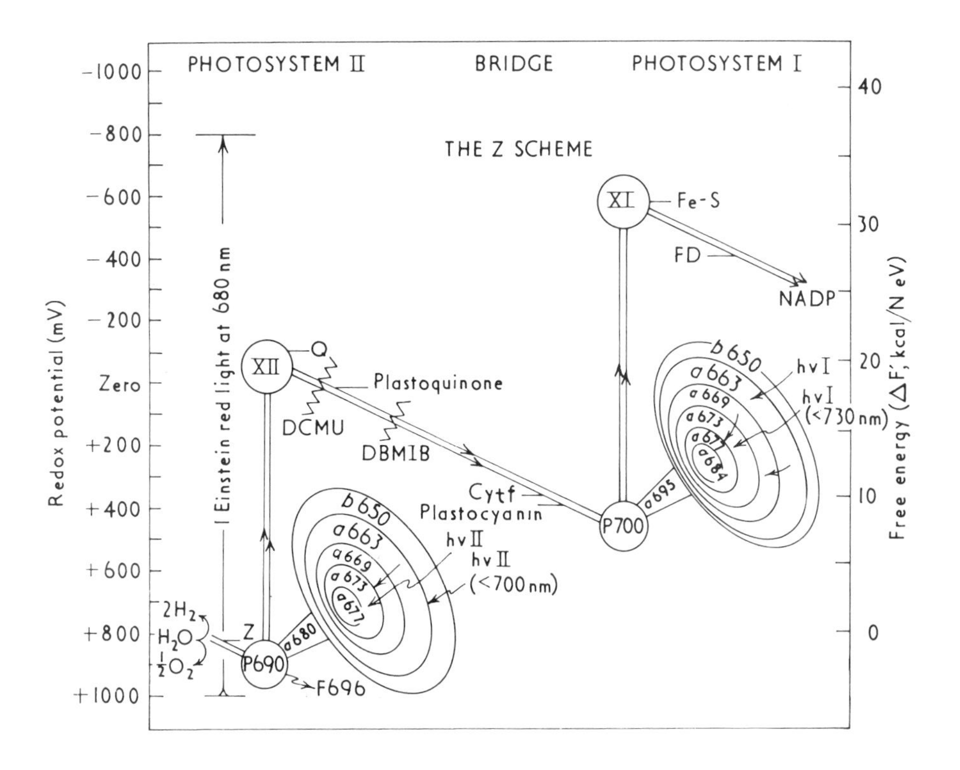


**Fig. 4.18** The Z-scheme. This is both an energy diagram and a map of electron flow but is not intended to be taken too literally. Redox potentials (left) are like football league tables (Section 2.13). Under given conditions, a team at the top of the league is likely to beat a team below it. Similarly, compounds with negative redox potentials tend to reduce compounds with more positive potentials. Thus, the transfer of electrons from water to $NADP^+$ (involving the cooperation of two photosystems) leads to the creation of the prospective hydrogen donor ($NADPH_2$) which is subsequently capable of reducing $CO_2$ to *$CH_2O$*. The energy needed for this electron transfer (right) is seen to be about 52 kcal or 26 kcal per electron (on a molar basis) and enters the scheme at two points. [Similarly, the difference (left) of approximately 1200 mV between the $O_2$ (+815) and hydrogen (−420) electrodes (Table 2.3) can be related (right) to a free energy change of approximately 57 kcal (Sections 2.9 and 2.13)]. The two trumpets represent light-harvesting arrays, or antennae, of chlorophyll molecules (Section 4.10) which funnel photon excitation energy into two reaction centres (*P690* and *P700*, see Section 4.12). Because of the nature of the Chl in these antennae (*a*695, for example, designates Chl *a* with an absorption maximum at 695 nm) PSI is preferentially excited by light of relatively long wavelengths (< 730 nm) and PSII by relatively short wavelengths (< 700 nm). Other known or suspected electron carriers (Section 4.9) and inhibitors such as DCMU and DBMIB (Section 4.13) occupy their probable positions or sites of action in the energy diagram. Two electrons (indicated by the two parallel lines) pass from carrier to carrier on their route from $H_2O$ to $NADP^+$. Each photon (quantum) entering each photosystem can lift one electron to a higher, or more reducing, level so that in order to transfer 2 electrons through the

## 4.23 The Z-scheme

As stated in Section 4.8 and as considered further in Chapter 5, the Z-scheme of Hill and Bendall postulates two light reactions in photosynthesis. These combine to lift electrons in two stages from water (the ultimate donor) to $NADP^+$ (the final acceptor). The two photosystems concerned (Section 4.13) are linked by a series of carriers which constitute a 'down-hill' bridge so that electrons follow a 'Z'–shaped path. Fig. 4.18 is therefore a map which shows the direction of electron flow and the intermediates (Section 4.4, 4.9) which lie between $H_2O$ and $NADP^+$. It is also an energy diagram showing the differences in redox potential along this route and the energy which is expended or made available as electrons move up and down between potentials. Oxygen is evolved in the reaction which occurs in PSII which are preferentially driven by shorter wavelength light. In turn, $NADP^+$ is reduced in PSI which preferentially uses longer wavelength light. Light-harvesting Chl antennae funnel the excitation energy photons to the reaction centres (Section 4.12). The electrons displaced by the excitation of the primary donor in reaction centre II are replaced by electrons from water in a reaction believed to be catalysed by a manganese-protein (Section 4.9). They pass, via Q and plastoquinone to the carriers cyctochrome *f* and plastocyanin which are associated with PSI. Here they fill the negatively charged holes created by excitation (oxidation) of *P700*. Electrons from this reaction centre of PSI travel via X, ferredoxin and the flavoprotein (Section 4.9) to $NADP^+$. Some of the electrical energy made available by the down-hill flow of electrons is conserved in a chemical form as ATP. In this sense, there is sufficient energy for ATP formation at two 'sites' and the use of an inhibitor which blocks the oxidation of plastoquinone and various donors and acceptors which interact with carriers in the bridge between the two photosystems supports this view. Accordingly, one 'site' lies in the bridge and the other between $H_2O$ and the reaction centre of PSII.

[It may be noted that the term 'site', which once implied that energy could be conserved as electrons passed between specific acceptors is now somewhat ambiguous if it is accepted that ATP formation is driven by a protonmotive force (pmf) (Sections 5.9, 5.10). Thus phosphorylation occurs within a compartment following generation of a pmf. The pmf itself is not localized, only the mechanism for

indicated pathway, 4 quanta are required (Section 3.9). Since 4 electrons are transferred for each molecule of $O_2$ released ($2H_2O + 2NADP \rightarrow 2NADPH_2 + O_2$), the whole sequence has to be operated twice and the quantum requirement then becomes 8 per $O_2$ (3.9). The energy content of one quantum mole of red light at 680 nm (left) is seen to be approximately equivalent to 1800 mV (left axis) or 42 kcal (right axis, see also Section 3.5). Some of the energy released, as electrons flow 'down-hill' across the 'bridge' between the two photosystems, is conserved (Chapter 5) as chemical energy in the form of ATP.

its generation. For example pyocyanine is now thought to return electrons from the top of PSI to *P700* and this pathway would bypass the natural 'site' in the bridge between the two photosystems. However, this artificial cofactor manufactures its own 'site' in the sense that it picks up a proton from the outside of the membrane as it receives electrons from X in PSI and then releases a proton to the inner compartment as it is oxidized by *P700*. A pmf is therefore generated as a consequence of light-excitation of PSI even though carriers such as plastoquinone, cytochrome f and plastocyanin are not involved].

Present evidence would locate the water splitting enzyme (manganese-protein) and hence $O_2$ evolution, at the inside of the membrane together with *P690*, *P700*, cytochrome *f* and plastocyanin. Conversely Q, ferredoxin and the flavoprotein would be towards the outside.

The Z-scheme was formulated by Hill & Bendall in 1960. It was based, in part, on the demonstration by Emerson *et al.* in 1958 that although far-red light alone was ineffective in photosynthesis (the 'red drop', Section 4.21) it greatly enhanced (Section 4.22) the effectiveness of light at shorter wavelengths. It became an experimental reality in 1962 when Duysens and his colleagues in Leiden showed that far-red light predominantly excited photosystem I (Section 4.13) and oxidized cyt *f* whereas red light at shorter wavelengths predominantly excited photosystem II and reduced cyt *f*. In addition, Hoch achieved an experimental (rather than a physical) separation of the two photosystems in 1963 and showed that whereas far-red light was very efficient in bringing about reaction in PSI and inefficient in driving PSII the converse was true of red light.

Fig. 4.19 illustrates the Z-scheme by analogy. The gentleman on the left is delivering a photon of relatively short wavelength red light to PSII. The projectile which he launches into the air is an electron on its way to the primary acceptor of PSII. Projectiles (electrons) running down-hill operate an ATP generator in the centre of the picture. On the right, the first process is repeated with the delivery of a photon to PSI and this time the projectile finds its way via X (top right) to the $NADP^+$ basket.

## 4.24 $P/e_2$ ratios

The precise amount of ATP synthesized for the passage of 2 electrons through the Z-scheme (the $P/e_2$ ratio) has proved difficult to determine. This is because there is some 'basal' electron transport in the absence of phosphorylation and because it seemed reasonable to assume that this should be subtracted in any computation.

**Fig. 4.19** The Z-scheme by analogy.

Unfortunately the basal rate is itself inhibited by ATP and the question of what should be subtracted becomes a matter for argument. The degree of interaction may also become very complex. Thus the basal rate may result from incomplete coupling as protons leak out of the thylakoid compartment (Section 4.2) while the rate of leakage is a function of the pH gradient across the thylakoid membrane and is therefore a function of light intensity (which increases this gradient) and ATP synthesis (which diminishes it). In-vivo, or in intact isolated chloroplasts, factors which diminished ATP consumption would therefore increase proton leakage and basal electron transport. This is seen by Heber as a control mechanism whereby 'flexible coupling' of this nature can vary the $P/e_2$ ratio between 1 and 2. In general it seems that the values will be nearer the lower of these two limits although a value of 1.5 would be needed to meet the requirement of the RPP pathway where 3 ATPs are consumed for every 2 molecules of NADPH reoxidized. Cyclic photophosphorylation (Section 5.4) and photophosphorylation linked to oxygen reduction (Section 5.6) might both add to this metabolic flexibility.

## 4.25 Energy conservation in the Z-scheme

In the Z-scheme, 8 quanta combine to bring about the transport of 4 electrons from 2 molecules of water so that one molecule of $O_2$ is evolved and 2 molecules of $NADP^+$ are reduced. A fraction of the energy associated with electron transport is also conserved as ATP.

As indicated above (Section 4.24), the $P/e_2$ ratios (molecules of ATP formed per 2 electrons transported) are not known with certainty and may be as variable *in vivo* as they have been reported to be *in vivo*. Nevertheless, it seems unlikely that in adequately coupled chloroplasts (Section 5.11) the $P/e_2$ ratio will be less than 1 or greater than 2. If the free energy change ($\Delta F'$) associated with the oxidation of NADPH is put at $-52$ kcal and free energy of hydrolysis of ATP (to ADP + Pi) is taken as $-10.8$ kcal this indicates a maximum energy conservation of approximately 148 kcal $[(2 \times 52) + (4 \times 11)]$ and a minimum of 126 $[(2 \times 52) + (2 \times 11)]$ for each gram molecule of $O_2$ evolved. Taking light at 680 nm, which is close to the red absorption maximum of chlorophyll the energy input is 8 Einsteins at 42 kcal per Einstein (336 kcal) giving values of energy conservation in the range of 37.5% (126/336) to 44% (148/336). If the quantum requirement is raised to 10 as a consequence of cyclic (Section 5.5) or pseudocyclic (Section 5.6) electron transport, this conservation value could fall to about 35% (148/420). Overall, therefore, the Z-scheme is clearly energetically feasible in the sense that the energy input is about three times the output. On the basis of 2 $NADPH_2 + 3$ ATP to meet the energy requirement of the RPP pathway (at a cost of 112–120 kcal per $(CH_2O)$ that cycle will operate with a $\Delta F'$ of 17–25 kcal and an 'efficiency' as high as 87% (other thermodynamic aspects are considered in Section 5.14). It may be noted that the actual energy requirement, at the molecular level, will be raised by 50% as a consequence of $O_2$ inhibition of photosynthesis—Chapter 13.

## 4.26 Quantum requirement of the Z-scheme (Section 4.25)

As indicated the Z-scheme demands an input of one photon (one quantum) for *each* electron moved through *each* photosystem. Actual measurements show almost perfect agreement with this requirement so that, e.g. 8 photons would be required per $O_2$ evolved. Thus quantum requirements of unity have been recorded for the transport of one electron to *P700*, for the photoxidation of ferrocytochrome c by PSI chloroplast preparations in 710 nm light, for the transfer of one electron from water to dichlorphenolindophenol by 630–660 nm light, or from water to ferricyanide in 640 nm light or for the transfer of one electron from ascorbate to $NADP^+$ in 700 nm light. Similarly two quanta were required to transfer two electrons from $H_2O$ to $NADP^+$ in 640–678 nm light.

# Chapter 5
# The Formation of ATP; the Generation of Assimilatory Power



## SUMMARY


The photochemical events in photosynthesis culminate in the generation of the 'assimilatory power' (ATP + reduced NADP) which is required to 'drive' carbon assimilation. When the chloroplast is illuminated electrons normally pass from $H_2O$ through a series of electron carriers (Chapter 4) to $NADP^+$ which becomes reduced. Electron transport brings about the establishment of a proton gradient and a separation of electrical charges which, in turn, facilitate the esterification of ADP and Pi to yield ATP. This process of photophosphorylation can be coupled to
(a) *Non-cyclic electron transport* (from water to $NADP^+$ or, in non-physiological systems, from a variety of donors to a variety of acceptors). In this process electrons are raised to higher energy levels by light excitation of chlorophyll but otherwise follow a 'down-hill' thermochemical gradient as they pass from carrier to carrier.
(b) *Pseudocyclic electron transport* (in which electrons donated by water are recombined with $O_2$ to form water after passing through an appropriate sequence of carriers).
(c) *Cyclic electron transport* (in which electrons in photosystem I are raised to a higher energy level as a consequence of light excitation of chlorophyll and are then returned to a lower energy level in the same photosystem after travelling down a thermochemical gradient).

In all of these processes the 'down-hill' passage of electrons and its associated phosphorylation is essentially analogous to respiratory-chain phosphorylation. In both respiratory-chain and photophosphorylation the electrons are originally lifted to the 'top of the hill' by chlorophyll excitation although in the case of animal respiration, the final 'run down the slope' (in which electrons separated from $H_2O$ in photosynthesis are finally recombined with $O_2$ to re-form water in respiration) may be separated from the original photochemical events by a long food chain.

## 5.1 Oxidative phosphorylation

In respiration, the H – O bonds which were separated in photosynthesis are re-formed and energy is released. As in photosynthesis, nicotinamide coenzymes play an important role. In photosynthesis, electrons are transferred from water to $NADP^+$ then $NADPH_2$ reduces $CO_2$ to the level of $CH_2O$. In respiration, electrons are passed from $CH_2O$ to NAD and $NADH_2$ donates electrons, via cytochromes, to $O_2$ so that water is reformed. [There are, in fact, several routes which eventually lead to the complete oxidation of carbohydrate to $CO_2$ and $H_2O$ and not all involve $NAD^+$. Nevertheless, the principle is the same, i.e. electrons are passed to a carrier which becomes reduced and subsequently the reduced carrier is reoxidized, with molecular $O_2$ acting as the terminal electron acceptor].

In intermediary metabolism, reduced coenzymes serve as hydrogen donors in a number of reductive processes so that all of the hydrogen from $CH_2O$ is not immediately restored to $O_2$ and the transfer of a fraction may be deferred more or less indefinitely. In this way $NADH_2$, $NADPH_2$, etc., may serve as electron and energy donors in respiration as in photosynthesis. Similarly, ATP is first generated and then utilized in a variety of metabolic reactions and life processes. The generation of ATP is an energy conservation mechanism. Part of the energy which would otherwise be released during the passage of electrons from substrate to oxygen is used to bring about the synthesis of ATP from ADP and Pi. Most of the respiratory energy conservation process occurs in a sub-cellular organelle called the mitochondrion and it does so during the transfer of hydrogens from $NADH_2$ to $O_2$. [Again, reduced carriers other than $NADH_2$ are involved in some reactions but the mechanism is similar]. The process of oxidative (or respiratory chain) phosphoryl-ation may therefore be summarized by the overall (and generalized) Equations 5.1 and 5.2.

$$NADH_2 + \tfrac{1}{2}O_2 \rightarrow NAD + H_2O \qquad \text{Eqn. 5.1}$$

and

$$3ADP + 3Pi \rightarrow 3ATP + 3H_2O \qquad \text{Eqn. 5.2}$$

As indicated, if electron transport is fully coupled to ATP formation as many as 3 molecules of ATP may be formed during the passage of 2 electrons from $NADH_2$ to $O_2$. If the $\Delta F'$ value for $NADH_2$ oxidation is taken as – 52 kcal and ATP hydrolysis as – 7 kcal this represents

$$\frac{(3 \times 7)}{52} \times 100 = \text{approximately } 40\,\% \text{ conservation}$$

When photophosphorylation was discovered by Arnon, Allen and Whatley, and independently by Frenkel, in 1954, biochemists had already been familiar with these broad concepts for several years. As will be seen below, however, this familiarity at first made it more, rather than less, difficult to grasp the concepts involved. Full clarification only followed the emergence of the Z-scheme.

## 5.2 The need for assimilatory power

Until the 1930's it was generally assumed that reduction of $CO_2$ was directly linked to excitation of chlorophyll. The first challenge to this proposition came from work by van Niel on photosynthetic bacteria. The green sulphur bacteria could be grown, for example, in a mixture of hydrogen sulphide ($H_2S$) and $CO_2$ and could photosynthesize according to the overall Equation 5.3.

$$CO_2 + 2H_2S \xrightarrow{h\nu} CH_2O + H_2O + 2S \qquad \text{Eqn. 5.3}$$

Van Niel pointed out the striking similarity between this equation and that (Equation 5.4) for green plant photosynthesis

$$CO_2 + 2H_2O \xrightarrow{h\nu} CH_2O + H_2O + O_2 \qquad \text{Eqn. 5.4}$$

Moreover chemosynthetic bacteria were known which could reduce $CO_2$ in the dark using hydrogen derived from the oxidation of a substrate and clearly in such circumstances (as Lebedev had pointed out in 1921) there was no role for excited Chl. This led van Niel to postulate a generalized bacterial and higher plant equation for photosynthesis,

$$CO_2 + 2H_2A \xrightarrow{h\nu} CH_2O + 2A + H_2O \qquad \text{Eqn. 5.5}$$

In this overall process, light excitation of Chl led to the generation of a reducing entity *H* and an oxidizing entity *OH*.

$$4H_2O \xrightarrow{h\nu} 4H + 4OH \qquad \text{Eqn. 5.6}$$

In both plants and bacteria *H* was then used to reduce $CO_2$ according to Equation 5.7.

$$CO_2 + 4H \rightarrow CH_2O + H_2O \qquad \text{Eqn. 5.7}$$

In higher plants the $OH$ discharged to give oxygen and water

$$4OH \rightarrow O_2 + 2H_2O \qquad \text{Eqn. 5.8}$$

whereas the bacteria it was used to oxidize a substrate as in

$$2H_2S + 4OH \rightarrow 4H_2O + 2S \qquad \text{Eqn. 5.9}$$

This scheme was well received for its apparent agreement with Warburg's requirement for 4 quanta per $CO_2$ but the implication that Chl was more concerned with photolysis of water than reduction of $CO_2$ made less impact.

At about this time Robert Hill published the first of his papers which were to change irrevocably the face of photosynthesis. In it, he reported that isolated chloroplasts could produce $O_2$ in the light by reducing compounds (artificial oxidants) other than $CO_2$, e.g.

$$2Fe^{3+} + H_2O \rightarrow 2Fe^{2+} + 2H^+ + \tfrac{1}{2}O_2 \qquad \text{Eqn. 5.10}$$

The photolysis of water thus progressed from a hypothesis to an experimental reality. The nature of the natural oxidant (hydrogen acceptor) remained unknown.

In 1943 Ruben incorporated some earlier suggestions of Thimann and of Lipmann in a proposal which was to prove remarkably accurate. This postulated a requirement for ATP as well as a need for the reductant $H$. The ATP was to be used in two ways, to provide the energy for a carboxylation and to assist in the reduction of the carboxyl group by a preliminary conversion to the more easily reduced carboxyl phosphate. The photosynthetic reductant $H$ in this hypothesis was to have the same redox potential as the nicotinamide coenzymes which are unable to bring about the unassisted reduction of carboxyl to aldehyde. The high energy phosphate esters (ATP) were to be generated by direct or indirect recombination of $H$ and $OH$. This hypothesis, which Rabinowitch (for once lacking his usual perspicacity) found 'improbable' was soon to be placed on a firm experimental basis as the evolving Benson–Calvin cycle (Reductive Pentose Phosphate Pathway) called for the utilization of $NADPH_2$ and ATP in precisely the same ways as Ruben had suggested. With this, and the demonstration by Arnon and his colleagues (Fig. 2.4) that chloroplasts could reduce $CO_2$ to $CH_2O$ in the dark using 'assimilatory power' ($ATP + NADPH_2$) Ruben's notion became fact.

In one respect, however, Ruben's hypothesis is no longer tenable. As in many later schemes, it seemed reasonable that ATP might be formed indirectly by reoxidation of $H$ and at least this had the merit of utilizing known pathways if $H$ in the form of $NADPH_2$ could be transmitted to the mitochondria. When Arnon, Allen and Whatley discovered cyclic photophosphorylation in 1954, however, it was achieved by chloroplast preparations that were largely free of mitochondria.

Moreover it could be demonstrated under anaerobic conditions which would not support oxidative photophosphorylation. Photophosphorylation associated with direct recombination of *H* and *OH* remained a possibility and indeed this principle remains embodied in the present concept of cyclic photophosphorylation (below). In August 1957 however the Arnon laboratory stood the world of photosynthesis on its head by announcing non-cyclic photophosphorylation (Equation 5.11).

$$NADP^+ + ADP + Pi + H_2O \xrightarrow{h\nu} NADPH_2 + ATP + H_2O + \tfrac{1}{2}O_2 \qquad \text{Eqn. 5.11}$$

Here was the apparent converse of oxidative phosphorylation in which ATP was formed as $NADH_2$ was reoxidized with a concomitant uptake of $O_2$ and formation of $H_2O$ [Water donates electrons to $NADP^+$ in Equation 5.11 above, the $H_2O$ which reappears on the right of the equation is released during the condensation of ADP and Pi]. While the photosynthetic world at large had therefore been prepared to accept cyclic photophosphorylation and $NADP^+$ reduction without demur it was entirely unprepared for the revelation that ATP formation might be associated with the reduction of $NADP^+$ rather than its reoxidation. This, at first glance, was so diametrically different to respiratory-chain phosphorylation that even the recognition that the one was light-driven and the other substrate-driven failed to bring much enlightenment. What was even more difficult to understand was the fact that $NADP^+$ reduction was faster when it was coupled to ATP formation than when ADP and Pi were omitted. As Hill and Bonner pointed out, this seemed to eliminate the possibility that ATP formation derived from a recombination of *H* and *OH*. If such a recombination occurred independently of phosphorylation it was difficult to see why the presence of ADP and Pi should make it go faster. [Chemiosmosis was not yet formulated]. If such a recombination depended on the presence of ADP and Pi the rate of formation of *H* and *OH* would be slowed by the extent to which *H* and *OH* reunited in the phosphorylation process.

This enigma was only resolved when Hill and Bendall postulated the Z-scheme. Photophosphorylation was then no longer the 'reductive' process that it had at first seemed but one in which ATP was formed, as in oxidative phosphorylation, by electrons following a downhill (oxidative) path through a series of carriers including cytochromes.

## 5.3 Cyclic photophosphorylation

What was at first called 'photosynthetic phosphorylation' soon became 'cyclic photophosphorylation' in order to distinguish it from the subsequently discovered variant in which $NADP^+$ or ferricyanide functioned as electron acceptors.

Cyclic photophosphorylation (Fig. 5.1) did not require $O_2$ uptake or lead to $O_2$ evolution and could be represented simply as

$$ADP + Pi \xrightarrow{h\nu} ATP + H_2O \qquad \text{Eqn. 5.12}$$

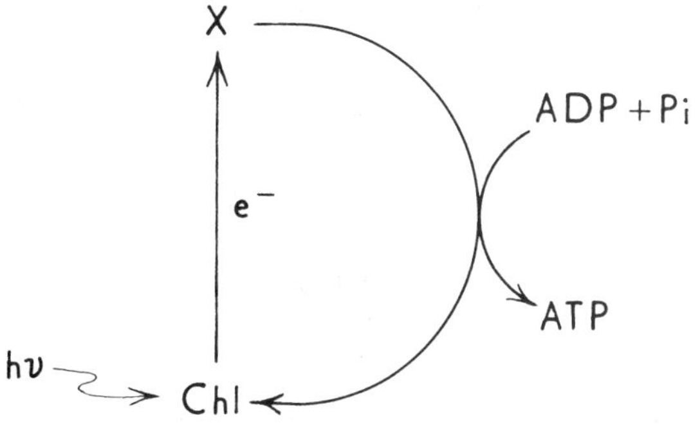

**Fig. 5.1** Cyclic photophosphorylation (or electron transport) as originally postulated by Arnon *et al.*

The term 'cyclic' relates to a mechanism of electron flow based by Arnon and his colleagues, on an idea by Lewis and Lipkin in which photochemical excitation led to electron loss. In their scheme Arnon *et al.* envisaged photoxidation of Chl as the primary event. This was to be followed by the transfer of the excited electron to a co-factor and then back via a series of carriers to fill the 'hole' in the oxidized Chl. The electron accordingly followed a cyclic route during which some of the energy it dissipated was conserved in ATP formation. At the time of its discovery a great deal of importance was attached to rates (which at first seemed to some, to be too low) and to the nature of the co-factor. The latter centered on the difference between 'non-physiological' or artificial co-factors and possible natural co-factors such as menadione (the equivalent of which was believed to occur as vitamin K within the chloroplast). At present these are no longer important issues. Good rates of phosphorylation have been achieved with a list of substances so long that it could easily be extended to include marmite and the dust from some laboratory floors. Whether many of these can function as co-factors of true cyclic photophosphorylation is another matter. Some such as methyl viologen (paraquat) undoubtedly bring about pseudocyclic photophosphorylation (below) but the precise role of others has not warranted examination. Of the 'non-physiological' co-factors only two can be more or less safely excluded from the pseudocyclic mode of operation. These are PMS (phenazine methosulphate, or more properly, methyl phenazonium methosulphate) and pyocyanine (one of its photoxidation products). [Even here there is some uncertainty. Pure pyocyanine undoubtedly catalyses photophosphorylation by higher plant chloroplast but may be less active in algal systems. Conversely, PMS has been used widely in experiments with higher plant chloroplasts but has been reported to be inactive if strict precautions are taken to avoid its photoxidation to pyocyanine]. In terms of the Z-scheme, pyocyanine short-circuits electron flow by returning electrons from some site near bound-ferredoxin to some point near *P700* (Fig. 5.2). Cyclic photophosphorylation is therefore primarily associated with PSI and the pyocyanine catalysed reaction will work in the presence of DCMU (an inhibitor of PSII) if steps are taken to prevent over-reduction. [In this and other

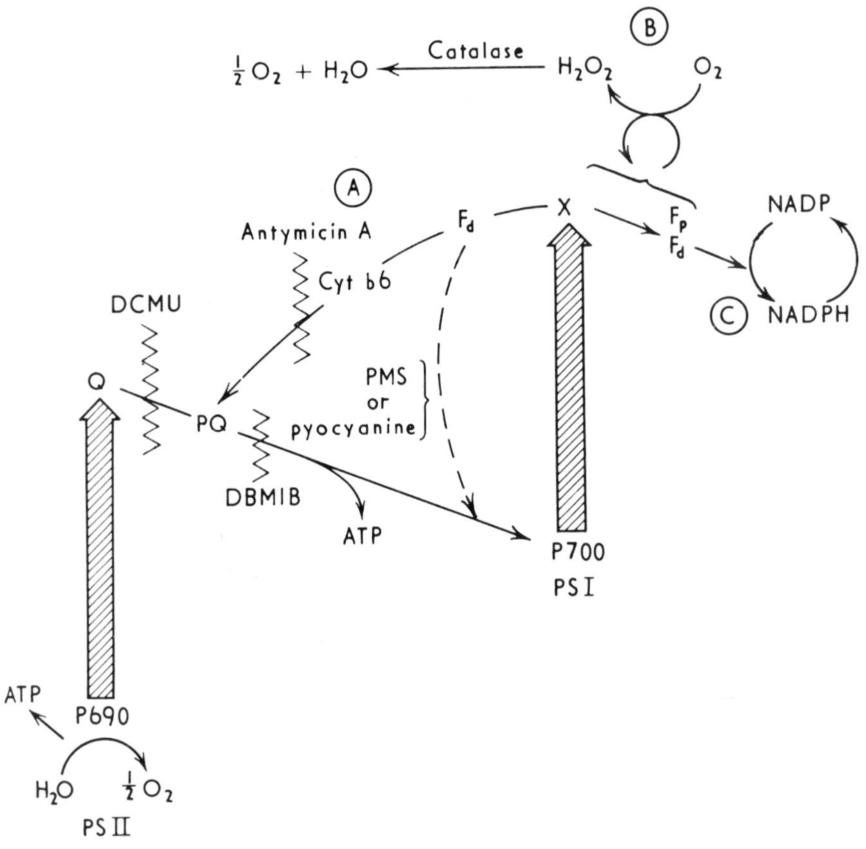


**Fig. 5.2** Possible routes of electron transport in chloroplasts. A, cyclic; B, pseudocyclic (Mehler reaction) and C, noncyclic. See Fig. 4.13 for details of pigment systems, and carriers of Z-scheme. In the presence of catalase (not a chloroplast component) the $H_2O_2$ produced is broken down to $O_2$ and water.

situations, the way in which an oxidation/reduction system is 'poised' is of importance because carriers have the dual function of electron acceptor and electron donor. If PMS becomes too reduced its electron accepting ability is slowed, a fact which might conceivably be an alternative explanation for its failure to 'work' under strictly anaerobic conditions—see above].

## 5.4 Cyclic photophosphorylation catalysed by ferredoxin

Ferredoxin is able to catalyse cyclic photophosphorylation under both aerobic and anaerobic conditions. As with pyocyanine, there is a 'poising' problem under anaerobic conditions. This time, however, it appears that ferredoxin cannot cycle until it is mostly reduced. Consequently it is possible to observe a lag in the onset of cyclic photophosphorylation which can be diminished by any of a number of treatments which speed the initial reduction (such as introducing some electrons from water or from an ascorbate/dichlorophenolindophenol couple). As with pyocyanine, however, there is also the danger of over-reduction (electrons from water could compete with electrons from cyclic flow to reduce plastoquinone). For this reason, CMU and DCMU (inhibitors of photosystem II) accelerate cyclic photophosphorylation by ferredoxin at low concentration but as their concentration is raised, inhibition sets in. (If 'poising' were not a problem, cyclic photophosphorylation should function in the complete absence of any contribution from photosystem II).

With ferredoxin as catalyst, cyclic photophosphorylation is favoured by the long wavelength red light which is optimal for photosystem I and is selectively inhibited by antimycin A (an antibiotic which inhibits electron transfer between cytochromes *b* and *c* in respiratory chain phosphorylation). The latter suggests that electrons from ferredoxin may re-enter the bridge between the two photosystems through a link (cytochrome *b* 563) which is more sensitive to antimycin A than the route followed by electrons from water. The fact that DBMIB inhibits both cyclic and non-cyclic photophosphorylation suggests that plastoquinone is common to both pathways (Fig. 5.2).

## 5.5 Cyclic photophosphorylation in-vivo

Whether or not cyclic photophosphorylation occurs in-vivo during normal functioning of non-cyclic photophosphorylation remains to be certainly established. It has been suggested by Arnon that $NADPH_2$ may poise the system like DCMU. Thus if the reoxidation of $NADPH_2$ by $CO_2$ were curtailed for any reason the accumulating $NADPH_2$ would favour cycling. While there seems little doubt, however, that reduced ferredoxin can be induced to react with other carriers in the chain the evidence in favour of direct oxidation by molecular oxygen (below) seems more compelling. Either process could contribute to ATP formation in excess of that associated with $NADP^+$ reduction.

## 5.6 Pseudocyclic photophosphorylation

As noted above, the term 'cyclic' relates to the proposition that the electrons involved in this process pursue a closed route through photosystem I, falling back into the reaction centre from which they were initially emitted. In this way the cyclic system neither consumes nor evolves oxygen and electron transport within it can not be detected on this basis. Similarly, however, conventional methods of measuring $O_2$ can not be used to follow a process in which there is simultaneous uptake and evolution of oxygen in equal quantities. At one stage such sequences were called 'pseudocyclic' because they could not be readily distinguished from a 'true' cycle in which $O_2$ does not participate. In Section 5.8 they are considered as examples of non-cyclic electron transport in which molecular oxygen functions as the electron acceptor (i.e. as Mehler reactions) (Fig. 5.3).

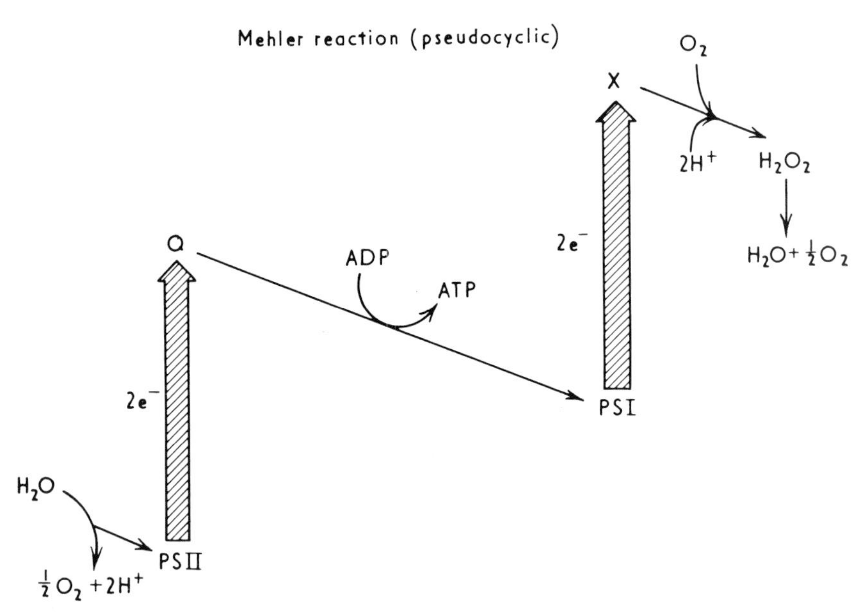


**Fig. 5.3** Pseudocyclic electron transport. Although there is no $O_2$ evolution this is not a true cycle as in Fig. 5.1. Instead $O_2$ evolution is masked by equimolar $O_2$ consumption but, nevertheless, electrons traverse both photosystems and are finally used to reduce $O_2$ rather than $NADP^+$.

## 5.7 Non-cyclic photophosphorylation

As the name suggests, this is a form of photophosphorylation associated with the transport of electrons from a donor (such as $H_2O$) to an acceptor (such as $NADP^+$ or ferricyanide). With $H_2O$ and $NADP^+$ it involves both photosystems (Section 4.13 and Fig. 5.2 for further details of noncyclic electron flow). However it can be driven by PSI alone if PSII is poisoned with DCMU and electrons are donated from a reductant of sufficiently low potential to feed directly into the Z-scheme between PSII and PSI. The blue dye, 2,6-dichlorophenolindophenol (DCIP) which is often used in visual demonstrations of the Hill reaction (Equation 5.13) can be made to function in this fashion (Fig. 5.4).

$$\text{Dye (blue, oxidized)} + H_2O \rightarrow \text{Dye } H_2 \text{ (colourless, reduced)} + \tfrac{1}{2}O_2 \qquad \text{Eqn. 5.13}$$

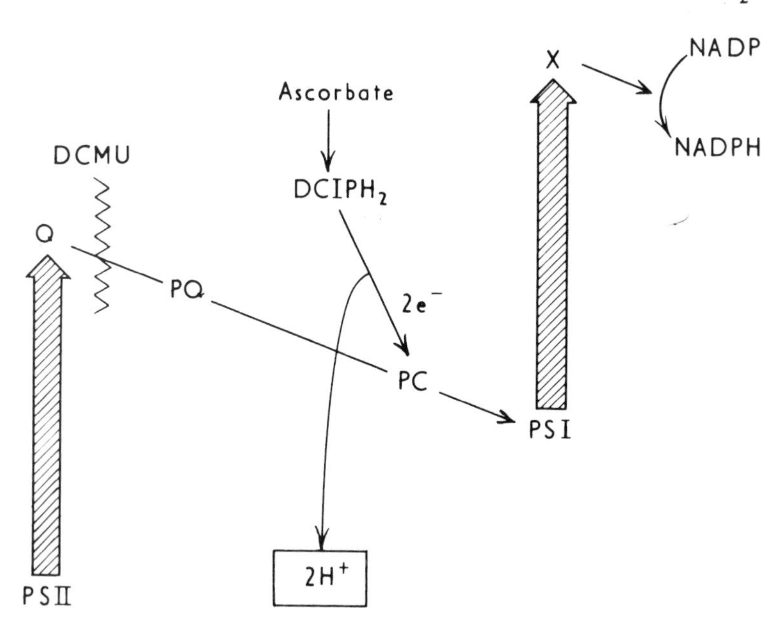


**Fig. 5.4** Non-cyclic electron transport involving one photosystem. Transport from PSII is blocked by DCMU and electrons are donated by an ascorbate/DCIP couple.

If added in catalytic quantities together with ferricyanide (another oxidant which by itself will support non-cyclic photophosphorylation) it will capture electrons near PQ and hand them directly to ferricyanide, bypassing some of the carriers in the 'bridge' between the two photosystems and thereby diminishing ATP formation. If, however, it is used in catalytic quantities together with ascorbate (as an electron donor) and, if PSII is blocked by the addition of DCMU, it will pass electrons from ascorbate via the 'bridge' and PSI to $NADP^+$. For this reason ATP is formed even though PSII is inoperative. [This approach was first used by Losada, Whatley and Arnon to demonstrate that non-cyclic photophosphorylation could be divided, by chemical manipulation into two parts: the photolysis of water and the subsequent transport of electrons through an ATP generating system to $NADP^+$. It now appears that the splitting of water is also associated with a phosphorylation site (Section 5.9). In the case of ascorbate-DCIP the plastoquinone site is bypassed and electrons are donated into the chain at plastocyanin. KCN (blocks at plastocyanin) is inhibitory while DBMIB (plastoquinone antagonist) or DCMU have no effect on this donor system. $DCIPH_2$ is considered to serve as an artificial carrier of protons across the membrane generating the protonmotive force required for phosphorylation (Section 5.9). Ascorbate-TMPD (n-tetramethyl-p-phenylenediamine) is an effective artificial donor system to PSI but one which is non-phosphorylating since TMPD only donates electrons and thus does not shuttle protons across the membrane].

## 5.8 The Mehler Reaction

In 1951 Mehler reported a variant of the Hill reaction in which molecular oxygen acts as the final oxidant. The Hill reaction leads to the evolution of half a molecule of oxygen for every molecule of water consumed

$$H_2O + A \xrightarrow{h\nu} AH_2 + \tfrac{1}{2}O_2 \qquad \text{Eqn. 5.14}$$

whereas the Mehler reaction leads to a correspondingly large $O_2$ consumption i.e.

$$H_2O + O_2 \xrightarrow{h\nu} H_2O_2 + \tfrac{1}{2}O_2 \qquad \text{Eqn. 5.15}$$

which is equivalent to

$$H_2O + \tfrac{1}{2}O_2 \xrightarrow{h\nu} H_2O_2 \qquad \text{Eqn. 5.16}$$

In the presence of excess catalase (an enzyme which facilitates the dismutation of hydrogen peroxide to water and $O_2$) no net $O_2$ exchange is observed (Fig. 5.3).

$$H_2O + O_2 \rightarrow H_2O_2 + \tfrac{1}{2}O_2 \qquad \text{Eqn. 5.17}$$

$$H_2O_2 \rightarrow H_2O + \tfrac{1}{2}O_2 \qquad \text{Eqn. 5.18}$$

And, indeed, if Equations 5.17 and 5.18 are added no net change is found although electrons have passed through the Z-scheme.

A great many natural and artificial electron acceptors will act as Mehler reagents including methyl viologen (paraquat) and flavin coenzymes such as FMN (flavin mononuleotide). These may be regarded as low potential and readily auto-oxidizable Hill oxidants

$$H_2O + A \rightarrow AH_2 + \tfrac{1}{2}O_2 \quad \text{Eqn. 5.19}$$

$$AH_2 + O_2 \rightarrow A + H_2O_2 \quad \text{Eqn. 5.20}$$

$$\text{SUM} \quad H_2O + O_2 \rightarrow H_2O_2 + \tfrac{1}{2}O \quad \text{Eqn. 5.21}$$

Although these equations illustrate the essential features of the Mehler reaction they may be misleading if it is inferred that the transfer of $H_2$ to $O_2$ is a simple process that proceeds in the same way regardless of the nature of the oxidant.

For example, the above sequence suggests that 2 electrons and 2 protons (i.e. 2 hydrogens) are transferred to $O_2$ to yield peroxide. Normally however, a single electron may be donated to $O_2$ to form the superoxide radical $O_2^-$

$$O_2 + e^- \rightarrow O_2^- \quad \text{Eqn. 5.22}$$

In the presence of an enzyme called superoxide dismutase (SOD) $O_2^-$ is converted to the peroxide radical ($O_2^{2-}$) and hydrogen peroxide

$$2O_2^- \rightarrow O_2^{2-} + O_2 \quad \text{Eqn. 5.23}$$

$$O_2^{2-} + 2H^+ \rightarrow H_2O_2 \quad \text{Eqn. 5.24}$$

but the superoxide radical can oxidize compounds such as ascorbate

$$AH_2 + 2O_2^- \rightarrow A + 2O_2^{2-} + 2H^+ \quad \text{Eqn. 5.25}$$

Thus the transfer of two electrons from water can simultaneously lead to the oxidation of ascorbate and a net $O_2$ uptake of $(2 - \tfrac{1}{2}) = 1\tfrac{1}{2}O_2$

$$H_2O + 2O_2 \rightarrow 2H^+ + 2O_2^- + \tfrac{1}{2}O_2 \quad \text{Eqn. 5.26}$$

$$AH_2 + 2O_2^- \rightarrow A + 2H^+ + 2O_2^{2-} \quad \text{Eqn. 5.27}$$

$$2O_2^{2-} + 4H^+ \rightarrow 2H_2O_2 \quad \text{Eqn. 5.28}$$

$$\text{SUM} \quad H_2O + AH_2 + 1\tfrac{1}{2}O_2 \rightarrow 2H_2O_2 + A \quad \text{Eqn. 5.29}$$

[In the presence of excess catalase this reaction would proceed with a net uptake of $\tfrac{1}{2}O_2$ per ascorbate oxidize].

## 5.9 Chemiosmosis and the establishment of an electrochemical potential difference or protonmotive force.

Light excitation of chlorophyll promotes an electron to a new high energy state (Section 4.16) initiating an electron flow. From water, the ultimate electron donor, electrons flow through the photosystems and their associated carriers to NADP (Section 4.9). Each oxidation/reduction step in this sequence behaves in a characteristic fashion. Some compounds, like plastoquinone, can accept and donate two hydrogens. Others, like cytochrome *f*, and ferredoxin can only accept one electron. NADP, the terminal acceptor can accept two electrons and one proton. Electron transport, therefore, involves a series of consecutive reactions in which protons are

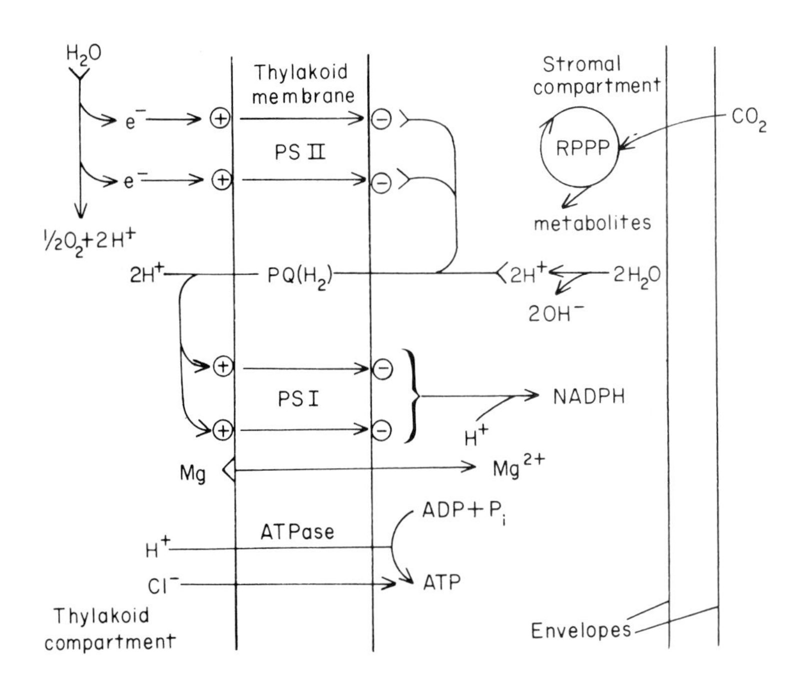


**Fig. 5.5** Consequences of illumination. The pigments and electron transport chain are housed in the thylakoid membrane. On illumination, electrons are donated to excited chlorophyll in photosystem II (PS II). The transfer of electrons to photosystem (PS I) via plastoquinone (PQ) requires protons that are taken up from the thylakoid compartment and released into the stroma. The stromal pH rises and $Mg^{2+}$ moves as a counter ion. [ATP] increases and [ADP] and [Pi] fall as $H^+$ is discharged through the ATPase in photophosphorylation. A reductant formed by donation of electrons from PS I will become available for the direct or indirect reduction of ferredoxin, NADP, and disulphide groups. Metabolite concentration will slowly increase. (RPPP, reductive pentose phosphate pathway).

released into, or taken up from, the immediate environment. As Hind and Jagendorf showed, a weakly buffered medium in which envelope-free chloroplasts are suspended becomes relatively alkaline upon illumination and relatively acid upon darkening. The Mitchell chemiosmotic hypothesis, as it is applied to photosynthesis seeks to correlate these facts and to ascribe to them a central role in ATP synthesis. It is proposed that protons which are taken up at some points in the electron transport chain are taken up exclusively from the outside of a thylakoid membrane and that protons which are put down are released exclusively on the inside of a thylakoid membrane. As a result of electron transport through the carrier chain a proton gradient is therefore established across the membrane. Protons are not, in this sense, passed directly across the membrane but simply released into the inner space (and taken up from the outer space) as a consequence of electron transport. The net effect is that the inner space becomes more acid and the outer space more alkaline (Figs. 5.5 and 5.6).

In non-cyclic electron flow two sites are proposed for generating the proton gradient, the splitting of water inside the thylakoid compartment and the oxidation-reduction of plastohydroquinone-plastoquinone. For example, studies with hydro-

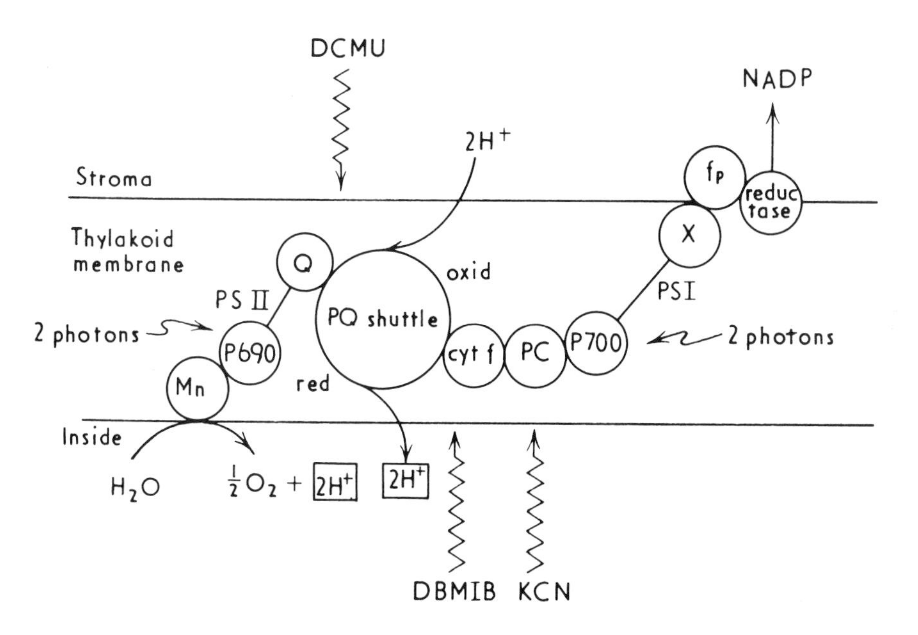


**Fig. 5.6** Another view of the mechanism and events shown in Fig. 5.5 but also giving some indication of the physical distribution of the major carriers and the points of action of inhibitors such as DCMU.

philic and hydrophobic acceptors and electron donors to PSI and PSII (and also experiments with antibodies to specific carrier proteins) indicate PSII is located towards the inner face and PSI on the outer surface of thylakoid membrane. Plastoquinone, a mobile carrier in the membrane, would accept protons from the stromal space as it is reduced and release protons inside the thylakoids as the plastohydroquinone is oxidized (Fig. 5.6). Photophosphorylation dependent only on PSII is revealed by blocking electron flow between PSII and PSI with KCN (at plastocyanin) or DBMIB (at plastoquinone) and adding benzoquinone or phenylenediamines as electron acceptors. Cyclic electron flow in the absence of artificial catalysts is also thought to involve electron transport through plastoquinone (Fig. 5.2 and Section 5.4).

Upon illumination the ΔpH between the thylakoid compartment and the stroma is about 3–3.5 pH units and may be determined by measuring the uptake of a weak base. This is sufficiently large that light/dark changes in the external pH with a suspension of thylakoids can be measured readily with a conventional glass electrode. In weakly buffered reaction mixtures containing 50 $\mu$g Chl/ml, pH shifts of 0.2–0.3 units are quickly established upon exposure to near-saturating light. On darkening there is a slower return to something near the starting pH and alternating rises and falls in pH in response to illumination and darkening can be continued almost indefinitely. The extent of the pH shift is a function of electron transport. It can be increased by the addition of co-factors which increase electron transport without themselves materially affecting the pH as a consequence of reduction. The pH of the medium becomes more acid if the illumination is decreased and returns to its original value if full light is restored. Detergents, and other treatments which are believed to render the membrane more permeable to protons (and, therefore, presumably accelerate leak-back) can, in most circumstances, discharge the pH shift. The fact that the pH does not continue to rise indefinitely upon illumination but, instead, quickly attains a steady state is itself attributed to an increasing back flow of protons through the membrane as the proton concentration in the inner space increases. The back-pressure exerted by the increased concentration of protons is assumed to exert a braking effect on electron transport, i.e. it would presumably become more difficult to release a proton into a high concentration of protons (inside) or to pick one up from a low concentration (outside). The acceleration of electron flow by uncouplers is seen as a response to the relief of this back-pressure as the proton gradient is discharged.

The gradient which results from electron transport has an electrical as well as a chemical component. An accumulation of positively charged protons will tend to produce an imbalance of electrical charge which will be offset, to a larger or smaller extent, by consequent inward or outward movement of charge compensating anions or cations. The protonmotive force (pmf) is the sum of the electrical component ($\Delta\psi$, the extent of the positive charge inside the thylakoids) and the chemical component ($Z\Delta pH$), where $Z = RT/f = 60$ mv

Thus, $pmf = \Delta\psi + Z\Delta pH$ Eqn. 5.30

A variant of this aspect of Mitchell's hypothesis has been proposed by Witt and his colleagues who conclude that charge separation is an immediate rather than a secondary consequence of the primary photochemical events. Thus, within 20 ns of illumination, the thylakoid membrane has become electrically energized. A force-field is therefore established upon Chl excitation and at the end of this stage a negative charge (electron) is located at the outer surface and a positive charge (hole) at the inner surface, giving rise to a potential difference ($\Delta\psi$) across the membrane. At the outer surface, two electrons from photosystem II then reduce plastoquinone and two from photosystem I reduce $NADP^+$ (Fig. 5.5). Each of these reductions is accompanied by the uptake of protons from the outer space. At the inner surface protons are released as PQ is oxidized. Although the initial charge separation is discharged by this time, an electrical field or charge imbalance persists because postively charged protons have been moved from the outside (which therefore, remains negative) to the inside (which therefore, remains positive).

## 5.10 The mechanism of ATP formation

In photosynthesis, the work of Jagendorf, Hind, Neumann and Uribe led to the conclusion that a light-generated high-energy intermediate in ATP synthesis might be simply a pH gradient. In a famous 'acid-bath' experiment Jagendorf and Uribe showed that an artificially generated gradient could bring about ATP synthesis in the dark. This work derived from and reinforced Mitchell's chemiosmotic hypothesis in which ATP synthesis is associated with the discharge of a proton gradient through a reversible ATPase. In simplest terms, the sequence of events is as follows
(a) Electron transport within a membrane establishes a proton gradient across that membrane (Section 5.9).
(b) ATP synthesis occurs by reversal of ATP hydrolysis via a membrane-localized, reversible, vectorial ATPase.
(c) Each molecule of water split off during the esterification process of ATP synthesis is released in a vectorial fashion, as a proton to one side of the membrane and an hydroxyl ion to the opposite side of the membrane, thus neutralizing the proton gradient established by electron transport (Fig. 5.7). In these terms, the proton gradient serves as a sink for hydrogen ions and hydroxyl ions, pulling the unfavourable equilibrium in the direction of ATP synthesis. Even in this simple-

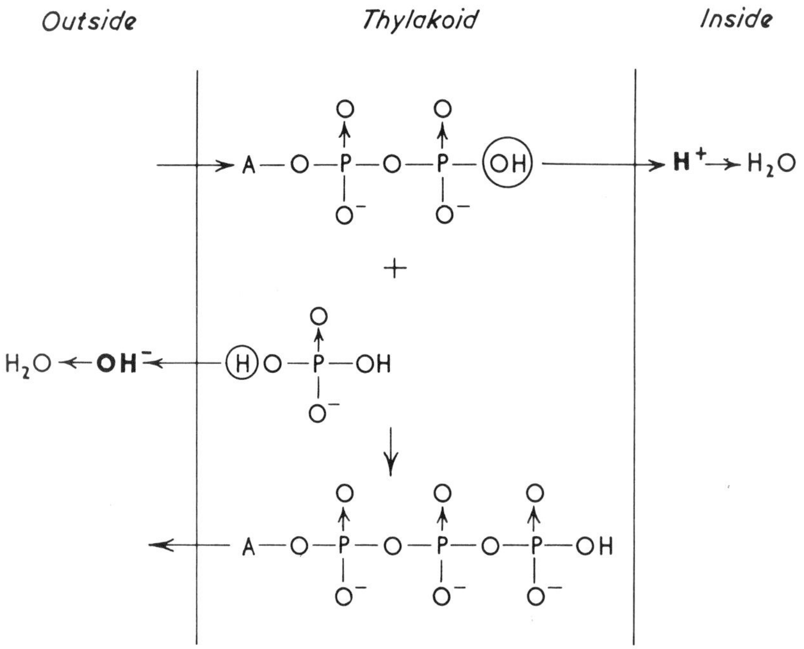

**Fig. 5.7** Simplified concept of ATP formation. Electron transport establishes a proton gradient (Fig. 5.5) which serves as a sink for protons and hydroxyl ions released in esterification thus displacing the unfavourable equilibrium position of this reaction.

minded picture the envisaged movement of ions has, of course, an electrical as well as a chemical component and the discharge of the chemical component which follows the recombination of $H^+$ and $OH^-$ outside the membrane is associated with a corresponding rectification of electrical imbalance. Since the first experiments and explanations, increasing weight has been attached to the notion that phosphorylation may be driven by a protonmotive force (i.e. a proton gradient a charge separation or any combination of both). Thus phosphorylation has been reported to have been observed without the establishment of a pH gradient or under conditions where an inadequate pH-gradient has been supplemented by a charge separation brought about by substituting $K^+$ and $H^+$. Conversely, uncoupling (Section 5.11) has been achieved without discharge of the pH gradient.

Especially in its later forms the chemiosmotic hypothesis readily accommodates a great many experimental observations. There is no doubt, for example, that a pH gradient is quickly established in illuminated chloroplasts and that the discharge of such a gradient is associated with ATP formation. Acceleration of electron transport by uncouplers or by ADP + Pi (photosynthetic control) is seen as a relief of the back pressure exerted by the gradient and its associated charge separation. At a relatively early stage Avron isolated a coupling factor (now designated $CF_1$) from chloroplasts (Section 5.12). If this compound is removed by washing chloroplasts with EDTA, electron transport is uncoupled from phosphorylation. If it is added back, phosphorylation and the effect produced by ADP and Pi on electron transport are restored. Racker showed that $CF_1$ had latent ATPase activity. The coupling factor in mitochondria, bacteria and chloroplasts is very similar. It has two components, an ATPase component (designated $F_1$ in bacteria and mitochondria and $CF_1$ in chloroplasts) and a membrane component (designated $CF_0$). Membranes are normally impermeable to protons. The $CF_0$ component provides a channel (Fig. 5.8) for conducting protons across the membrane through the ATPase during

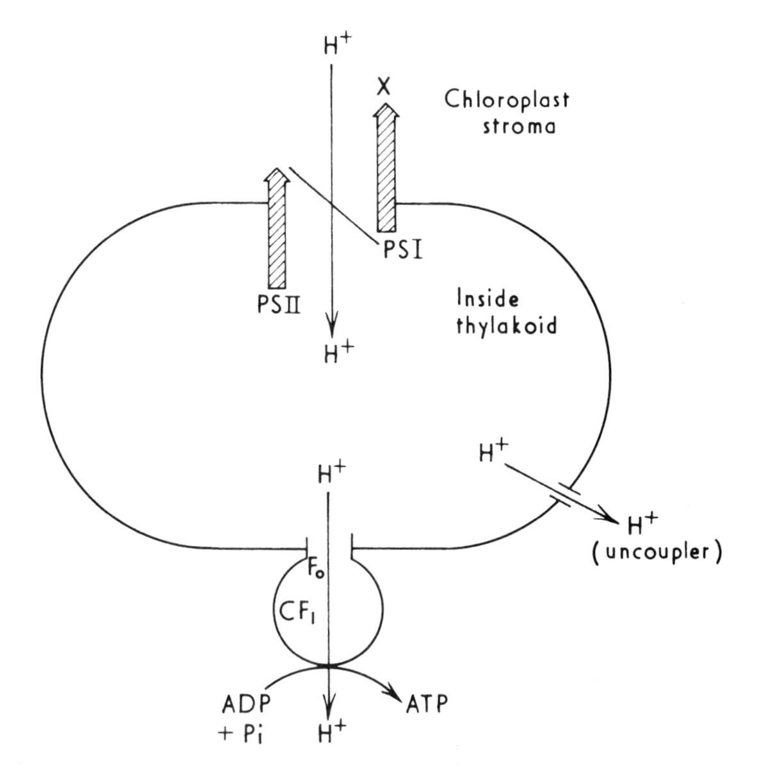


**Fig. 5.8** Illustration of the electron transport linked proton uptake into the inner compartment of the thylakoids. The proton gradient across the thylakoid membrane can by dissipated by transport through the coupling factor which facilitates ATP synthesis or artificially by uncouplers. $CF_1$ is the ATPase component of the coupling factor and $CF_0$ forms a channel in the membrane through which protons are transported.

phosphorylation. The total coupling factor complex in chloroplasts is referred to as $CF_1$-$CF_0$ (Section 5.12 and Fig. 5.8).

The actual mechanism of ATP synthesis may well be much more complicated than Fig. 5.7 implies. For example Dixon has suggested a variant in which Pi moves as $H_1PO_4^{2-}$ into a phosphorylation compartment. Protons which have accumulated in this compartment as a consequence of electron transport encourage the inward movement both by virtue of the positive charge and by forming a sink in which $H_1PO_4^{2-}$ is partially converted to $H_2PO_4^-$ (Fig. 5.9). External $ADP^{3-}$ also exchanges $ATP^{3-}$ within the coupling factor complex in the thylakoid membrane. The $ATP^{3-}$ then dissociates externally (to $H^+$ and $ATP^{4-}$) thereby facilitating this movement. In this way, proton accumulation brings about an increase in the concentration of ADP and Pi at the phosphorylation site. Moreover, phosphate might move after covalent attachment to a carrier which, following protonation in an acid compartment would become a more powerful phosphorylating agent than $H_1PO_4^{2-}$.

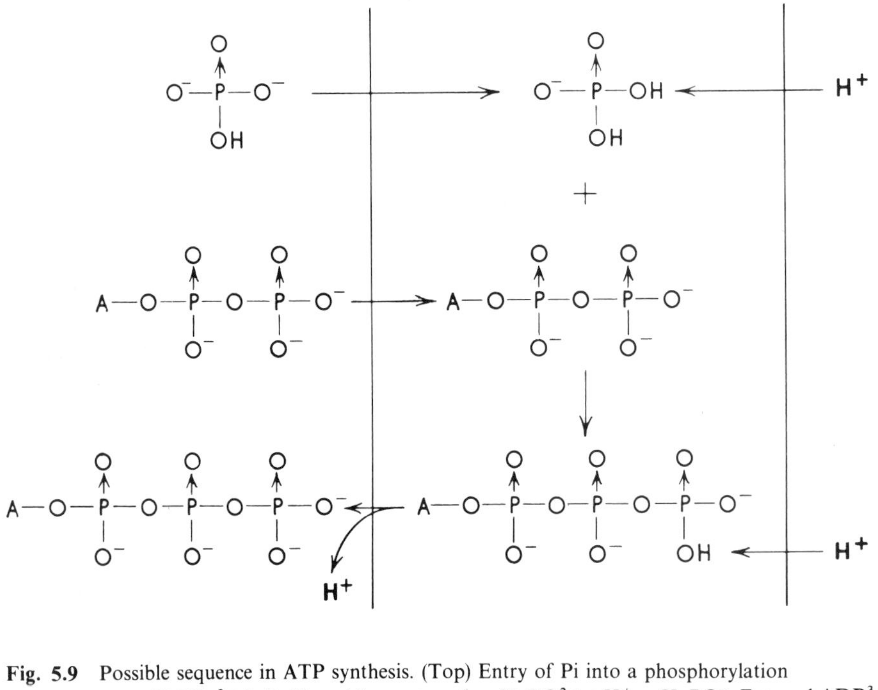

**Fig. 5.9** Possible sequence in ATP synthesis. (Top) Entry of Pi into a phosphorylation compartment as $H_1PO_4^{2-}$ is facilitated by protonation $H_1PO_4^{2-} + H^+ \rightarrow H_2PO_4^-$ External $ADP^{3-}$ exchanges with internal $ATP^{3-}$ which dissociates externally ($ATP^{3-} \rightarrow ATP^{4-} + H^+$) also facilitating this exchange.

[In any event, diffusion gradients for ATP and ADP must be set up between the stroma and the outer surface of the thylakoids in order that ATP might travel from its site of synthesis to its site of consumption in the reactions of the photosynthetic carbon cycle].

ATP is a 'high energy' compound because of its internal molecular environment in which there is electrostatic repulsion between adjacent phosphate groups. Synthesis of ATP from ADP and Pi would therefore be favoured in an environment in which over-riding electrostatic forces facilitated the condensation of the two reactants. Bennun has suggested that electrostatic complex formation involving $CF_1$ (Section 5.12) might facilitate energy conservation (see also Chapter 2).

## 5.11 Coupling, uncoupling and photosynthetic control

In-vivo, electron transport is 'coupled' to phosphorylation (i.e. the two processes are linked). If coupling were complete (or extremely 'tight') there would be no electron transport in the absence of ATP formation. In-vitro, although the rate of electron transport may be increased by a factor of 4–6 in the presence of ADP and Pi, there is always some basal or non-phosphorylating electron transport, presumably because of proton leakage (Sections 5.9, 5.10). A great many compounds can act as uncouplers of photophosphorylation, decreasing the ATP/2e ratio and increasing the rate of non-phosphorylating electron transport. In terms of the Mitchell hypothesis and its variants the back pressure of the proton gradient imposes a restraint on electron transport so that anything which acts by removing this restraint might be expected to accelerate electron transport. Thus detergents which make holes in the membrane will discharge the gradient by permitting rapid back diffusions. Arsenate and thiophosphate act as uncouplers in the presence of ADP (and are competitive with respect to Pi) suggesting that they form unstable derivatives of ADP which spontaneously decompose. Compounds such as $NH_3$ which act as proton-sinks ($NH_3 + H^+ \rightarrow NH_4^+$) and antibiotics, such as nigericin and gramicidin D, which are believed to facilitate exchange of $H^+$ for $K^+$ will also tend to discharge the chemical component of the $H^+$ gradient thereby accelerating electron transport and decreasing photophosphorylation. The mechanism of action of these compounds is not always as straightforward as it might appear at first glance and, in many cases, may still be imperfectly understood. For example $NH_3$ acts as a sink for protons according to the reaction $NH_3 + H^+ \rightarrow NH_4^+$ but whereas this would decrease 'H' it would not alter the charge contribution to the total pmf. If, on the other hand, $NH_3$ diffuses into the thylakoid compartment and $NH_4^+$ diffuses out both the chemical and electrical components of $H^+$ will be discharged. Several uncouplers may function, in part, by initiating changes of this nature. Some movement of charge compensating ions (inward movement of anions such as $Cl^-$ and /or outward movement of cations such as $Mg^{2+}$) must also occur in all normal circumstances because the magnitude of the electrical component of the proton gradient would otherwise become too high to be tenable. Careful heat treatment of chloroplasts will accelerate both influx and efflux of protons, leaving the gradient untouched or increased but inhibiting photophosphorylation. Emmett and Walker attributed this to a heat-induced increase in the permeability of the membrane which permits more rapid movement of charge compensating ions so that the electrical component of the $H^+$ gradient is discharged.

When both are present, ADP and Pi also accelerate electron transport presumably by discharging part of the $H^+$ gradient through the reversible ATPase. When all of the ADP is esterified, electron transport slows. The acceleration and subsequent retardation constitute 'photosynthetic control' which is analogous to 'respiratory control' in mitochondria.

## 5.12 Coupling factor

As already noted (Section 5.10) Avron made the important discovery (in 1963) that when envelope-free chloroplasts are treated with EDTA they become uncoupled and at the same time lose a 'coupling factor' ($CF_1$). When added back, in the presence of $Mg^{2+}$ the coupling of ATP synthesis to electron transport was restored. Subsequently coupling factors have been extensively studied by Avron, Racker, McCarty, Nelson and others. Brief heating, incubation with trypsin, or reduction by high concentrations of sulfhydryl reagents (such as dithiothreitol) converts free $CF_1$ to a highly active $Ca^{2+}$-dependent ATPase, implying a role for $CF_1$ in photophosphorylation. When bound to the thylakoid membrane it is attached to a specific hydrophobic protein $CF_0$ thus comprising an ATPase complex ($CF_0$-$CF_1$) which undergoes conformational changes in the light or when subjected to a reasonably large pH gradient. In its relation to the Mitchell Chemiosmotic Hypothesis (Section 5.11) $CF_0$ is regarded as a channel which specifically directs protons to $CF_1$, the catalyst of ATP formation. The complete $CF_0$-$CF_1$ complex, when reconstituted in artificial phospholipid vesicles, catalyses ATP-Pi exchange and synthesises ATP if subjected to a transmembrane pH gradient.

## 5.13 Thermodynamics of ATP formation

The standard free energy of hydrolysis of ATP to ADP + Pi (Sections 2.4, 2.11) depends to some extent on the presence or absence of $Mg^{2+}$ but is usually put at about $-7$ kcal. This relates to equimolar concentrations of ATP, ADP and Pi and the actual phosphorylation potential (the free energy of ATP formation) will vary with the concentration of reactants and products, according to Equation 5.31

$$\Delta F = \Delta F' + RT \ln \frac{[\text{ATP}]}{[\text{ADP}][\text{Pi}]} \qquad \text{Eqn. 5.31}$$

and will be largest when most of the adenylate is present as ATP. Since envelope-free chloroplasts can readily convert almost all of any added ADP to ATP the phosphate potential (as measured by the observed equilibrium in a phosphorylation mixture) indicates the actual energy requirement. Such measurements by Kraayenhof give values of about 14 kcal when inserted in the equation above. The corresponding electrical potential may be calculated (Section 2.13) from the relationship

$$\Delta F' = -\text{n}fE' \text{ or } E' = \frac{-\Delta F'}{\text{n}f}$$

and for a 2-electron reaction and a $\Delta F'$ of 14 kcal, this becomes

$$\frac{14}{2 \times 23} = 300\,\text{mV (approx.)} \qquad \text{Eqn. 5.32}$$

which may be compared with the potential drop between plastoquinone and cyt $f$ (Fig. 4.18) which is of corresponding magnitude. [The total potential drop between the two photosystems is about 450 mV for 2 electrons]. ATP formation at such a 'site' (Section 4.23) is therefore energetically feasible. The corrresponding calculation for the second 'site' (Section 4.23) is speculative because it has not yet proved possible to assign firm redox values to Y and Z (Section 4.23) and reaction centre II (Section 4.12). However, if only 70% of the energy (1800 mV/quantum mole) of a photon of 680 nm light entered the electron transport chain and the potential difference between water and the primary electron acceptor in photosystem II is about 800 mV then an initial fall in potential of about 460 mV could be envisaged. This would again be more than adequate to meet the 300 mV drop demanded by a phosphorylation potential of 14 kcal.

## 5.14 Relationship between proton transport, electron transport and energy.

The chemiosmotic hypothesis predicts the influx of one proton for every electron passing each site and an efflux of 2 $H^+$ per ATP synthesized. This would give $P/2e^-$ ratios of 1.0 at each site. Actual measurements support the notion of 1 proton per 1 $e^-$ equivalence but determinations of $H^+$/ATP ratios range from 2–4. In the Z-scheme two phosphorylation sites are proposed (Section 5.9) which would result in $4H^+$ transported per two electrons (Fig. 5.5). This could give rise to 1, 1.33, or 2 molecules of ATP according to which values of $H^+$/ATP ratios prove to be the more realistic. As seen in 5.13 a phosphorylation potential of 14 kcal is equivalent to 300 mV for a two electron process and thus, in turn (on the basis of 1 pH unit $\equiv$ 60 mV) is equivalent to a $\Delta$pH of 5.0. Recorded values of $\Delta$pH lie in the 2.5–3.5 range and this would imply the need for both a proton gradient and a membrane potential. [e.g. a $\Delta$pH of 3.3 (= 200 mV) and a membrane potential of 100 mV]. In calculations of this nature it is important to note that the energy available depends not only on the redox potential difference but also on the number of electrons involved. Thus 3 electrons (and therefore $3H^+$, given a $1e^-$, $1H^+$ equivalence) will afford half as much energy again as 2 electrons if lowered through the same potential. Thus if $3H^+$ are needed to bring about the synthesis of one ATP then a $\Delta$pH of 3.3 units (= 200 mV) would suffice without any additional contribution made by a membrane potential. If a membrane potential as high as 100 mV did exist the required $\Delta$pH would fall to 1.7. [Although the contribution made by membrane potential remains a matter for interpretation, values as high as 100 mV would not seem to be improbable]. The transport of $2e^-$ through the Z-scheme gives $4H^+$ per 4 quanta (Fig. 5.6). This could yield 2 ATP/4 quanta. The transport of electrons through the Z-scheme has a fixed quantum requirement. Thus each photochemical event involves the movement of one electron and demands an input of one quantum. NADP accepts electrons from water after they have traversed the entire sequence of carriers therefore the reduction of NADP in this process requires 4 quanta, as 2 electrons are passed through the 2 photosystems. The quantum requirement for ATP generation is uncertain, however, because it depends on how many protons are moved into the thylakoid compartment for every electron transported and how many ATP's are generated by this proton gradient. If 2 protons are released at the inside of the thylakoid membrane as water donates 2 electrons to PSII (and 2 more as PQ donates 2 electrons to PSI) then the passage of 2 electrons through the Z-scheme will induce the release of $4H^+$ within the thylakoid compartment. The relationship between electron transport and ATP formation then depends on how many of these protons must be discharged through the ATP generator (Mitchell's ATPase) in order to form one ATP from ADP + Pi.

In the RPP pathway 3 molecules of ATP are required for each molecule of $CO_2$ fixed. As Table 5.1 indicates the quantum requirement for the synthesis of 3 molecules of ATP would be 6, 9 and 12 for $H^+$/ATP ratios of 2, 3 and 4 respectively. However, since 2 molecules of $NADPH_2$ are also required the minimum quantum requirement must be at least 8, dictating the values of 8, 9 and 12 given in the last column of Table 5.1. With an $H^+$/ATP ratio of 2, one molecule of ATP *in excess* of the immediate requirement of the RPP pathway would be generated for every 2 molecules of NADPH formed. For the higher quantum requirements it would not matter

**Table 5.1.** Minimum quantum requirements (QR) for generation of assimilatory power and $CO_2$ assimilation in the RPP pathway. The values are based on 1 proton transported per quanta absorbed, with $H^+/ATP$ ratios ranging from 2 to 4; and with a minimum of 3 ATP, 2 NADPH required per $CO_2$ fixed.

| $H^+$ required per ATP synthesized | $H^+$ or quanta required/3 ATP | QR and $H^+$ transported/ 2 NADPH | Minimum QR per $CO_2$ assimilated |
|---|---|---|---|
| 2 | 6 | 8 | 8 |
| 3 | 9 | 8 | 9 |
| 4 | 12 | 8 | 12 |

whether the ATP *deficit* was made good by cyclic electron transport or non-cyclic electron transport. For example, with an $H^+/ATP$ ratio of 4 there would be a deficit of one ATP for every two molecules of NADP reduced. The generation of an extra molecule of ATP would require 4 protons and this gradient could be established by the passage of 2 electrons through 2 sites (non-cyclic, costing 4 quanta) or 4 electrons through one site (cyclic, also costing 4 quanta). If however, cyclic electron transport led to the generation of more ATP than non-cyclic (if, e.g. cyclic electron transport also led to the transport of 4 protons per 2 electrons, now costing only 2 quanta) the overall quantum requirement for $CO_2$ fixation would be 10.

## 5.15 The reduction of nicotinamide adenine dinucleotide phosphate (NADP)

FERREDOXIN.

In his early work Robert Hill showed that isolated chloroplasts would evolve $O_2$ in the presence of artificial oxidants *or in the presence of aqueous extracts from acetone treated leaves*, thus implying the presence 'of soluble hydrogen acceptors for the chloroplasts which were derived from the plant itself'. Hill & Scarisbrick had used muscle methaemoglobin with ferrous oxalate in chloroplast systems. The chloroplasts reduced the ferrous oxalate. This in turn reduced methaemoglobin and the resulting haemoglobin combined with oxygen. The appearance of oxyhaemoglobin allowed a spectroscopic assay of the reaction. These observations led Hill and his colleagues to the demonstration of a protein fraction in acetone leaf-extracts which would reduce methaemoglobin (Section 4.9).

The demonstration that chloroplasts could also reduce $NAD^+$ and $NADP^+$ came a decade later when it was independently observed by several workers. A protein factor was again implicated. In 1957 Arnon, Whatley and Allen isolated an 'NADP$^+$-reducing factor' and in 1958 San Pietro and Lang partially purified a 'photosynthetic pyridinine nucleotide reductase'. By 1960 there was no real doubt that these various extracts all contained the same protein. Chemically this proved very similar to bacterial ferredoxin and Arnon found it 'appropriate' to call the chloroplast protein 'ferredoxin' in preference to its other names.

### FERREDOXIN—NADP REDUCTASE.

Following further purification it became clear that ferredoxin was, itself, entirely incapable of reducing NADP$^+$, rather that it was simply an electron carrier of low potential. [The ability of chloroplasts to reduce low-potential artificial electron carriers such as methyl viologen (paraquat) had been independently demonstrated by Jagendorf and Avron and by Hill and Walker. Methyl viologen has a potential of about $-0.450$ V, ferredoxin about $-0.430$ V and NADP$^+$ about $-0.320$ V.] Ferredoxin will also react with molecular oxygen (supporting pseudocyclic photophosphorylation) and with other electron carriers of the Z-scheme (cyt *b*?) to support soluble it is normally bound to the thylakoids, so that a mixture of washed thylakoids, ferredoxin and NADP$^+$ will support a rapid Hill reaction when illuminated.

The reductase is a typical flavoprotein with absorption peaks at 275, 385 and 456 nm. Its affinity for NADP$^+$ (Km $= 9.8 \times 10^{-6}$M) is so much greater than its affinity for NAD$^+$ (Km $= 3.7 \times 10^{-3}$M) that it is virtually NADP$^+$–specific under physiological conditions. The enzyme will, however, function as a transhydrogenase (NADPH + NAD$^+ \rightleftharpoons$ NADH + NADP$^+$) and a diaphorase (oxidation of $NADPH_2$ by a number of electron acceptors such as ferricyanide and quinones).

### THE MECHANISM OF NADP REDUCTION

Electrons are normally passed from water via the carriers of the Z-scheme to ferredoxin. This donates electrons to the flavoprotein reductase which in turn reduces NADP$^+$ (which carries 2 electrons and 1 proton in the reduced state). In order to transfer electrons from ferredoxin to NADP$^+$ a separate flavoprotein, 'ferredoxin–NADP$^+$ reductase' was also required. Although this enzyme can be made cyclic photophosphorylation. The physiological role of the reductase in pseudocyclic electron flow and $NADPH_2$ reoxidation remains to be established but, in principle, it would seem that chloroplasts have a potential mechanism for energy dissipation which is sometimes ascribed to photorespiration (Chapter 13).

## GENERAL READING


1 ARNON D. I. (1969) Role of ferredoxin in photosynthesis. *Naturwissenchaften.*, **56**, 295–305.
2 AVRON M. (1977) Energy transduction in chloroplasts. *Ann. Rev. Biochem.*, **46**, 143–155.
3 BEALE S. I. (1978) $\delta$ -Aminolevulinic acid in plants: Its biosynthesis, regulation, and role in plastid development. *Ann. Rev. Plant Physiol.*, **29**, 95–120.
4 BICKFORD E. D. & DUNN S. (1972) *Lighting for Plant Growth.* The Kent State University Press. 221 pp.
5 BISHOP N. I. (1971) Photosynthesis: the electron transport system of green plants. *Ann. Rev. Biochem.*, **40**, 197–226.
6 BLACK C. C. (1973) Photosynthetic carbon fixation in relation to net $CO_2$ uptake. *Ann. Rev. Plant Physiol.*, **24**, 253–286.
7 BLINKS L. R. (1954) The photosynthetic function of pigments other than chlorophyll. *Ann. Rev. Plant Physiol.*, **5**, 93–111.
8 BOARDMAN N. K. (1970) Physical separation of the photosynthetic photochemical systems. *Ann. Rev. Plant Physiol.*, **21**, 115–140.
9 BOARDMAN N. K. (1977) Comparative photosynthesis of sun and shade plants. *Ann. Rev. Plant Physiol.*, **28**, 355–377.
10 BOYER P. D., CHANCE B., ERNSTER L., MITCHELL P., RACKER E. & SLATER E. C. (1977) Oxidative phosphorylation and photophosphorylation. *Ann. Rev. Biochem.*, **46**, 955–1026.
11 BROWN J. S. (1972) Forms of chlorophyll *in vivo. Ann. Rev. Plant Physiol.*, **23**, 73–86.
12 BUTLER W. L. (1978) Energy distribution in the photochemical apparatus of photosynthesis. *Ann. Rev. Plant Physiol.*, **29**, 345–378.
13 CHENIAE G. M. (1970) Photosystem II and $O_2$ evolution. *Ann. Rev. Plant Physiol.*, **21**, 467–498.
14 CRAMER W. A. & WHITMARSH J. (1977) Photosynthetic cytochromes. *Ann. Rev. Plant Physiol.*, **28**, 133–172.
15 GOEDHEER J. C. (1972) Fluorescence in relation to photosynthesis. *Ann. Rev. Plant Physiol.*, **23**, 87–112.
16 HALL D. O. (1976) The coupling of photophosphorylation to electron transport in isolated chloroplasts. In: *The Intact Chloroplast*, ed. J. Barber. Elsevier/North-Holland Biomedical Press. The Netherlands. pp. 135–170.
17 HATCH M. D. & SLACK C. R. (1970) Photosynthetic $CO_2$-fixation pathways. *Ann. Rev. Plant Physiol.*, **21**, 141–162.
18 HEBER U. (1974) Metabolite exchange between chloroplasts and cytoplasm. *Ann. Rev. Plant Physiol.*, **25**, 393–421.
19 HILL R. (1965) The biochemists' green mansions. The photosynthetic electron-transport chain in plants. *Essays Biochem.*, **1**, 121–151.
20 HINKLE P. C. & MCCARTY R. E. (1978) How cells make ATP. *Sci. Am.*, **238**, 104–123.
21 JACKSON W. A. & VOLK R. J. (1970) Photorespiration. *Ann. Rev. Plant Physiol.*, **21**, 385–432.
22 JOLIOT P. (1965) Cinétiques des réactions liées a l'émission d'oxygéne photo-synthétique. *Biochim. Biophys. Acta.*, **102**, 116–134.
23 JUNGE W. (1977) Membrane potentials in photosynthesis. *Ann. Rev. Plant Physiol.*, **28**, 503–536.
23 KELLY G. J. & LATZKO E. (1976) Regulatory aspects of photosynthetic carbon metabolism. *Ann. Rev. Plant Physiol.*, **27**, 181–205.
25 KIRK J. T. O. (1971) Chloroplast structure and biogenesis. *Ann. Rev. Biochem.*, **40**, 161–196.
26 MYERS J. (1971) Enhancement studies in photosynthesis. *Ann. Rev. Plant Physiol.*, **22**, 289–312.
27 RASCHKE K. (1975) Stomatal Action. *Ann. Rev. Plant Physiol.*, **26**, 309–340.
28 REBEIZ C. A. (1973) Protochlorophyll and chlorophyll biosynthesis in cell-free systems from higher plants. *Ann. Rev. Plant Physiol.*, **24**, 129–172.
29 RIED A. (1972) Improved action spectra of light reaction I and II. Proc. of the 2nd International

Congress on Photosynthesis Research Stresa 1971, Vol. I. Ed. G. Forti, M. Avron and A. Melandri. Dr. W. Junk, N. V., The Hague. pp. 763–772.
30 SCHWARTZ M. (1971) The relation of ion transport to phosphorylation. *Ann. Rev. Plant Physiol.*, **22**, 469–484.
31 SIMONIS W. & URBACH W. (1973) Photophosphorylation *in vivo*. *Ann. Rev. Plant Physiol.*, **24**, 89–114.
32 THORNBER J. P. (1975) Chlorophyll-proteins: Light-harvesting and reaction center components of plants. *Ann. Rev. Plant Physiol.*, **26**, 127–158.
33 TREBST A. (1974) Energy conservation in photosynthetic electron transport of chloroplasts. *Ann. Rev. Plant Physiol.*, **25**, 423–458.
34 WALKER D. A. & CROFTS A. R. (1970) Photosynthesis. *Ann. Rev. Biochem.*, **39**, 389–428.
35 ZELITCH I. (1975) Pathways of carbon fixation in green plants. *Ann. Rev. Biochem.*, **44**, 123–145.

## References

BARBER J. (ed.) (1976) *The Intact Chloroplast – Topics in photosynthesis*, vol. 1, Elsevier/North-Holland Biomedical Press, The Netherlands.

CLAYTON R. K. (1970) *Light and Living Matter*, vol. 1. McGraw-Hill, New York.

DOUCE R. & JOYARD J. (1979) Structure and function of the plastid envelope. *Adv. Bot. Res.* **7**, 1–117.

GIBBS M. (ed.) (1971) *Structure and Function of Chloroplasts*. Springer-Verlag, Berlin, Heidelberg, New York.

GIESE A. C. (1964) *Photophysiology*. Academic Press, New York.

GOVINDJEE (ed.) (1975) *Bioenergetics of Photosynthesis*. Academic Press, New York.

HATCH M. D. & BOARDMAN N. K. (eds.) (1981) *The Biochemistry of Plants. A Comprehensive Treatise*, vol. 8. Academic Press, New York.

HILL R. & BENDALL F. (1960) Function of the two cytochrome components in chloroplasts: a working hypothesis. *Nature, London*, **186**, 136–137.

LAVOREL J. & ETIENNE A. L. (1977) *In vivo* chlorophyll fluorescence. In: *Primary Processes of Photosynthesis*, (ed. J. Barber) pp. 203–268. Elsevier/North-Holland Biomedical Press, The Netherlands.

MCCARTY R. E. (1979) Roles of a coupling factor for photophosphorylation in chloroplasts. *Ann. Rev. Plant Physiol.*, **30**, 79–104.

MCELROY W. D. & GLASS, M. (eds.) (1961) *Light and Life*. The John Hopkins Press, Baltimore.

MITCHELL P. (1966) Chemiosmotic coupling in oxidative and photosynthetic phosphorylation. *Biol. Rev. Cam. Philos. Soc.*, **41**, 445–502.

OLSON J. H. & HIND, G. (eds.) (1976) Chlorophyll-proteins, reaction centers and photosynthetic membranes. Brookhaven Symposia in Biol. no. 28. Brookhaven Nat. Lab. Assoc. Universities Inc. Upton, New York.

RABINOWITCH E. I. (1945, 1951, 1956) *Photosynthesis and Related Processes*, vol. I, II. Wiley (Interscience), New York.

SAN PIETRO A. GREER F. A. & ARMY T. J. (eds.) (1967) *Harvesting the Sun*. Academic Press, New York.

TREBST A. & AVRON M. (eds.) (1977) *Photosynthesis I. Photosynthetic Electron Transport and Photophosphorylation*. Encyclopaedia of Plant Physiology (New Series) vol. 5. Springer-Verlag, Berlin, Heidelberg, New York.

WALKER D. A. (1979) *Energy, Plants and Man*. Packard Publishing Ltd., Chichester.

WITT H. T. (1971) Coupling of quanta, electrons, fields, ions and phosphorylation in the functional membrane of photosynthesis. *Quarterly Rev. Biophys.*, **4**, 365–477.

# PART B

# Chapter 6
# The Reductive Pentose Phosphate Pathway and Associated Reactions



## SUMMARY

There is only one *primary* carboxylation mechanism in living organisms. This is primary in the sense that all else derives from it. It is the avenue through which all organic carbon is ultimately derived from carbon dioxide. It has been called the Calvin Cycle, the Benson–Calvin Cycle, the Photosynthetic Carbon (Reduction) Cycle or (as here) the *Reductive Pentose Phosphate* (RPP) *pathway* (Fig. 6.1). It is common to $C_3$, $C_4$, and CAM plants, although the last two also have additional auxiliary or preliminary fixation mechanisms.

The RPP pathway has four principal features:

1 *Carboxylation*, in which $CO_2$ is joined to the acceptor, ribulose 1,5-bisphosphate (RBP) to give two molecules of 3-phosphoglycerate (PGA). This reaction has a highly favourable equilibrium position and is catalysed by RBP carboxylase, a complex enzyme with a high affinity for $CO_2$.

2 *Reduction*, in which PGA is reduced, at the expense of assimilatory power, to triose phosphate.

3 *Regeneration*, in which five molecules of triose phosphate are rearranged to regenerate three molecules of the $CO_2$-acceptor ($5C_3 \rightarrow 3C_5$).

4 *Autocatalysis*. For every three molecules of $CO_2$ which enter the RPP pathway one molecule of triose phosphate ($C_3$) constitutes product. Although this can be

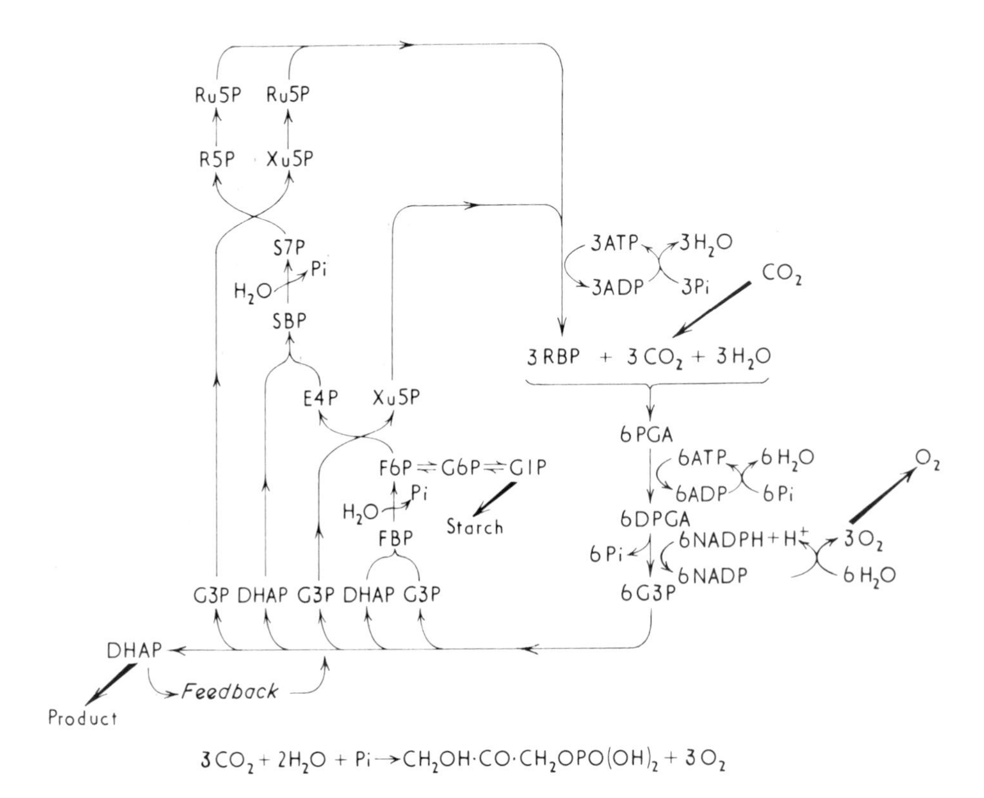


**Fig. 6.1.** The Reductive Pentose Phosphate Pathway. On the right, 3 molecules of RBP combine with 3 molecules of $CO_2$ and 3 molecules of water to give 6 molecules of PGA. These are phosphorylated at the expense of ATP and the resulting DPGA is reduced by NADPH to G3P. Part of this is converted to its isomer DHAP. Aldol condensation of these 2 triose phosphates gives a molecule of FBP which undergoes hydrolysis to F6P. This hexose phosphate is also the precursor of G6P and G1P which, after further transformation, gives rise to starch. The F6P also enters the first transketolase reaction donating a 2-carbon unit to G3P to form Xu5P and E4P. The process of condensation, phosphorylation and 2-carbon transfer is repeated yielding, SBP, S7P and two more molecules of pentose phosphate respectively. All 3 molecules of pentose monophosphate are finally converted to Ru5P which is phosphorylated to RBP. The entire cycle can be divided into 3 phases. The initial carboxylation is followed by reduction to triose phosphate. 5 of these $C_3$ molecules are rearranged to regenerate 3 ($C_5$) molecules of $CO_2$-acceptor. It should be noted that the cycle consumes 9 molecules of ATP and 6 molecules of NADPH in the formation of one triose phosphate product (which can also feed back into the cycle to promote autocatalytic acceleration). In total, 5 molecules of $H_2O$ are consumed in the cycle proper and 3 are released in the generation of assimilatory power. If the triose phosphate product were hydrolysed to give free triose in a reaction consuming 1 molecule of $H_2O$ there would be no net Pi consumption and the entire sequence would simplify to the classic overall equation for photosynthesis (Eqn. 1.1).

converted to starch, sucrose etc., it can also re-enter the regenerative phase of the pathway thus leading to an autocatalytic build-up of intermediates. In this way the pathway can function as a 'breeder' reaction producing more $CO_2$-acceptor than it consumes. This unique autocatalytic feature is the basis of plant growth.
Although sucrose is sometimes regarded as the ultimate carbohydrate end-product of photosynthesis it is, in reality, the major transport metabolite in many plants and is synthesized in the cytoplasm from exported triose phosphate. Conversely, leaf starch is formed within the chloroplast and occurs in the majority of higher plant leaves as a transitory storage product. It is most abundant when photosynthesis exceeds metabolic consumption of assimilates and is then accumulated by day and remobilized by night.

## 6.1 Photosynthetic carbon fixation

Photosynthesis involves the 'fixation' or incorporation of atmospheric $CO_2$ into an organic carbon molecule within the plant. All known fixation processes in photosynthesis are carboxylations, i.e. $CO_2$ is joined to a pre-existing acceptor (R) in such a way that a new carboxyl group ($R \cdot COO^-$) is formed:

$$R + CO_2 \rightarrow R \cdot COO^- \qquad \text{Eqn. 6.1}$$

## 6.2 Regeneration

If photosynthetic $CO_2$ fixation is to continue unchecked the acceptor (R) must be replenished as rapidly as it is consumed. There must therefore be an associated mechanism which brings about the regeneration of the $CO_2$ acceptor (Section 6.13).

## 6.3 Autocatalysis

Most higher plant processes (and most living organisms) depend, directly or indirectly, on compounds derived from photosynthetic carbon assimilation. Originally, however, there must be one basic process which is entirely self-sufficient in the sense that it can produce *and increase* its own substrate. A reaction which can produce more substrate than it utilizes (a breeder reaction) is inevitably autocatalytic because, within limits, its rate must increase with increasing substrate concentration. The need for autocatalysis, though self-evident is frequently neglected and therefore merits emphasis. Thus if a carboxylation mechanism (i.e. the carboxylation plus the

associated replenishing reactions) could only regenerate precisely the same quantity of substrate that it utilized, then the plant would have no capacity for growth and no ability to adjust its rate of photosynthesis to match improving conditions (such as daily increases in light intensity and temperatures). The green plant must therefore contain at least one mechanism which permits photosynthetic product to be converted into additional substrate. The only known mechanism which meets this essential criterion is the Benson–Calvin cycle or reductive pentose phosphate pathway (RPP pathway) (See Section 6.13).
[Ref. **55**]

## 6.4 Energetics

A carboxylation costs, in energy, about as much (+7 kcal) as the formation of ATP from ADP and Pi. [The change in *enthalpy* calculated from the empirical bond and resonance energies is about +3 kcal/mol. Calculation of $\Delta F'$ also needs to take into account the *entropy* change (Section 2.3) which is significant if $CO_2$ is in the gaseous form but, in any event, the calculated value can be inaccurate by as much as 5 kcal]. Carbon dioxide is a 'well-adjusted' molecule with lots of resonance possibilities and the energy required to take it apart exceeds, by about 7 kcal, the energy recovered in the formation of the new carboxyl group. Because of this energy requirement, some 'carboxylations' (e.g. the formation of oxaloacetate from pyruvate and $CO_2$, Eqn. 6.2)

$$CH_3 \cdot CO \cdot COOH + CO_2 \rightarrow HOOC \cdot CH_2 \cdot CO \cdot COOH \qquad \text{Eqn. 6.2}$$

are so unfavourable energetically that they may be entirely discounted as feasible biological carboxylating mechanisms. Others (e.g. the reductive carboxylation of pyruvate to malate, Eqn. 6.3)

$$CH_3 \cdot CO \cdot COOH + NADPH_2 + CO_2 \rightarrow HOOC \cdot CHOH \cdot CH_2 \cdot COOH + NADP^+ \qquad \text{Eqn. 6.3}$$

may be regarded as less unfavourable but are still most unlikely to result in the formation of new carboxyl groups. This follows from the extremely low concentration of $CO_2$ in the atmosphere (about 300 ppm or 0.03 %) and the need for $CO_2$ to diffuse into the leaf (i.e. the internal concentration is presumed to be 7 $\mu$mol or less in the absence of $CO_2$-concentrating mechanisms, Chapter 14). The major photosynthetic carboxylation mechanism(s) must operate within these constraints and would clearly benefit from equilibrium constants such that the accumulating product did not quickly slow the rate of reaction. At present only two major photosynthetic carboxylations (Eqn. 6.4 & 6.5) have been clearly established and both proceed with a decrease in free energy of approximately 7–8 kcal.

$$PEP + CO_2 + H_2O \rightarrow \text{oxaloacetate} + Pi \quad [\Delta F' = -7\,\text{kcal}] \qquad \text{Eqn. 6.4}$$

$$RBP + CO_2 + H_2O \rightarrow 2 \times \text{3-phosphoglycerate} \quad [\Delta F' = -8\,\text{kcal}] \qquad \text{Eqn. 6.5}$$

Both substrates, phosphoenolpyruvate (PEP) and ribulose 1,5-biphosphate (RBP), may be regarded as high energy compounds with sufficient built-in energy to drive the requisite carboxylation to virtual completion.

Two more aspects of the energetics of carboxylation may also be noted. In the RPP pathway it might be asked why a sequence has evolved which carboxylates RBP rather than Ru5P since both molecules are 'high energy' in the sense that they are unable to form the 5-membered rings which would increase their stability. (Accordingly carboxylation of Ru5P would also involve a large decrease in free energy and the fact that one product was glycerate could be corrected by subsequent phosphorylation). From an energetic standpoint the answer could lie in the fact that RBP formation will pull the reactions of the regenerative phase in the desired direction. The Ru5P kinase reaction also plays an important role in regulation (Chapter 9).

Phosphoenolpyruvate also provides an example of the different ways in which a 'high energy' compound may be exploited. Thus direct hydrolysis of PEP yields about 14 kcal (Eqn. 6.6).

$$PEP + H_2O \rightarrow \text{pyruvate} + Pi \quad [\Delta F' = -14\,\text{kcal}] \qquad \text{Eqn. 6.6}$$

In glycolysis about 7 kcal of this 14 is conserved as ATP (Eqn. 6.7)

$$PEP + ADP \rightarrow \text{pyruvate} + ATP \quad [\Delta F' = -7\,\text{kcal}] \qquad \text{Eqn. 6.7}$$

but the reaction still proceeds with a $\Delta F'$ of about $-7$ kcal, making it very difficult to reverse.

Alternatively about 7 kcal of the energy potential of PEP can be used to 'drive' an equally favourable carboxylation with a $\Delta F'$ of about $-7$ kcal (Eqn. 6.8).

$$PEP + CO_2 + H_2O \rightarrow \text{oxaloacetate} + Pi \quad [\Delta F' = -7\,\text{kcal}] \qquad \text{Eqn. 6.8}$$

If both ATP formation and carboxylation are achieved, as they are in the reaction catalysed by PEP carboxykinase (Eqn. 6.9), the free energy change falls to near zero and this reaction becomes an energetically feasible mechanism for the entry of phosphoenolpyruvate into reversed glycolysis (gluconeogenesis)

$$\text{Oxaloacetate} + ATP \rightleftarrows PEP + CO_2 + ADP \quad [\Delta F' = 0] \qquad \text{Eqn. 6.9}$$

In $C_4$ photosynthesis, PEP regeneration could (in theory) be achieved by the direct consumption of both high energy groups of ATP (Eqn. 6.10)

$$\text{pyruvate} + \text{ATP} \rightarrow \text{PEP} + \text{AMP} + \text{Pi} \quad [\Delta F' = 0] \qquad \text{Eqn. 6.10}$$

thus overcoming the unfavourable equilibrium position of the reaction catalysed by pyruvic kinase (reverse of Eqn. 6.7) and indeed it was the search for the catalyst of this reaction (which is known in certain bacteria) which led to the elucidation of the pyruvate phosphate dikinase reaction (Eqn. 6.11). Here both high energy groups of ATP are also utilized

$$\text{pyruvate} + \text{Pi} + \text{ATP} \rightarrow \text{PEP} + \text{AMP} + \text{PPi} \quad [\Delta F' = +7\,\text{kcal}] \qquad \text{Eqn. 6.11}$$

but this reaction is, in itself, as energetically unfavourable as the reverse of reaction 6.7 ($\Delta F' = +7$, in the direction of PEP synthesis). It is linked, however, to the hydrolysis (Eqn. 6.12) of pyrophosphate(PPi).

$$\text{PPi} + H_2O \rightarrow \text{Pi} + \text{Pi} \quad [\Delta F' = -7\,\text{kcal}] \qquad \text{Eqn. 6.12}$$

so that in effect the combined equation (6.11 plus 6.12) has a $\Delta F'$ of near zero and is therefore freely reversible (Section 2.5). In order that this becomes biologically as well as energetically feasible it is necessary that the pyrophosphatase is present in ample amounts and that it has a very low *K*m (high affinity for PPi). If this were not so, the first stage (Eqn. 6.11) could not readily generate enough product to allow displacement of its unfavourable equilibrium position by the operation of the PPiase sink (Eqn. 6.12).
[Refs. 5, 7, 17, 18, 19, 20, 22, **28, 32, 53**]

## 6.5 Affinity for $CO_2$

Just as a carboxylation will be favoured by a large equilibrium constant [as Table 2.1 shows a high *K* is equivalent to a large negative free energy change ($\Delta F'$) and indicates that Equations 6.4 and 6.5 (above) will go to near completion in the forward direction, as written] so the catalyst (the carboxylase) will be favoured by a high affinity for $CO_2$.

At suitably low substrate concentrations, the rate of an enzyme catalysed reaction can be doubled by doubling the substrate concentration [S]. At very low substrate levels the Michaelis–Menten equation:

$$v = \frac{V\text{max}[S]}{K\text{m} + [S]} \qquad \text{Eqn. 6.13}$$

essentially becomes:

$$v = \frac{V\max[S]}{Km} \qquad \text{Eqn. 6.14}$$

where [S] becomes almost negligible in relation to the magnitude of $Km$. Thus a linear relationship between v and [S] will be approached but never quite achieved at low [S]. If the substrate concentration is increased indefinitely, however, the active sites through which the enzyme exerts its catalytic function will become saturated. If rate is plotted against substrate concentration [S] the relationship is therefore linear at very low concentrations but as saturation is approached the relationship will depart from linearity as the advantage derived from adding more substrate diminishes. (Fig. 6.2)

The complete curve is therefore part of a rectangular hyperbola which approaches but never quite reaches (or reaches at infinitely high [S]) a theoretical maximum called the $V$max (maximum velocity). This value is characteristic for a given enzyme at a given enzyme concentration. So is the substrate concentration (the Michaelis Constant or $Km$) which gives half the maximum velocity. The $Km$ has the added advantage that it is independent of enzyme concentration (Fig. 6.2).

[In practice, the maximum velocity is not determined by increasing the substrate concentration to infinity because this is obviously not practical. Instead, advantage is often taken of the fact that a rectangular hyperbola becomes a straight line if the values of rate (v) and [S] are plotted as their reciprocals (1/v and 1/[S]). When [S] is infinitely high, 1/[S] is infinitely low (i.e. 1/[S] becomes zero and therefore the intercept on the vertical axis (Fig. 6.3) equals 1/$V$max. Similarly, on a reciprocal plot, 1/$Km$ is the value which would give twice this intercept (1/2 $V$max) and is therefore numerically equal to the negative of the value at which the extrapolated curve cuts the horizontal axis)].

The $Km$ is inversely related to the affinity of an enzyme for its substrate. In short, if an enzyme needs a lot of substrate to work at a fast rate (a high $Km$) then it is said to have a low affinity. Clearly if a carboxylase is called upon to function well in the presence of low concentrations of $CO_2$ it will be at an advantage if it has a high affinity (low $Km$) for this substrate. Again, the enzymes which catalyse Equations 6.4 and 6.5 meet this criterion. PEP carboxylase (the catalyst in Eqn. 6.4) has been known to have a high affinity for $CO_2$ /bicarbonate since it was first characterized in 1954. RBP carboxylase, which was discovered about the same time, constituted a considerable problem for many years because it seemed at first that it needed as much as 6 % $CO_2$ to work at half maximal velocity and this was incompatible with its proposed role *in vivo*. More recently, it has become clear that the enzyme was first assayed in a relatively inactive state and that provided appropriate activation

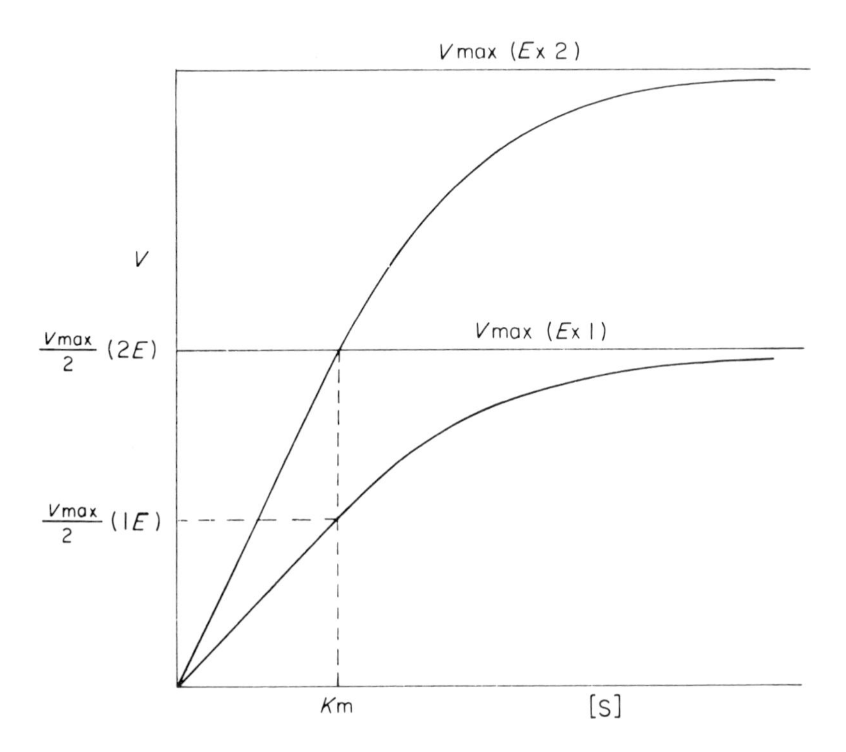


**Fig. 6.2.** Velocity (rate) of an enzyme catalysed reaction as a function of substrate concentration [S]. Two concentrations of enzyme ($E \times 1$, and $E \times 2$) are depicted. The maximum rate (*V*max) is doubled when the enzyme concentration is doubled but the substrate *K*m needed to give half maximum velocity is unchanged.

procedures are employed the affinity for $CO_2$ is no less high than that of PEP carboxylase.

[In theory the *V*max is as important in this regard as the affinity. In other words it would not matter how much substrate an enzyme needed in order to work at an appropriate rate provided that the organism contained sufficient enzyme. In practice there may not be enough space to house the required catalyst. Even as it is, ribulose bisphosphate carboxylase is the major stromal protein and may account for as much as 50% of the leaf protein. This may reflect an evolutionary increase brought about by the need to maintain photosynthetic activity in the face of the falling $CO_2$ concentration and increasing $O_2$ concentration brought about by photosynthetic activity.]
[Refs. 7, 17, 18, 19, 26, **34, 39, 42, 44, 53, 54**]

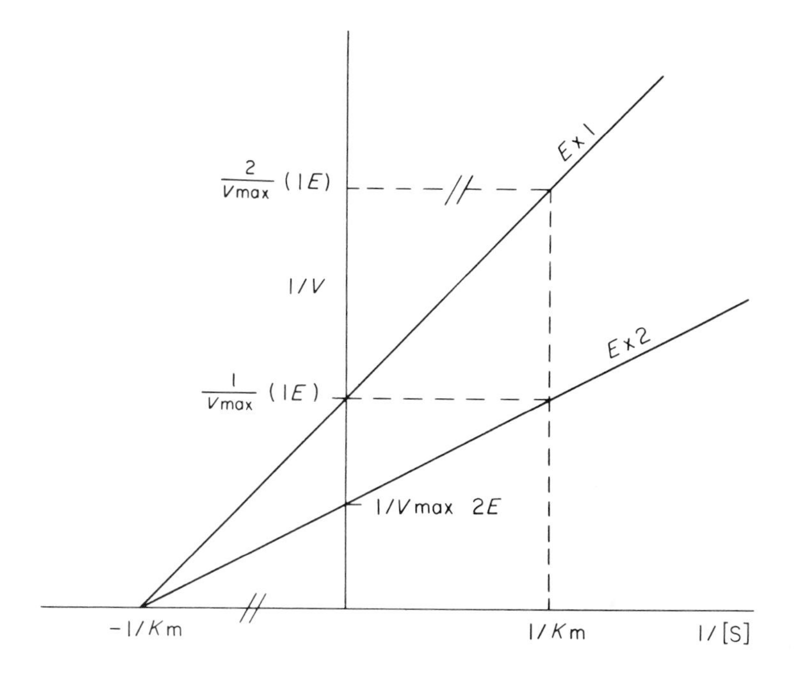


**Fig. 6.3.** Double reciprocal plot of Fig. 6.2 converts hyperbolic sections to a more convenient straight-line relationship. When $[S] = \infty$, $1/[S] = 0$ so that the intercepts with the vertical axis give the values of $1/V\text{max}$, for each enzyme concentration. (Theoretically the maximum velocity is attained in the presence of infinitely high substrate). The $K\text{m}$ could be derived as in Fig. 6.2 by dropping a vertical line from the values equal to the reciprocals of half these maximum velocities but it is simpler to extrapolate to the point of interception with the negative horizontal axis which gives the same numerical value (disregarding the negative sign) of $1/K\text{m}$.

## 6.6 Carboxylation characteristics summarized

The major carboxylation mechanism in photosynthesis might be expected to meet four criteria

1 It must be able to regenerate its $CO_2$-acceptor.
2 It must work autocatalytically, as a breeder reaction, producing more potential substrate than it utilizes.
3 Its carboxylation reaction should have a very favourable equilibrium position to offset the low $[CO_2]$ in the environment.
4 Its carboxylase should have a high affinity for $CO_2$ to permit effective function in low $[CO_2]$.

Only the Reductive Pentose Phosphate pathway satisfies all four requirements. [Ref. **55**]

## 6.7 Formulation of the reductive pentose phosphate pathway

The RPP pathway was elucidated in the 1950's by Calvin, Benson and others in Berkeley, California. The importance of their findings was acknowledged by the award of the Nobel Prize to Calvin and the full sequence of reactions is often referred to as the Calvin Cycle or Benson–Calvin cycle (in an attempt to do justice to the contributions of Benson). In recent years it has become unfashionable to refer to metabolic pathways in this way, perhaps because of the difficulty of giving adequate credit to the many research workers who have contributed to their elucidation. Neutral terms like 'photosynthetic carbon cycle' are acceptable but relatively unspecific because they could also be applied to the entire $C_4$ pathway (Chapter 10). RPP pathway is used here because it is precise and descriptive.

Calvin and his colleagues took advantage of two newly developed techniques, radioactive tracers and paper chromatography. The long-lived radioactive tracer, $^{14}C$, had been discovered by Samuel Ruben and Martin Kamen in 1940 and became available in large quantities in 1945 as a by-product of nuclear reactors. Martin and Synge had completed much of their work on chromatography which was to win them the Nobel Prize in 1952 and paper chromatography had been used for the separation of amino acids as early as 1944. [Chromatography based on icing-sugar columns was first used by the Russian botanist Tswett to separate chlorophylls, and associated pigments, in leaf extracts. The name Tswett is also the Russian for 'colour' and therefore particularly apt for the discoverer of a process for separating coloured compounds. The term 'chromatography' is now, however, applied equally to the separation of non-coloured compounds]. Together, $^{14}C$ and paper chromatography constituted remarkable new tools with which to attack the formidable problem of photosynthetic carbon assimilation. The availability of $^{14}C$ allowed Calvin and his colleagues to label $CO_2$ and then to follow, with *relative* ease the appearance of radioactivity in newly formed compounds when $^{14}CO_2$ was fed, in the light, to the green alga *Chlorella*. Calvin is a chemist and reputedly used *Chlorella* (this most famous of photosynthetic organisms—see Sections 3.9 & 4.10) because it more closely resembled a green solution than did a cabbage. Be this as it may, *Chlorella* also lent itself remarkably well to experiments in which it could be uniformly and intensely illuminated and then run into hot ethanol, in order to bring biologically catalysed reactions to a sudden halt. [Gaffron and his colleagues in Chicago used a similar approach, employing *Scenedesmus* and there are obvious advantages in using very large populations of uniformly grown algae rather than small and variable samples of higher plant leaves in initial experiments of this nature].

After brief exposure to $^{14}CO_2$ in the light, samples of 80% ethanol extract were

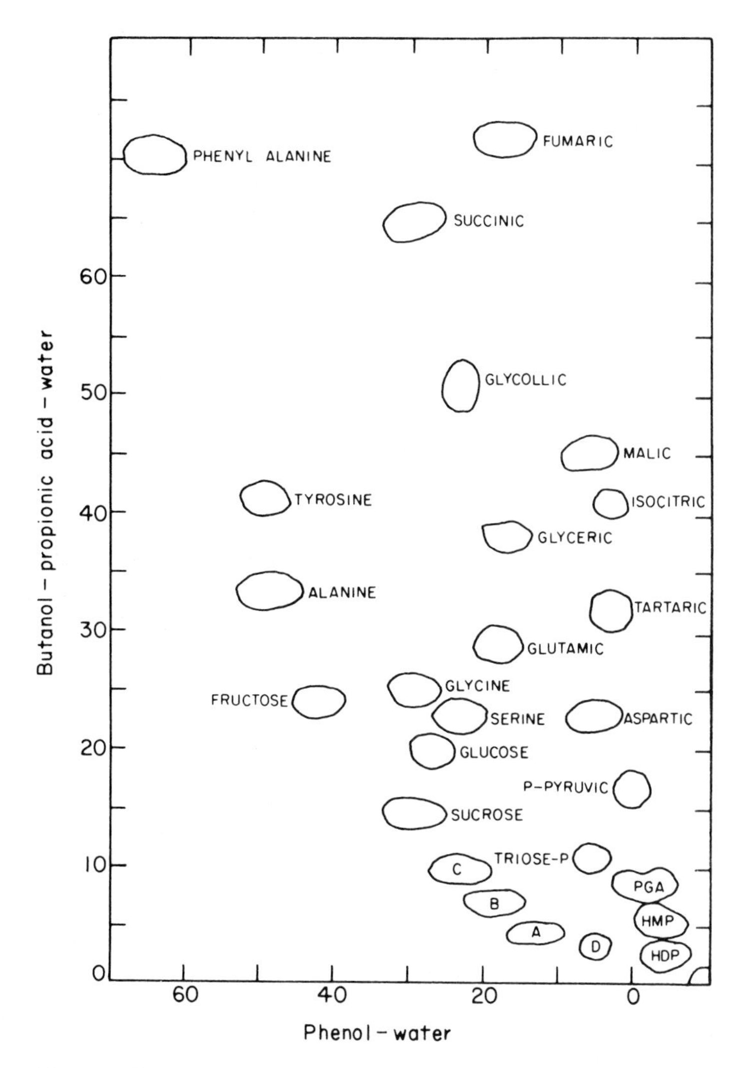


**Fig. 6.4.** Chromatography Map. When 'run' in the same solvents under precisely the same conditions each radioactive compound applied to the source (bottom right) tends to move to the same spot and tentative identification can therefore be based on position. Similarly the radioactivity in each spot indicates the concentration of the labelled compound concerned. (From Benson *et al* 1950)

dried on the corners of large sheets of filter paper and an appropriate mixture of solvents allowed to flow through the paper as it hung from a trough in a sealed and temperature controlled tank. The original 'spot' of extract separated first of all into a chain of spots and then into a two dimensional pattern of spots after the paper was dried, rotated through 90° and developed with a second solvent mixture. In the hands of workers in Calvin's laboratory like Martha Kirk this is a precise and reproducible technique which allows compounds to be identified by their position relative to the origin (the location of the original spot). It also allows their concentration to be determined by their $^{14}C$ content. [Initially, of course, each compound must be chemically identified before it can be assigned a position on a chromatography map and appropriate allowance made for large pools of initially unlabelled compounds]. Together, the availability of two dimensional chromatography and $^{14}C$ permitted analysis within weeks which would previously have taken years, even if sufficient material had been available to permit conventional analysis (Fig. 6.4).
[Refs. 3, 4, 6, **33, 35, 38, 50, 52**]

### 6.8 Kinetic studies

As the period of photosynthesis was shortened, fewer and fewer compounds came to contain $^{14}C$. When their radioactivity was plotted as a fraction of the total $^{14}C$ recovered, only one (3-phosphoglycerate) displayed a negative slope which allowed a reasonable extrapolation to 100% at zero time (Fig. 6.5). [In theory, the product of the carboxylation would contain virtually all of the radioactivity after an infinitely short time interval, before subsequent conversions had been given an opportunity to occur]. The fact that the first-formed product contained 3 atoms of carbon per molecule implied that it may have been formed by addition of $CO_2$ to a 2-carbon acceptor. As pentose phosphates became implicated, it became increasingly clear that the $CO_2$-acceptor was not a $C_2$ but a $C_5$ compound and that a $C_5 + C_1$ addition was followed (or perhaps accompanied) by separation of a $C_6$ intermediate into two molecules of $C_3$ product.
[Refs. 3, 4, 6]

### 6.9 Transients

The identity of the $CO_2$-acceptor (ribulose 1,5-diphosphate) was strongly suggested by transient changes in the concentration of this compound and by concomitant

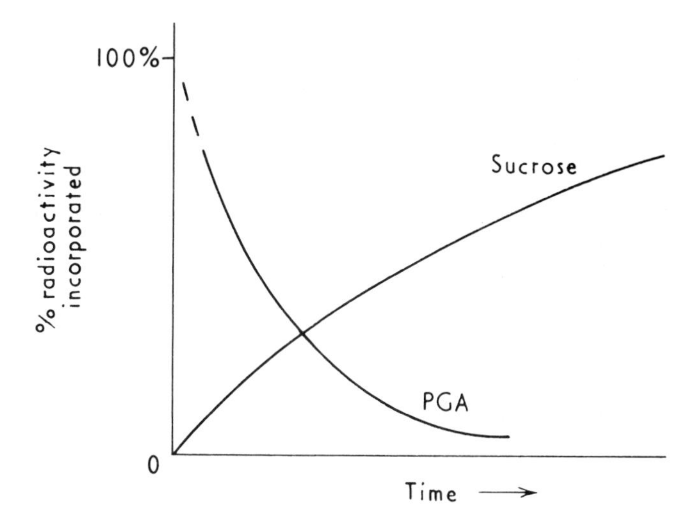


**Fig. 6.5.** As the time of exposure to $^{14}CO_2$ in the light is lengthened, less and less radioactivity is found in PGA but more and more in end-products of photosynthesis, like sucrose. PGA is the only compound which shows this negative slope when $^{14}CO_2$ is fed to *Chlorella* and a reasonable interpolation shows that it would be the only compound labelled after the shortest possible time of exposure. (After Calvin & Bassham)

changes in PGA. [It should be noted that the acceptor is now called ribulose 1,5-*bis*phosphate (RBP) to emphasize the fact that the phosphate groups are attached at different points within the molecule rather than joined together in a pyrophosphate (Pi-Pi) structure as the term '*di*phosphate' might imply]. When, for example, the $CO_2$ concentration was suddenly lowered, PGA fell and RBP increased (Fig. 6.6). Conversely in passing from light to dark, PGA rose and RBP fell (Fig. 6.7). The first response (Fig. 6.6) was precisely that which would be expected if RBP were the acceptor and PGA the carboxylation product, since the acceptor would clearly be spared and the product diminished if $CO_2$ were limiting. The second response (Fig. 6.7) implies that the acceptor must be regenerated from the product. Thus if some of the PGA underwent light-driven conversion to RBP, this reaction would cease in the dark and PGA would accumulate (and RBP diminish) as residual acceptor underwent carboxylation.
[Refs. 3, 4, 6].

## 6.10 Intra-molecular labelling

When labelled intermediates were chemically degraded and the distribution of radioactivity within molecules was determined it finally became possible to trace the

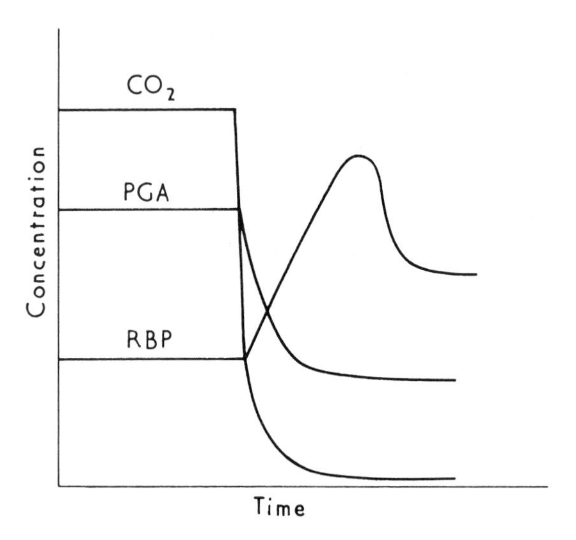


**Fig. 6.6.** Transients following change in $CO_2$ concentration. When the $CO_2$ concentration was abruptly decreased during illumination, the end-product of $CO_2$-fixation (PGA) also decreased but the $CO_2$-acceptor (RBP) rose at first (because it was being consumed at a much slower rate) and fell subsequently as its regeneration was, in turn, retarded.

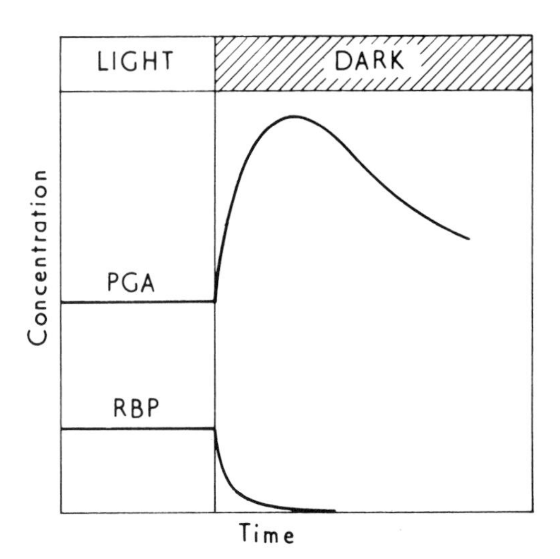


**Fig. 6.7.** Light/Dark Transients. PGA rose at first because it continued to be produced but not reduced. (The subsequent fall in PGA can be attributed to non-reductive metabolism). RBP falls because of continued consumption and arrested regeneration (after Calvin & Bassham)

path of carbon throughout the entire sequence of the RPP pathway. The initial massive incorporation of $^{14}C$ into the carboxyl of PGA (Table 6.1) leaves little doubt that this is the point at which radioactivity is first introduced. Similarly the distribution of label within the other intermediates listed in Table 6.1 can be most easily explained by a rearrangement of five 3-carbon units according to Fig. 6.8. In this sequence (now an accepted part of the RPP pathway, see Section 6.11 & Fig. 6.1) two $C_3$ units (on the right) condense (in a specific, head to head fashion dictated by their structure) to form a $C_6$ compound. This donates a $C_2$ segment to a $C_3$ acceptor, creating a new $C_5$ molecule and leaving a $C_4$ remnant. This residual $C_4$

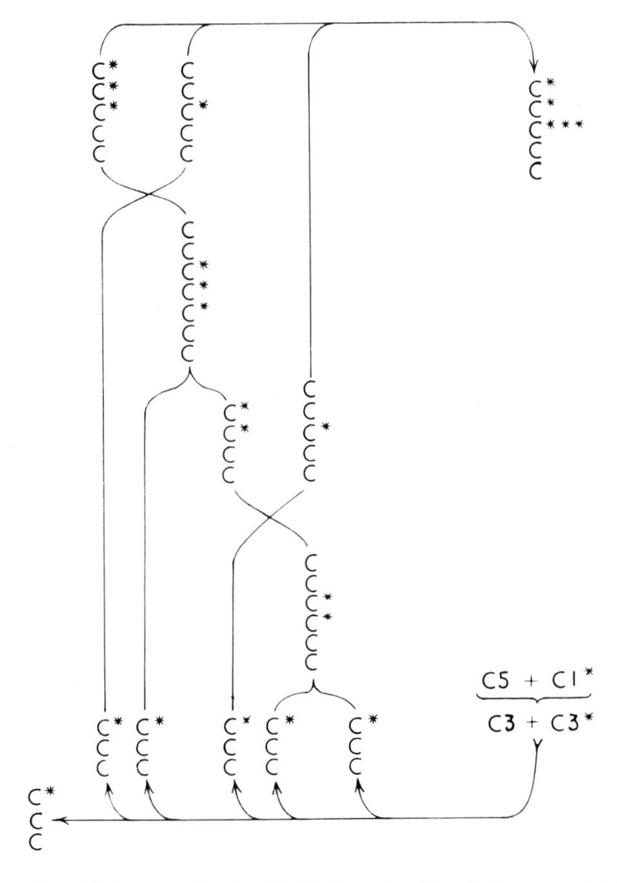


**Fig. 6.8.** Intra-molecular distribution of radioactivity amongst RPP pathway metabolites. Initial fixation of $^{14}CO_2$ into PGA and the subsequent reactions of the RPP pathway would initially distribute the radioactivity within the molecules as shown in Fig. 6.12. Similar distribution was observed in practice (after Calvin & Bassham).

condenses with another $C_3$ to form a $C_7$. A $C_2$ transfer from the $C_7$ molecule to a $C_3$ acceptor finally creates two further $C_5$ intermediates. If the radioactivity within these compounds is now totalled, it will be seen that the hexose ($C_6$) and the heptulose ($C_7$) will have their central carbons equally labelled whereas the central carbon of the pentose would contain by far the greater part of the $^{14}C$ recovered. It must be borne in mind that this sequence is part of a cycle which has been allowed to run for a short period in a particular experiment. Because of the cyclic nature of the whole process, label will then spread, in turn, from the pentose derivatives to the triose derivatives. Accordingly the values actually recorded do not exactly match the distribution which is predicted by Fig. 6.8 but they are sufficiently close to preclude any reasonable alternative based on known reactions (see Fig. 6.8).
[Refs. 3, 4, 6].

**Table 6.1.** Distribution of $^{14}C$ within intermediates of the RPP pathway.

| Carbon | PGA | Hexose | Heptose | Pentose |
|---|---|---|---|---|
| 1 | 82* | 3 | 2 | 11* |
| 2 | 6 | 3 | 2 | 10* |
| 3 | 6 | 43* | 28* | 69*** |
| 4 | | 42* | 24* | 5 |
| 5 | | 3 | 27* | 3 |
| 6 | | 3 | 2 | |
| 7 | | | 2 | |

## 6.11 The enzymes

Although the combined use of radioactive carbon and two-dimensional chromatography was crucial in elucidating the principal features of the path of carbon in photosynthesis it was essential that the partial reactions concerned and their enzymic catalysts were already known or could be demonstrated to be real. Of these, pride of place must go to the carboxylation itself.

### 1 Ribulose bisphosphate carboxylase oxygenase.

This is probably the most plentiful, and arguably the most important, enzyme known to man. Its presence in extracts of *Chlorella* was first shown by Quayle, Fuller, Benson and Calvin in 1954. It was originally called 'carboxydismutase' in order to convey the nature of the reaction which it catalyses (see Eqns. 6.15).

$$
{}^{*}CO_2 + \begin{array}{c} CH_2OPO(OH)_2 \\ | \\ C{=}O \\ | \\ HCOH \\ | \\ HCOH \\ | \\ CH_2OPO(OH)_2 \end{array} \longrightarrow \begin{array}{c} CH_2OPO(OH)_2 \\ | \\ O{=}\overset{+}{C}(HO){-}COH \\ | \\ C{=}O \\ | \\ HCOH \\ | \\ CH_2OPO(OH)_2 \end{array} + H_2O^{*} \rightleftarrows \begin{array}{c} CH_2OPO(OH)_2 \\ | \\ HCOH \\ | \\ HO\overset{+}{C}{=}O \\ HO^{*}C{=}O \\ | \\ HCOH \\ | \\ CH_2OPO(OH)_2 \end{array} \qquad \text{Eqn. 6.15a}
$$

ribulose 1,5-bisphosphate (RBP) — intermediate — 2(PGA)

$$
\downarrow O_2^{\blacklozenge}
$$

Eqn. 6.15b

$$
\begin{array}{c} CH_2OPO(OH)_2 \\ | \\ HO^{\blacklozenge}O^{\blacklozenge}COH \\ | \\ C{=}O \\ | \\ HCOH \\ | \\ CH_2OPO(OH)_2 \end{array} + H_2O^{*} \longrightarrow \begin{array}{c} CH_2OPO(OH)_2 \\ | \\ O^{\blacklozenge}{=}COH \end{array} + \begin{array}{c} HO^{*}C{=}O \\ | \\ HCOH \\ | \\ CH_2OPO(OH)_2 \end{array} + H_2O^{\blacklozenge}
$$

intermediate — 2-phosphoglycolate — 3-phosphoglycerate

---

E.C. 4.1.1.39
$Km\ (CO_2) = 12\ \mu M$
$Km\ (O_2) = 250\ \mu M$
$Km$ (RBP) $= 40\ \mu M$. Concentration of RBP in chloroplasts: 0.2–0.6 mM
$\Delta F' = -8.4$ kcal (carboxylase reaction)
mol. wt. = 550 000. 8 large subunits (55 000 mol. wt.) and 8 small subunits (15 000 mol. wt.)*

---

* Protein size is commonly expressed in daltons. One dalton is the mass of one hydrogen atom and Avogadro's Number of hydrogen atoms have a combined mass of almost exactly 1g. One gram molecule (mol) of RBP carboxylase would weigh 550 kg.

Briefly, this involves the simultaneous addition of $CO_2$ and water and an intramolecular oxidation/reduction reaction (or dismutation). Thus, the addition of a $C_1$ molecule ($CO_2$) to a $C_5$ acceptor leads to the formation of two $C_3$ products (3-phosphoglycerate) one containing an hydrogen ion from the water and the other an hydroxyl ion from the water (Eqn. 6.15a.). At one stage it seemed possible that (at least in-vivo) one of the two $C_3$ units might become more reduced (yielding triosephosphate) by some direct reductive process but this was never substantiated. There is now no doubt that the in-vitro reaction yields only two molecules of 3-phosphoglycerate, one of which contains the newly fixed carbon in its carboxyl group (carbon number one, as opposed to carbon number three which carries the phosphate). What has become clear subsequently is that oxygen may compete with $CO_2$ and that the carboxylase may also act as an oxygenase (Section 6.16). In this event, one molecule of 3-phosphoglycerate is produced as before whereas the remaining $C_2$ fragment becomes 2-phospho*glycolate* (Eqn. 6.15b). It is the RBP oxygenase reaction which is the basis of the photorespiratory carbon oxidation cycle. This is discussed at length in Chapter 13 and is integrated with the RPP pathway in Figs. 6.11 and 6.12. Because $CO_2$ and $O_2$ compete for the same site on the enzyme, the rates of the two reactions are determined by the concentrations of the two gases. Thus when $CO_2$ is high and $O_2$ is low, the carboxylase reaction would be predominant. The actual rates of the two reactions also depend on other factors such as pH and temperature but as a rough guide, the carboxylase activity would be three to five times that of the oxygenase activity under in-vivo conditions (i.e. 250 $\mu$M $O_2$, 10 $\mu$M $CO_2$).

As first described, the carboxylase appeared to have a very high *K*m (Section 6.5) with respect to $CO_2$, requiring the equivalent of something like 6% in the gas phase to attain half maximal velocity. This was something of an enigma because it was very difficult to understand how a carboxylase with such a low affinity for $CO_2$ could function effectively in the low concentrations of $CO_2$ which must exist at the site of carboxylation. [If $CO_2$ enters the organism by unassisted diffusion from an atmosphere (air) containing only 0.03% (300 parts per million) there must be appreciably less than this value at the carboxylation site]. This led to speculation about $CO_2$ concentration devices but more recently it has been shown that, if an appropriate activation procedure is employed (in which the enzyme is preincubated with $Mg^{2+}$ and bicarbonate at high pH), the *K*m ($CO_2$) falls to a very low level.

The activation and carboxylation process is illustrated in Eqns. 6.15c and 6.15d

ACTIVATION

$$\underset{\text{(inactive)}}{E+CO_2} \overset{\text{slow}}{\rightleftarrows} \underset{\text{(inactive)}}{E-CO_2} + Mg^{2+} \overset{\text{fast}}{\rightleftarrows} \underset{\text{(active)}}{E-CO_2-Mg} \qquad \text{Eqn. 6.15c}$$

CARBOXYLATION

$$E-CO_2-Mg+RBP+{}^*CO_2 \rightleftarrows E-CO_2-Mg+{}^*PGA+PGA \qquad \text{Eqn. 6.15d}$$

The inactive form of the enzyme first combines with a molecule of $CO_2$ in a slow, reversible reaction. When this enzyme-$CO_2$ complex subsequently reacts with $Mg^{2+}$ in a second reversible step the enzyme-$CO_2$-Mg complex is rapidly formed and it is only this form of the enzyme which is fully active (Eqns. 6.15c & d). The $Mg^{2+}$ is necessary for the activation of the enzyme but not for the carboxylase reaction. The actual carboxylation of RBP occurs when a second molecule of $CO_2$ reacts with the enzyme (Eqn. 6.15d). The molecule of $CO_2$ which binds to the enzyme during activation (Eqn. 6.15c) does not participate in the carboxylase reaction and it is most likely that the two molecules of $CO_2$ bind at different sites on the enzyme. It has been established that $CO_2$ and not $HCO_3^-$ is the species responsible for both the activation process and the actual carboxylase reaction. Although the alkaline environment in the stroma of an illuminated chloroplast would keep most of the carbon as the bicarbonate ion, the equilibrium between $CO_2$ and $HCO_3^-$ is maintained by the enzyme carbonic anhydrase which catalyses the reaction $H^+ + HCO_3^- \rightleftarrows CO_2 + H_2O$. This would ensure that $CO_2$ removed by the carboxylase reactions is rapidly replaced from the pool of bicarbonate.

If $CO_2$ and $Mg^{2+}$ are removed, the active complex dissociates to the inactive form so that the concentrations of $CO_2$ and $Mg^{2+}$ determine the activation state of the enzyme. The activation process is pH-dependent and has an alkaline optimum suggesting that the activation may involve the formation of a carbamate ion when the enzyme-$CO_2$ complex is formed. Activation increases the maximum velocity and, as noted above, increases the affinity of the enzyme for $CO_2$, resulting in greatly increased carboxylase activity, particularly at air levels of $CO_2$. As the activation state of the enzyme is altered, the oxygenase activity is increased and decreased in the same fashion as the carboxylase activity. Substrate inhibition of the enzyme by high levels of RBP has been reported but it now appears that this results from chemical decomposition of the RBP to form inhibitory compounds. Certain other chloroplast metabolites, notably NADPH, ferredoxin and 6-phosphogluconate have been reported to increase the activity of RBP carboxylase or to alter its state of activation (Chapter 9). It is difficult to assess the role such metabolites might play *in vivo* particularly when it is remembered that the concentration of many of these metabolites, relative to the amount of enzyme, may be much lower than in the *in-vitro* situation where the effects are observed. It can be calculated that the concentration of active binding sites of RBP carboxylase in chloroplasts may be as high as 3–4 mM. This results from the large amount of this protein in chloroplasts plus the presence of eight binding sites on the enzyme molecule. Many of the metabolites in chloroplasts are present at concentrations well below this figure and their effect on the carboxylase is therefore uncertain.
[Refs. **34, 42, 47, 49**]

2 PHOSPHOGLYCERATE KINASE.

This enzyme was already well-known in glycolysis, where the reaction which it catalyses (Eqn. 6.16) runs in the direction of PGA formation. In photosynthesis it gives rise to DPGA (1,3-diphosphoglycerate or glycerate 1,3-bisphosphate) at the expense of ATP.

$$\begin{array}{l} \mathrm{C}(=\mathrm{O})\mathrm{OH} \\ | \\ \mathrm{HCOH} \\ | \\ \mathrm{CH_2OPO(OH)_2} \end{array} + \mathrm{ATP} \xrightarrow{Mg^{2+}} \begin{array}{l} \mathrm{C}(=\mathrm{O})\mathrm{OPO(OH)_2} \\ | \\ \mathrm{HCOH} \\ | \\ \mathrm{CH_2OPO(OH)_2} \end{array} + \mathrm{ADP} \qquad \text{Eqn. 6.16}$$

3-phosphoglycerate (PGA) — 1,3-diphosphoglycerate (DPGA)

---

E.C. 2.7.2.3
*Km* (PGA) = 0.5 mM Concentration of PGA in chloroplasts = 3–5 mM.
*Km* (ATP) = 0.1 mM Concentration of AMP + ADP + ATP in chloroplasts = 1–3 mM.
$\Delta F' = +4.5$ kcal
mol. wt. = 47 000

---

This is the only step in the RPP pathway with a large positive $\Delta F'$ value (indicating that the equilibrium position lies towards the formation of PGA). Phosphoglycerate kinase is not light-activated in chloroplasts and is apparently not altered in an allosteric manner by other metabolites. For the reaction to operate in the direction of DPGA synthesis, as required for photosynthesis, a high ratio of substrates to products is therefore required i.e. [PGA] [ATP]/[DPGA] [ADP] must be high. Such a situation does occur in chloroplasts where DPGA is only present in very small amounts (largely as a result of its removal by the glyceraldehyde phosphate dehydrogenase reaction, see Equation 6.17) and the concentration of PGA is relatively high resulting in very high ratios of [PGA] [DPGA]. As suggested by the ratio above, the reaction is also sensitive to the concentrations of ATP and ADP and any decrease in [ATP] or increase in [ADP] may decrease the rate of DPGA formation (Chapter 9).

For photosynthesis to continue at high rates, the pools of intermediates must be maintained within the chloroplast as the pathway operates as a cycle. In the case of PGA, there is a great potential for loss of intermediates as this metabolite is rapidly transported by the chloroplast phosphate transporter and is present in the chloroplast at relatively high concentrations. Loss of PGA to the cytoplasm is prevented by the high pH in the chloroplast stroma which maintains most of the PGA as a trivalent anion ($PGA^{3-}$) whereas only the divalent anion ($PGA^{2-}$) is transported. Thus the chloroplast is able to maintain relatively high concentrations of PGA which favour its reduction whilst preventing loss of this intermediate to the cytoplasmic compartment. (see Section 8.23 & p. 239).

### 3 TRIOSEPHOSPHATE DEHYDROGENASE [NADP$^+$].

In glycolysis an NAD-specific enzyme catalyses the oxidation of glyceraldehyde 3-phosphate to DPGA. In photosynthesis an essentially similar reaction (Eqn. 6.17) is catalysed by an NADP-specific enzyme but it proceeds in the reductive direction and Pi is released rather than incorporated.

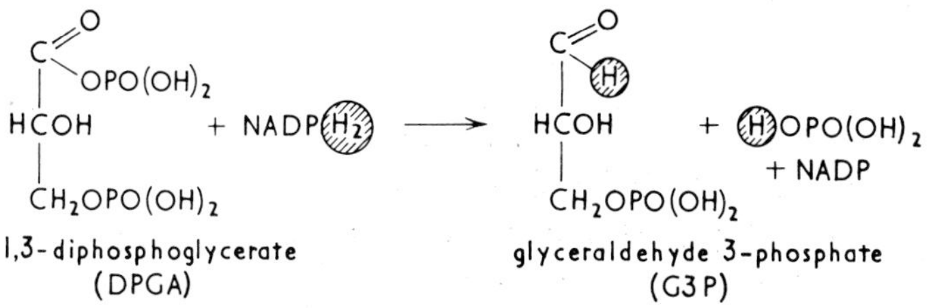

$$\begin{array}{c} C(=O)OPO(OH)_2 \\ | \\ HCOH \\ | \\ CH_2OPO(OH)_2 \end{array} + NADP(H_2) \longrightarrow \begin{array}{c} C(=O)(H) \\ | \\ HCOH \\ | \\ CH_2OPO(OH)_2 \end{array} + (H)OPO(OH)_2 + NADP \qquad \text{Eqn. 6.17}$$

1,3-diphosphoglycerate (DPGA) — glyceraldehyde 3-phosphate (G3P)

---

E.C. 1.2.1.13
$Km$ (DPGA) = 1 μM Concentration of DPGA in chloroplasts = very low, unknown
$Km$ (NADPH) = 4 μM Concentration of total NADPH + NADP = 0.1–0.5 mM
$\Delta F' = -1.5$ kcal
mol. wt. = 600 000

---

The enzyme is light-activated by the ferredoxin/thioredoxin system (Chapter 9) and, *in vitro* by dithiothreitol. A sigmoidal increase in activity is produced by NADPH with maximal rates occurring at between 0.5 and 1.5 mM NADPH. ATP acts similarly in the range 1–6 mM.

### 4 TRIOSEPHOSPHATE ISOMERASE

This is another enzyme, known in glycolysis, which catalyses (Eqn. 6.18) the conversion of glyceraldehyde 3-phosphate (G3P) to its isomer, dihydroxyacetone-phosphate (DHAP).
Glyceraldehyde 3-phosphate and its isomer, dihydroxyacetone phosphate are known collectively as triose phosphate. In the RPP pathway 2/5 of the triose

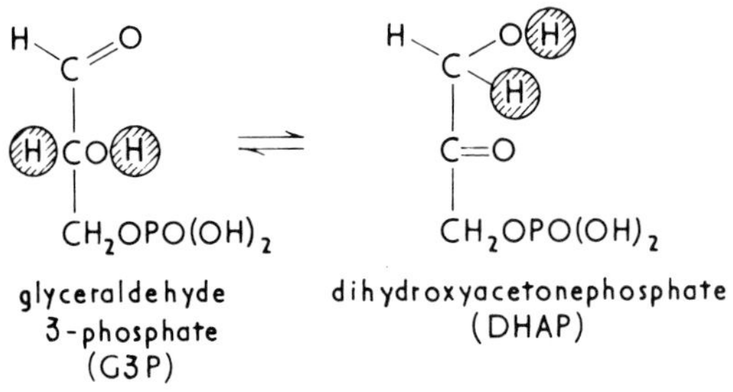

H C=O | HCOH | $CH_2OPO(OH)_2$ ⇌ H C OH H | C=O | $CH_2OPO(OH)_2$

glyceraldehyde 3-phosphate (G3P) — dihydroxyacetonephosphate (DHAP)

Eqn. 6.18

E. C. 5.3.1.1
$Km$ (DHAP) = 1.1 mM
$Km$ (G3P) = 0.3 mM. Concentration of DHAP + G3P in chloroplast = 0.3–0.4 mM
$\Delta F' = -1.8$ kcal.
mol. wt. = 53 000

phosphate is utilized in the form of DHAP. The equilibrium for this reaction (Eqn. 6.18) lies to the right and at equilibrium 22/23 of the total triose phosphate will be present as DHAP. The enzyme is inhibited by low concentrations of phosphoglycolate.

## 5 Aldolase (FBP aldolase).

Yet another enzyme of glycolytic fame. In photosynthesis, the reaction, (Eqn. 6.19) actually brings about the aldol (*ald*ehyde-alcoh*ol*) condensation which gives the enzyme its name rather than the reverse separation of fructose 1,6-bisphosphate (FBP) into its component triose phosphates (G3P and DHAP).

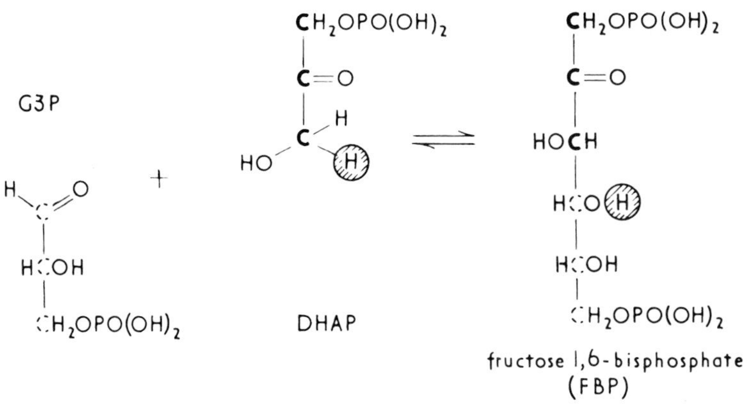

G3P: H C=O | HCOH | $CH_2OPO(OH)_2$ + DHAP: $CH_2OPO(OH)_2$ | C=O | HO C H H ⇌ $CH_2OPO(OH)_2$ | C=O | HOCH | HCOH | HCOH | $CH_2OPO(OH)_2$

fructose 1,6-bisphosphate (FBP)

Eqn. 6.19

E.C. 4.1.2.13
$K$m (FBP) = 20 $\mu$M
$K$m (G3P) = 0.3 mM
$K$m (DHAP) = 0.4 mM
$\Delta F' = -5.5$ kcal
mol. wt. = 150 000

The highly negative $\Delta F'$ for the formation of FBP from DHAP and G3P suggests that the equilibrium for this reaction is strongly towards FBP formation but because the reaction has two substrates and one product, the equilibrium position is greatly influenced by the concentration of the compounds involved. For example, with an initial concentration of G3P plus DHAP of 1 mM, 76% will be converted to FBP before equilibrium is reached, whereas with a starting concentration of 0.1 mM, only 42% will be converted to FBP at equilibrium. The situation is altered by the fact that the G3P and DHAP are not present in equal amounts because the equilibrium of the triose phosphate isomerase reaction favours DHAP by 22:1. This effectively draws the equilibrium back towards G3P and DHAP e.g. for 0.1 mM G3P plus DHAP, only 33% will be as FBP if the G3P and DHAP are also at the equilibrium dictated by the triose phosphate isomerase reaction.

## 6 Fructose bisphosphatase (FBPase).

This catalyses the hydrolysis of FBP to fructose 6-phosphate (F6P) and free Pi (Eqn. 6.20).

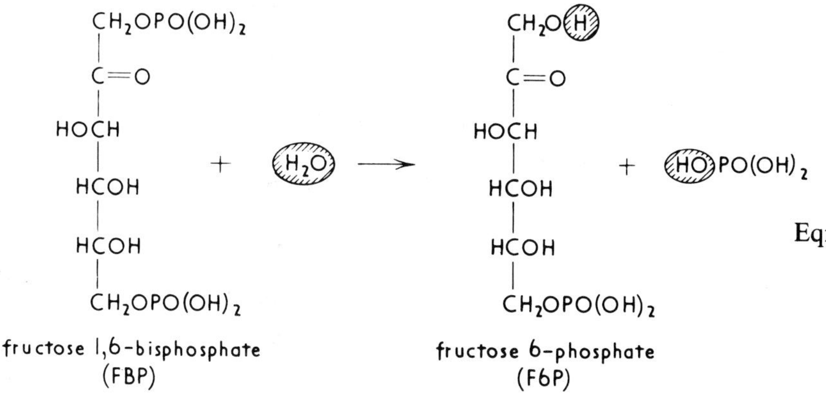

$$\begin{array}{c} CH_2OPO(OH)_2 \\ | \\ C{=}O \\ | \\ HOCH \\ | \\ HCOH \\ | \\ HCOH \\ | \\ CH_2OPO(OH)_2 \end{array} + H_2O \longrightarrow \begin{array}{c} CH_2OH \\ | \\ C{=}O \\ | \\ HOCH \\ | \\ HCOH \\ | \\ HCOH \\ | \\ CH_2OPO(OH)_2 \end{array} + HOPO(OH)_2 \qquad \text{Eqn. 6.20}$$

fructose 1,6-bisphosphate (FBP) fructose 6-phosphate (F6P)

E.C. 3.1.3.11
$Km\ (Mg^{2+}) = 3$ mM
$Km$ (FBP) = 0.2 mM. Concentration of FBP in chloroplasts = 0.1–0.3 mM
$\Delta F' = -4.0$ kcal
mol. wt. = 160 000

This hydrolysis is the second Pi-releasing step in the RPP pathway and it is catalyzed by a specific chloroplast FBPase which is different from the cytoplasmic enzyme. The large negative $\Delta F'$ value indicates an equilibrium in favour of F6P formation. Activity is more pronounced (up to 250 $\mu$mol $mg^{-1}$ Chl $h^{-1}$) in the absence of electron sinks such as $CO_2$ than in their presence (100 $\mu$mol $mg^{-1}$ Chl $h^{-1}$). Experiments with isolated chloroplasts, leaf protoplasts and whole leaf extracts have shown that the activity of the chloroplast FBPase (measured at pH values around 8.0) is much higher following illumination. The activation can also be achieved with the purified enzyme by preincubation with reducing agents such as DTT or reduced ferredoxin. Under these conditions, the activation is accompanied by an increase in the number of available sulphydryl groups showing that an actual reduction of the protein has occurred. The mechanism of light activation is also discussed in Chapter 9. Activation results in a decrease in the pH optimum and an increase in the affinity for FBP and $Mg^{2+}$. These changes in activation status together with the alteration of the stromal environment indicate that the enzyme will have very low activity in the dark.

## 7 TRANSKETOLASE.

This enzyme also plays a role in respiration (in the Oxidative Pentose Phosphate pathway). In photosynthesis, it is involved in two reactions. In the first (Eqn. 6.21), it transfers a $C_2$-unit (glycoaldehyde-transketolase addition complex called dihydroxyethylthiamine pyrophosphate) from the top of F6P to carbon 1 of G3P yielding xylulose 5-phosphate and erythrose 4-phosphate ($C_6 + C_3 \rightarrow C_5 + C_4$).

Eqn. 6.21

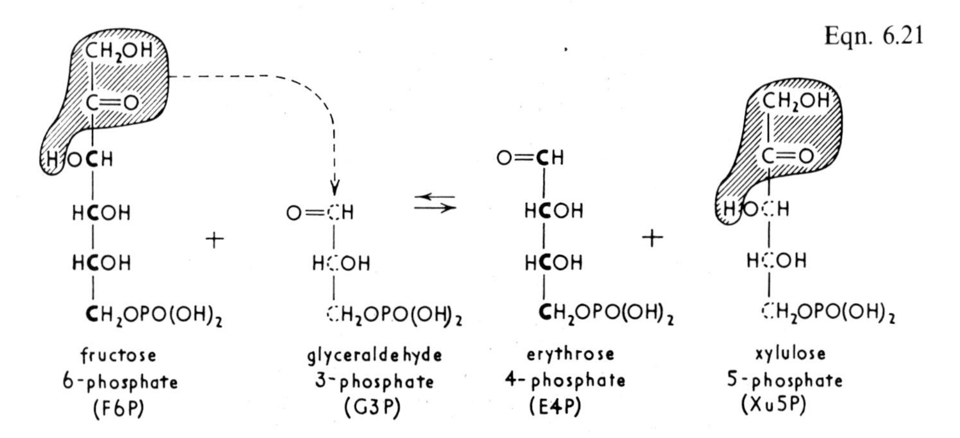

E.C. 2.2.11
$K$m (F6P) not known = Concentration of F6P in chloroplasts = 0.6–1.5 mM
$K$m (G3P) not known = Total DHAP plus G3P in chloroplast = 0.3–0.4 mM
$\Delta F' = 1.47$ kcal
mol. wt. = 140 000

At one time it was believed that the $C_2$ unit which is transferred might be an important source of glycolate in photorespiration (Chapter 13) but this view no longer attracts wide support and the derivation of phosphoglycolate from RBP (Eqn. 6.15b) is not disputed.

### 8 Aldolase.

The second aldol condensation (cf. Eqn. 6.19 above) joins erythrose 4-phosphate from the first transketolase reaction (see Eqn. 6.21 above) to DHAP, yielding sedoheptulose 1,7-bisphosphate ($C_4 + C_3 \rightarrow C_7$).

$$\underset{\substack{\text{erythrose}\\\text{4-phosphate}\\\text{(E4P)}}}{\begin{array}{c} H\diagdown C{=}O \\ | \\ HCOH \\ | \\ HCOH \\ | \\ CH_2OPO(OH)_2 \end{array}} + \underset{\text{DHAP}}{\begin{array}{c} CH_2OPO(OH)_2 \\ | \\ C{=}O \\ | \\ HO{-}C{-}H \\ H \end{array}} \rightleftharpoons \underset{\substack{\text{sedoheptulose}\\\text{1,7-bisphosphate}\\\text{(SBP)}}}{\begin{array}{c} CH_2OPO(OH)_2 \\ | \\ C{=}O \\ | \\ HOCH \\ | \\ HCOH \\ | \\ HCOH \\ | \\ HCOH \\ | \\ CH_2OPO(OH)_2 \end{array}} \qquad \text{Eqn. 6.22}$$

### 9 Sedoheptulose bisphosphatase (SBPase).

The second phosphatase reaction (see Eqn. 6.20 above) releases phosphate from carbon 7 of SBP leaving sedoheptulose 7-phosphate (S7P).

E.C. 3.1.3.37
Concentration of SBP in chloroplasts = 0.2–1 mM
$\Delta F' = -4.0$ kcal (by analogy with reaction 6.20)
$K$m = 0.24 mM, mol wt. 50 000 approx.

The enzyme from spinach is highly specific for SBP and requires a divalent cation ($Mg^{2+}$ or $Mn^{2+}$) for activity. Like FBPase, it can be activated by DTT or by the ferredoxin-thioredoxin system. It appears to be present in amounts which are lower than any other enzyme of the RPP pathway with maximum recorded rates of 60 $\mu$mol. $mg^{-1}$ Chl $h^{-1}$. Preliminary evidence indicates that, unlike FBPase, metabolites of the RPP pathway may be involved in its activation *in vivo*.

$$\begin{array}{c} CH_2OPO(OH)_2 \\ | \\ C{=}O \\ | \\ HOCH \\ | \\ HCOH \\ | \\ HCOH \\ | \\ HCOH \\ | \\ CH_2OPO(OH)_2 \end{array} + H_2O \longrightarrow \begin{array}{c} CH_2OH \\ | \\ C{=}O \\ | \\ HOCH \\ | \\ HCOH \\ | \\ HCOH \\ | \\ HCOH \\ | \\ CH_2OPO(OH)_2 \end{array} + HO\,PO(OH)_2 \qquad \text{Eqn. 6.23}$$

sedoheptulose 1,7-bisphosphate (SBP) — sedoheptulose 7-phosphate (S7P)

10 TRANSKETOLASE.

The second transketolase reaction (cf. Eqn. 6.21 above) transfers a $C_2$ unit as before, to G3P. However, since the donor is now S7P the other product (ribulose 5-phosphate) is also a pentose ($C_3 + C_7 \rightarrow C_5 + C_5$).

Eqn. 6.24

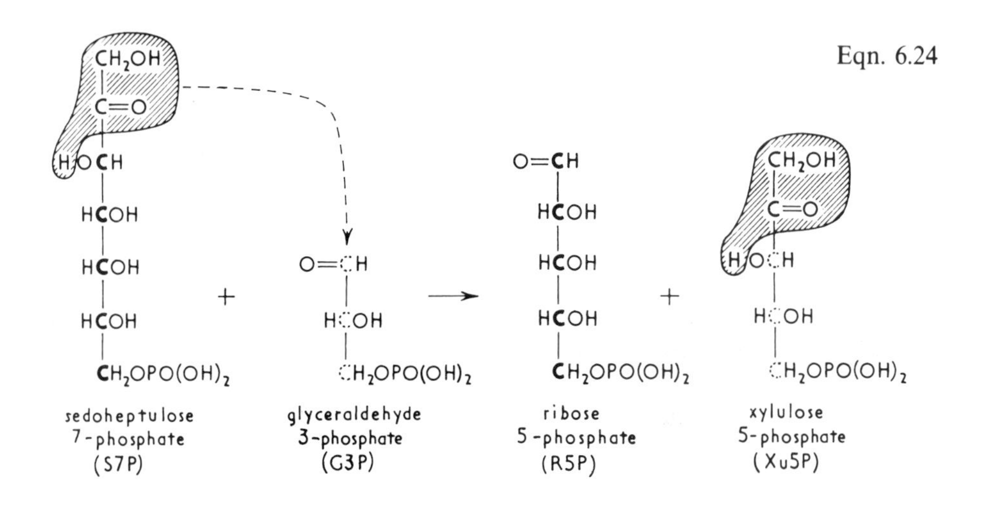

## 11 RIBOSE PHOSPHATE ISOMERASE.

Converts ribose 5-phosphate (R5P) to its isomer, ribulose 5-phosphate (Ru5P) in a freely reversible reaction (Eqn. 6.25)

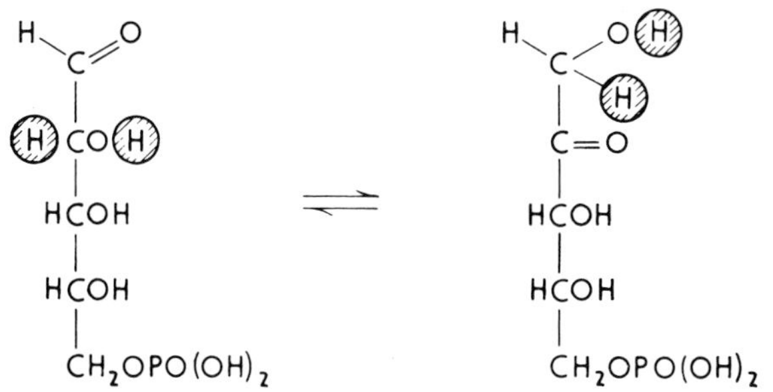

$$\begin{array}{c} \text{H}\diagdown\text{C}\diagup\!\!\!=\text{O} \\ | \\ \text{(H)CO(H)} \\ | \\ \text{HCOH} \\ | \\ \text{HCOH} \\ | \\ \text{CH}_2\text{OPO(OH)}_2 \end{array} \rightleftharpoons \begin{array}{c} \text{H}\diagdown\text{C}\diagup\text{O(H)} \\ \diagdown\text{(H)} \\ | \\ \text{C=O} \\ | \\ \text{HCOH} \\ | \\ \text{HCOH} \\ | \\ \text{CH}_2\text{OPO(OH)}_2 \end{array} \quad \text{Eqn. 6.25}$$

ribose 5-phosphate (R5P) ribulose 5-phosphate (Ru5P)

E.C. 5.3.1.6
$Km$ (R5P) = 2.0 mM
$\Delta F'$ = 0.64 kcal
mol. wt. = 54 000

## 12 RIBULOSE PHOSPHATE 3-EPIMERASE.

Like Equation 6.25 this is also an isomerization but substrate and product are epimers which differ only in the orientation of the hydrogen and hydroxyl attached to carbon 3. The substrate is xylulose 5-phosphate (Xu5P), from the two transketolase reactions (Eqns. 6.21 & 6.24) and the product is again ribulose 5-phosphate.

E. C. 5.1.31
$Km$ (Xu5P) = 0.5 mM
$\Delta F'$ = −0.13 kcal
mol. wt. = 46 000

The reaction is freely reversible as indicated by the low $\Delta F'$ value, and at equilibrium the ratio of Xu5P/Ru5P varies from about 1 to 3.

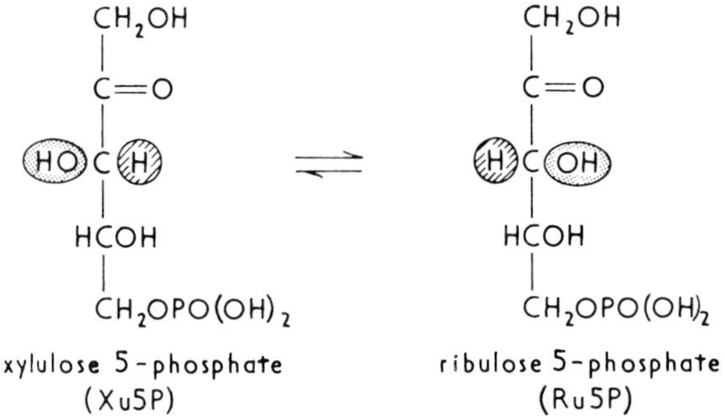

$$\begin{array}{ccc} \mathrm{CH_2OH} & & \mathrm{CH_2OH} \\ | & & | \\ \mathrm{C{=}O} & & \mathrm{C{=}O} \\ | & & | \\ \mathrm{HO\,C\,H} & \rightleftharpoons & \mathrm{H\,C\,OH} \\ | & & | \\ \mathrm{HCOH} & & \mathrm{HCOH} \\ | & & | \\ \mathrm{CH_2OPO(OH)_2} & & \mathrm{CH_2OPO(OH)_2} \end{array} \qquad \text{Eqn. 6.26}$$

xylulose 5-phosphate (XuP5) ribulose 5-phosphate (Ru5P)

13 PHOSPHORIBULOKINASE

This enzyme catalyses the last reaction of the RPP pathway converting Ru5P produced by Equations 6.25 and 6.26 to ribulose 1,5-bisphosphate (RBP) at the expense of ATP.

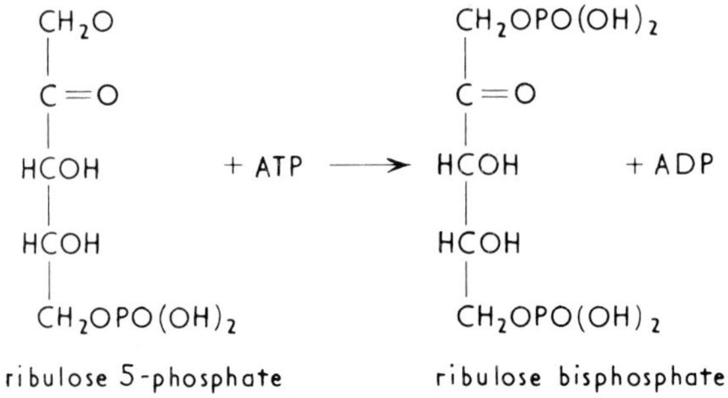

$$\begin{array}{ccccc} \mathrm{CH_2O} & & & \mathrm{CH_2OPO(OH)_2} & \\ | & & & | & \\ \mathrm{C{=}O} & & & \mathrm{C{=}O} & \\ | & & & | & \\ \mathrm{HCOH} & + \mathrm{ATP} & \longrightarrow & \mathrm{HCOH} & + \mathrm{ADP} \\ | & & & | & \\ \mathrm{HCOH} & & & \mathrm{HCOH} & \\ | & & & | & \\ \mathrm{CH_2OPO(OH)_2} & & & \mathrm{CH_2OPO(OH)_2} & \end{array} \qquad \text{Eqn. 6.27}$$

ribulose 5-phosphate ribulose bisphosphate

---

E. C. 2.7.1.19
$Km$ (Ru5P) = 0.2 mM. Concentration of total pentose monophosphate in chloroplast = 0.1 mM
$Km$ (ATP) = 0.1 mM Concentration of AMP + ADP + ATP in chloroplasts = 1–3 mM
$\Delta F' = -5.2$ kcal.
mol. wt. = 240 000

---

This reaction utilizes one third of the ATP required for $CO_2$ fixation and, as the $\Delta F'$ value suggests, the equilibrium position is strongly towards RBP formation. There is good evidence for light activation of this enzyme *in vivo* and an activation of the enzyme *in vitro* by DTT once again suggests that a photosynthetically produced reductant may be responsible. Phosphoribulokinase is also regulated by the energy charge of the adenylate system i.e. by the concentrations of ATP, ADP and AMP.

## 6.12 Free energies

Standard 'biological' free energy changes ($\Delta F'$) differ from those sometimes used by the physical chemist ($\Delta F^{\circ}$) because they are calculated with some regard for biological environments (e.g. $\Delta F'$ is calculated for pH 7.0 whereas $\Delta F^{\circ}$ relates to pH 0). There are often quite large differences in published $\Delta F'$ values. These derive from difficulties in measurement, differences in assumptions and even errors in calculation. Even greater disparities derive from the uses to which these values are put and the conclusions which are drawn from them. The student is always reminded that a free energy change (which is, in itself, a thinly disguised equilibrium constant) will only indicate something about the feasibility of a reaction and nothing about the rate. As previously noted (Section 2.9) the oxidation of $H_2$ by molecular $O_2$ proceeds with a decrease in energy of about 56 kcal but although these gases can be induced to combine with explosive violence they can be stored in each others company indefinitely without detectable reaction. Although the $\Delta F$ value as such tells us nothing about rate it does however carry certain implications. Malic dehydrogenase is an active and widely distributed enzyme but its $\Delta F'$ value is so strongly positive in the direction malate → oxaloacetate that it is impossible to envisage this reaction occurring *in this direction* in the cell except in a metabolic context like the tricarboxylic acid cycle (in which its unfavourable equilibrium position is offset by build-up of substrates and removal of end-products). If, on the other hand, such an unfavourable equilibrium position can be so displaced it may be asked what, if anything, can be deduced from $\Delta F$ values. The answer is that they do serve to focus attention on metabolic problems even when used in a subjective fashion. Enzymes such as Ochoa's malic enzyme (Eqn. 6.3) were once thought to be possible candidates in photosynthetic carbon fixation but the relatively unfavourable $\Delta F$ values of reactions that they catalysed were one element in the continuing search for other mechanisms and the establishment, for example, of the role of PEP carboxylase (Eqn. 6.8) in CAM (Chapter 15).

Bassham has attached great importance to $\Delta F$ values as indicators of 'regulated enzymes'. Again, however, there are some dangers in the way in which this could be interpreted. It is true that a reaction with a large $\Delta F$ value will tend to go to near completion if the enzyme concerned is not subject to some form of regulation. This is not to say that a reaction with a relatively small $\Delta F$ value may not play a role in the way in which carbon flow is regulated. In this instance the catalyst may not be subject to regulatory control but the equilibrium position of the reaction itself, the forces of mass action, may be of importance.

Table 6.2 lists $\Delta F'$ values derived by Bassham and Krause. The equilibrium position of the aldolase reactions is anomalous because it is a reaction with two

**Table 6.2.** Free energy changes in the RPP pathway (after Bassham & Krause, 1969).

| Equation | $\Delta F'$ (kcal) |
|---|---|
| 6.15 (RBP → PGA) | −8.4 |
| 6.16 + 6.17 (PGA → DPGA → G3P) | +4.3 |
| 6.18 (G3P → DHAP) | −1.8 |
| 6.19 (G3P + DHAP → FBP) | −5.2 |
| 6.20 (FBP → F6P) | −3.4 |
| 6.21 (F6P + G3P → E4P + Xu5P) | +1.5 |
| 6.22 (E4P + G3P → SBP) | −5.6 |
| 6.23 (SBP → S7P) | −3.4 |
| 6.24 (S7P + G3P → R5P + Xu5P) | +0.1 |
| 6.25 (R5P → Ru5P) | +0.5 |
| 6.26 (Xu5P → Ru5P) | +0.2 |
| 6.27 (Ru5P → RBP) | −5.2 |

$\Delta F'$ = Standard 'biological' free energy change (pH 7.0)

substrates and one product. The condensation will, therefore, be more and more favoured as the concentration of triose phosphates is increased. The large negative $\Delta F'$ value for RBP carboxylase (−8.4 kcal) is in accord with its role in assimilating carbon from atmospheres containing only about 0.03% (300 ppm) $CO_2$. Similarly this value suggests that the carboxylase is regulated but although it is undoubtedly affected by many metabolites the evidence that it is inactivated in the dark is much less compelling than was once believed (Chapter 9). The bisphosphatases also have relatively large $\Delta F'$ values (−3 to −4 kcal) and it seems clear that fructose bisphosphate undergoes light activation (Chapter 9). The fact that this enzyme works much better in the alkaline environment of the illuminated stroma makes it feasible that both bisphosphatases may be entirely inactive in the dark (Chapter 9).

The $\Delta F'$ value of +4.3 for Equation 6.16 + 6.17 (PGA → DPGA → G3P) given in Table 6.2 is less favourable than is often quoted elsewhere (−3.0 kcal is popular and Fig. 9.6 is based on a value of 1.2 kcal). There is general agreement, however, that the value for reaction in Equation 6.16 (PGA → DPGA) is about +4.5 kcal and this leaves the overall reduction of PGA to G3P unfavourable by anything from +1.2 to +4.3 kcal. What is clear about this aspect is that the reaction in Equation 6.16 itself is the most unfavourable in the entire RPP pathway and therefore particularly susceptible to changes in [ATP], [ADP] and [PGA] (see Section 9.6). Apart from the reaction in Equation 6.27 (Ru5P → RBP), which strongly favours RBP formation, the other reactions are all freely reversible. Even relatively small $\Delta F'$ values may still be large enough to be significant, however, and for example, the fact

that the equilibrium position between the two triose phosphates favours DHAP probably account for the preponderance of this compound amongst exported metabolites (Chapter 8).
[Refs. 7, 18, 19, **32**]

## 6.13 The operation of the RPP pathway

In its operation, the RPP pathway can be divided into several parts. Initially there is the carboxylation of RBP and, in Fig. 6.1, this is represented in triplicate so that all other stages of the cycle may be followed as they occur. This also emphasizes the fact that for every three molecules of $CO_2$ which enter the cycle one molecule of triose phosphate can be released as product ($3 \times C_1 \rightarrow 1 \times C_3$). Secondly, there is the reduction. This is the only reductive reaction in the main pathway of photosynthetic carbon assimilation. Accordingly, it utilizes all of the $NADPH_2$ which is consumed in the RPP pathway. In theory, the reduction of PGA to triose phosphate could occur in one step but in practice the overall energy gap is diminished by an input of ATP in the reaction catalysed by PGA kinase (Eqn. 6.16) (Fig 6.9).

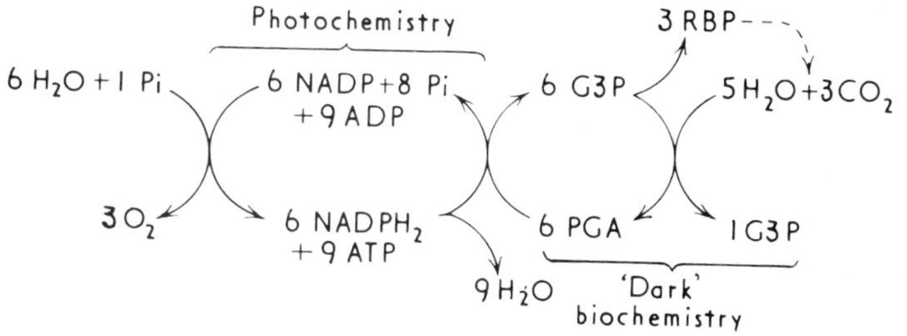

Photochemistry
3 RBP
$6\ H_2O + 1\ Pi$
$6\ NADP + 8\ Pi + 9\ ADP$
6 G3P
$5\ H_2O + 3\ CO_2$
$3\ O_2$
$6\ NADPH_2 + 9\ ATP$
6 PGA
1 G3P
$9\ H_2O$
'Dark' biochemistry

**Fig. 6.9** Utilization of 'assimilatory power' ($NADPH_2$ + ATP) in the RPP pathway.

Once triose phosphate has been formed by reduction of PGA (Eqns. 6.16–6.18) all of the remaining 10 reactions (constituting the so-called 'sugar phosphate shuffle') are concerned with the regeneration of RBP, the $CO_2$-acceptor, and the majority simply constitute a device by which five triose phosphates are 'rearranged' to give three pentose phosphates ($5 \times C_3 \rightarrow 3 \times C_5$).

Equations 6.25 and 6.26 ensure that all of the regenerated pentose phosphate is in the form of ribulose 5-phosphate. Finally ribulose 5-phosphate kinase (Eqn. 6.27) utilizes ATP to convert Ru5P to RBP and the cycle is complete.
[Refs. 3, 6, **55**]

## 6.14 Utilization of assimilatory power

As discussed in Section 6.13, and as shown in Fig. 6.1 assimilatory power is used at three points. The bulk of the ATP (six molecules for every three molecules of carbon dioxide fixed) goes in the phosphorylation of PGA (Eqn. 6.16) and may be regarded as a contribution to the subsequent reduction of DPGA to triose phosphate (Eqn. 6.17) which utilizes six molecules of $NADPH_2$ (one for each triose formed). By far the greater part of the total assimilatory power is therefore employed in the reduction of PGA and, in principle, most (5/6) of the ensuing triose phosphate is then reoxidized to PGA (via the cycle) as carboxylation proceeds. Only a relatively small fraction (3 ATP's or about 5% of the assimilatory power) enters the cycle in Equation 6.27 and this input may be regarded as a means of simultaneously maintaining the impetus of the cycle and an appropriate working concentration of RBP. [The free energy of RBP formation from Ru5P (Table 6.2) is about −5 kcal indicating that the equilibrium position of Equation 6.27 is well towards RBP—see Section 2.5]. Otherwise, the introduction of the second phosphate group at this particular point seems to serve no good purpose. [RBP is a high energy compound in the sense that ring structure which would lead to increased resonance stability is precluded by the phosphate at carbon 5 and this must contribute to the fact that its carboxylation, which in itself is an endergonic or energy requiring process (Section 6.4) nevertheless leads to a decrease in free energy (energy release) of about 8 kcal. The introduction of a second phosphate at carbon 1 does not, however, contribute to this effect and there would not seem to be any thermodynamic advantage, other than that indicated above, in phosphorylation followed by carboxylation rather than carboxylation followed by phosphorylation (see Section 6.4)].
[Refs. 5, 22].

## 6.15 Feedback

Although five out of the six molecules of triose phosphate formed in the RPP pathway (Fig. 6.1.) are consumed in the regeneration of the $CO_2$-acceptor (RBP) the remainder constitutes product. As such it may have several fates including export to the cytosol (Chapter 8) and starch synthesis within the chloroplast. In some circumstances, however, it may also feed back into the main pathway (see Section 6.3). In this way it is theoretically possible to double the concentration of RBP for every five turns of the cycle, if the cycle is depicted as it is in Fig. 6.10. [Six molecules of PGA are formed by the carboxylation of three molecules of RBP and, of these, five are used in the regeneration of the $CO_2$-acceptor. Thus after five revolutions, an additional five molecules of $C_3$ would have accumulated. If these were returned to

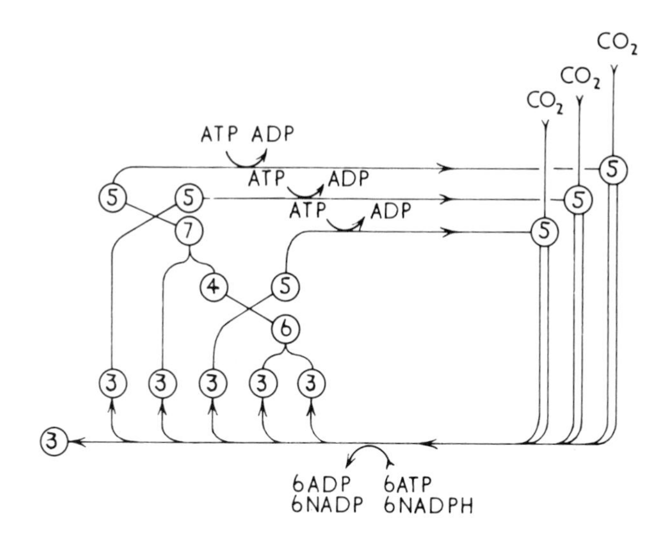


**Fig. 6.10.** The RPP pathway. A still further simplified version of Fig. 6.1 to emphasize carbon traffic.

the cycle, instead of being removed as product, a further three molecules of RBP would be formed].
[Ref. **55**].

## 6.16 Integration of the RPP pathway and the photorespiratory carbon oxidation cycle

Photorespiration is considered at more length in Chapter 13 but there is now good evidence that photorespiration can be regarded as a cyclic sequence of events, which integrate with the RPP pathway (Figs. 6.11 and 6.12). According to this concept RBP carboxylase also functions as an oxygenase. In high $CO_2$ or low $O_2$ the carboxylase action predominates but in air a proportion of the RBP is split, in an $O_2$-consuming reaction which yields one molecule of phosphoglycerate and one molecule of phosphoglycolate ($HOOC{\cdot}CH_2OPO(OH)_2$) instead of two molecules of PGA. The phosphoglycolate is hydrolysed and, after leaving the chloroplast as glycolate, enters the peroxisome where it is oxidized to glyoxylate ($CHO{\cdot}COOH$) before undergoing transamination to glycine ($CH_2NH_2COOH$). Passing into the mitochondrion, two molecules of glycine form serine ($CH_2OH{\cdot}CHNH_2COOH$) according to Equation 6.28.

$$2\,\text{glycine} + H_2O \rightarrow \text{serine} + CO_2 + NH_3 + 2H^+ + 2e^-. \qquad \text{Eqn. 6.28}$$

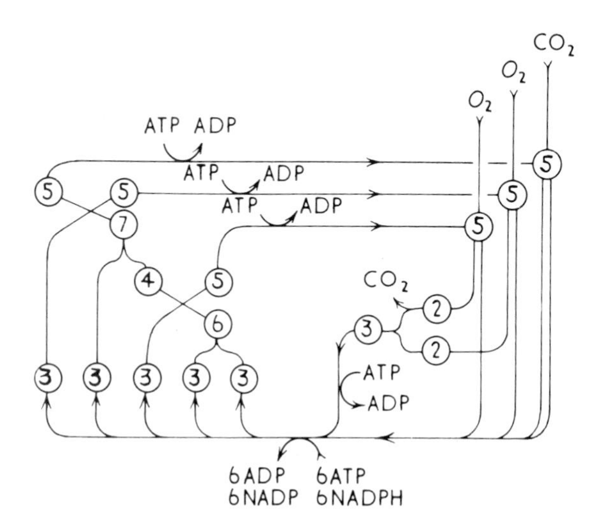


**Fig. 6.11.** Integration of the RPP pathway and photorespiratory cycle at the $CO_2$-compensation point. Two out of every 3 molecules of RBP are oxidized, each yielding 1 molecule of PGA and 1 molecule of phosphoglycolate, a 2-carbon compound. In this outline diagram most of the partial reactions in which these 2-carbon units engage are disregarded but it is seen that a condensation of two 2-C compounds leads to the generation of a new 3-carbon compound which re-enters the cycle. It should be noted that there is no net product under these conditions (see Fig. 6.12 for details)

This reaction is the source of the $CO_2$ and $NH_3$ released in photorespiration, whereas $O_2$ is consumed both in the initial oxygenase reaction and in the subsequent oxidation of glycolate. The serine passes to the peroxisome where it gives hydroxypyruvate ($CH_2OH{\cdot}CO{\cdot}COOH$) which is reduced to glycerate ($CH_2OH{\cdot}CHOH{\cdot}COOH$). This moves into the chloroplast where it is phosphorylated to PGA and, once there, equilibrates with that already formed and retained in the stroma (Fig. 6.12).

The proposed sequence accounts for $^{14}CO_2$ labelling which suggest that glycolate is derived from an intermediate of the RRP pathway by a mechanism which leads to the incorporation (into the carboxyl of glycolate) of one molecule of $^{18}O_2$. It explains the fact that photorespiration is repressed by high $[CO_2]$ or low $[O_2]$ and that $^{18}O_2$ incorporated into glycolate subsequently appears in glyoxylate,

acceptor ($5C_3 \rightarrow 3C_5$) but the sixth 3-carbon unit (which would otherwise constitute product) is lost, i.e. 2 out of the 3 carbons which would go into product are never fixed because they are displaced by $O_2$ and the third is released in the conversion of glycine to serine ($2C_2 \rightarrow C_3 + CO_2$). Boxes around structural formulae together with numbers in parentheses indicate how many of each of these molecules are involved in this sequence as depicted in Fig. 6.11

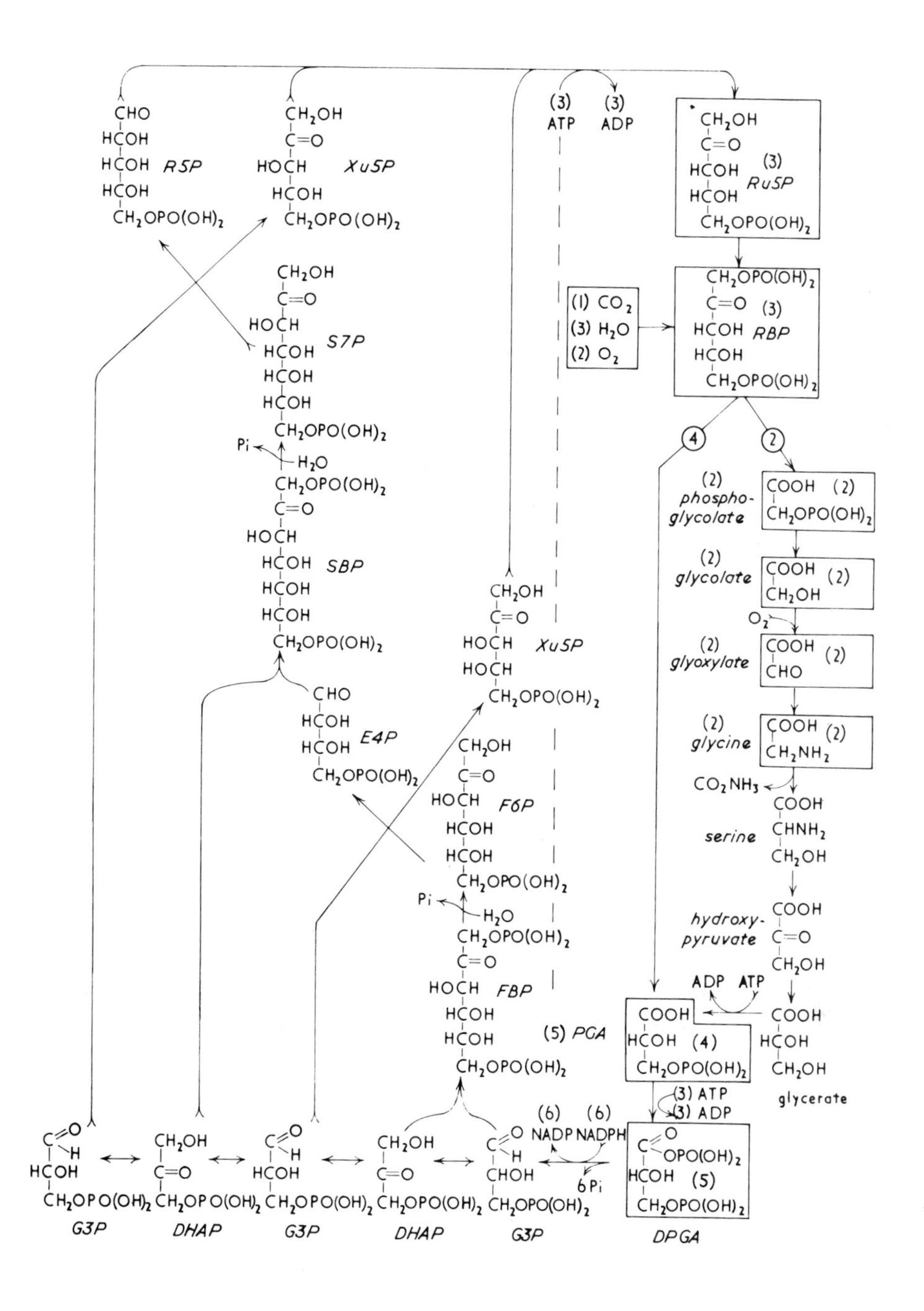


**Fig. 6.12.** The Integrated Pathway in detail. The reactions to the left of the dotted line are exactly the same as those depicted in outline in the regenerative phase of Fig. 6.1 or 6.10. Similarly the carboxylation and reduction of PGA to triose phosphate involves precisely the same reactions as in Figs. 6.1 and 6.10 (i.e. as in the RPP pathway proper) but the flow of carbon through these reactions is changed. Thus only one RBP is carboxylated instead of 3 so only 4 molecules of PGA are formed at this stage. The 2 molecules of phosphoglycolate which are formed instead, in the oxygenase reaction enter the sequence glycolate→ glyoxylate→ glycine→ serine→ hydroxypyruvate→ glycerate. Finally the glycerate is phosphorylated, restoring a fifth molecule of PGA so that 5 molecules of triose phosphate can be formed. These can regenerate the $CO_2$-

glycine, serine, hydroxypyruvate and PGA. It is consistent with the fact the glycolate formation from sugar monophosphates of the RPP cycle requires ATP and that in the presence of $^{18}O_2$ the enrichment of PGA (at the compensation point) is only about 1/5 of the enrichment of glycolate (as Figs. 6.11 and 6.12 show, only one out of every 5 molecules of PGA formed at the compensation point is produced via glycolate and serine).

The ammonia which is released in this sequence could be several fold greater than that undergoing primary assimilation and must therefore be reassimilated if the plant is not to lose all of its nitrogen. Studies with $N^{15}$ show that reassimilation does occur. It is suggested that glutamine ($HOOC{\cdot}CHNH_2(CH_2)_2{\cdot}CONH_2$) is formed in the cytoplasm by the action of glutamine synthetase (Eqn. 6.29).

$$\begin{matrix} COOH \\ | \\ (CH_2)_2 \\ | \\ CHNH_2 \\ | \\ COOH \end{matrix} + NH_3 + ATP \longrightarrow \begin{matrix} CONH_2 \\ | \\ (CH_2)_2 \\ | \\ CHNH_2 \\ | \\ COOH \end{matrix} + ADP + Pi \qquad \text{Eqn. 6.29}$$

The glutamine then enters the chloroplast where the GOGAT reaction (Eqn. 6.30) leads to the formation of new glutamate ($HOOC{\cdot}CHNH_2(CH_2)_2COOH$)

$$\begin{matrix} COOH \\ | \\ (CH_2)_2 \\ | \\ CO \\ | \\ COOH \end{matrix} + \begin{matrix} CONH_2 \\ | \\ (CH_2)_2 \\ | \\ CHNH_2 \\ | \\ COOH \end{matrix} + 2H^+ + 2e^- \longrightarrow \begin{matrix} COOH \\ | \\ (CH_2)_2 \\ | \\ CHNH_2 \\ | \\ COOH \end{matrix} + \begin{matrix} COOH \\ | \\ (CH_2)_2 \\ | \\ CHNH_2 \\ | \\ COOH \end{matrix} \qquad \text{Eqn. 6.30}$$

Ferredoxin acts as the reductant in this reaction. The glutamate formed can re-enter the glutamine synthetase reaction (see Fig. 13.6) or give rise to 2-oxoglutarate ($HOOC{\cdot}CO(CH_2)_2COOH$) in one of the transamination reactions (Eqn. 6.31) which is responsible for the conversion of glycolate to glycine

$$\text{glutamate} + \text{glyoxylate} \rightarrow \text{glycine} + \text{2-oxoglutarate.} \qquad \text{Eqn. 6.31}$$

For purposes of stoichiometry, at least, the other molecule of glycine in this sequence is derived from serine by a second transamination (Eqn. 6.32)

$$\text{glyoxylate} + \text{serine} \rightarrow \text{glycine} + \text{hydroxypyruvate.} \qquad \text{Eqn. 6.32}$$

The overall picture could be very neat, even to the extent that the oxidation of glycine to serine is sufficiently well coupled to yield three molecules of ATP which is more

than enough for reassimilation of ammonia into glutamine if the plant is not to lose all of its nitrogen. [Alternatively part of the reductive power from glycine oxidation may be shuttled to the peroxisome for reduction of hydroxypyruvate to glycerate, see Fig. 13.6].
[Refs. **41, 43, 56**]

### 6.17 Carbon traffic through the RPP pathway and the photorespiratory pathway

Precisely how many molecules of RBP are subject to oxygenation is a matter for debate. For every three molecules of $CO_2$ which are fixed in the RPP pathway six molecules of PGA are formed (Fig. 6.10). Five of these are recycled to RBP and only one constitutes net product in the form of triose phosphate. If, as a result of oxygenation, one molecule of phosphoglycolate and five molecules of PGA were formed (i.e. if net fixation was diminished by one third) only 1.5 atoms of carbon would be derived from the photorespiratory pathway for every 15 going through the main cycle (i.e. a diversion of only 10%). If two molecules of RBP underwent oxygenation for every one carboxylated (as illustrated in Fig. 6.11) then one fifth (20% of the carbon traffic) of all of the PGA formed would be derived from serine. This could be regarded as the principal chain of events at the $CO_2$ compensation point and would represent a situation in which there was no net product. If the flow of carbon through serine exceeded 20% of the total flow through both pathways this could only occur at the expense of net consumption of pre-formed products or metabolites. In these circumstances the intermediates of the RPP pathway would be depleted unless the status quo were maintained by utilization of reserves (e.g. by starch degradation).
[Refs. **43, 56**]

### 6.18 Starch synthesis

Starch is often regarded as a characteristic end-product of photosynthesis and there is no doubt that it frequently accumulates in chloroplasts by day and is mobilized by night (Sections 8.1 & 9.7). The extent of accumulation varies considerably, however. Some species such as snow-drop (*Galanthus nivalis*) do not appear to form starch at all and starch is not an important product in the leaves of several major crop species such as wheat. Tobacco has the reputation of being a formidable starch former, yet certain varieties accumulate very little if they are grown under optimal conditions.

Massive accumulation of starch and a decreased ability to degrade all of the starch synthesized during the preceding day is often a symptom of some deficiency in growing conditions or a change in a function such as that associated with the initiation of flowering. Many species will not accumulate starch, except in the guard cells, if grown in relatively low light, even if this is applied continuously. For reasons of this sort it is clearly difficult to generalize but at least in many instances it seems clear that the maximum photosynthetic production of metabolites outpaces the rate of utilization and that in these circumstances excess product can often be stored as transitory starch within the stroma (starch is often formed faster and degraded less readily in isolated leaves from which soluble products can not be exported to distant sinks). The role of Pi in regulating starch synthesis is considered more fully in Chapter 9. Stromal starch is believed to be formed primarily from triose phosphate released from the RPP pathway. (Most of this triose phosphate will be derived by direct reduction of PGA but some small contribution will be made by carbon which finds its way back into the PGA pool after traversing the photorespiratory pathway). The term 'released' is used advisedly because triose phosphate produced in this manner may be exported from the chloroplast or utilized in the regenerative aspect of the RPP pathway as well as being consumed in starch synthesis. If retained within the chloroplast the initial reactions are the same as those involved in sucrose synthesis in the cytoplasm i.e. a proportion of the triose phosphate undergoes aldol condensation to FBP which is then hydrolysed to F6P. This is then converted to its isomer, G6P, by hexose phosphate isomerase (Eqn. 6.33).

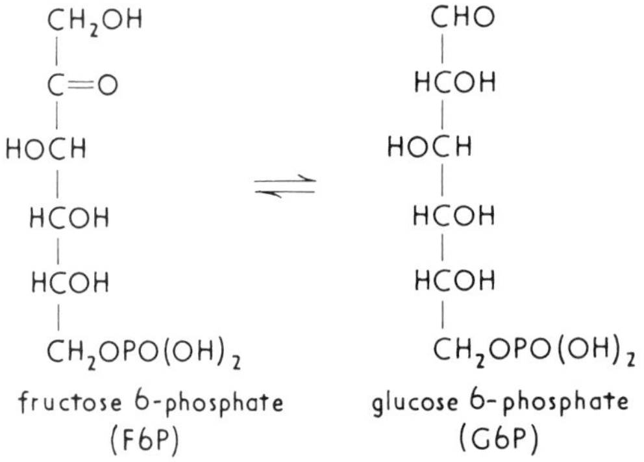

$$\begin{array}{c} CH_2OH \\ | \\ C{=}O \\ | \\ HOCH \\ | \\ HCOH \\ | \\ HCOH \\ | \\ CH_2OPO(OH)_2 \\ \text{fructose 6-phosphate} \\ \text{(F6P)} \end{array} \rightleftharpoons \begin{array}{c} CHO \\ | \\ HCOH \\ | \\ HOCH \\ | \\ HCOH \\ | \\ HCOH \\ | \\ CH_2OPO(OH)_2 \\ \text{glucose 6-phosphate} \\ \text{(G6P)} \end{array} \qquad \text{Eqn. 6.33}$$

In a reaction (Eqn. 6.34) catalysed by phosphoglucomutase, G6P becomes G1P.

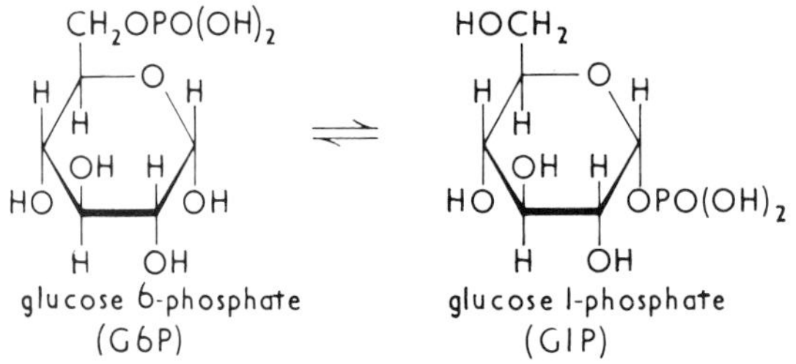

$$\begin{array}{c} CH_2OPO(OH)_2 \\ \text{H} \quad \text{O} \quad \text{H} \\ \text{H} \\ \text{OH} \quad \text{H} \\ \text{HO} \qquad\qquad \text{OH} \\ \text{H} \quad \text{OH} \\ \text{glucose 6-phosphate} \\ \text{(G6P)} \end{array} \rightleftharpoons \begin{array}{c} HOCH_2 \\ \text{H} \quad \text{O} \quad \text{H} \\ \text{H} \\ \text{OH} \quad \text{H} \\ \text{HO} \qquad\qquad OPO(OH)_2 \\ \text{H} \quad \text{OH} \\ \text{glucose 1-phosphate} \\ \text{(G1P)} \end{array} \qquad \text{Eqn. 6.34}$$

At equilibrium, a mixture of the above enzymes would yield hexose phosphates in the proportions of approximately

$$\begin{array}{ccccc} \text{F6P} & \rightleftarrows & \text{G6P} & \rightleftarrows & \text{G1P} \\ 10 & & 20 & & 1 \end{array}$$

Although these reactions are therefore freely reversible, the overall equilibrium may not be unimportant in determining the distribution of carbon between starch and pentose monophosphate in the illuminated chloroplast. Both F6P and TP (triose phosphate) are substrates for the first transketolase reaction (Eqn. 6.21) and this will, in turn, influence the amount of TP entering the second aldolase condensation (Eqn. 6.22) and the second transketolase reaction (Eqn. 6.24). An active sink for G1P would therefore tend to deflect carbon towards starch. It seems most probable that this sink must be provided by ADPglucose pyrophosphorylase which catalyses Eqn. 6.35 and is activated by a high ratio of [PGA]/[Pi]. Such a ratio will be greatest

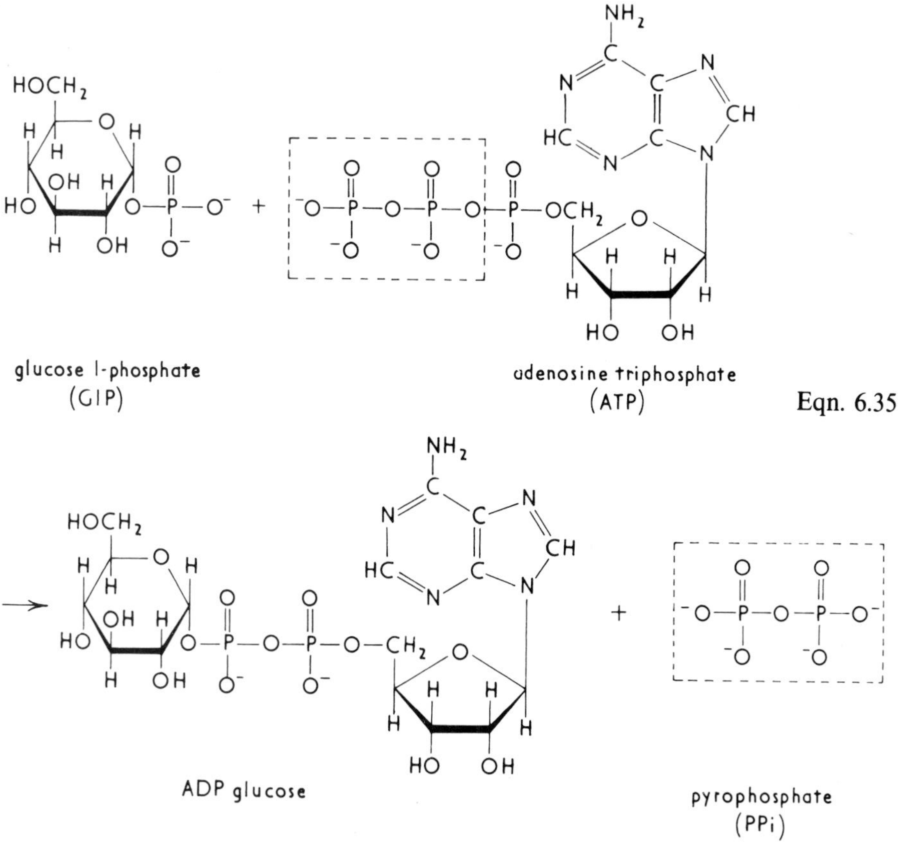

glucose 1-phosphate (G1P) + adenosine triphosphate (ATP) Eqn. 6.35

→ ADP glucose + pyrophosphate (PPi)

when low external Pi decreases triose phosphate and PGA export. Thus, maximal sink activity should, in theory, coincide with maximal supply of starch precursor. Finally, starch synthetase transfers glucosyl units from ADPglucose to suitable primers (such as maltose, maltotriose, other oligosaccharides and dextrins), extending their chains to form unbranched $\alpha(1 \rightarrow 4)$-linked polymers such as amylose (Eqn. 6.36).

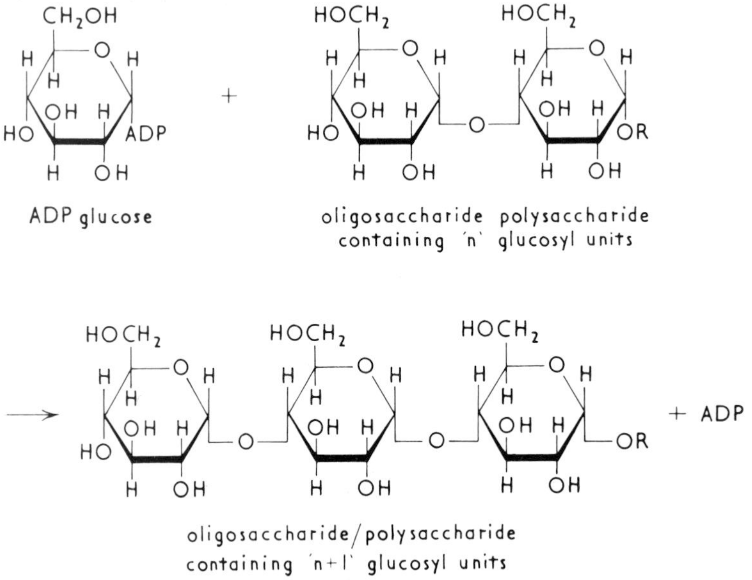

ADP glucose + oligosaccharide polysaccharide containing 'n' glucosyl units — Eqn. 6.36

→ oligosaccharide/polysaccharide containing 'n+1' glucosyl units + ADP

Amylose may also be formed by starch phosphorylase (Eqn. 6.37) but though the possibility that this enzyme may contribute to synthesis cannot be entirely precluded the reaction is readily reversible and a role in degradation is currently regarded as more feasible.
[Refs. 21, **48**]

### 6.19 Starch degradation

In general, starch in storage tissues is believed to be degraded by the combined action of several enzymes. Starch phosphorylase catalyses the phosphorolysis of α-1,4 glucosyl chains yielding G1P (Eqn. 6.37).

$$\alpha-1{,}4\ (\text{glucosyl})_{n+1} + \text{Pi} \rightleftarrows \alpha-1{,}4\ (\text{glucosyl})_{n} + \text{G1P} \qquad \text{Eqn. 6.37}$$

α-amylase attacks $\alpha(1 \rightarrow 4)$ linkages of amylose (which is composed of straight chains and gives a characteristic blue colour with iodine) or amylopectin (which has $\alpha - 1,6$ linked branches and gives a purple colour with iodine). The end-products are short-chain dextrins together with maltotriose, maltose and some glucose. β-amylase attacks only the non-reducing ends of chains, splitting off pairs of glucosyl units as maltose. *R*-enzyme attacks the 1,6 branching points of amylopectin making this component of starch susceptible to further attack by β-amylase or phosphorylase. *D*-enzyme (disproportionating enzyme) can transfer groups of glucosyl units between short-chain dextrins, producing a mixture of longer and shorter chain dextrins and glucose.

In leaves, it seems likely that β-amylase is cytoplasmic and may only be concerned with starch degradation of senescent chloroplasts or those which have taken on the

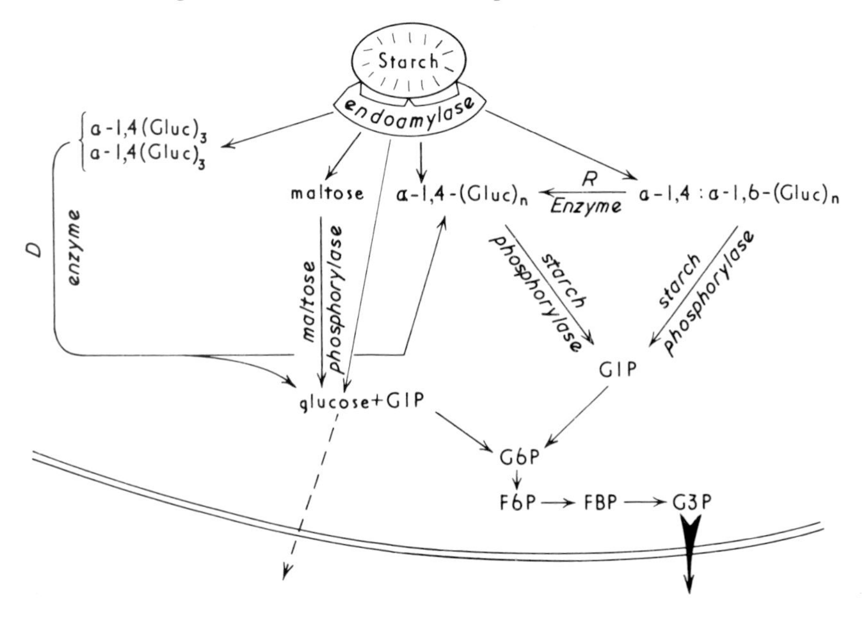


**Fig. 6.13.** Starch degradation in chloroplasts. The starch grain is attacked by an endoamylase which releases branch-chain dextrins, straight-chain dextrins, oligosaccharides (such as maltotriose and maltose) and some glucose. *D*-enzyme catalyses reactions such as maltotriose + maltotriose → maltopentose + glucose. *R*-enzyme attacks 1, 6-branches in amylopectin. Maltose phosphorylase and starch phosphorylase respectively catalyse the phosphorolysis of maltose and longer $\alpha(1 \rightarrow 4)$ glucans yielding G1P which can undergo further conversion to triose phosphate. Glucose is slowly exported via a relatively specific translocator (Section 8.17) and triose phosphate is rapidly exported via the Pi translocator (after Okita *et al*, 1979). Internal degradation of triose phosphate to PGA could yield the ATP consumed in the conversion of F6P → FBP. Since all of the starch which is accumulated by day can sometimes be degraded during the following night it seems likely that regulation of starch metabolism does not reside entirely in the synthetic reaction.

characteristics of amyloplasts. A stromal endoamylase which has somewhat different characteristics from other plant α-amylases (having no requirement for $Ca^{2+}$ and lacking heat stability) is believed to initiate starch degradation in chloroplasts (Fig. 6.13). With the help of *R*-enzymes and *D*-enzymes this produces substrates which readily undergo phosphorolysis. G1P is then converted to triose phosphate via the following sequence (Eqn. 6.38) which demands the availability of phosphofructokinase activity.

$$G1P \rightarrow G6P \rightarrow F6P \rightarrow FBP \begin{array}{l} \nearrow G3P \\ \searrow DHAP \end{array} \qquad \text{Eqn. 6.38}$$

This triose phosphate can, of course, be exported from the chloroplast via the Pi translocator (Section 8.23). High Pi favours degradation in three ways, by initiating phosphorolysis, inactivating ADPglucose pyrophosphorylase (by providing a low [PGA]/[Pi] ratio) and by accelerating triose phosphate export. Kraminer (unpublished) has, in fact, shown that an increase in Pi will switch synthesis to degradation in illuminated chloroplasts.

The fate of maltose is somewhat uncertain. It seems unlikely that it will be directly exported and maltase (which would catalyse its hydrolysis to glucose) has not been detected. Degradation by maltose phosphorylase to glucose and G1P is possible.
[Refs. **36, 40, 45, 45a, 46, 48a, 51**]

## 6.20 Regulation of starch synthesis and degradation

This topic is discussed elsewhere (Section 9.7) but it may be noted here that much remains to be resolved. It is not clear, for example, how much degradation occurs in the light but it seems equally clear that some must occur unless the endoamylase is completely suppressed. De Fekete and Vieweg have suggested that maltose may inhibit amylase activity. Maltose is also a good primer for starch synthase and they argue that a high level of maltose would not only promote synthesis directly but could limit amylase activity to a point at which it actually furthered net synthesis by providing increased substrates for starch synthase.

Fructose bisphosphatase is inactivated in the dark (Section 9.2) and this would prevent the futile cycling (Eqn. 6.39) which would otherwise occur in the presence of an active phosphofructokinase.

phosphofructokinase ATP ⟲ FBP / F6P ⟳ Pi FBPase Eqn. 6.39

Similarly, the RPP pathway could not proceed in the light in the presence of a highly

active phosphofructokinase and any activity in the illuminated stroma would result in the formation of a wasteful sink for ATP.
[Refs. 10, **36**]

## 6.21 Sucrose synthesis

Although sucrose was once believed to be formed in the chloroplasts it is normally entirely absent from the products of $^{14}CO_2$ fixation by isolated chloroplasts (see also, Fig. 6.14). Recent work on enzyme distribution and on product formation also strongly suggest a cytoplasmic origin (Fig. 9.9). The penultimate reaction in this sequence is catalysed by sucrose phosphate synthetase. This enzyme facilitates the transfer of a glucose moiety from UDPglucose to F6P (cf. Eqn. 6.36) and the sucrose phosphate formed is hydrolysed by a specific phosphatase to give free sucrose. The formation of UDPglucose is analogous to the formation of ADP glucose in starch synthesis (Section 6.18, Eqn. 6.35).
[Refs. 10, **33a, 37, 49b**]

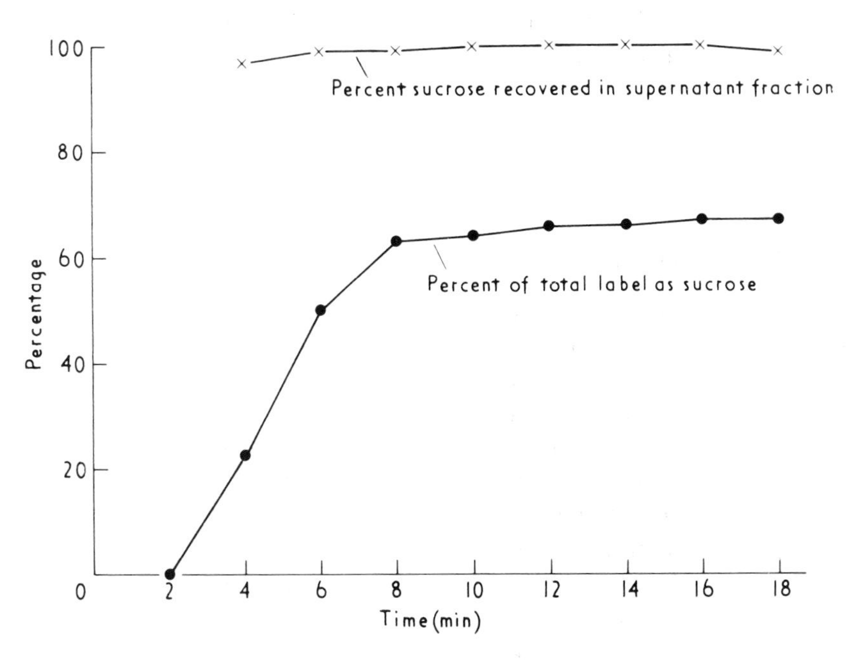


**Fig. 6.14.** Accumulation of sucrose in the cytoplasm of protoplasts during illumination in $^{14}CO_2$. It can be seen that the quantity of sucrose in the chloroplast was more or less unchanged and consistent with slight contamination during fractionation. [This is entirely consistent with the view that sucrose synthesis is a cytoplasmic event because sucrose is not transported across the inner envelope (Chapter 8) at anything other than negligible rates (S. P. Robinson, unpublished].

## 6.22 Up the carbon path

In this chapter we have described the RPP pathway as it is presently understood. It must be admitted, however, that the following account (which was written by Dominic Recaldin and was first printed in the New Scientist on the 5 December, 1968) puts the whole thing in a somewhat different light.

It was in the early '50s, I remember, when I had my first real test. I hadn't been on the job—Cell Security, that is—more than six months when the call came through. I had been expecting it for some time. Bad news travels fast, and the rumours that we were losing production secrets were getting around a little too quickly for my liking.

'He wants to see you', said a tiny voice in my ear. 'Give me fifteen minutes.' I slipped outside and grabbed an empty solute molecule. We took the reticulum to the nuclear boundary. I flipped open my cell-security card and waved it at the guard. He was reading a comic, and picking his teeth with dedication.

'You're expected', he said. He did not look up. It seemed churlish to disturb him.

The chief was staring bleakly at an open file as I walked in. It looked as if it was empty, but I couldn't be sure.

'What are you doing about this leak?'

A fair question, I thought.

'No leads yet, I'm afraid. I was wondering whether you had any thoughts.'

'Cell security is your job. And the way things are going you won't have a job soon. Neither will any of us. The Enemy's very hot on comparative biochemistry. If he cracks our secrets he cracks the lot. *Scenedesmus, Chlorella*, even the big boys like spinach. Nobody's safe'.

As I left him and went down ill-humouredly to the plant for another abortive look around, I mulled over what little we did know. For almost a year now we had been losing vital information about our photosynthetic mechanism for making sugars. Pilfering was the obvious answer. Perhaps They had found a way to smuggle out intermediates at various points along the production line and were piecing them together sequentially. I soon knocked that idea on the head. I was down on the shop floor for six weeks checking every inventory, counting every glucose unit we made. There was never anything missing.

Then there were the periodic exterminations. Small batches of us would be packed like cattle into a tiny container in glaring light, and then plunged into boiling alcohol—a technique pretty horrible even to Them. They were analysing us of course. Probably using two-dimensional chromatography, the big, new discovery then. But it wouldn't have done Them a bit of good, even so. They had no idea how to arrange the intermediates in the proper order. We were well aware, too, of their pathetic attempts to extract information out of our enzymes. But everybody knows that enzymes are trained from birth never to give valuable information once they are outside the cell. 'Name, number, rank' and that's all. Nevertheless, they were getting information out of us and the question remained: How?

I found myself standing morosely at the packing bench next to old George.

'Hello, George. How's it going?'

George lumbered off with a dozen crates of starch granules and stacked them in the delivery bay. 'Must be getting old', he said. 'Used to be able to stack them fourteen at a time. Only do twelve now.'

'Perhaps they're getting heavier, George'. I ventured.

'Never thought of that', he said.

Neither had I 'til that moment. It was an idea, of sorts.

'Do we have any old stock anywhere, George? Last year's?'

'Upstairs.'

'Give me one of those crates', I said.

I took it upstairs and put it on a pair of scales. Then I got out a crate of the previous year's stock and compared it. It wasn't as heavy. There wasn't much in it, but it definitely wasn't as heavy. It didn't seem very important, and yet I felt certain I was on to something. But I just couldn't see how it tied in with the problem. I distinctly remember I was about to forget the whole thing when my brain made a deductive click that felt like a blow on the head. I was so excited that I had to dial the sickbay number three times before my trembling fingers could manage the extension number.

'Sick bay.'

'Security here. Which department has had the greatest number of sickness absentees in the past year?'

'I wondered when someone was going to ask me that. It's funny. The carbon dioxide absorbing unit's got a terrible record. Can't understand it.'

I nearly passed out.

'Was there any particular illness they complained of? Any symptoms that stood out?'

'That's odd too. Nearly all of them showed signs of—well, I feel a bit silly saying this but—they all seemed to be suffering from radiation sickness. It's nonsense, I know, but the symptoms—'.

I slammed down the telephone and leaped up in the air in one jubilant movement. Then I ran downstairs, grabbed hold of George and waltzed him round the floor singing 'The Merry Widow'. After thirty-two breathless bars I raced upstairs again and called control.

'Chief? Security. Would you believe . . . . . . . . . . . . radiocarbon tracers?'

'Would I believe what?'

'That's how They're finding out. They feed us carbon dioxide with the carbon radioactive. Then they analyse our products, and follow the dilution of radioactivity. Then all they need do is assemble the intermediates in order of increasing dilution and They've solved it.'

'I can't understand a word you're saying.'

'Chief? Can I suggest a piece of counter-intelligence?'

'Put it in your report.'

The report was a masterpiece, even though I say it myself. And I was really rather proud of our counter-move. What we did, in brief, was manufacture a whole series of daft enzymes that would do all sorts of crazy things to sugar molecules. We had 3-carbon sugars going to 5-carbon sugars; 4-carbon sugars going to 7-carbon sugars and back to 5-carbon sugars, the whole thing was a riot. My *piece de resistance* was to arrange the entire scheme in the form of a hideously complex cycle—They are very fond of cycles—picking up carbon dioxide at one point and producing sugar (eventually!) at another. It worked beautifully. In six months They were quite convinced They had the carbon pathway all figured out. They all patted themselves on the back and dished out Nobel Prizes. Nobody bothered us after that. They think They've unlocked the secret of the cell, so They're happy. We've gone back to our old simple method of carbon fixation, so we're happy. With so much happiness in the world we should all be smiling!

Although this was described as an anthropomorphic fantasy it could also be regarded as a cautionary tale. When the RPP pathway was first elucidated there seemed little left to learn about photosynthetic carbon assimilation. An eminent

scientist is reputed to have said 'there are problems in photosynthesis for sure but they are not those associated with the path of carbon'. The characterization of photorespiration and $C_4$ photosynthesis and the recognition of the ecological role of CAM have shown such optimism to be misplaced. At the present time, it is even more difficult to imagine that whole new metabolic pathways remain to be discovered but it is no less certain that a full understanding of the relationship between the chloroplast and its cellular environment will not be forthcoming for many years. The statement that 'enzymes are trained from birth never to give valuable information once they are outside the cell' is also particularly apposite. There is no doubt that highly purified enzymes are essential for some research purposes. Equally it is clear that the availability of highly purified RBP carboxylase did little to resolve the seeming inadequacy of its characteristics until the early work of Pon on its $Mg^{2+}$ requirement was remembered. At the moment sucrose phosphate synthetase seems similarly inadequate and it may be supposed that the resolution of this particular problem will require the attention of the cell-biologist as well as the enzymologist.
[Refs. **47, 54**]

### General References


1 ARNON D. I. (1961) Cell-free photosynthesis and the energy conversion process. In *Light and Life*. W. D. McElroy and B. Glass (eds.), pp. 489–569. Baltimore: The John Hopkins Press.
2 ARNON D. I. (1967) Photosynthetic activity of isolated chloroplasts. *Physiol. Rev.* **47**, 317–58.
3 BASSHAM J. A. & CALVIN M. (1957) *The Path of Carbon in Photosynthesis*. pp. 1–104. Englewood Cliffs, N. J: Prentice-Hall Inc.
4 BENSON A. A. & CALVIN M. (1950) Carbon dioxide fixation by green plants. *Ann. Rev. Plant Physiol.* **1**, 25–40.
5 CALVIN M. & PON N. G. (1959) Carboxylations and decarboxylations. *J. Cell. Comp. Physiol.* **54**, 51–74.
6 CALVIN M. & BASSHAM J. A. (1962) *The Photosynthesis of Carbon Compounds*, pp. 1–127. New York: Benjamin.
7 CONN E. E. & STUMPF P. K. (1976) *Outlines of Biochemistry*, pp. 1–629. New York: John Wiley & Sons, Inc.
8 FRANCK J. & LOOMIS W. E. (eds.). (1949) *Photosynthesis in Plants*, pp. 381–401. Ames: Iowa State College Press.
9 GIBBS M. Photosynthesis. (1967) *Ann. Rev. of Biochem.* **36**, Para II, 757–84.
10 GIBBS M. (ed.). (1971) *Structure and Function of Chloroplasts*, pp. 1–286. Berlin: Springer-Verlag.
11 GOODWIN T. W. (ed.). *Biochemistry of Chloroplasts.* (1967) Vol. II. Proceedings NATO Advanced Study Institute, Aberystwyth, 1965, pp. 1–706. London: Academic Press.
12 HATCH M. D. & SLACK C. R. (1970) Photosynthetic $CO_2$-fixation pathways. *Ann. Rev. Plant Physiol.* **21**, 141–62.
13 HATCH M. D., OSMOND C. B. & SLATYER R. O. (eds.). Photosynthesis and Photorespiration. *Proceedings of the Conference at A.N.U., Canberra 1970.* New York: Wiley Interscience.
14 JENSEN R. G. & BAHR J. I. (1977) Ribulose 1,5-bisphosphate carboxylase/oxygenase. *Ann. Rev. Plant Physiol.* **28**, 379–400.

15 KOK B. & JAGENDORF A. T. (eds.). (1963) *Photosynthetic Mechanisms in Green Plants.* Washington D.C: Nat. Acad. Sci. 1145.

16 KELLY G. J. & LATZKO E. (1976) Regulatory aspects of photosynthetic carbon metabolism. Gibbs M. (ed.). *Ann. Rev. Plant Physiol.* **27**, 181–205.

17 LEHNINGER A. L. (1970) *Biochemistry*, pp. 1–833. New York: Worth Publishers, Inc.

18 LEHNINGER A. L. (1971) *Bioenergetics*, pp. 1–245. Philippines: W. A. Benjamin, Inc.

19 NEILANDS J. B. & STUMPF, P. K. (1958) *Outlines of Enzyme Chemistry*, pp. 1–411. John Wiley & Sons Inc., New York.

20 PAULING L. (1960) *The Nature of the Chemical Bond*, pp. 1–644. Ithaca, New York: Cornell University Press.

21 PRIDHAM J. B. (ed.). (1974) Phytochemical Society Symposium on Plant Carbohydrate Biochemistry, Edinburgh (1972). London: Academic Press.

22 QUAYLE J. R. & FERENCI T. (1978) Evolutionary aspects of autotrophy, *Microbiological Reviews* pp. 251–73.

23 RABINOWITCH E. I. (1956) *Photosynthesis and Related Processes.* Vol. 2, Part I, pp. 537 *et seq.* New York: Wiley Interscience.

24 ROBINSON S. P. & WALKER D. A. (1981) Photosynthetic carbon reduction cycle (Calvin cycle) (including fluxes and regulation). In *The Biochemistry of Plants. A comprehensive treatise.* (eds. P. K. Stumpf and E. E. Conn.) Vol. 8 (Photosynthesis). London: Academic Press.

25 SAN PIETRO A., GREER F. A. & ARMY T. J. (eds.). (1967) *Harvesting the Sun*, pp. 1–342. London: Academic Press.

26 SIEGELMAN H. W. & HIND G. (eds.). (1978) In *Photosynthetic Carbon Assimilation.* Proceedings of a symposium held at Brookhaven National Laboratory, Upton, New York, 1978, pp. 1–445. New York: Plenum Press.

27 STILLER M. (1962) The path of carbon in photosynthesis. *Ann. Rev. Plant Physiol.* **13**, 151–70.

28 *Symposia of the Soc. for Exp. Biol. V. Carbon Dioxide Fixation & Photosynthesis.* (1951).

29 WALKER D. A. & CROFTS A. R. (1970) Photosynthesis. *Ann. Rev. Biochem.* **39**, 389–428.

30 ZELITCH I. (1971) *Photosynthesis, Photorespiration and Plant Productivity.* pp. 1–347. London: Academic Press.

31 ZELITCH I. (1975) Pathways of carbon fixation in green plants. *Ann. Rev. Biochem.* **44**, 123–45.

## Specific citations

32 BASSHAM J. A. & KRAUSE G. H. (1969) Free energy changes and metabolic regulation in steady-state photosynthetic carbon reduction. *Biochim. Biophys. Acta*, **189**, 207–21.

33 BENSON A. A., BASSHAM J. A., CALVIN M., GOODALE T. C., HAAS V. A. & STEPKA W. (1950) The path of carbon in photosynthesis. V. Paper chromatography and radioautography of the products. *J. Amer. Chem. Soc.* **72**, 1710–18.

33a BIRD I. F., CORNELIUS M. J., KEYS A. J., WHITTINGHAM C. P. (1974) Intracellular site of sucrose synthesis in leaves. *Phytochemistry*, **13**, 59–64.

34 CALVIN M. & PON N. G. (1959) Carboxylations and decarboxylations. *J. Cell. Comp. Physiol.* **54**, 51–74.

35 CONSDEN R., GORDEN A. H. & MARTIN A. J. P. (1944) Qualitative analysis of proteins: A partition chromatography method using paper. *Biochem. J.* **38**, 224–32.

36 De FEKETE M. A. R. & VIEWEG G. H. (1974) Starch metabolism: Synthesis versus degradation pathways. In *Plant Carbohydrate Biochemistry.* Proc. Phytochem. Soc. Symp., Edinburgh 1973 (ed. J. B. Pridham), pp. 127–44. London: Academic Press.

37 EVERSON R. G., COCKBURN W. & GIBBS M. (1967) Sucrose as a product of photosynthesis in isolated spinach chloroplasts. *Plant Physiol.* **42**, 840–4.

38 GAFFRON H., FAGER E. W. & ROSENBERG J. L. (1951) Intermediates in photosynthesis. Formation

and tranformation of phosphoglyceric acid. In *Symposia of the Soc. for Exp. Biol. V. Carbon Dioxide Fixation & Photosynthesis*, p. 261–283.

39 HEATH O.V.S. (1951) Assimilation by green leaves with stomatal control eliminated. In *Symposia of the Soc. for Exp. Biol. V. Carbon Dioxide Fixation & Photosynthesis*, pp. 94–115.

40 HELDT H. W., CHON C. J., MARONDE E., HEROLD A., STANKOVIC Z. S., WALKER D. A., KRAMINER A, KIRK M. A. & HEBER U. (1977) Role of orthophosphate and other factors in the regulation of starch formation in leaves and isolated chloroplast. *Plant Physiol* **59**, 1146–55.

41 KEYS A. J., BIRD I. F., CORNELIUS M. J., LEA P. J., WALLSGROVE R. M. & MIFLIN B. J. (1978) Photorespiratory nitrogen cycle. *Nature*, **275**, 741–3.

42 LORIMER G. H., BADGER M. R. & ANDREWS T. J. (1976) The activation of ribulose-1,5-bisphosphate carboxylase by carbon dioxide and magnesium ions. Equilibria, kinetics, a suggested mechanism and physiological implications. *Biochemistry*, **15**, 529–36.

43 LORIMER G. H., WOO K. C., BERRY J. A. & OSMOND C. B. (1978) The $C_2$ photorespiratory carbon oxidation cycle in leaves of higher plants: Pathway and consequences. In: Photosynthesis 77. *Proceedings of the Fourth International Congress in Photosynthesis*, September 1977. Hall, D. O., Coombs, J. & Goodwin, T. W. (eds.), pp. 311–22. London: Biochemistry Society

44 POINCELOT R. P. (1974) Uptake of bicarbonate ion in darkness by isolated chloroplast envelope membrane and intact chloroplast of spinach. *Plant Physiol.* **54**, 520–6.

45 OKITA T. W., GREENBERG E., KUHN D. N. & PREISS J. (1979) Subcellular localization of the starch degradative and biosynthetic enzymes of spinach leaves. *Plant Physiol.* **64**, 187–92.

46 PEAVEY D. G., STEUP M. & GIBBS M. (1977) Characterization of starch breakdown in the intact spinach chloroplast. *Plant Physiol.* **60**, 305–8.

47 PON N. G. (1959) Studies on the carboxydismutase system and related materials. Ph. D. Thesis, University of California, Berkeley

48 PREISS J. & LEVI C. (1978) Regulation of α-1, 4-glucan metabolism in photosynthetic systems. In: Photosynthesis 77. *Proceedings of the Fourth International Congress in Photosynthesis*, September *1977*, (eds.) D. O. Hall, J. Coombs and T. W. Goodwin, pp. 457–68. London: Biochem. Soc.

49 QUAYLE J. R., FULLER R. C., BENSON A. A. & CALVIN M. (1954) Enzymatic carboxylation of ribulose diphosphate. *J. Amer. Chem. Soc.* **76**, 3610–1.

49a ROBINSON S. P., MCNEIL P. H. & WALKER D. A. (1979) Ribulose bisphosphate carboxylase—lack of dark inactivation of the enzyme in experiments with protoplasts. *FEBS Letters*, **97**, 296–300.

49b ROBINSON S. P. & WALKER D. A. (1979) The site of sucrose synthesis in isolated leaf protoplasts. *FEBS Letters*, **107**, 295–299.

50 RUBEN S. & KAMEN M. D. (1940) Radioactive carbon of long half-life. *Phys. Rev.* **57**, 549.

51 THOMAS M., RANSON S. L. & RICHARDSON J. A. (1973) *Plant Physiology* London: Longman.

52 TSWETT M. (1906). See Rabinowitch, E. I. *Photosynthesis and Related Processes*. Vol. I, p. 402. New York: Interscience, (1945).

53 WALKER D. A. (1962) Pyruvate carboxylation and plant metabolism. *Biol. Rev.* **37**, 215–56.

54 WALKER D. A. (1973) Photosynthetic induction phenomena and the light activation of ribulose diphosphate carboxylase. *New Phytol.* **72**, 209–35.

55 WALKER D. A. (1974) Some characteristics of a primary carboxylating mechanism. *Phytochemical Society Symposium on Plant Carbohydrate Biochemistry, Edinburgh (1972)* (ed. J. B. Pridham) pp. 7–26. London: Academic Press.

56 WALKER D. A. (1978) Report of Symposium 7 (Regulation of Metabolism). In: Photosynthesis 1977. *"Proceedings of the Fourth International Congress on Photosynthesis", September 1977.* (eds), D. O. Hall, J. Coombs, and T. W. Goodwin, pp. 489–93. London: Biochemistry Society

# Chapter 7
# Induction



## SUMMARY


Induction is a fundamental aspect of photosynthesis. When whole plant leaves, protoplasts and chloroplasts are taken from the dark and brightly illuminated, photosynthetic carbon assimilation does not commence immediately but only after a lag or induction period. Several minutes may elapse before the rate approaches maximum. Conversely most of the 'photochemical' events such as electron transport proceed at more or less maximal rate from the outset. Induction is therefore primarily associated with the RPP pathway. It occurs in many circumstances in which stomata are not involved.

The Osterhout–Hass hypothesis explains induction in two ways: light-activation of catalysts or building-up of substrates. The former implies that some of the enzymes of carbon assimilation are inactive in the dark and are 'switched on' in the light. The latter now incorporates autocatalysis (Section 6.3)—the ability of the cycle to generate more intermediates than it consumes—and implies that one or more of these intermediates is depleted in the dark and is not instantly reformed in the light. Photosynthetic electron transport and photophosphorylation cease immediately in the dark so that reactions which depend upon assimilatory power will also cease more or less immediately whereas others, such as the carboxylation, could continue for some time leading to depletion of subsequent pools of intermediates.

Because of the possibility of allosteric interactions it is clear that there could be massive interplay between activation of enzymes and the pool sizes of various intermediates. Contemporary research implicates both aspects of the Osterhout–Hass hypothesis and recognizes the likelihood of such interaction. It is also important to note that improved catalysis may result either from activation of enzymes (switching from an inactive form in the dark to an active form in the light) or from an improved environment (an enzyme which is already fully active may work better in the relatively alkaline environment of the illuminated stroma). Again, both of these effects may complement one another.

It is our view that induction is most closely associated with two aspects of the RPP pathway. The conversion of PGA to DPGA is a freely reversible but unfavourable reaction which is driven, in photosynthesis, by high concentrations of PGA and is particularly susceptible to decreases in the [ATP]/[ADP] ratio. The lag in $O_2$ evolution is therefore primarily related to the increase in [PGA] which occurs during the first minutes of illumination. (Oxygen evolution involves the transfer of electrons from $H_2O$ to $NADP^+$ and this acceptor is regenerated by the reduction of PGA, via DPGA, to triose phosphate). Similarly the lag in $CO_2$-fixation seems to be primarily associated with an increase in hexose monophosphate. The regenerative phase of the RPP pathway consumes triose phosphate in four reactions simultaneously (two catalysed by transketolase and two by aldolase). Of all the cycle intermediates, triose phosphate is also the most readily lost by export. The rearrangement of five molecules of triose phosphate to give three molecules of pentose monophosphate therefore involves considerable readjustment of pool sizes in the light (since triose phosphate formation from PGA ceases immediately in the dark and almost certainly reverses). The re-establishment of appropriate pools may well be compounded by the changing equilibrium position of the aldolase reaction, in which two reactants yield one product by condensation, and by the initially low activity of the bisphosphatases.

Induction is largely artificial in the sense that there must be very few situations in which plants or leaves pass more or less immediately from relatively long periods of darkness into bright light. It is of interest to investigators because it has provided so many insights into the mechanisms of regulation and the inter-relationship between the chloroplast and its cellular environment.

## 7.1 Induction in whole plants and leaves

When green leaves are strongly illuminated after a period of darkness, photosynthetic oxygen evolution and $CO_2$-fixation do not normally reach a maximum for several minutes (Fig. 7.1). This initial lag has been called 'induction' (or 'simple induction' by Rabinowitch) in order to distinguish it from a variety of transient fluctuations in gaseous exchange ('gulps' and 'bursts') which occur under some conditions. Simple induction is not an artefact of measurement. It occurs in species which lack stomata and it can be readily observed in experiments with isolated

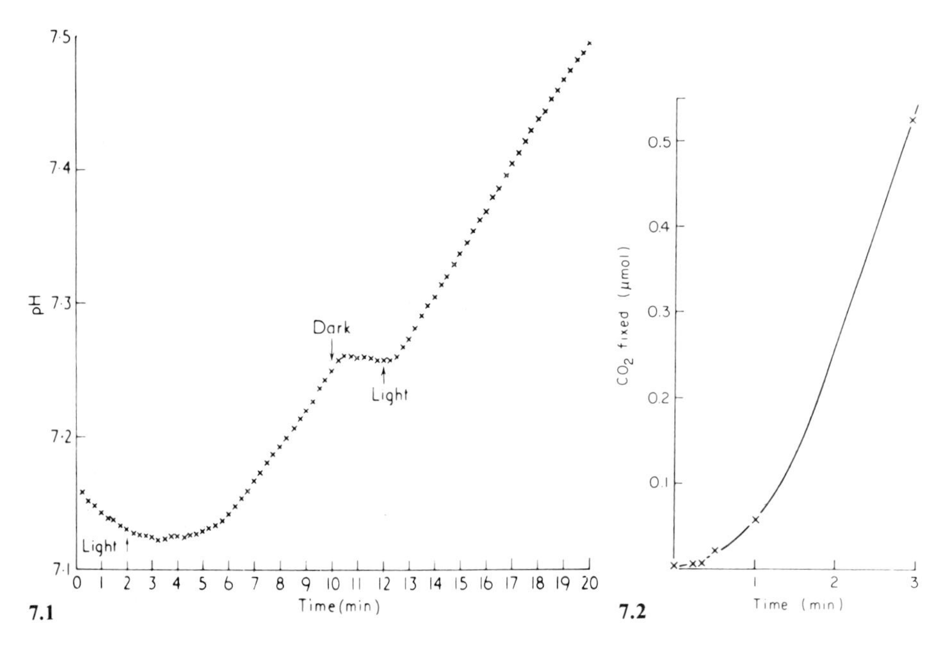


**Fig. 7.1.** Induction in *Lemna*. Whole plants illuminated at 20°C in bicarbonate solution (approximately $10^{-3}$M). As in the original experiments of Blinks and Skow (1938) the pH increases as $CO_2$ is taken up from solution. Note long initial lag which is not repeated after a short dark interval.

**Fig. 7.2.** Induction in chloroplasts *in vivo* (after Heber and Willenbrink, 1964). The curve shows the kinetics of $^{14}CO_2$-fixation by chloroplasts within a leaf. The chloroplasts were subsequently isolated in non-aqueous media from the freeze-dried tissue.

chloroplasts and protoplasts. It is an intrinsic and fundamental feature of photosynthesis.

Rabinowitch also distinguished between 'short' and 'long' induction. At normal temperatures, the former was normally complete within two to five minutes. The latter was a secondary, slow acceleration which might last several hours and for this reason could not be certainly attributed to photosynthesis as such.

Induction in whole plants and leaves has been measured by infra-red gas analysis, by manometry, by fluorescence spectrometry and by $O_2$ electrodes. In aquatic plants it can also be followed with a pH electrode as $CO_2$-uptake from dilute bicarbonate brings about increases in alkalinity (Fig. 7.1). Accumulation of $^{14}C$ in chloroplasts within the leaf has been followed by non-aqueous fractionation and shown to proceed with the same lag that characterizes the onset of $O_2$ evolution by the intact tissue (Fig. 7.2). Induction is relatively independent of light intensity but strongly dependent on temperature suggesting that it is more strongly related to biochemical events which normally double in rate for every 10°C rise in temperature ($Q_{10} = 2$) than to truly photochemical events which are independent of temperature ($Q_{10} = 1$) (Fig. 7.3). Induction is diminished if re-illumination follows a relatively short dark

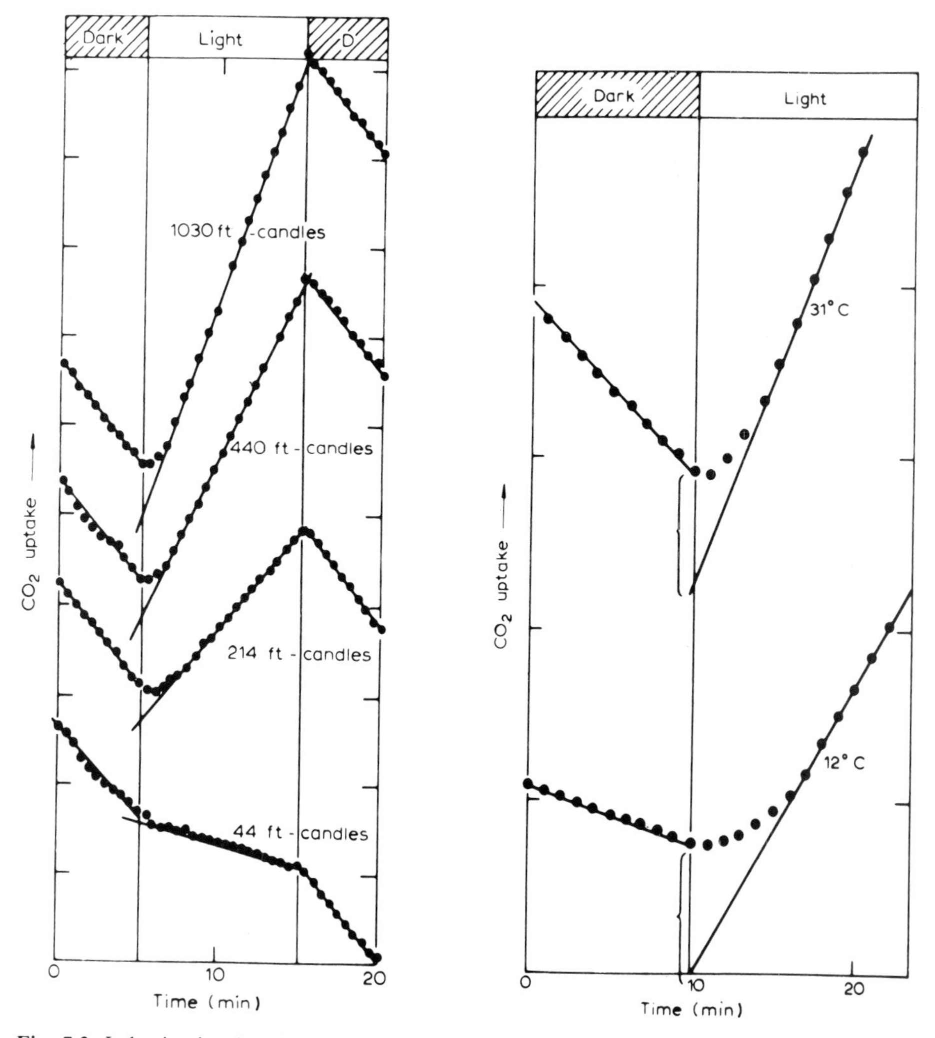


**Fig. 7.3.** Induction in wheat leaves (after McAlister, 1937). Changes in [$CO_2$] were followed by an infrared method. In these experiments the initial lag was independent of light intensity (left). Fig. on right shows the extension of the lag at a lower temperature.

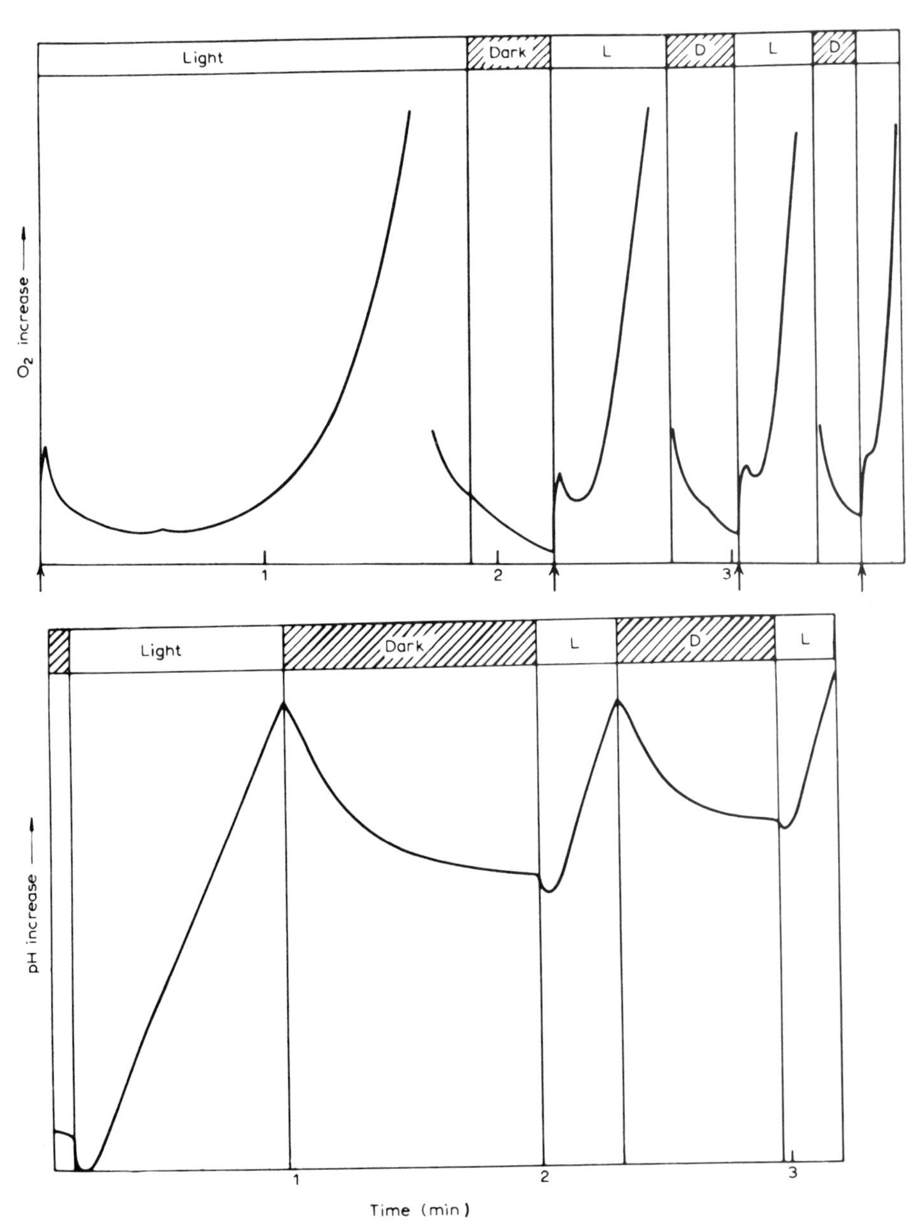


**Fig. 7.4a.** Induction in *Ricinus* leaf (after Blinks and Skow, 1938a). Oxygen evolution was followed with an electrode held against the leaf surface with agar. The duration of the preceding dark periods were (from left to right) 10 minutes, 34 seconds, 21 seconds, and 13.5 seconds respectively. **b.** Induction in pond lily (after Blinks and Skow, 1938b). Changes in pH, associated with uptake and release of $CO_2$, measured with a glass electrode.

interval (e.g. Figs. 7.1, 7.9). This suggests that some constraint is lifted during illumination which is not immediately reimposed in the dark.
[Refs. 1, 4, 5, 7, 9, **22**]

## 7.2 Historical aspects

Induction periods which lasted several minutes at room temperature were first observed by Osterhout and Hass in experiments with *Ulva* at Woods Hole in 1918 and by Warburg (using *Chlorella*) in 1920. In 1938 Blinks and Skow made continuous records of lags in oxygen evolution from *Ricinus* leaves and in pH changes associated with the onset of photosynthesis in waterlily (Fig. 7.4). Steeman-Nielsen also observed induction in photosynthetic oxygen evolution from *Fucus*. Probably the most extensive and useful measurements during this period were made by McAlister and Myers who followed $CO_2$ uptake by wheat leaves using a spectrographic (infra-red) method (Fig. 7.3).
[Refs. **14, 15, 28, 29, 30, 36, 46**]

## 7.3 The role of stomata

When higher plants are taken from the dark their stomata will normally be closed and will commence to open upon illumination. The time course of opening is often

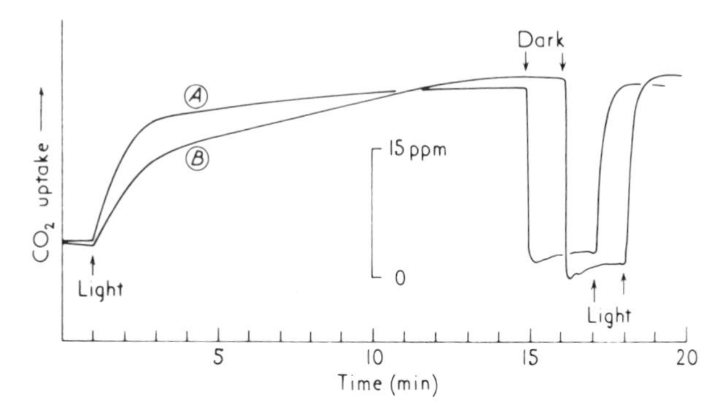


**Fig. 7.5.** $CO_2$ uptake by spinach, followed by infra-red gas analysis (Delaney, unpublished). Leaf discs (10 $cm^2$) were strongly illuminated (ca. 6000 ft. candles) at 20°C in air containing 300 ppm $CO_2$ flowing at 1 l/min. Curve *A*, disc taken from leaf previously illuminated in bright light for 1 h. Curve *B*, disc taken from plant in room light (ca. 150 ft. candles). As in Figs. 7.1 to 7.4 there is a lag of some minutes before the maximum rate is reached. It can be seen that the rate also increases more rapidly in the disc which has been pre-illuminated in strong, rather than weak light. After a brief dark interval, the lag is greatly diminished. Under these conditions part of the initial lag may be attributed to the relative slowness of the stomatal response.

very similar to the time course of induction and is similarly affected by temperature and (to some extent) by light intensity. Particularly when infra-red gas analysis is used to follow photosynthesis at air-levels of $CO_2$ (Figs. 7.3, 7.5) it is clearly difficult to distinguish between inductive effects which are independent of stomata and those occasioned by stomatal opening. Moreover, there seems little doubt that the rate of diffusion of $CO_2$ from the external atmosphere to the site of carboxylation will affect the rate of photosynthesis or that endo-diffusion of $CO_2$ will be affected by stomatal resistance. It is therefore not unreasonable to ask if there is an inductive lag in higher plant photosynthesis other than that brought about by stomatal opening. On the basis of comparative metabolism alone it would have to be concluded that induction is a normal feature of photosynthesis. Certainly, many studies of induction have

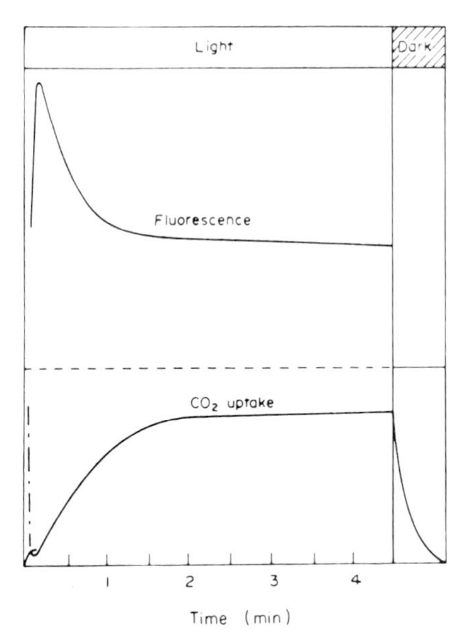


**Fig. 7.6.** Induction in wheat (after McAlister and Myers, 1940). Simultaneous measurement of $CO_2$ uptake and fluorescence showing a commonly observed 'mirror-image' relationship.

---

exchange. All of the species examined showed the characteristic initial lag which is longer in leaves taken from the dark rather than from the light. Other consistent features include the greatly decreased lag following a brief dark interval despite increased post-illumination oxygen uptake (Walker, 1976a). (a) Sunflower. Left, from light; right, from dark. (b) Spinach. Left, from light, right, from dark. The oscillations in (a) are not always seen but have been observed in several species in a number of occasions, usually following illumination and are reminiscent of the results of Van der Veen (1948) and others. It is entirely possible that these reflect over-compensation (or 'hunting') by control mechanisms such as those discussed in Section 7.13.

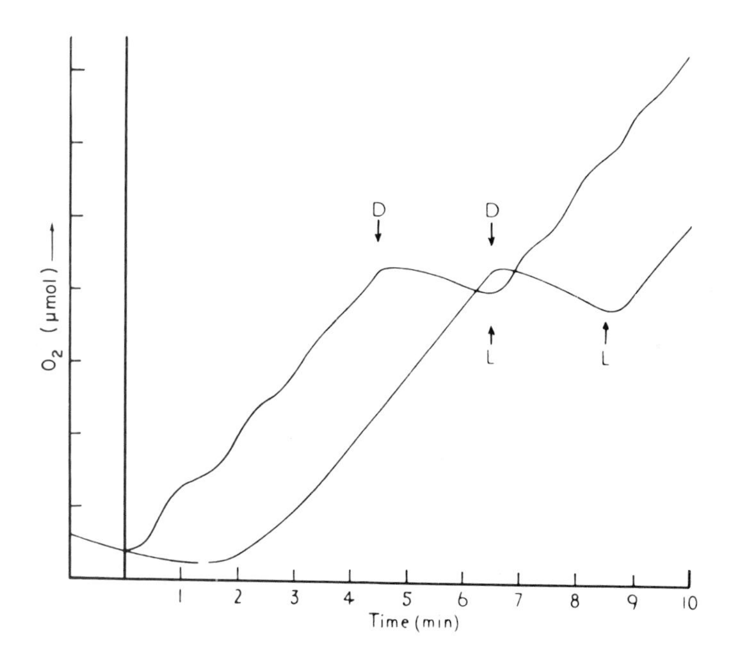


Fig. 7.7a

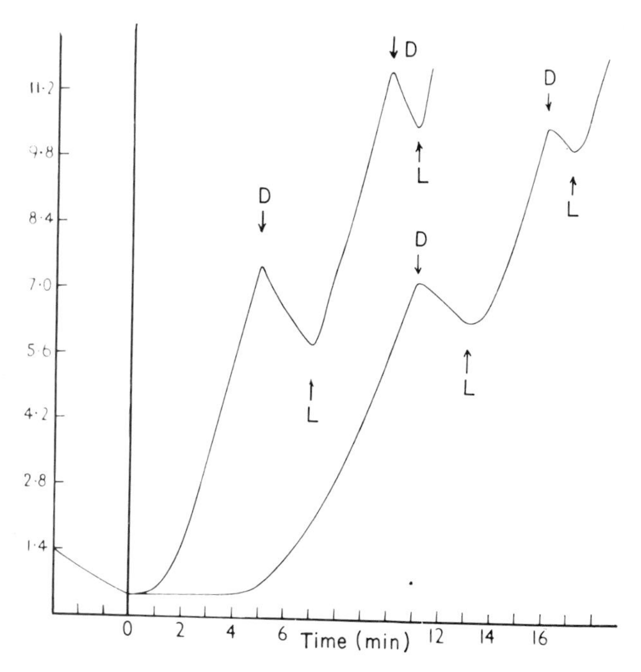


Fig. 7.7b

**Fig. 7.7.** Examples of induction in leaves. Leaf discs (10 cm²) were illuminated in a small (3 ml) chamber containing an oxygen electrode and 0.2 ml of 1.0 M bicarbonate on a filter paper disc. The concentration of $CO_2$ in the gas phase during the course of the experiment is indeterminate (balanced as it is between uptake by the leaf and release from the unbuffered bicarbonate solution) but is believed to be saturating and is sufficient to maintain rapid rates of $CO_2$ evolution (250–350 µmol $mg^{-1}$ Chl $h^{-1}$). Silicone rubber impressions of the leaf surface made according to the method of Zelitch immediately after illumination showed no evidence of stomatal opening but it is assumed that at these concentrations closed stomata would offer only minimal resistance to gaseous

been made with algae which do not possess stomata. Moreover, stomata open well in moderate light and yet some induction can still be observed in passing from moderate to saturating light. Closed stomata would be expected to interfere with the evaluation of induction by gaseous exchange measurements but scarcely to interfere with the onset of photosynthesis within the leaf. It is of interest therefore that fluorescence in whole leaves also shows an initial rise which may be taken as an indication of impaired photosynthesis (Fig. 7.6). Induction has also been observed in saturating $CO_2$ (Fig. 7.7) and in leaf discs from which the epidermis has been stripped (Fig. 7.8) and in experiments with isolated chloroplasts (Fig. 7.9) and protoplasts (Fig. 7.10).
[Refs. **10, 29, 36, 38a, 46, 47**]

### 7.4 Lack of induction in photochemistry

Many of the so-called photochemical events in photosynthesis have an enzymic component and cannot be expected to behave as entirely 'pure' photochemical reactions. Some are slower than others by several orders of magnitude. Although it is therefore difficult to generalize it seems clear that events such as pH shifts (Section 5.9), photophosphorylation (Sections 5.3–5.7) and the Hill reaction (Section 5.2) reach their maximum rates within seconds of the onset of illumination (Figs. 7.11 & 7.12) and that induction cannot therefore be attributed to the slow onset of reactions of this nature. This is not to say that the rate of generation of assimilatory power (Section 5.2) may not increase with time under certain conditions but that, if it does,

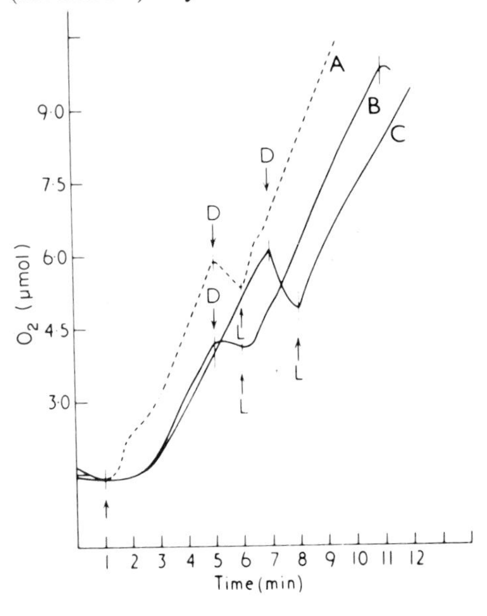


**Fig. 7.8.** As for Fig. 7.7. Tobacco. A from light, C from dark, B as for C but with upper epidermis stripped from the leaf.

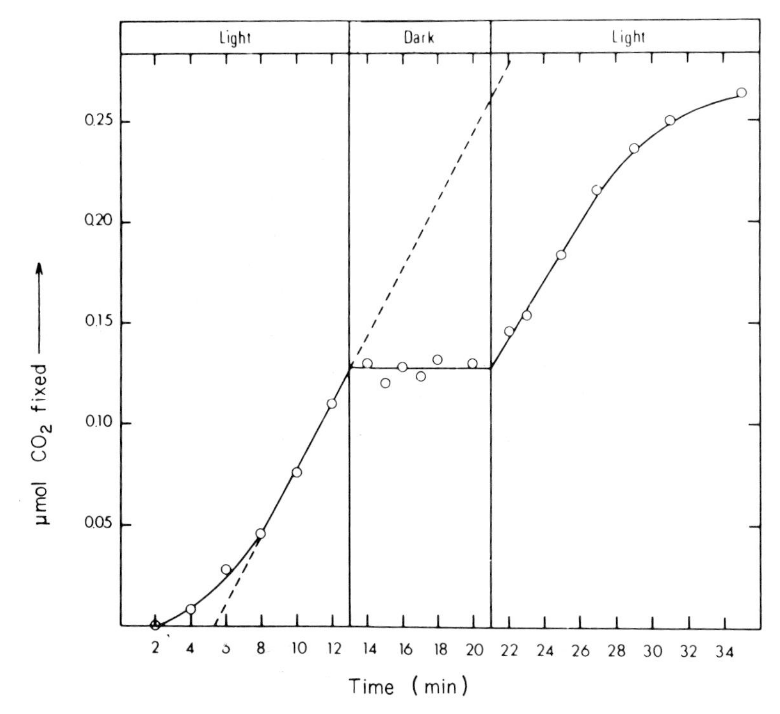


Fig. 7.9a

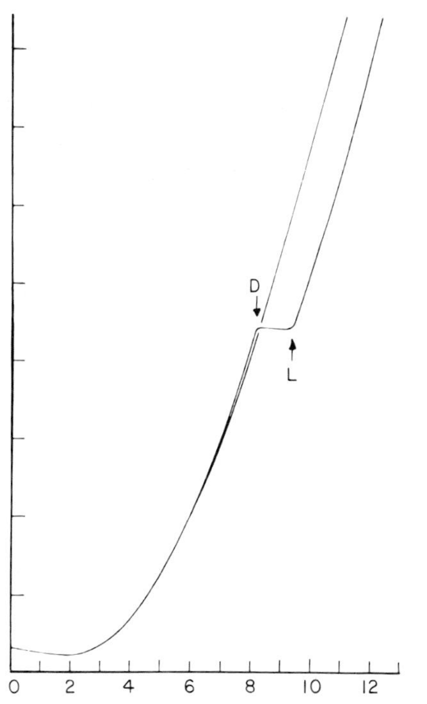


Fig. 7.9b

**Fig. 7.9a.** Absence of induction in isolated pea chloroplasts following a brief dark interval (after Walker, 1965). Initially several minutes elapsed before $CO_2$ fixation reached its maximal rate but following a break in illumination, fixation was resumed without detectable lag. There was no evidence of appreciable dark fixation of $CO_2$ following cessation of illumination.
**b.** Unchanged rate following a dark interval. Two identical reaction mixtures each containing pea chloroplasts and orthophosphate ($10^{-3}$ M). One subjected to a dark interval as indicated. Traces were recorded simultaneously.

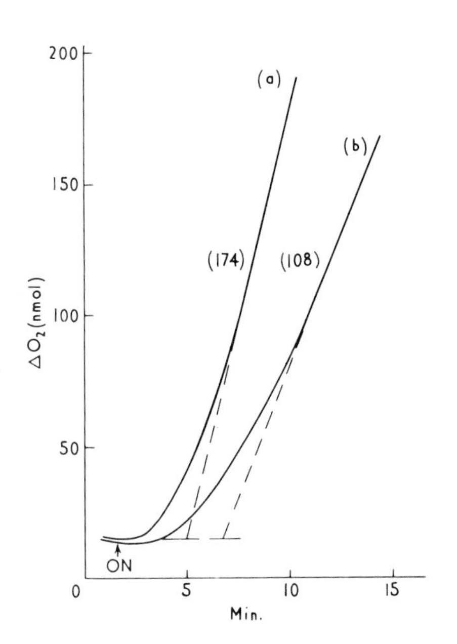


**Fig. 7.10.** Induction in wheat protoplasts comparing lag and final rate in protoplasts isolated from (a) illuminated and (b) darkened leaves. Temp. 20°C. Rates in parentheses in $\mu$mol. $mg^{-1}$ Chl $h^{-1}$ (Edwards unpublished).

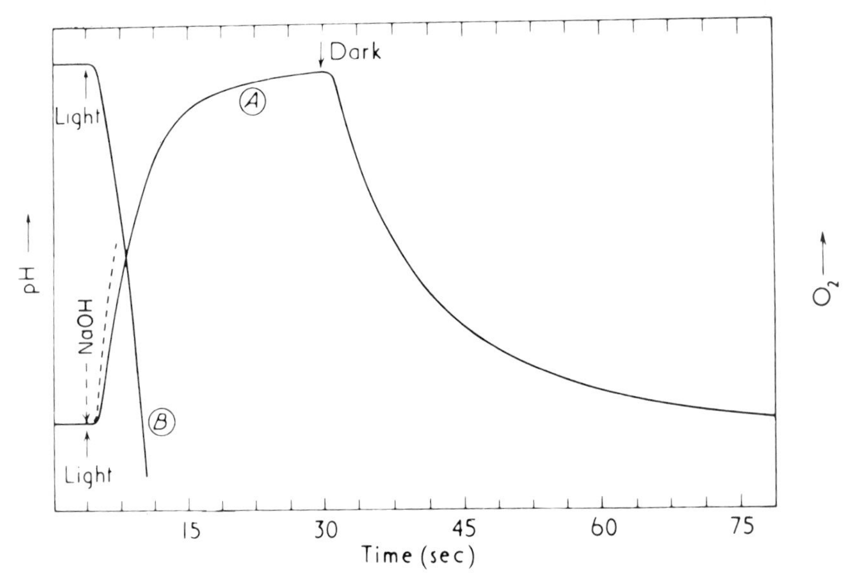


**Fig. 7.11.** Curve *A* Light-generated pH shift. The broken line records the change in pH which followed the addition of NaOH to osmotically shocked chloroplasts in the dark and is included to show the response time of the apparatus. The lower trace shows the response of the same chloroplasts to illumination in the presence of pyocyanine. The reagent is believed to return electrons from the 'top' (reducing) end of photosystem I to some point between photosystems I and II. The pH shift is associated with electron transport and this figure shows that it can be discerned as rapidly as the pH meter can respond.
Curve *B* Ferricyanide-dependent pH shift. Ferricyanide accepts electrons but not protons so that the solution becomes more acid as the reaction proceeds. The reaction mixture contained nigericin to inhibit the converse pH shift directly associated with electron transport. Again it is seen that ferricyanide reduction commences as rapidly as the instrument can record. (After Walker, 1976a)

the original constraint is likely to be imposed by non-photochemical events, such as a sub-optimal rate of supply of ADP etc.
[Ref. 7]

## 7.5 Induction *in vitro*

The first unambiguous records of induction in isolated chloroplasts can be attributed to Gibbs and his colleagues. Initially these were entirely based on measurements of $^{14}CO_2$ incorporation but when Walker and Hill reported the advent of chloroplast systems capable of exhibiting high rates of $CO_2$-dependent $O_2$ evolution, these also displayed the same initial lags (Figs. 7.9 & 7.13). In-vitro induction is characterized by the same dependence on temperature and relative

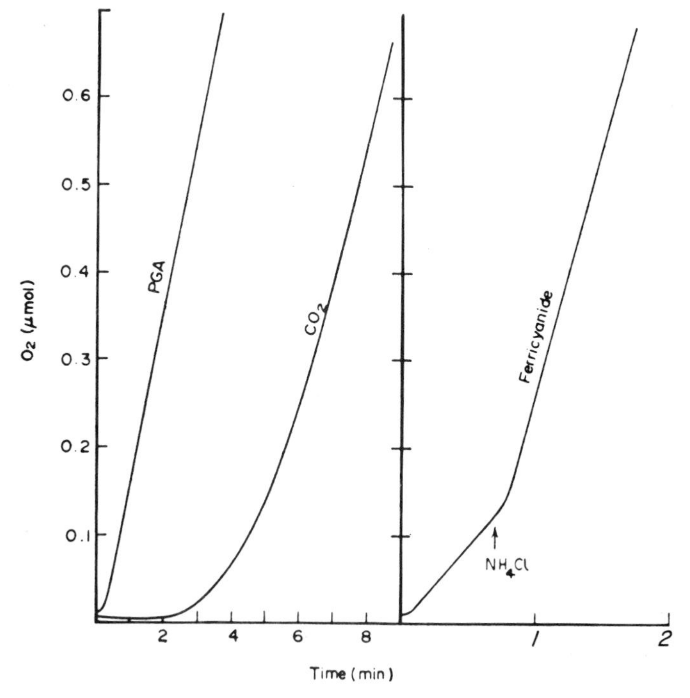


**Fig. 7.12.** Absence of induction in Hill reaction. With ferricyanide the maximum rate of oxygen evolution was achieved within seconds of the onset of illumination ($NH_4Cl$ added as indicated to show acceleration in rate due to uncoupling). With 3-phosphoglycerate the lag was longer than with ferricyanide but still very short. Under these particular conditions the maximal rate with $CO_2$ (included for comparison) was not approached until several minutes had elapsed. The same chloroplasts were used for all three determinations but in the ferricyanide experiment they were osmotically shocked in the reaction vessel. (Walker, 1973)

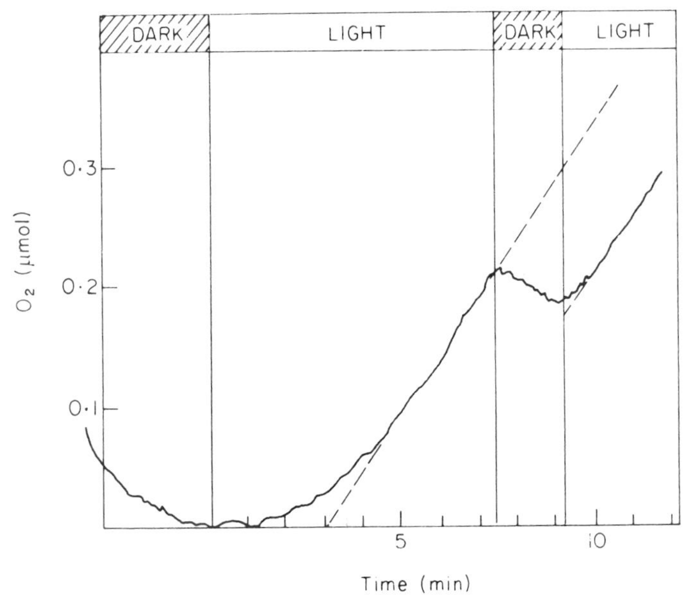


**Fig. 7.13.** First polarographic recording of $CO_2$-dependent $O_2$ evolution by isolated chloroplasts. Note similarity between kinetics of $O_2$-evolution and those of $CO_2$-fixation (Fig. 7.9a). The dark uptake of $O_2$ was largely artefactual and could be explained by oxidation of components of the reaction mixture. (For further details see Walker & Hill, 1967).

independence of light intensity which is shown by the parent tissue (cf. Figs. 7.3 & 7.14ab). It should be noted that induction may well be exaggerated *in vitro* because of the relative proportions of chloroplast and surrounding media. In the leaf, the proportions of chloroplast to cytosol are of the order of 1 to 1 and export of metabolites will quickly affect the immediate environment in which the plastid finds itself. At 100 μg of chlorophyll per ml of reaction mixture (proportions which cannot be greatly exceeded in ordinary vessels without impeding illumination) the ratio of chloroplast to medium is closer to 1/1000 and build-up of exported metabolite must inevitably take longer.
[Ref. **37, 41**]

## 7.6 The molecular basis of induction

On the basis of their first observations, Osterhout and Hass suggested that induction might be based on the building-up of intermediates or the light activation of catalysts. Contemporary explanations are also based on these proposals although there is now more awareness of the possibility of interaction between the two. Thus

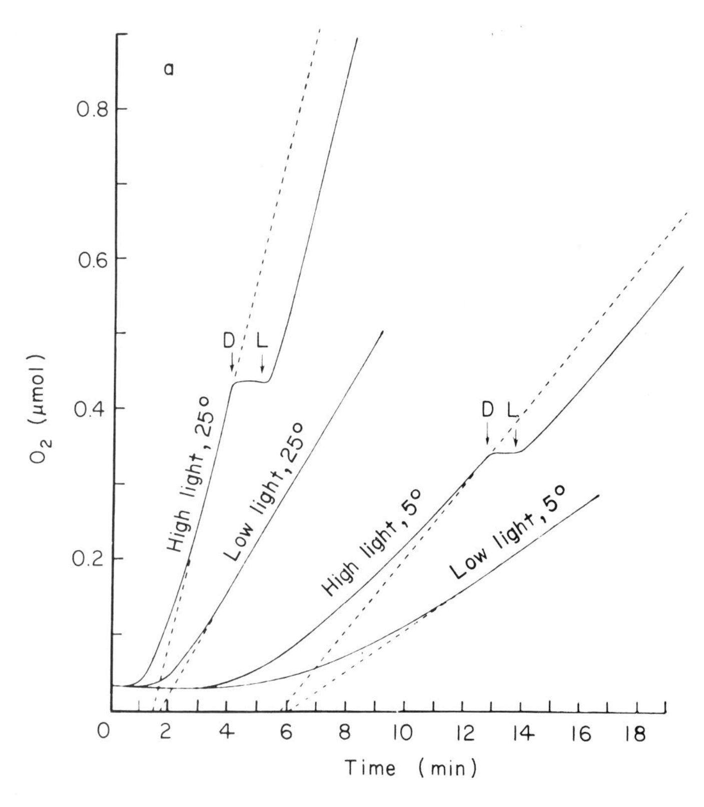


**Fig. 7.14a.** Induction in $O_2$ evolution by isolated chloroplasts. Lags are seen to be independent of light intensity but dependent on temperature. One electrode vessel was at 5°C, the other at 25°C. High light traces were recorded simultaneously immediately prior to low light measurement. (After Walker *et al*, 1973).

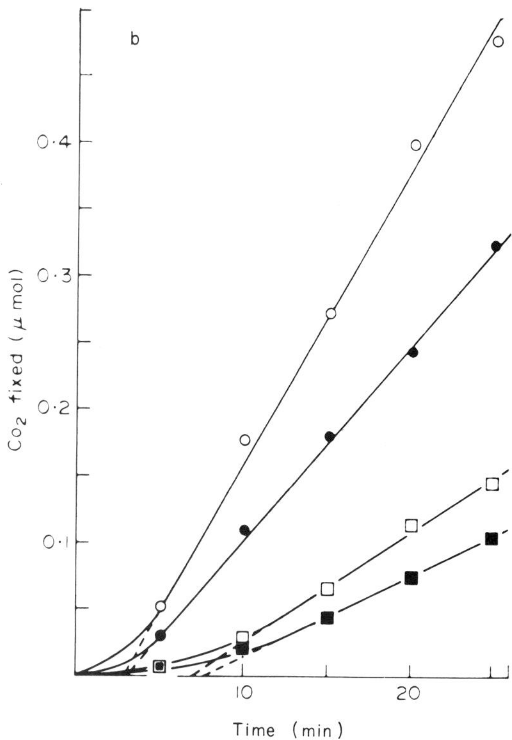


**Fig. 7.14b.** Induction in $CO_2$ fixation by isolated chloroplasts. Lags seen to be independent of light intensity but dependent on temperature. Open symbols, high light intensity; closed symbols, low light intensity; circles 25°C; squares, 10°C. (After Baldry *et al*, 1966).

the gradual accumulation of a vital intermediate may also bring about allosteric activation of an enzyme concerned in its formation or consumption.

In a more precise restatement of the Osterhout–Hass hypothesis, Rabinowitch also talked of 'a deficiency of this acceptor (the $CO_2$-acceptor) at the beginning of the light period and an autocatalytic adjustment of its concentration in the light to the level needed to maintain photosynthesis at the rate corresponding to the prevailing

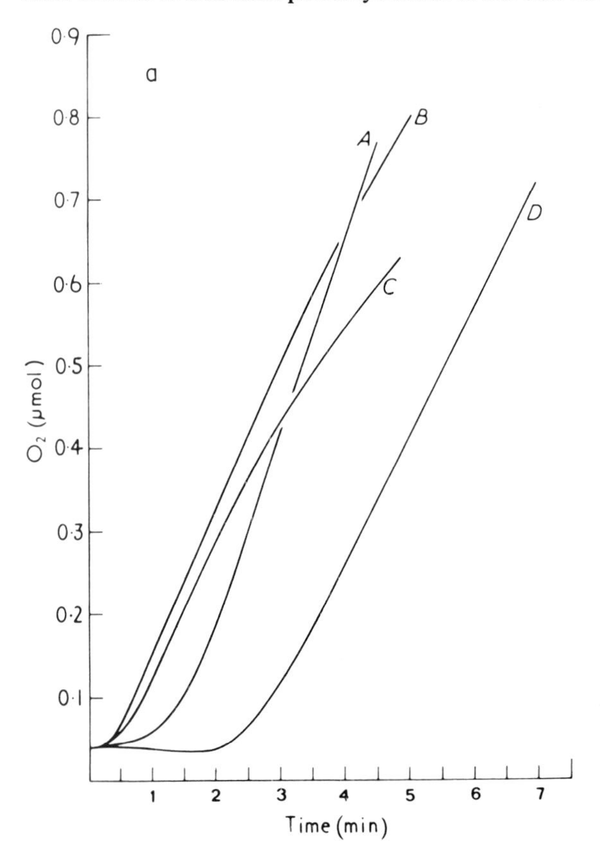


**Fig. 7.15a.** Shortening of induction by compounds released from illuminated chloroplasts and by the addition of cycle intermediates. Curve *D*, $CO_2$-dependent $O_2$ evolution by chloroplasts prepared from leaves harvested after 12 h darkness. Curve *C*, as for *D* but the chloroplasts were resuspended in medium in which other chloroplasts from the same preparation had been illuminated for 30 min and then removed by centrifugation. Curves *A* and *B*, as for *D* but R5P added to *A*, PGA was added to *B*. Note that the lag is shortened by compounds released from illuminated chloroplasts in much the same way as it is by added PGA, indicating a major contribution by this compound. R5P also shortens the lag but the kinetics are then more like those exhibited by chloroplasts from pre-illuminated leaves. The lower and declining rate which results from the addition of PGA at the outset is characteristic. (After Walker, 1976a)

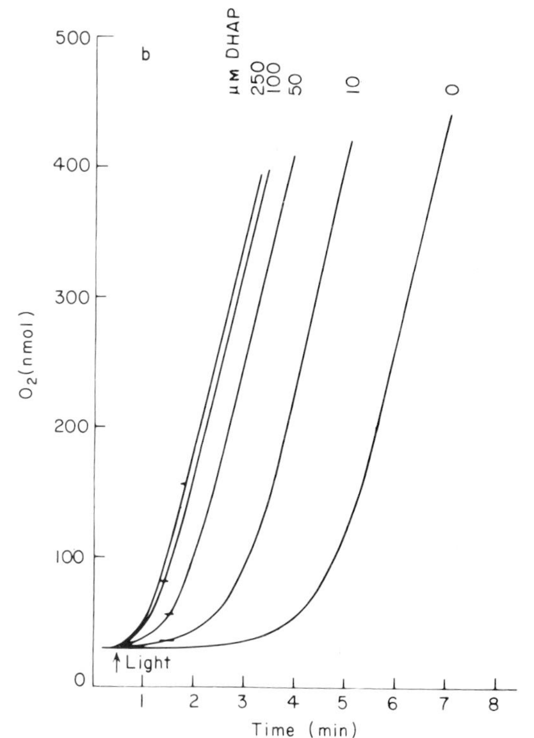


**Fig. 7.15b.** Shortening of induction by catalytic quantities of DHAP. The horizontal bars show the $O_2$ which could be derived by quantitative conversion of DHAP to PGA (via the appropriate partial reactions of the RPP pathway) and its subsequent reduction back to triose phosphate. (After Leegood and Walker, 1980)

light intensity'. This, in turn, incorporates the important principle of autocatalysis (Section 6.3) and the fact that the period of acceleration reflects a situation in which triose phosphate is preferentially fed back into the cycle rather than being converted into starch or exported into the cytosol.
[Refs. 4, **30**]

## 7.7 The contribution of autocatalysis

Although the difficulty of clearly separating the effect of substrate concentration from that of enzyme activation has already been noted (Section 7.6) there is good evidence that the level of intermediates of the RPP pathway in isolated chloroplasts is usually too low to support maximal photosynthesis and that the ensuing lag represents the time which must elapse before their full steady-state concentration is achieved.
[Refs 1, 5, 9]

This may be summarized as follows:

(a) Once the full rate has been reached the initial lag is not repeated, or is greatly shortened, following a short dark interval (Figs. 7.1, 7.4, 7.7, 7.8, 7.9). [Refs. **37, 38, 39, 41, 42**] The same would result from light activation of catalysts but:

(b) If chloroplasts are added to a medium in which other chloroplasts (since removed) have been allowed to photosynthesize, the lag is also greatly diminished (Fig. 7.15). This strongly suggests that something has accumulated in the medium which can re-enter the new chloroplasts and shorten the lag. Again an effect on the enzymes concerned cannot be discounted but in these circumstances it must be secondary (i.e. activation must be promoted by accumulated product). [Ref. **12**]

(c) Analysis shows that a variety of cycle intermediates (particularly triose phosphates) appear in the medium in which chloroplasts have been illuminated. If these are added back to fresh chloroplasts, as pure substances, they all cause lag shortening (Fig. 7.15). They are not all equally effective in this regard, presumably because they do not all penetrate the chloroplast with equal rapidity (Chapter 8) but those which enter most easily (triose phosphates and PGA) cause the greatest

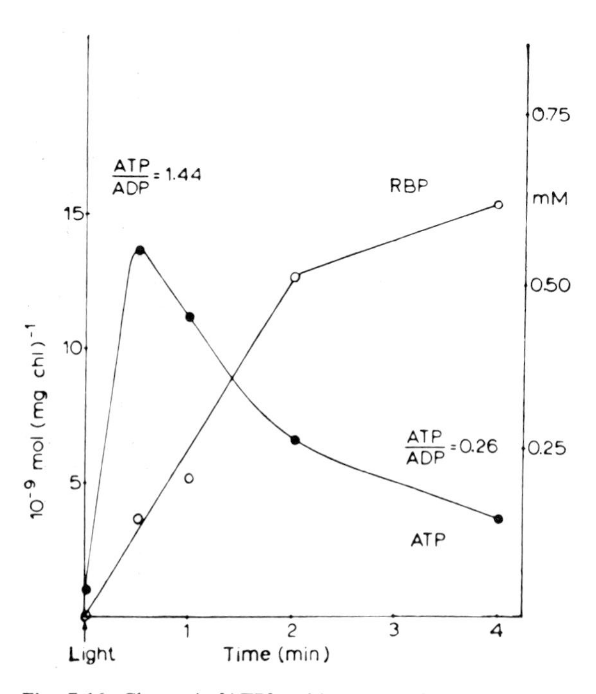


**Fig. 7.16.** Change in [ATP] and increase in [RBP] during induction. From Lilley *et al*, 1977.

shortening. The effectiveness of compounds such as pentose monophosphates which penetrate more slowly can be increased by pre-incubation prior to illumination. All can act catalytically. (Fig. 7.15b). [Refs. 6, **12, 13, 16, 38, 40, 41**].

(d) Analysis shows that some intermediates, particularly PGA, and hexose phosphates (Fig. 7.17) and RBP (Fig. 7.16) are initially low in chloroplasts and that they increase as induction terminates. [Ref. **26**]

(e) Orthophosphate, which promotes metabolite export does not appear to cause a significant inhibition of light activation of cycle enzymes (Section 7.8). [Refs. 2,8,**17**]

(f) As in whole tissues, induction is lengthened at low temperatures but is largely independent of light intensity. [Refs. **12, 42**]

(g) Induction is also considerably longer in chloroplasts isolated from dark stored tissue (Fig. 7.18). While this is consistent with the degradation of essential intermediates it is, of course, equally consistent with dark de-activation of enzymes. [Refs. 7, **17, 38, 41**]

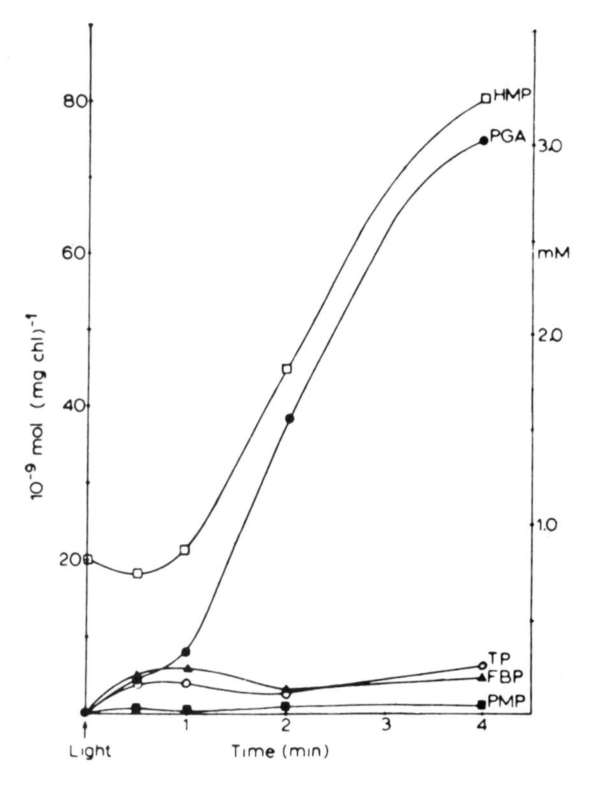


**Fig. 7.17.** Increase in stromal [PGA] and [hexose monophosphates] during induction. From Lilley *et al*, 1977. (In this experiment and in Fig. 7.16 the induction period lasted about 3 min).

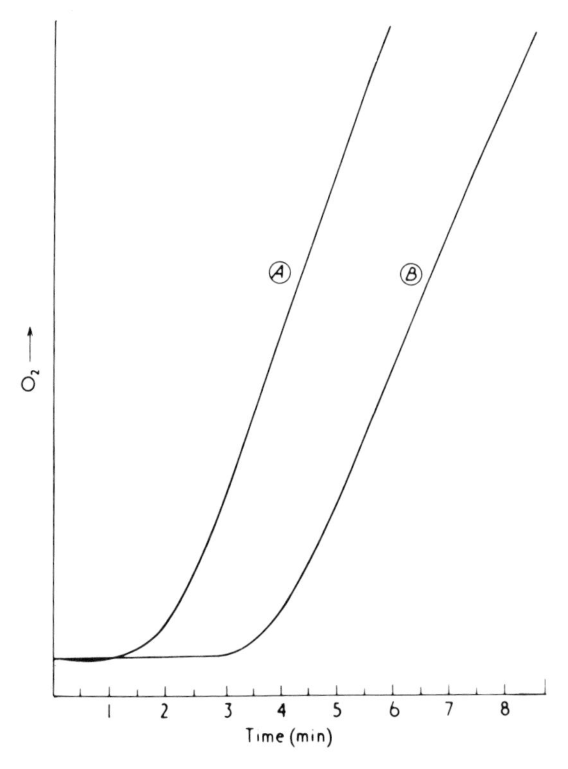


**Fig. 7.18.** Shortening of induction in chloroplasts by pre-illumination of the parent tissue. (cf. Figs. 7.8 and 7.10). Curve *A*, $CO_2$-dependent $O_2$ evolution by spinach chloroplasts isolated from leaves pre-illuminated for 30 min in white light at 2000 ft candles. Curve *B*, as for *A* but chloroplasts isolated at the same time from non-illuminated leaves. All of the leaves were harvested after 12 h darkness and randomised prior to extraction. (Walker, 1976a).

## 7.8 Light activation of catalysts

While the evidence listed above (Section 7.7) clearly implies that induction is correlated with the autocatalytic build-up of intermediates it is not suggested that the light activation of photosynthetic enzymes is an unimportant factor. Such activation is considered in more detail in Chapter 9. A distinction should be drawn, however, between light activation of enzymes and the creation, by illumination, of an environment which favours catalysis. For example the pH of the stroma is more alkaline in the light than it is in the dark and will therefore favour the action of enzymes with more alkaline pH optima. High pH may also contribute to the conversion of an enzyme from a less active to a more active state. Although these are

**Table 7.1.** Rates of $CO_2$-dependent oxygen evolution and the effect of illumination on RuBP carboxylase activity of wheat protoplasts

| Exp. | Oxygen evolution | RuBP carboxylase Dark | RuBP carboxylase Light | Ratio Light/Dark |
|---|---|---|---|---|
| 1 | 139 | 182 | 228 | 1.25 |
| 2 | 152 | 161 | 186 | 1.16 |
| 3 | 167 | 174 | 206 | 1.18 |
| 4 | 136 | 150 | 207 | 1.38 |
| 5 | 117 | 165 | 166 | 1.01 |
| 6 | 115 | 144 | 141 | 0.98 |

RBP carboxylase was measured in protoplast after 6 min preincubation in the dark or light under the same conditions as for oxygen evolution (2.5mM $NaHCO_3$, pH 7.6). In exp. 5 and 6, the wheat was kept in darkness for 24 h prior to isolating the protoplasts. Rates are expressed in $\mu$mol mg $Chl^{-1}$ $h^{-1}$. Similar lack of marked activation has been observed in experiment carried out in air levels of $CO_2$. The *initial* rates of $O_2$ evolution by wheat protoplasts during induction are extremely low and whereas the rates displayed by the isolated carboxylase are commensurate with the final observed rates of photosynthesis (above) they are greatly in excess of those which would be required to meet the needs of carboxylation during induction. (After Robinson *et al*, 1979)

intrinsically different effects the result could be the same and if both are involved they might well be additive. However, it is difficult to explain the virtual absence of photosynthetic carbon assimilation during the first minute or two of illumination on this basis. Other things being equal, a zero rate would call for completely inactive enzymes and although substantial changes in activity may occur upon illumination, activities in the dark (of RBP carboxylase for example) are often substantially in excess of those needed to maintain the low rates observed during induction (Table 7.1)

Other enzymes, however, are undoubtedly deactivated during the dark, *in vivo*, and their reactivation in the light would, therefore, be expected to contribute to induction. The residual induction (of 30 seconds or less) seen in the presence of high concentrations of TP (Fig. 7.15b) may well be attributable to this cause. Nevertheless, it seems clear that the light activation of FBPase, for example, is far too rapid (Fig. 7.19) to account for the longer delays in the onset of photosynthesis which still occur in many circumstances. In experiments of this sort it is inescapable that, within 30 seconds of illumination, the activities of this and other enzymes of the RPP

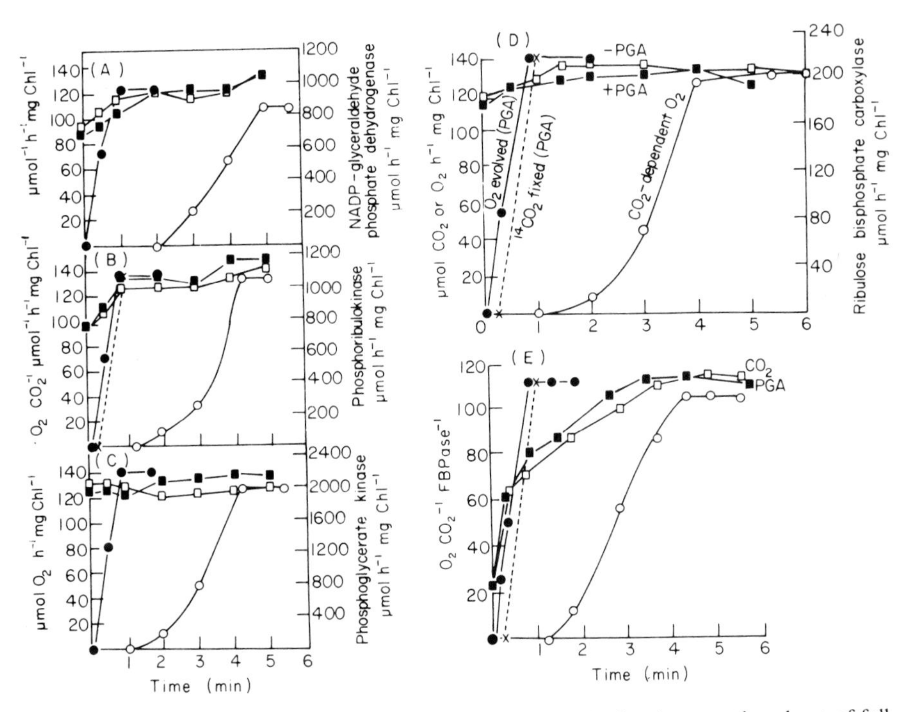

**Fig. 7.19** (A)–(E). An experiment showing a clear separation in time between the advent of full enzyme activity (as a result of light activation) and the advent of full photosynthesis (as a consequence of autocatalysis). Rates of photosynthesis by wheat chloroplasts in relation to the light activation of NADP-dependent glyceraldehyde phosphate dehydrogenase (A), phosphoribulokinase (B), phosphoglycerate kinase (C), RBP carboxylase (D) and FBPase (E). Open symbols depict $O_2$ evolution (O) and enzyme activity (□) in the presence of 10 mM $NaHCO_3$. Closed symbols show $O_2$ evolution (●), enzyme activity (■) and, in (B), $^{14}CO_2$-fixation (- - -), with 10 mM PGA and 10mM $NaHCO_3$. Light activation was measured by simultaneously removing samples and assaying them in the absence of dithiothreitol. Each curve shows that the highest measured rates of $CO_2$-fixation or $O_2$ evolution can be sustained by each of the enzymes prior to detectable $O_2$ evolution in those mixtures containing $CO_2$ alone. (After Leegood and Walker, 1980)

pathway are sufficiently high to support maximal photosynthesis and yet this potential is not realized for a further 2–3 minutes unless exogenous metabolites (such as PGA or DHAP) are added.
[Refs. **20, 21, 23, 24, 25a,b,c, 27, 32, 33a,b**].

## 7.9 Effects of light intensity and temperature

As already noted (Sections 7.1, 7.5) induction is largely independent of light intensity but the initial lag is increased as the temperature is lowered (Figs. 7.3 & 7.14). On a purely theoretical basis this is entirely consistent with the key role proposed for the concentration of cycle intermediates. If for example the rate of photosynthesis were lowered by lowering the light intensity the steady-state substrate concentration needed to maintain the new decreased rate would be smaller and likewise the time taken to reach this new rate would be diminished. Accordingly the lag would remain unchanged if the rate were directly proportional to substrate concentration. Conversely, at a lower temperature, more substrate would be needed to maintain a given rate and the time needed to reach that higher substrate level would be increased. Suppose, for example, that in the steady-state a decrease in temperature of 10°C slowed the RPP pathway by one half. Presumably the concentration of intermediates needed to maintain this slower rate would be the same as before but the time needed to build up this concentration in a mixture illuminated at the lower temperature from the outset would be increased. In fact the extension of the lag by low temperatures is sometimes surprisingly long and the final rate surprisingly high

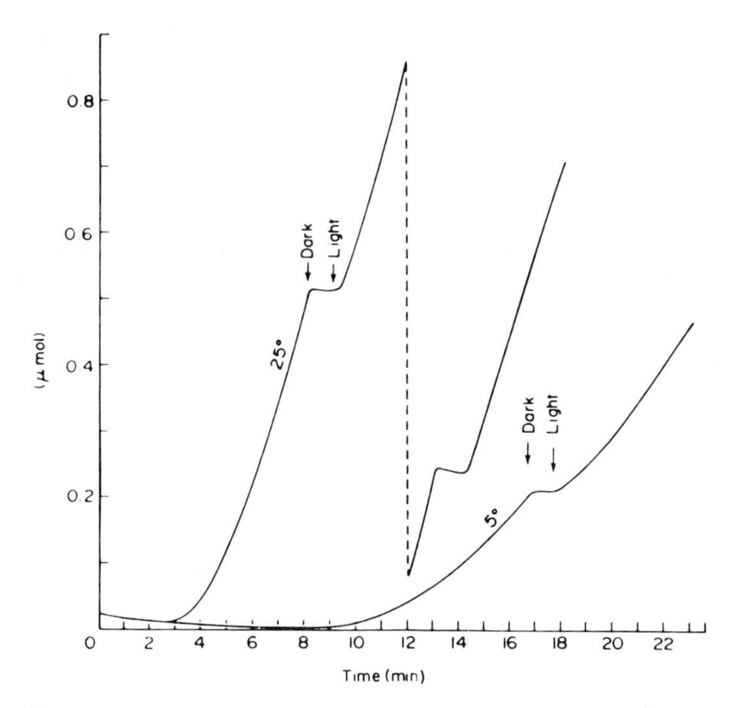


**Fig. 7.20.** Induction dependent on temperature but showing markedly anomalous behaviour at lower temperature with respect to extension of lag and maximal rate. One electrode vessel was at 25°C, the other at 5°C. Despite the substantially diminished lag observed after the dark interval, the time taken for resumption of the near maximal rate was also longer than usual at both temperatures. (After Walker *et al.*, 1973).

(Fig. 7.20). The former may be accounted for in part by the high $Q_{10}$ values observed at low temperatures in photosynthesis. Thus at temperatures below 15°C the increase in rate for a 10°C rise in temperature may greatly exceed the value of about 2 which is typical of many metabolic processes. This has been attributed to the autocatalytic nature of the RPP pathway (Section 6.3) and the fact that feed-back will also increase as the temperature is raised so that the enzymes concerned will be simultaneously presented with a higher substrate as well as a more favourable

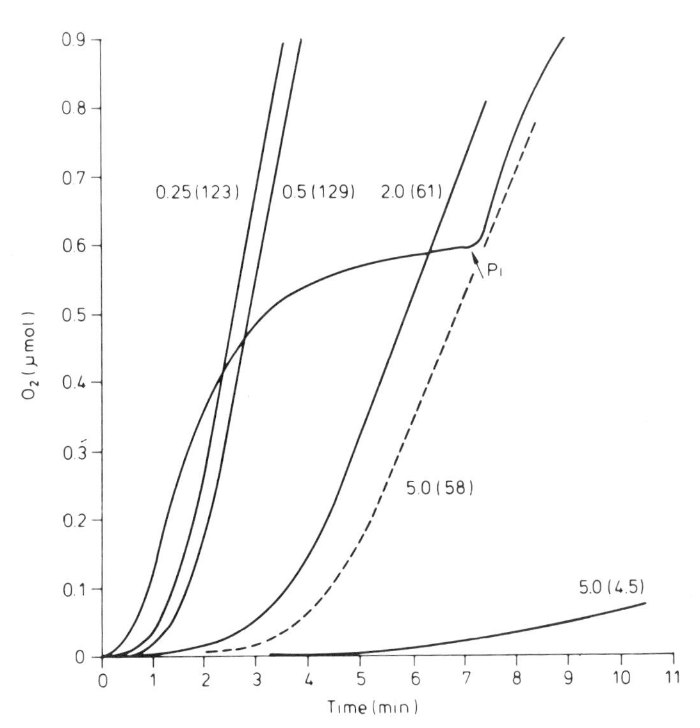


**Fig. 7.21a.** Inhibition of photosynthetic carbon assimilation in isolated spinach chloroplasts by orthophosphate. Kinetics of $CO_2$-dependent $O_2$ evolution in presence of increasing quantities of exogenous Pi. Characteristic induction period is shortest in the absence of added Pi but photosynthesis then soon declines as in Fig. 8.9. Small increments in [Pi] lengthen the lag and either increase the rate or leave it unchanged. Larger quantities of Pi cause lag extension and rate depression. Pi values (mM) are recorded against each trace and the figures in parentheses are the rates in $\mu$mol mg$^{-1}$ Chl h$^{-1}$ (at 20°C). Broken line shows extent of alleviation of Pi inhibition (at [Pi] of 5 mM) by the inclusion of 10 mM PPi. These previously unpublished assays are essentially the same as those first reported by Cockburn *et al*, (1968) but were carried out in a medium containing no exogenous $Mg^{2+}$ and sufficient catalase to minimize secondary effects associated with peroxide formation.

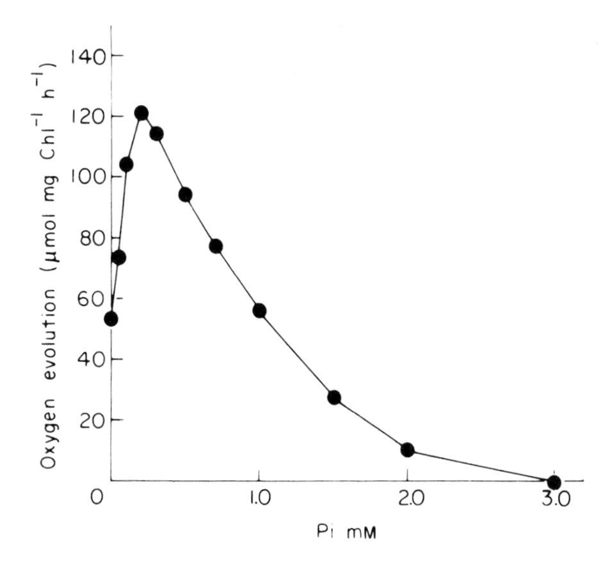


**Fig. 7.21b.** Rate of $CO_2$-dependent $O_2$ evolution by pea chloroplasts as a function of [Pi]. To the left of the peak, Pi is limiting, a parameter which is believed to manifest itself by its effect on the [ATP] / [ADP] ratio (Section 7.13). To the right Pi is inhibitory, enforcing excess export of triose phosphate via the Pi-translocator (Section 8.23). (From Robinson and Walker, 1978)

temperature. On this basis, a decrease in temperature from 10° C to 5° C might be expected to extend induction rather more than would be expected from the temperature response of most metabolic processes. Conversely the relatively high steady-state rates observed at low temperatures might follow if accumulation of substrate continued beyond the level reached at higher temperatures i.e. if low temperature were to some extent offset by the accumulation of more substrate. [Refs. **11, 34, 42**]

## 7.10 Induction and orthophosphate

Inhibition of in-vitro photosynthesis by high concentrations of orthophosphate has been known since some of the first work on isolated chloroplasts by Arnon and his colleagues. Only subsequently did it become apparent that the kinetics of Pi inhibition were complex and could be attributed, at least in part, to an extension of the induction period (Section 8.14). Thus as the external [Pi] is increased the initial lag is extended (Fig. 7.21). At relatively low concentrations the lag extension is not accompanied by a decrease in rate but as the [Pi] is increased and the lag is extended still further, the final rate is also depressed. [Ref. 7, **17**]

## 7.11 Reversal of orthophosphate inhibition by cycle intermediates

At concentrations of Pi in the region of 10 mM, lag extension and rate depression may be so pronounced that photosynthesis is more or less entirely repressed (Section 8.14). The inhibition may then be largely reversed by the addition of certain intermediates of the RPP pathway. The two triose phosphates and PGA cause more

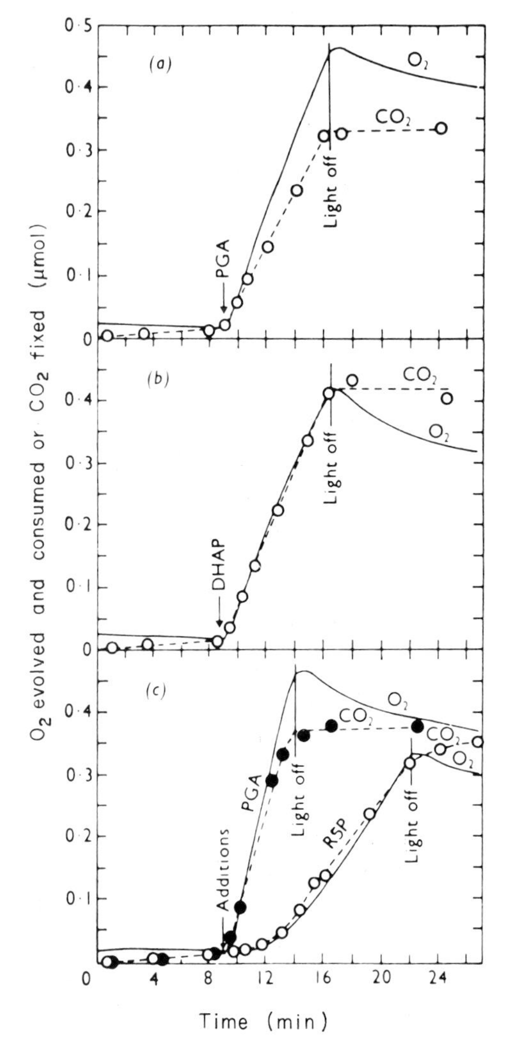


**Fig. 7.22.** Reversal of Pi inhibition by intermediates of the RPP pathway. Continuous lines $O_2$, broken lines $CO_2$. Each reaction mixture initially contained sufficient Pi to prolong induction indefinitely. PGA, R5P and DHAP added as indicated. (From Cockburn *et al*, 1968)

or less immediate reversal whereas R5P produces its response only after a relatively long lag. Fructose bisphosphate also reverses Pi inhibition after a lag but probably as a result of external cleavage to triose phosphate. With triose phosphate and R5P, the kinetics of reversal are the same for $CO_2$ fixation as they are for $O_2$ evolution (Fig. 7.22) implying that the entire RPP pathway is set in motion once again. If PGA is added at the outset together with high Pi, $O_2$ evolution rapidly follows and $CO_2$-fixation increases gradually suggesting that some reduction of PGA may occur independently of $CO_2$-fixation. If the addition of PGA is delayed however, the rate of $O_2$ evolution consequent upon its addition is increased and there is an immediate onset of $CO_2$-fixation (Fig. 7.22), again implying that the high Pi has in some way impeded the normal operation of the cycle and that PGA has reversed this effect (see Section 7.14). Compounds such as RBP, which do not cross the chloroplast envelope (Chapter 8) are ineffective.
[Refs. **18, 33**].

### 7.12 The nature of orthophosphate inhibition

Orthophosphate is a competitive inhibitor of RBP carboxylase and part of the observed inhibition by Pi may be attributed to this cause. However, if induction reflects the time that must elapse before the autocatalytic build-up of intermediates is complete and, if the addition of intermediates which can enter the chloroplast shortens induction as it does (Section 7.7c), it is reasonable to suppose that Pi might extend induction by impeding autocatalysis (Chapter 6). This possibility led to the concept and characterization of the Pi-translocator (Section 8.23) and it is now accepted that Pi produces most of its response by enforced export of triose phosphate from the cycle. Fig. 7.23 illustrates the inter-relationships between [Pi], metabolite movement and induction . Thus, during induction, triose phosphate is preferentially retained in the cycle, bringing about autocatalytic acceleration. In the steady-state, triose phosphate is exported in exchange for external Pi. If Pi is increased, enforced transport extends induction and Pi is much more inhibitory, for this reason, than if it is added after the steady-state has been achieved. [It may be noted that the conceptual interrelationships between autocatalysis, transport and the Pi-translocator are many and varied and that each contributed to the elucidation of the other as did long standing cooperation between the laboratories of Heber (Dusseldorf) Heldt (Munich) and Walker (London and Sheffield). It should not be supposed however that the concept of controlled permeability followed direct measurements. Indeed a 1970 review suggested that 'A direct obligatory exchange between orthophosphate (outside) and sugar phosphate (inside) could account for

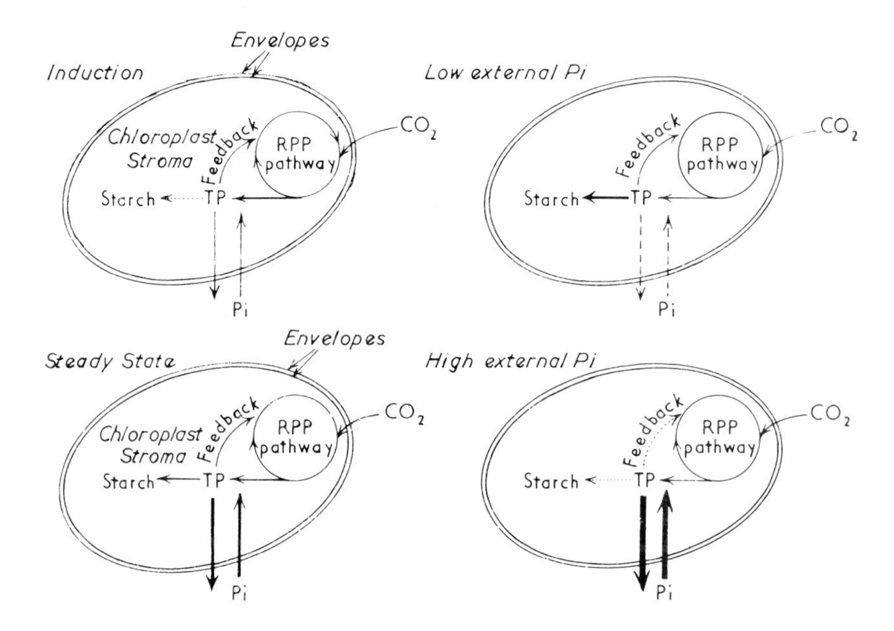


**Fig. 7.23.** Proposed relationship between induction, Pi and metabolite movement. In normal concentrations of external Pi, triose phosphate formed in the RPP pathway will either (a) feed back into the cycle, (b) undergo further conversion to starch etc. in the stroma or (c) be exported. At the onset of illumination the lag which is observed will reflect the time taken for autocatalytic feedback to raise the concentration of cycle intermediates to the steady-state level dictated by the prevailing light intensity etc. During this period feedback will be favoured by rapid utilization within the cycle. As the steady-state is approached relative surplus will increasingly permit export and internal storage. In low external Pi (which it is believed can be achieved experimentally in some leaf tissue by feeding mannose [33]) export will be diminished and surplus triose phosphate diverted into starch synthesis etc. In high external Pi, obligatory export will diminish the proportion of triose phosphate available for feedback and prolong induction beyond the normal period of 1–3 min (see Fig. 7.21). In-vivo less extreme fluctuations in Pi than those illustrated will influence the balance between storage within the chloroplast and export to the cytoplasm.

the inhibition of photosynthesis by orthophosphate and its reversal by sugar phosphate'].

It will be seen that the inter-relationship between Pi, triose phosphate and rate may also become rather more complicated than this simple picture indicates. Thus external Pi will not only encourage export of triose phosphate but will also interfere with re-entry. In many circumstances, there will also be competition for triose phosphate between internal consumption and export. In the first instance the amount of triose phosphate available for export will be inversely related to the effectiveness of the aldol condensation (in which triose phosphate is converted to

fructose bisphosphate—Section 6.11) but subsequently the reactions catalysed by transketolase (Section 6.11) will also act as triose phosphate sinks. Both external triose phosphate and external PGA can inhibit photosynthesis by isolated chloroplasts in some circumstances. In relatively high concentration exogenous triose phosphate tends to inhibit $O_2$ evolution rather than $CO_2$-fixation and is believed to act in this way by exchanging with PGA which would otherwise be reduced internally. Both PGA and triose phosphate will tend to deplete the chloroplast of Pi.
[Refs. **3**, **9**, **20**, **25**, **31**].

### 7.13 Induction in $O_2$ evolution and the role of PGA

At one stage it was believed that pentose monophosphates accumulated during induction but more recent work by Lilley *et al* has shown that although ribulose *bis*phosphate increases during the lag the most spectacular rises occur in PGA and hexose monophosphates (Fig. 7.17). Triose phosphate accumulates in the medium to a much larger extent than PGA but remains at a low level within the stroma. It has also been shown in experiments with reconstituted chloroplasts that there is little or no reduction of PGA to triose phosphate in the presence of pentose monophosphates or fructose bisphosphate. This is because ribulose 5-phosphate, in the presence of its kinase (Section 6.11), acts as a powerful sink for ATP (Section 9.6). This in turn produces an unfavourable ATP/ADP ratio which inhibits PGA reduction (Fig. 7.24) partly, perhaps, by allosteric regulation and partly by its effect on the DPGA ⇄ PGA equilibrium. Accordingly reduction of PGA and its associated $O_2$ evolution will not occur to a significant extent while [PGA] is low and [Ru5P] is high (Fig. 7.25). The build-up of PGA which is necessary to overcome the unfavourable ATP/ADP ratio is almost certainly helped by the fact that, at the alkaline pH which obtains in the illuminated stroma, PGA carries one more negative charge than triose phosphate and is therefore less readily exported than triose phosphate. This is reflected by the fact that the triose phosphate in the stroma does not rise appreciably during induction (although it accumulates in the medium) and that there is about 10 times as much PGA inside the chloroplast as there is outside.

The rate of PGA reduction by chloroplast extracts is dependent on the ratio [PGA][ATP]/[ADP] within the physiological range of concentrations of these metabolites. During RBP-dependent oxygen evolution by reconstituted chloroplasts, the lag produced by addition of ADP (Fig. 7.26) can be largely eliminated if this ratio is sufficiently increased by the addition of PGA (Fig. 7.27) or ATP (Fig. 7.28). In the presence of NADPH, DPGA is reduced to triose phosphate and the steady-state concentration of DPGA is likely to be low. Thus it is the concentrations of PGA,

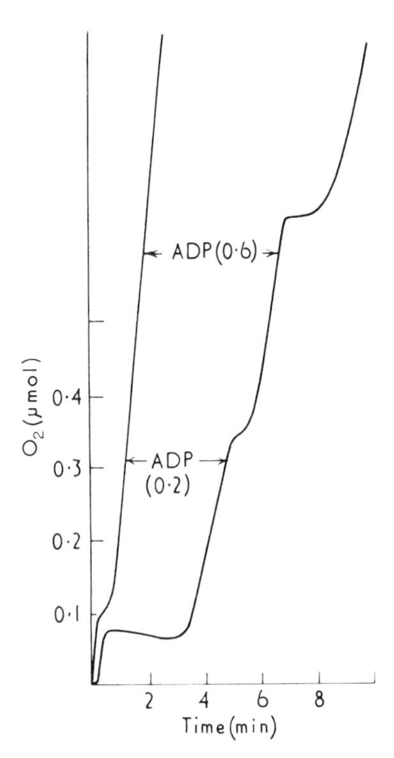


**Fig. 7.24.** In the reconstituted chloroplast system, R5P gives rise to $O_2$ evolution according to the reaction sequence shown in Fig. 7.25. During this process, however, it serves as an ATP sink and the ADP so generated blocks the conversion of PGA to DPGA (Fig. 7.25). This experiment shows that in the presence of an added ATP generator (which complements endogenous photophosphorylation) $O_2$ evolution is sustained regardless of the presence of R5P or the addition of ADP. Conversely, photophosphorylation alone is unable to make good the unfavourable ATP/ADP ratio which is initially caused by the presence of R5P and subsequently by the addition of ADP. (Walker, unpublished).

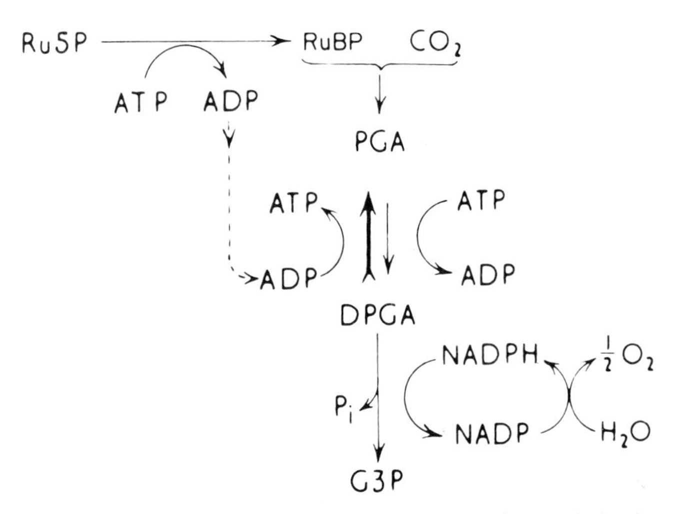


**Fig. 7.25.** Diagramatic representation of inhibition of $O_2$ evolution by an ATP sink. In this instance Ru5P → RBP consumes ATP and the resulting ADP inhibits PGA → DPGA.

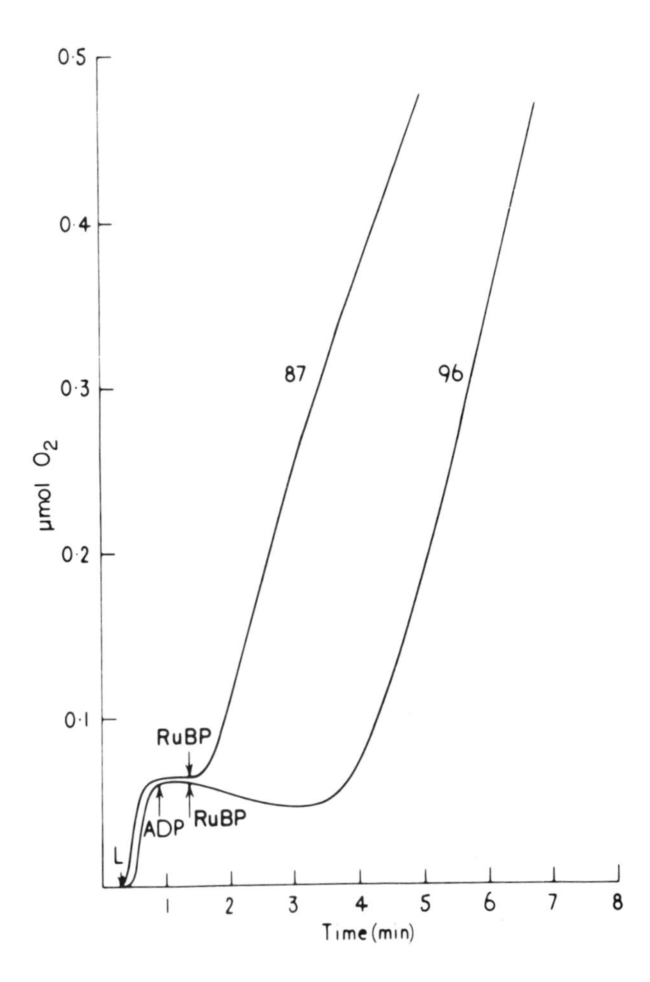


**Fig. 7.26.** Simulation of induction in reconstituted chloroplast system to show very short lag with RBP as a substrate (despite need to build up PGA) and pronounced lag extension caused by additional ADP. Each reaction medium contained catalytic ADP, and $NADP^+$, plus $NaHCO_3$, $MgCl_2$ ferredoxin, broken chloroplasts, and stromal extract. After the added $NADP^+$ was reduced, 1 μmol. of ribulose bisphosphate (RBP) was added. In the lower trace, 1 μmol of ADP was added prior to RBP, resulting in a lag of 3 min before the maximum rate of oxygen evolution was achieved. Rates of oxygen evolution (μmol $mg^{-1}$ Chl $h^{-1}$) are given alongside the traces. (From Robinson and Walker, 1978)

ATP and ADP which determine the rate of PGA reduction. Figs. 7.29 and 7.30 show the results of an experiment with the reconstituted chloroplast system which was designed to mimic the conditions within the intact chloroplast during induction. After the added $NADP^+$ had all been reduced, ATP, ADP, and PGA were added

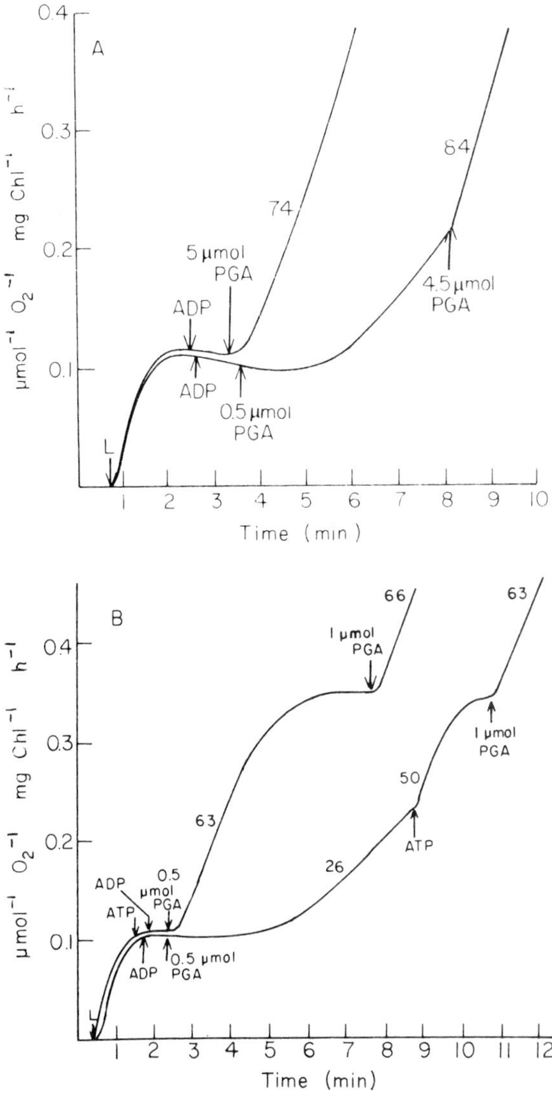


**Fig. 7.27 and 7.28.** Oxygen evolution by reconstituted chloroplasts with PGA. Reaction mixture similar to that in Fig. 7.26 except that $NaHCO_3$ was omitted. **Fig. 7.27.** Reversal of ADP inhibition by increased concentrations of PGA. ADP was added to give a final concentration of 1 mM as indicated. The lag produced by ADP is greatly reduced at higher PGA concentrations (upper trace) **Fig. 7.28.** Reversal of ADP inhibition by ATP. ATP was added to a final concentration of 5 mM (upper trace) and ADP was added to 1 mM (both traces) as indicated. The lag is diminished by ATP. Rates of oxygen evolution ($\mu$mol. $mg^{-1}$ Chl $h^{-1}$) are given alongside the traces.

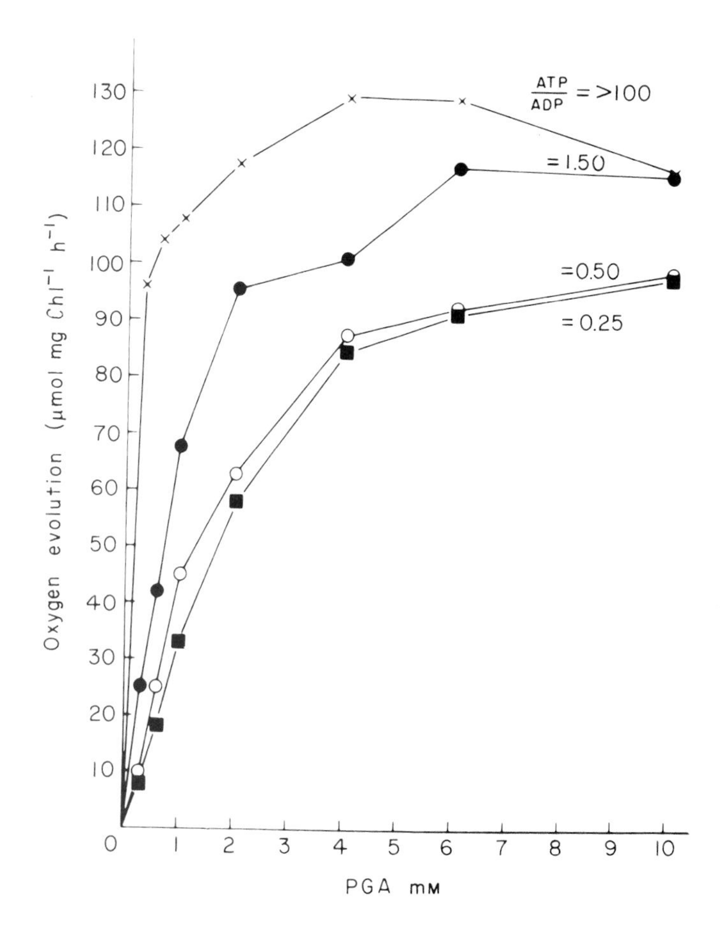


**Fig. 7.29.** Effect of ATP/ADP ratio on the rate of PGA-dependent oxygen evolution by reconstituted chloroplasts at different PGA concentrations. Reaction mixtures as for Fig. 7.27 except that ADP was initially omitted. ATP, ADP and PGA were added together once the NADP had been reduced. The rate of oxygen evolution was dependent on the ATP/ADP ratio and on the PGA concentration. (From Robinson and Walker, 1978)

simultaneously to give various concentrations of PGA and various ATP/ADP ratios with a fixed total concentration of adenine nucleotides (1.5 mM). As shown in Fig. 7.29, the maximum rate of oxygen evolution was dependent on the concentration of PGA and on the ATP/ADP ratio. As this ratio was decreased, higher concentrations of PGA were required to achieve a given rate of oxygen evolution. The time taken to reach the maximum rate is plotted as a function of [PGA] in Fig. 7.30. When all of the adenine nucleotides were added as ATP([ATP]/[ADP] > 100), the lag was

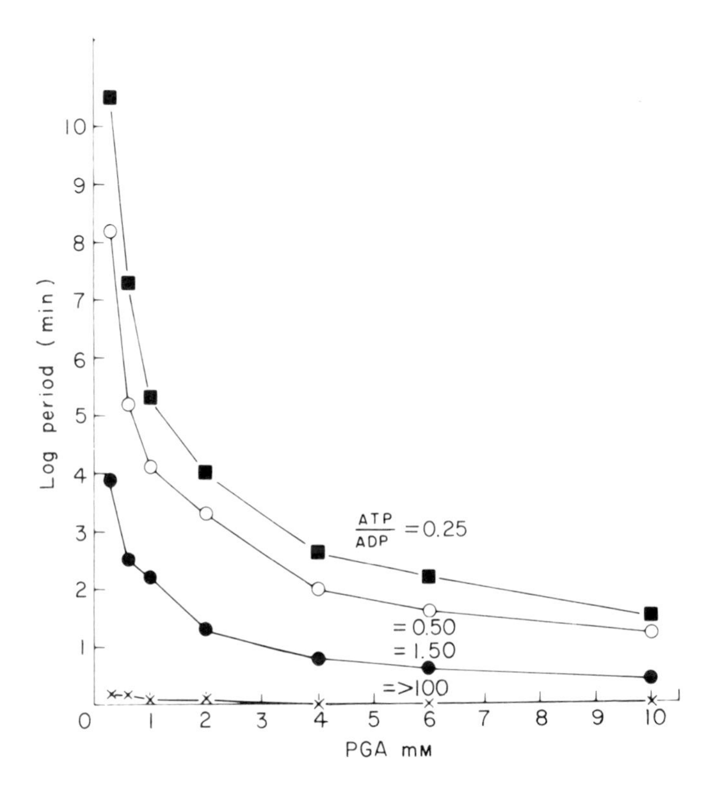


**Fig. 7.30.** Time taken to reach maximum rate as a function of PGA concentration and ATP/ADP ratio. The lag is increased by low ATP/ADP and at lower PGA concentrations. (From Robinson and Walker, 1978).

< 0.2 min. The lag increased as the ATP/ADP ratio was decreased, but for a given ratio it was dependent on the PGA concentration. Thus both the rate of oxygen evolution and the time taken to reach that rate were dependent on the relative concentrations of PGA, ATP and ADP.

The control of PGA reduction by [ADP] is one of several regulatory mechanisms (Chapter 9). If pentose monophosphates were to accumulate, the drain of ATP would ensure that pentose monophosphate formation from PGA would diminish until the normal steady-state ratios were re-established. Some limitation of pentose phosphate formation would be beneficial to the extent that it would diminish photorespiratory consumption. Hexose monophosphates are not readily exported from the chloroplast (Section 8.15) and presumably help to dampen any oscillations between export and internal consumption which might otherwise occur. Because the Pi-translocator (Section 8.23) has a high affinity for PGA the

induction period in $O_2$ evolution is readily shortened when this compound is added to chloroplasts. It is generally assumed that PGA then acts as the natural Hill oxidant, entering the chloroplast and undergoing reduction to triose phosphate. Recent experiments, however, imply that PGA can only enter the chloroplasts as $PGA^{2-}$ and that at alkaline pH (in which $PGA^{3-}$ predominates) there may be little or no PGA-dependent $O_2$ evolution as such. Even at high pH, however, PGA may shorten induction or reverse Pi inhibition (Section 8.7) by inhibiting the Pi-translocator (Section 8.23).
[Refs. **19, 26, 33, 35, 45**].

## 7.14 Induction in $CO_2$-fixation

If isolated chloroplasts are osmotically shocked in a solution at pH 8.0, containing $Mg^{2+}$, a sulphydryl reducing agent such as dithiothreitol and bicarbonate they will start to fix $CO_2$ in the dark as soon as they are provided with ribulose bisphosphate. (Endogenous RBP would be so diluted by the rupturing process normally employed that it could not maintain an easily detectable rate without supplementation). The fact that the same chloroplasts will not commence equally rapid $CO_2$-fixation on first illumination therefore implies that the carboxylase is more or less completely inactive in the environment provided by the darkened stroma or that the RBP available to it is, at first, insufficient to maintain the rates which are seen later. It might be imagined that some choice could be made between these alternatives on the basis of experiment but at present the answers which have been forthcoming are not unequivocal. For example a number of workers have reported light/dark modulation of carboxylase activity but in experiments with protoplasts from wheat, peas and spinach Robinson *et al* have found the difference in activity to be only marginal. Even Jensen *et al*, who attach great importance to light activation, have reported dark values of about 2/3 of those seen in the light (Fig. 7.31). Like Lilley *et al*, Jensen *et al* report light driven increases in RBP but found considerable quantities of RBP in the dark (i.e. concentrations considerably higher than those required to support half maximal activity of the isolated carboxylase). Coexistence of high [RBP] and significantly active carboxylase are precluded in theory by the large negative $\Delta F'$ value for the carboxylation which indicates that the reaction should proceed to completion. Similarly the conditions within the darkened stroma may not favour carboxylase action even though the carboxylase itself is fully active (Section 7.8). However this could not be explained in terms of [$Mg^{2+}$] because $Mg^{2+}$ action appears to be restricted to activation rather than catalysis and, in any case, recent work by Krause and Ben-Hayyim (Section 9.3) shows that $CO_2$ fixation is not significantly affected in isolated chloroplasts in which light-induced increases in

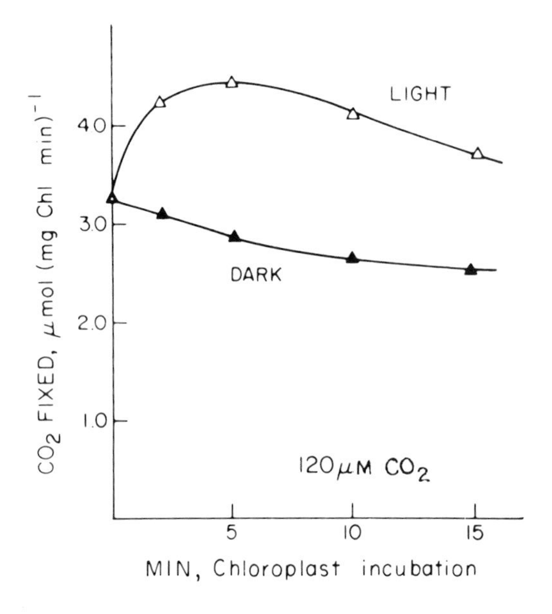


**Fig. 7.31.** Behaviour of chloroplast RBP carboxylase activity during incubation with $CO_2$. Chloroplasts were incubated at pH 7.8, 25°C, in the dark (▲) or light (△), and assayed for RBP carboxylase as indicated (After Jensen *et al*, 1978).

stromal [$Mg^{2+}$] are abolished by ammonia. Similarly it seems unlikely that the relatively acid pH of the darkened stroma could suppress carboxylase activity completely or that the response to alkaline pH upon illumination would be slow enough to make a material contribution to induction lags of several minutes.

At this time we are therefore bound to conclude that if the estimates of RBP by Jensen *et al* are correct then much of the RBP which they measure is not immediately available for carboxylation. [The *K*m for RBP is 20–30 μM which, on the basis of a stromal volume of 25 μl/mg Chl would be 0.50–0.75 nmol/mg Chl. In Fig. 7.32 which, incidentally, shows inhibition of $CO_2$-fixation by DL-glyceraldehyde, the initial RBP concentration exceeded this by an order of magnitude]. The increases in RBP which they and Lilley *et al* (1977) see upon illumination are then consistent with the notion that the lag in $CO_2$-fixation reflects the time required to build the concentration of the $CO_2$ acceptor to its steady-state working level. If enzyme activation is an important contributory factor in RBP accumulation the enzymes most likely to be concerned are the bisphosphatases and Ru5P kinase. For example, FBPase has a low activity in the dark and the kinetics of its activation at 20°C are *sometimes* similar to induction in $CO_2$-fixation. It is not easy to separate $O_2$

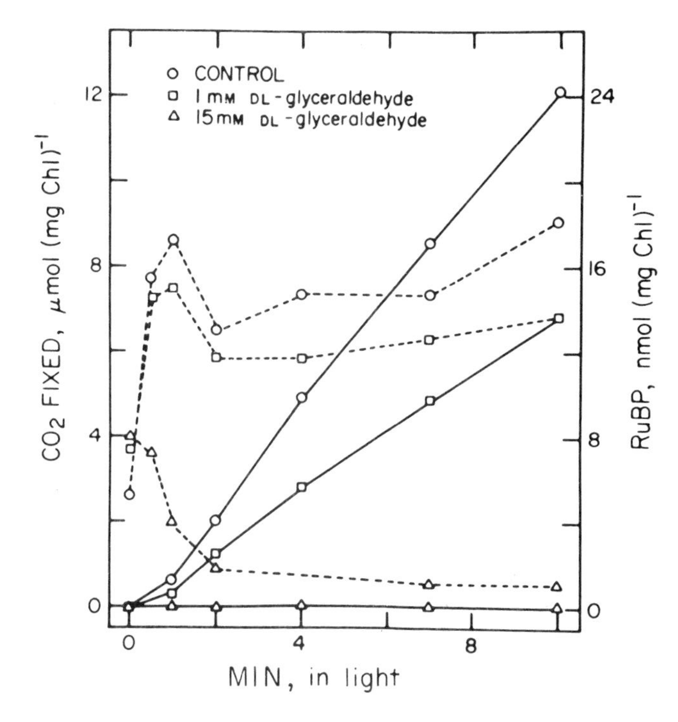


**Fig. 7.32.** The effect of DL-glyceraldehyde on $CO_2$-fixation and levels of chloroplast RBP. $CO_2$-fixation (———) determined in presence of zero (0), 1.0 mM (□), or 15.0 mM (△) DL-glyceraldehyde. RBP (- - -) was determined from similar media (Jensen *et al*, 1978).

evolution from $CO_2$-fixation in intact chloroplasts but when PGA is added at the outset, $O_2$ evolution soon follows whereas the lag in $CO_2$-fixation is not shortened by the same factor (Fig. 7.33).

On the other hand if the lag is extended by high [Pi] the kinetics of PGA-dependent $O_2$ evolution and FBPase activation are largely unchanged (Fig. 7.19) in accord with the present view that Pi inhibits by enforcing export of intermediates via the Pi-translocator (Sections 7.11, 8.14). If PGA is then added, after some minutes illumination, *both* $O_2$ evolution *and* $CO_2$-fixation start immediately (Fig. 7.22). The fact that exogenous PGA gives rise to more or less immediate $O_2$ evolution (an indication of similarly rapid formation of triose phosphate) implies that at the outset of illumination triose phosphate is not able to fill the RBP pool to an optimal level. Conversely the rapid onset of $CO_2$-fixation which follows PGA addition to a Pi inhibited system (Fig. 7.22) implies that any residual barrier to full activity (such as inadequate FBPase activity) has been overcome during illumination.
[Refs. **24, 26, 33, 34, 45**]

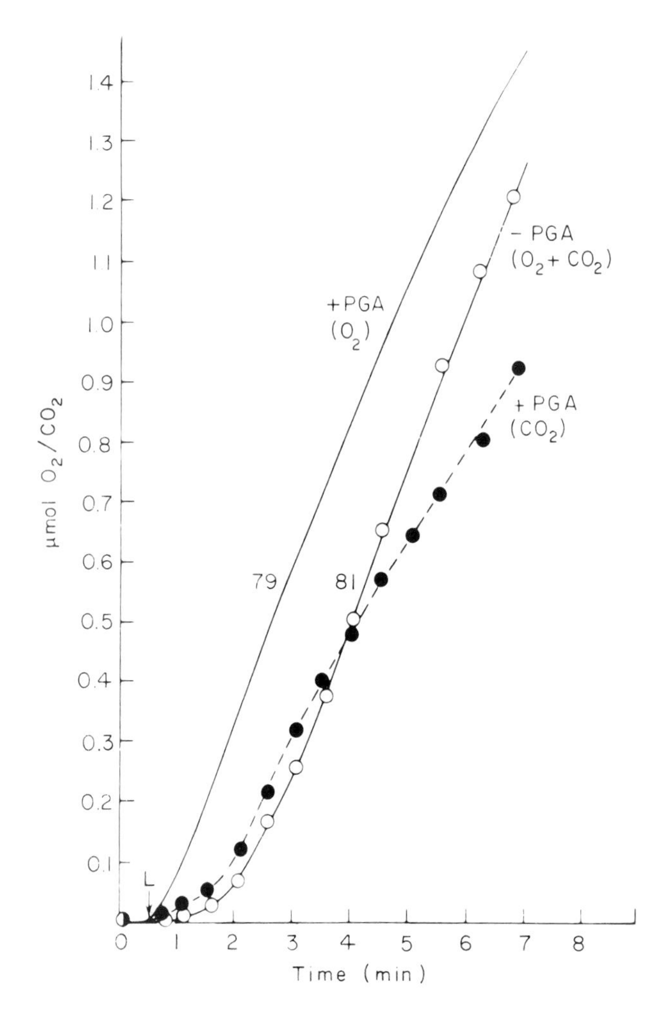


**Fig. 7.33.** Simultaneous measurement of oxygen evolution and $CO_2$-fixation by intact spinach chloroplasts in the presence and absence of PGA. Oxygen evolution is shown by continuous lines and incorporation of $^{14}CO_2$ into acid-stable products by circles. Although the lag in oxygen evolution is virtually eliminated by PGA, the lag in $CO_2$-fixation is only slightly decreased. (From Robinson and Walker, 1978)

## 7.15 Induction in the reconstituted chloroplast system

As noted above (Section 7.14) $CO_2$-fixation quickly starts in reconstituted chloroplast systems which are supplied with RBP or R5P. However, with FBP as substrate there is (at least in some circumstances) a distinct lag and this becomes

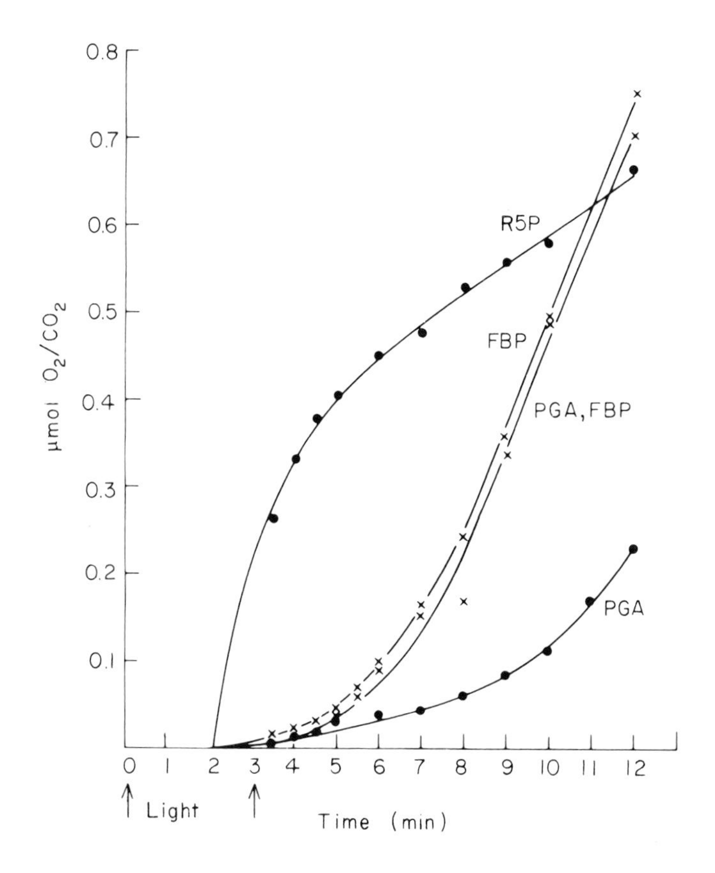


**Fig. 7.34.** $CO_2$-fixation in the reconstituted system. Substrates (as indicated) added after 2 min pre-illumination. With R5P there was a substantial fixation in the first minute. With FBP and PGA there was an evident lag (Walker, unpublished).

more marked with PGA or G3P or when exogenous substrate is omitted entirely (Figs. 7.34, 7.35 & 7.36).

In the reconstituted chloroplast system the presence of bicarbonate, $Mg^{2+}$ and dithiothreitol at alkaline pH should ensure that all of the enzymes concerned are fully activated. The observed lags in $CO_2$-fixation must therefore be associated with pool-filling and even though normal pool-sizes may be greatly affected by dilution in this system these lags are informative and point to the need to build up appropriate concentration of triose phosphate and hexose monophosphate.

Furbank and Lilley have recently studied $CO_2$ fixation in stromal extracts and have only observed non-linear kinetics in the presence of substrate amounts of

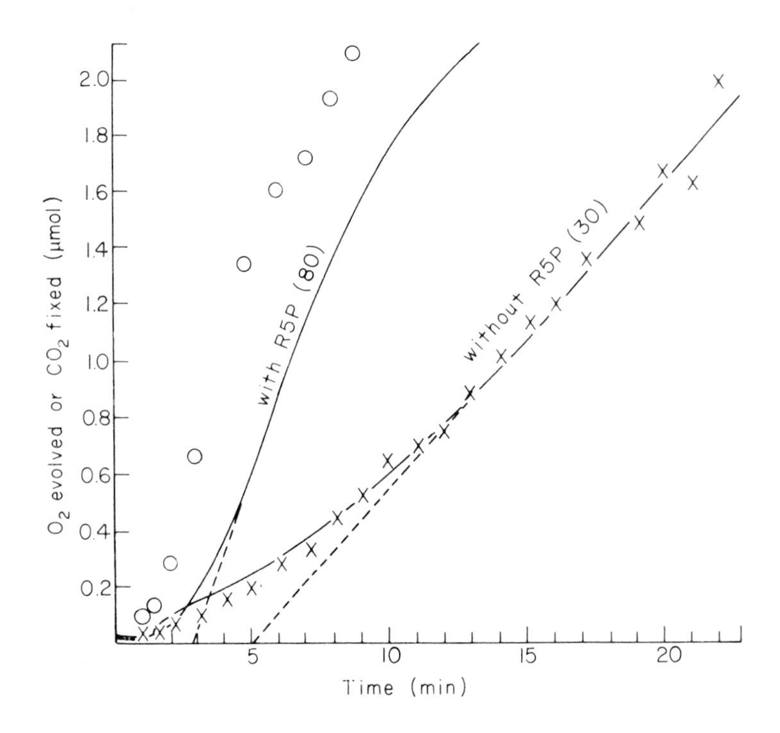


**Fig. 7.35.** Induction in a reconstituted chloroplast system. Both mixtures contained $NaHCO_3$. R5P was added to one mixture as indicated. Immediately prior to illumination $NaH^{14}CO_3$ was added to both and simultaneous measurements of $O_2$ and $CO_2$ carried out. The continuous lines are $O_2$ electrode traces, the circles (+ R5P) and the crosses (− R5P) record $^{14}CO_2$ fixed into acid stable products. Maximal rates are given in parentheses as μmol $CO_2$ or $O_2$ $mg^{-1}$ Chl $h^{-1}$. The difference in the kinetics of $O_2$ evolution and $CO_2$-fixation seen in the presence of R5P is caused by the creation of an unfavourable [ATP] / [ADP] ratio (Section 7.13). (From Walker and Lilley, 1974)

erythrose 4-phosphate. They have suggested therefore 'that in the absence of information on all relevant intermediates, care should be used in interpreting rate increases with time as being diagnostic of autocatalysis'. We are in complete agreement with this view and readily concede that the apparent autocatalytic acceleration recorded in Figs. 7.34 and 7.35 could only be made definitive by simultaneous measurements of all of the intermediates concerned. It may be noted, however, that the Furbank and Lilley system contained no thylakoids and depended entirely for a continuing supply of ATP on the presence of creatine phosphate-creatine phosphate kinase. Their system, therefore, differs in several respects from that employed in the experiments used in the derivation of Figs. 7.34 and 7.35. Perhaps the most important of these is that a favourably high [ATP]/[ADP] ratio is

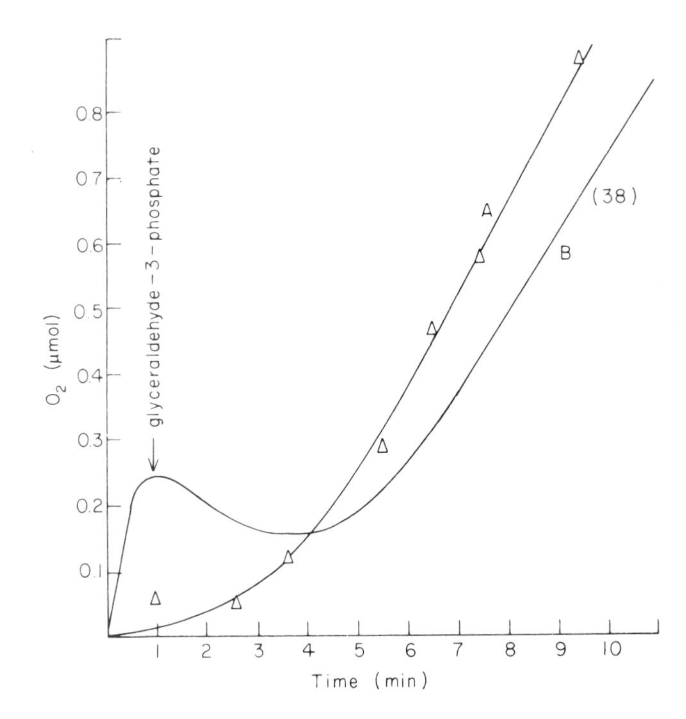


**Fig. 7.36.** Induction in a reconstituted chloroplast system. Simultaneous measurements of $O_2$ evolution (continuous trace) and $CO_2$-fixation (triangles) in the presence of G3P. (cf. Fig. 7.35). (From Stokes and Walker, 1976)

maintained by the inclusion of CP-CPK. Conversely, if ATP generation is left to the presence of illuminated thylakoids it is entirely possible to arrive at an experimental situation in which a given [ATP]/[ADP] ratio will maintain a linear rate of $O_2$ evolution (or $CO_2$ fixation) once a reasonable threshold of PGA has been obtained. However, it can be seen in Figs. 7.27 and 7.28 that $O_2$ evolution in these circumstances is determined by the concentration of ATP and ADP as well as by the concentration of PGA. In short, a build-up of PGA could readily manifest itself even if the [ATP]/[ADP] ratio were not maintained at a high value. What presently emerges from studies of this sort is that the reaction catalysed by PGA kinase (Section 6.11) plays a key role, both in induction and in the internal regulation of the RPP pathway (Chapter 9).
[Refs. **20, 33, 43, 44**]

## 7.16 Sequence of events in induction

(a) At the onset of illumination electron transport will commence at a rapid rate and associate vectorial movements of protons from the stroma tc the thylakoid compartment will cause the stroma to become more alkaline, shifting from about pH 7 to about 8.

(b) Anions such as $Cl^-$ will move into the thylakoid compartment and cations such as $Mg^{2+}$ will move out, in order to compensate for the charge imbalance caused by proton efflux. The $Mg^{2+}$ shift was thought at one stage to be important in relation to enzyme activation but it now seems likely that the stromal [$Mg^{2+}$] is always sufficiently high to satisfy the requirements of enzyme activation and catalysis.

(c) A proton pump (which facilitates $H^+/K^+$ exchange) is located in the chloroplast envelope and will help to maintain the pH difference between the stroma and the cytosol which would otherwise decay as protons diffused slowly into the chloroplast. The action of this pump may be affected by external $Mg^{2+}$ and it may also interact with the Pi-translocator.

(d) The stromal [ATP]/[ADP] ratio will rise rapidly as photophosphorylation commences at maximum rate and ATP consumption is, at first, limited by a deficiency of the substrates for the kinase reactions.

(e) The [NADPH]/[$NADP^+$] ratio will rise as reduction initially outstrips consumption. In the absence of a fully active sink for NADPH (before PGA reduction attains its maximum rate) electrons will be directed towards the reduction of thioredoxin etc. so that the entire stromal environment will become more reducing.

(f) At alkaline pH and through the action of thioredoxin and the like, disulphide groups on enzymes will become reduced. In consequence enzymes such as Ru5P kinase, triose phosphate dehydrogenase and the bisphosphatases will become activated. The precise contribution of such activation to induction remains to be established but FBPase and SBPase seem to have a particularly low activity in the dark and, if low activity is allied to low substrate concentration, their activation *may* make an important contribution to the early stages of induction. RBP carboxylase has also been regarded as a light activated enzyme and one which may be directly activated by NADPH. This view is not supported by experiments with protoplasts which show that RBP carboxylase is nearly as active in the dark as in the light.

(g) A number of enzymes of the RPP pathway (including FBPase) work better at the alkaline pH of the illuminated stroma. This is not a question of enzyme activation [see (f) above], simply one of optimum pH.

(h) The concentration of RBP, PGA and hexose monophosphates will increase. It seems likely that mobilization of starch may contribute directly to the latter in certain circumstances. As Ru5P increases the drain on ATP will increase and the [ATP]/[ADP] ratio will become less favourable to PGA reduction. PGA will build up further and be retained within the chloroplast by the alkalinity of the stroma. As Fig. 7.26 shows, even if the RBP carboxylase is fully active and supplied with adequate substrate, an induction lag in $O_2$ evolution can be observed if the [ATP]/[ADP] ratio is unfavourable. This inhibition by ADP can be offset by high [PGA] (Fig. 7.27). Similarly $CO_2$-fixation will commence very rapidly in a reconstituted chloroplast system supplied with FBP although $O_2$ evolution may be delayed indefinitely. In such systems induction in $CO_2$-fixation can be demonstrated but only in the absence of added substrate or in the presence of intermediates which precede FBP. Presumably the time taken to fill the various pools of intermediates must be exaggerated by dilution in these mixtures (the volume of the reconstituted system is about 1000 times greater, per mg Chl, than that of the intact chloroplast). Nevertheless, the enzymes concerned are fully activated. This indicates that, at least in these circumstances, induction is largely a matter of substrate build-up.

As the enzymes concerned become fully activated and the substrates of the cycle build up, photosynthesis will enter a steady-state. In the presence of saturating $CO_2$ and adequate Pi the rate will then be determined by light intensity and temperature.

Maximal photosynthesis will continue unchecked only in the presence of external Pi. If Pi becomes limiting less triose phosphate will be exported. High [PGA]/[Pi] ratios will favour the allosteric activation of ADPglucose pyrophosphorylase (Section 9.7) and if F6P could be consumed in starch synthesis at a sufficient rate to prevent a further decline in stromal [Pi] it ought in theory to be possible to divert triose phosphate from export to internal storage (as starch) without affecting the rate of photosynthesis. In practice, however, it seems probable that the F6P pool will tend to rise and, in the presence of ample triose phosphate this will lead to increased consumption of carbon in the first transketolase reaction.

Any tendency for pentose phosphate accumulation will then inevitably lead to a further decline in the [ATP]/[ADP] ratio (because of the Ru5P → RBP sink for ATP). In turn, this will suppress PGA reduction and Ru5P consumption will diminish its own regeneration. In short there is no way in which the regenerative phase of the RPP pathway can place a greater load on the available supply of ATP than it can bear. Only at optimal Pi is there no danger that regeneration will proceed at a rate which will fail to use the available assimilatory power to its best advantage.

## 7.17 Restoration of induction

Although induction is virtually absent after a brief interruption of illumination it is evident that it must eventually be restored after longer periods in the dark. It is of interest, therefore, that restoration occurs more quickly in whole leaves and in protoplasts than in isolated chloroplasts. Leegood has shown that, in wheat, FBPase is rapidly deactivated and reactivated in both chloroplasts and protoplasts. The difference in behaviour appears to reside in the level of metabolites which falls more rapidly in the protoplasts than in the chloroplasts. [Refs. **25c, 33b**]

## General References


1 GIBBS M. (1971) Carbohydrate metabolism by chloroplasts. In *Structure and Function of Chloroplasts* (ed. M. Gibbs), pp. 169–214. Berlin: Springer-Verlag.
2 HELDT H. W. (1976) Metabolite carriers of chloroplasts. In *Encyclopedia of Plant Physiology. Transport in Plant III* (eds. C. R. Stocking & U. Heber) New Series, Volume 3, pp. 137–43. Berlin: Springer-Verlag.
3 HEROLD A. & WALKER D. A. (1979) Transport across chloroplast envelopes—the role of phosphate. In: *Membrane Transport in Biology* (eds. G. Giebisch, D. C. Tosteson & J. J. Ussing). Vol. II, 412–439. Heidelberg: Springer-Verlag.
4 RABINOWITCH E. I. (1956) *Photosynthesis and Related Processes*, Vol. II, Pt. 2, p. 537 *et seq*. New York: Interscience.
5 WALKER D. A. (1973) Photosynthetic induction phenomena and the light activation of ribulose diphosphate carboxylase. *New Phytol.* **72**, 209–35.
6 WALKER D. A. (1974) Chloroplast and Cell—Concerning the movement of certain key metabolites etc. across the chloroplast envelope. *Med. Tech. Publ. Int. Rev. Sci. Biochem. Ser. 1*, (ed. D. H. Northcote) Vol. 11, pp. 1–49. London: Butterworths.
7 WALKER D. A. (1976a) $CO_2$-fixation by Intact Chloroplasts: Photosynthetic induction and its relation to transport phenomena and control mechanisms. In: *The Intact Chloroplast*. (ed. J. Barber) Chapter 7, pp. 235–78. Amsterdam: Elsevier.
8 WALKER D. A. (1976b) Plastids and Intracellular Transport. In *Encyclopedia of Plant Physiology. Transport in Plants III*, (eds. C. R. Stocking & U. Heber) New Series, Volume 3, pp. 85–136. Berlin, Springer-Verlag.
9 WALKER D. A. & CROFTS A. R. (1970) Photosynthesis. *Ann. Rev. Biochem.* **39**, 389–428.


## Specific Citations


10 BALASUBRAMANIAM S. & WILLIS A. J. (1969) Stomatal movements and rates of gaseous exchange in excised leaves of *Vicia faba*. *New Phytol.*, **68**, 663–75.
11 BALDRY C. W., BUCKE C. & WALKER D. A. (1966) Some effects of temperature on carbon dioxide fixation by isolated chloroplasts. *Biochim. Biophys. Acta*, **126**, 207–13.
12 BALDRY C. W., WALKER D. A. & BUCKE C. (1966) Calvin-cycle intermediates in relation to induction phenomenon in photosynthetic carbon dioxide fixation by isolated chloroplasts. *Biochem. J.*, **101**, 641–6.
13 BAMBERGER E. S. & GIBBS M. (1963) Studies on photosynthetic carbon dioxide fixation by whole spinach chloroplasts. *Plant Physiol.* Suppl. **38**, x.
14 BLINKS L. R. & SKOW R. K. (1938a) The time course of photosynthesis as shown by a rapid electrode method for oxygen. *Proc. Natn. Acad. Sci. U.S.A.*, **24**, 420–7.
15 BLINKS L. R. & SKOW R. K. (1938b) The time course of photosynthesis as shown by the glass electrode, with anomalies in the acidity changes. *Proc. Natn. Acad. Sci. U.S.A.*, **24**, 413–9.
16 BUCKE C., WALKER D. A. & BALDRY C. W. (1966) Some effects of sugars and sugar phosphates on carbon dioxide fixation by isolated chloroplasts. *Biochem. J.*, **101**, 636–41.

17 COCKBURN W., BALDRY C. W. & WALKER D. A. (1967) Some effects of inorganic phosphate on $O_2$ evolution by isolated chloroplasts. *Biochim. Biophys. Acta.*, **143**, 614–24.
18 COCKBURN W., WALKER D. A. & BALDRY C. W. (1968) Photosynthesis by isolated chloroplasts. Reversal of orthophosphate inhibition by Calvin cycle intermediates. *Biochem. J.* **107**, 89–95.
19 FLIEGE R., FLUGGE U-I, WERDAN K. & HELDT H. W. (1973) Specific transport of inorganic phosphate, 3-phosphoglycerate and triose phosphates across the inner membrane of the envelope in spinach chloroplasts. *Biochim. Biophys. Acta.*, **502**, 232–47.
20 FURBANK R. T. & LILLEY R. M. C. (1980) Effects of inorganic phosphate on the photosynthetic carbon reduction cycle in extracts from the stroma of pea chloroplasts. *Biochim. Biophys. Acta*, **592**, 65–75
21 HEBER U. & PERCELD P. (1977) Substrate and product fluxes across the chloroplast envelope during bicarbonate and nitrite reduction. In *Proceedings of the Fourth International Congress on Photosynthesis*, (eds. D. O. Hall, J. Coombs, T. W. Goodwin) p. 107–18. Biochem. Soc. London.
22 HEBER U. & WILLENBRINK J. (1964) Site of synthesis and transport of photosynthetic products within the leaf cell. *Biochim. Biophys. Acta.* **82**, 313–24.
23 JAGENDORF A. T. & HIND G. (1963) Studies on the mechanism of photophosphorylation. In *Photosynthetic Mechanism of Green Plants*, pp. 599–610. National Academy of Sciences, National Research Council Publications 1145.
24 JENSEN R. G., SICHER R. C. Jr. & BAHR J. T. (1978) Regulation of ribulose 1,5-bisphosphate carboxylase in the chloroplast. In Photosynthetic Carbon Assimilation. Basic Life Sciences Vol. II, "*Proceedings of a Symposium held at Brookhaven National Laboratory, New York, May 31–June 2, 1978.*" (eds. H. W. Siegelman & G. Hind), pp. 95–112. New York: Plenum Press.
25 KAISER W. & URBACH W. (1976) Rates and properties of endogenous cyclic photophosphorylation of isolated intact chloroplasts measured by $CO_2$ fixation in the presence of dihydroxyacetone phosphate. *Biochim. Biophys. Acta*, **423**, 91–102.
25a LEEGOOD R. C. & WALKER D. A. (1980) Autocatalysis and light activation of enzymes in relation to photosynthetic induction in wheat chloroplasts. *Arch. Biochem. Biophys.* **200**, 575–582.
25b LEEGOOD R. C. & WALKER D. A. (1980) Modulation of fructose bisphosphatase activity in intact chloroplasts. *FEBS Letters* **116**, 21–4.
25c LEEGOOD R. C. & WALKER D. A. (1981) Photosynthetic induction in wheat protoplasts and chloroplasts. Autocatalysis and light activation of enzymes. *Plant, Cell and Environment* **4**, 59–66.
26 LILLEY R. McC., CHON C. J., MOSBACH A. & HELDT H. W. (1977) The distribution of metabolites between spinach chloroplasts and medium during photosynthesis *in vitro*. *Biochim. Biophys. Acta*, **460**, 259–72.
27 LORIMER G. H., BADGER M. R. & HELDT H. W. (1978). The activation of ribulose 1, 5-bisphosphate carboxylase/oxygenase. In Photosynthetic Carbon Assimilation. Basic Life Sciences Vol. II, *Proceedings of a Symposium held at Brookhaven National Laboratory, Upton, New York, May 31–June 2, 1978.* (eds. H. W. Siegelman & G. Hind) pp. 43–59. New York: Plenum Press.
28 MCALISTER E. D. (1937) Time course of photosynthesis for a higher plant. *Smithson Misc. Coll.*, **95**, 24.
29 MCALISTER E. D. & MYERS J. (1940) The time course of photosynthesis and fluorescence, observed simultaneously. *Smithson, Misc. Coll.*, **99**, 6.
30 OSTERHOUT W. J. V. & HASS A. R. C. (1919) On the dynamics of photosynthesis. *J. Gen. Physiol.* **1**, 1–16.
31 PEAVEY D. G. & GIBBS M. (1975) Photosynthetic enhancement studied in intact spinach chloroplasts. *Plant Physiol*, **55**, 799–802.
32 ROBINSON S. P., MCNEIL P. H. & WALKER D. A. (1979) Ribulose bisphosphate carboxylase—lack of dark inactivation of the enzyme in experiments with protoplasts. *FEBS Letters*, **97**, 260–300.
33 ROBINSON S. P. & WALKER D. A. (1978) Regulation of photosynthetic carbon assimilation. In Photosynthetic Carbon Assimilation. Basic Life Sciences Vol. II, *Proceedings of a Symposium held at*

*Brookhaven National Laboratory, Upton, New York, May 31– June 2, 1978.* (eds. H. W. Siegelman & G. Hind) pp. 43–59. New York. Plenum Press.
33a ROBINSON S. P. & WALKER D. A. (1980) The distribution of metabolites between chloroplast and cytoplasm during the induction phase of photosynthesis in leaf protoplasts. *Plant Physiol.* **65,** 902–905.
33b ROBINSON S. P. & WALKER D. A. (1980) The significance of light activation of enzymes during the induction phase of photosynthesis in isolated chloroplasts. *Arch Biochem. Biophys.* **202**, 617–23.
34 SELWYN M. J. (1966) Temperature and photosynthesis II. A mechanism for the effects of temperature on carbon dioxide fixation. *Biochim. Biophys. Acta.* **126**, 214–24.
35 SLABAS A. R. & WALKER D. A. (1976) Transient inhibition by ribose-5-phosphate of photosynthetic $O_2$ evolution in a reconstituted chloroplast system. *Biochim. Biophys. Acta,* **430**, 154–64.
36 STEEMAN-NIELSEN E. (1942) Der Mechanismus der Photosynthese. *Dansk bot. Ark.*, **11**, 2.
37 TURNER J. F., BLACK C. C. & GIBBS M. (1962) The effect of light intensity on the triphosphopyridine nucleotide reduction, adenosine triphosphate formation and carbon dioxide assimilation in spinach chloroplasts. *J. Biol. Chem.*, **237**, 577–9.
38 WALKER D. A. (1964) Improved rates of carbon dioxide fixation by illuminated chloroplasts. *Biochem. J.* **92**, 22c–23c.
38a WALKER D. A. (1967) Photosynthetic activity of isolated pea chloroplasts. In *Biochemistry of the Chloroplast* (1965) NATO Adv. Study Inst. Aberystwyth. (Ed. T. W. Goodwin) Vol. 2, 53–69. New York: Academic Press.
39 WALKER D. A., BALDRY C. W. & COCKBURN W. (1968) Photosynthesis by isolated chloroplasts, simultaneous measurement of carbon assimilation and oxygen evolution. *Plant Physiol.* **43**, 1419–22.
40 WALKER D. A., COCKBURN W. & BALDRY C. W. (1967) Photosynthetic oxygen evolution by isolated chloroplasts in the presence of carbon cycle intermediates. *Nature*, **216**, 597–9.
41 WALKER D. A. & HILL R. (1967) The relation of oxygen evolution to carbon assimilation with isolated chloroplasts. *Biochim. Biophys. Acta.* **131**, 330–8.
42 WALKER D. A., KOSCLUKIEWICZ K. & CASE C. (1973) Photosynthesis by isolated chloroplasts: some factors affecting induction in $CO_2$-dependent oxygen evolution. *New Phytol.* **72**, 237–47.
43 WALKER D. A. & LILLEY R. MCC. (1974) Autocatalysis in a reconstituted chloroplast system. *New Phytol.* **73**, 657–662.
44 WALKER D. A. & SLABAS A. R. (1976) Stepwise generation of the natural oxidant in a reconstituted chloroplast system. *Plant Physiol.* **57**, 203–6.
45 WALKER D. A., SLABAS A. R. & FITZGERALD M. P. (1976) Photosynthesis in a reconstituted chloroplast system from spinach. Some factors affecting $CO_2$-dependent oxygen evolution with fructose-1, 6-bisphosphate as substrate. *Biochim. Biophys. Acta.* **440**, 147–62.
46 WARBURG O. (1920) Uber die Geschwindigkeit der photochemischen Kohlensaurezersetzung in lebenden Zellen II. *Biochem Z.*, **103**, 188–217.
47 WILLIS A. J. & BALASUBRAMANIAM S. (1968) Stomatal behaviour in relation to rates of photosynthesis and transpiration in *Pelargonium*. *New Phytol.* **67**, 265–85.

# Chapter 8
# Plastids and Intracellular Transport




SUMMARY

The $C_3$ chloroplast is principally an organelle which imports $CO_2$ and orthophosphate and exports triose phosphate at high rates. Although it accommodates a great many additional metabolic processes its *major* roles in anabolism and transport can therefore be summarized by the equation:

$$3CO_2 + 2H_2O + Pi \rightarrow \text{triose phosphate} + 3O_2$$

Assimilatory power (ATP and NADPH) cannot be directly exported at high rates because of the relative impermeability of the inner envelope to adenylates and total impermeability to nicotinamide nucleotides. These compounds can, however, be exchanged by participation in shuttle mechanisms (Fig. 8.1).


## 8.1 The chloroplast as a transporting organelle

Clearly the principal function of the chloroplast is to generate assimilatory power and to use this to reduce $CO_2$ to the level of carbohydrate. The whole plant, on the other hand, can only benefit from these activities if the chloroplast exports what it

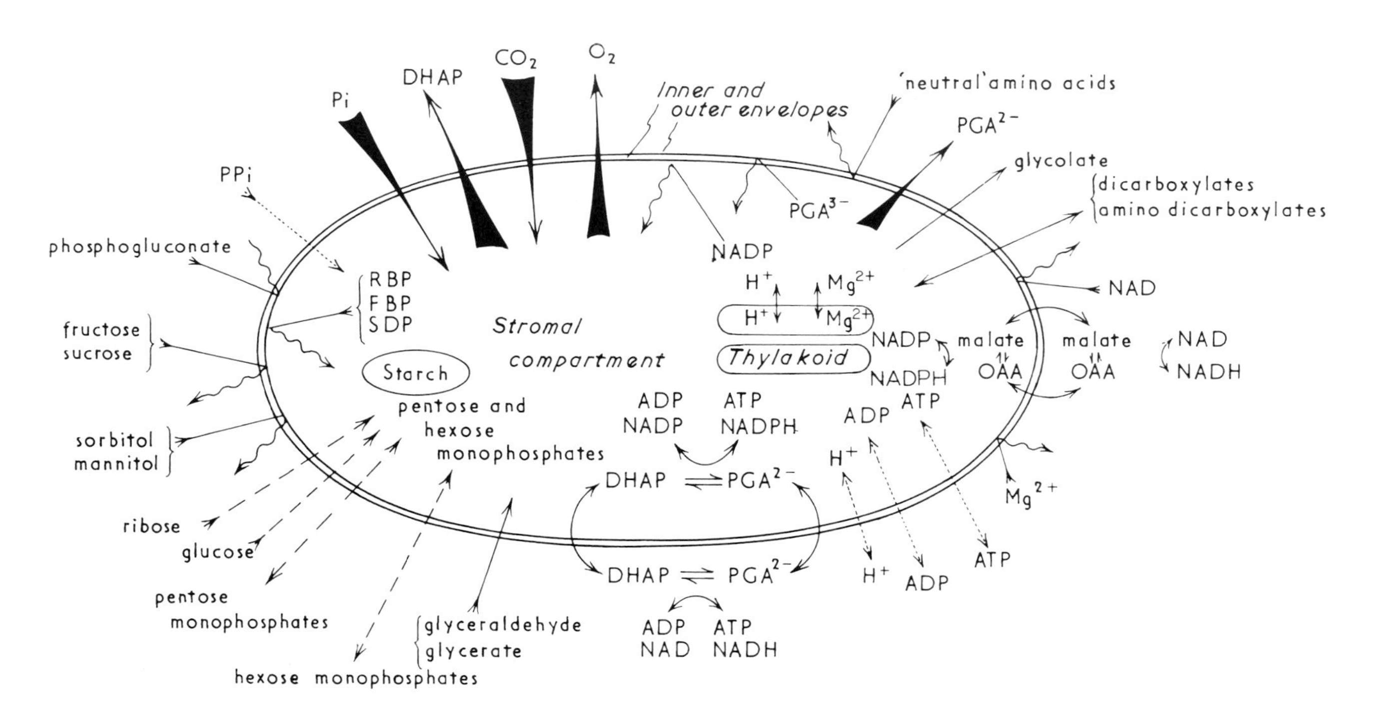


Fig. 8.1 summarizes the major metabolite movements. The 'boldness' of the arrows is intended to convey an impression of the rates of movement and the magnitude of the exchange. Thus much prominence is given to the stoichiometric exchange of Pi and triose phosphate via the phosphate translocator whereas direct movement of hexose monophosphate is believed to be very slow and unimportant.

manufactures. Similarly the chloroplast can only export elaborated carbon if it imports the wherewithal for its manufacture. In the long term the chloroplast must balance import and export and, in the short term, the transport of key metabolites must match the rate of photosynthesis. Nevertheless it is also imperative that the chloroplast continuously regenerates its $CO_2$ acceptor and that the RPP pathway is permitted to function as an autocatalytic sequence or breeder reaction (Chapter 6). In essence, these are conflicting roles and it is evident that the transport of metabolites, coenzymes and ions between the chloroplast and its immediate cellular environment must occur in a controlled fashion. During the past decade a considerable body of evidence has been amassed which indicates that the principal imports into the $C_3$ plastid are Pi and $CO_2$ and that the principal export is triose phosphate (Fig. 8.2). This evidence and that relating to the movement of other compounds is considered below. [The methods employed are also summarized in Table 8.1].

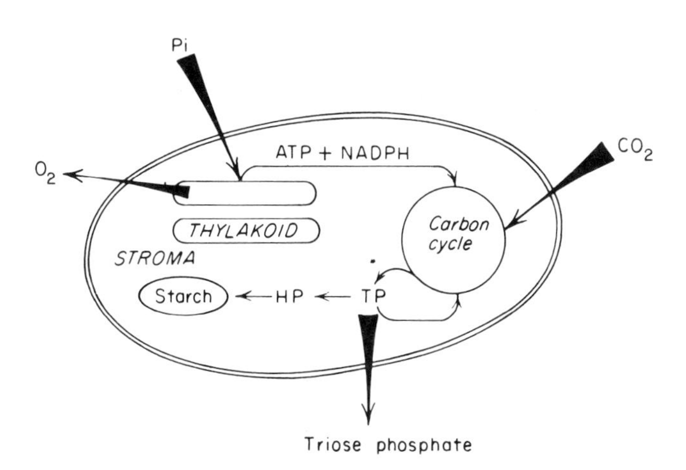


**Fig. 8.2.** Movement of principal metabolites across chloroplast envelopes in spinach. Carbon dioxide and orthophosphate enter and triose phosphate is exported. Oxygen escapes. Reduction of $NADP^+$ and formation of ATP from $ADP^+$ is associated with the thylakoids which house the pigments and other constituents of the photochemical apparatus. The RPP pathway or carbon reduction cycle is located in the stroma. Triose phosphate (TP) is made in the cycle and is available for export, recycling, and starch synthesis via hexose phosphate (HP).

## 8.2 The experimental basis of transport studies: fractionation of whole tissues

Radioactive carbon has been used extensively in the study of metabolic pathways and it might seem that the simplest way of studying transport might be to feed

**Table 8.1**

A. Information on distribution and flow of compounds *in vivo*
  I. Nonaqueous cell fractionation
    (a) Kinetic measurements of substrate levels and substrate distribution
    (b) Kinetic measurements of tracer distribution
  II. Aqueous cell fractionation
    Measurements of chloroplast contents in intact chloroplasts after 'fast' isolation

B. Information on metabolite distribution and flow *in vitro*
  I. Direct methods using aqueously isolated chloroplasts
    (a) Distribution of metabolites or tracer between chloroplasts and medium
    (b) Measurements of substrate concentration after centrifugation of chloroplasts through silicone oil
  II. Indirect methods using aqueously isolated chloroplasts
    (a) Response of chloroplast metabolism to the addition of metabolites
    (b) Measurement of the activity of 'cryptic' enzymes
    (c) Osmotic response of chloroplasts to additives

C. Distribution of metabolites, or tracer, between chloroplast and cellular environment using isolated protoplasts

$^{14}CO_2$ to an intact leaf and then macerate the tissue and see what is released from the chloroplast in the way of photosynthetic product. In practice this is more difficult than it seems for two main reasons. Firstly, a transport metabolite must ultimately be used in further metabolism and, since processes such as the RPP pathway and glycolysis share common intermediates, it is clearly difficult to decide whether compound X has been transported from chloroplast to cytoplasm, as such, or whether it is converted into Y for transport and back to X in the first stage of subsequent metabolism. Secondly, chloroplasts and other sub-cellular organelles are mostly very small and very fragile. If the contents of a shopping basket were put through a mincing machine, it would not be easy, without prior knowledge and the ability to read, to decide precisely what had been originally contained within each packet. Many cellular fractionation procedures are equally brutal. If, however, leaves are allowed to photosynthesize in the presence of $^{14}CO_2$ and then quickly frozen in liquid nitrogen prior to fractionation in non-aqueous solvents, there are reasonable grounds for believing that what was once associated with the chloroplast will remain associated with the chloroplast. The fact that some starch is found in extra-chloroplastic fractions, when this procedure is applied, indicates that it is by no means exact but it can nevertheless provide valuable indications of metabolite distribution. [Starch is one of the exceptional compounds which may be readily

**Plate 8.1a** Starch Pictures
In the winter of 1864 Sachs, the famous German plant physiologist, re-illuminated a starch-free leaf which was partly obscured by tin-foil and showed (by the starch–iodine reaction) that starch was only re-formed in the illuminated chloroplast. [In experiments of this sort leaves are freed of starch by keeping them in the dark for 24–48 hours. After illumination they are killed in boiling 80% ethanol, which extracts chlorophyll, and the white leaf tissue is washed in water and stained in a solution of iodine in potassium iodide. The starch-iodine complex is really blue (amylase is stained blue and amylopectin is stained purple) but in high concentration appears black]. Molisch did similar experiments in the 1920's but re-illuminated the leaf through a photographic negative and obtained a recognizable picture. Starch pictures of great intrinsic beauty can also be prepared in this way. The picture in Plate 8.1a which depicts 'Innocence' was first drawn in chalk by Paul-Pierre Prudhon (1758–1823). Here it is executed in starch in a *Pelargonium* leaf—a process which may even be thought to enhance its intrinsic quality.

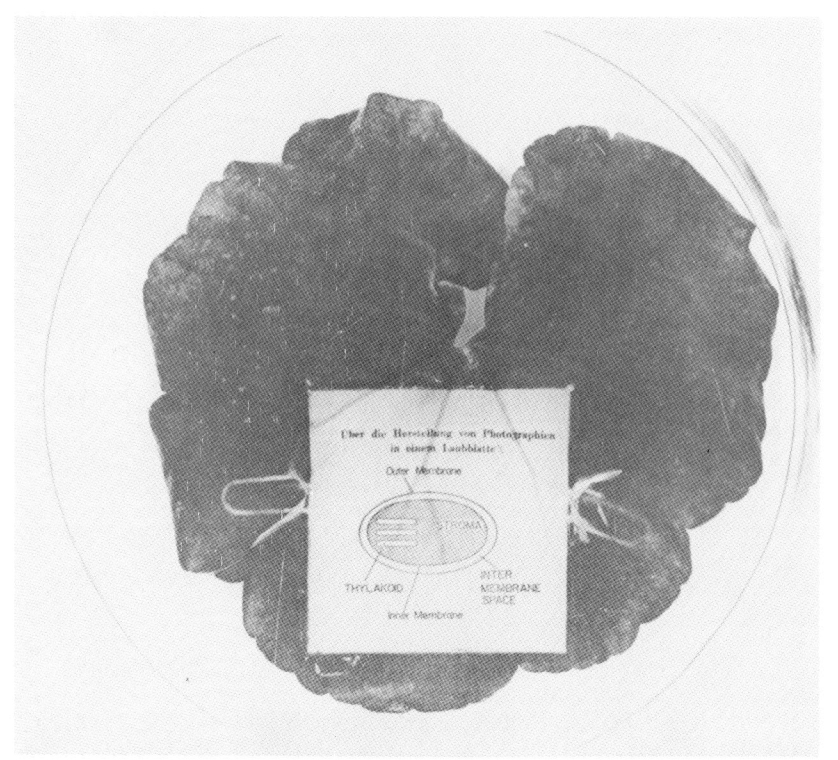


**Plate 8.1b** shows a diagram of Heldt's beneath a facsimile of the title of Molisch's original paper. The resolution achieved is what excites most interest at the present time. This is because the major export metabolite from the chloroplast is triose phosphate and the major transport metabolite which enters the vascular tissues is sucrose. If leaf discs are floated on sucrose solutions in the dark they produce starch uniformly throughout the tissue, i.e. light is not required for starch synthesis, only for photosynthesis. It may then be asked why internally generated sucrose which must move, in starch-picture experiments, through immediately adjacent darkened cells on its way to the vascular tissues, produces no blurring of the image. Sucrose itself can not re-enter the chloroplast but its degradation products must if externally applied sucrose can initiate starch synthesis in the dark. The problem may be simply a question of concentration. It may be that externally applied sucrose enters a different compartment to internally generated sucrose. Whatever the answer, it is of considerable relevance to metabolite transport. This would clearly be most effective if the transport metabolite were protected in some way, from degradation until it reached its destination. How this protection is achieved in leaves will only be known when it is possible to say why starch pictures can be made to exhibit such remarkable definition. Successive enlargements (**8.1c** & **d**) of the starch-picture allow us to see that the 'cut-off point' is at the cellular level, i.e. starch is formed profusely in illuminated cells but not at all in immediately adjacent cells. In the highest magnification the guard cells of the stomata (arrows) are clearly visible in between the dots made by the draughtsman in the original diagram, because these do not lose their starch in the dark and remain as minute pairs of black marks or 'parentheses' scattered at random. These provide a scale, one guard-cell long, against which the discrete nature of starch formation may be judged.

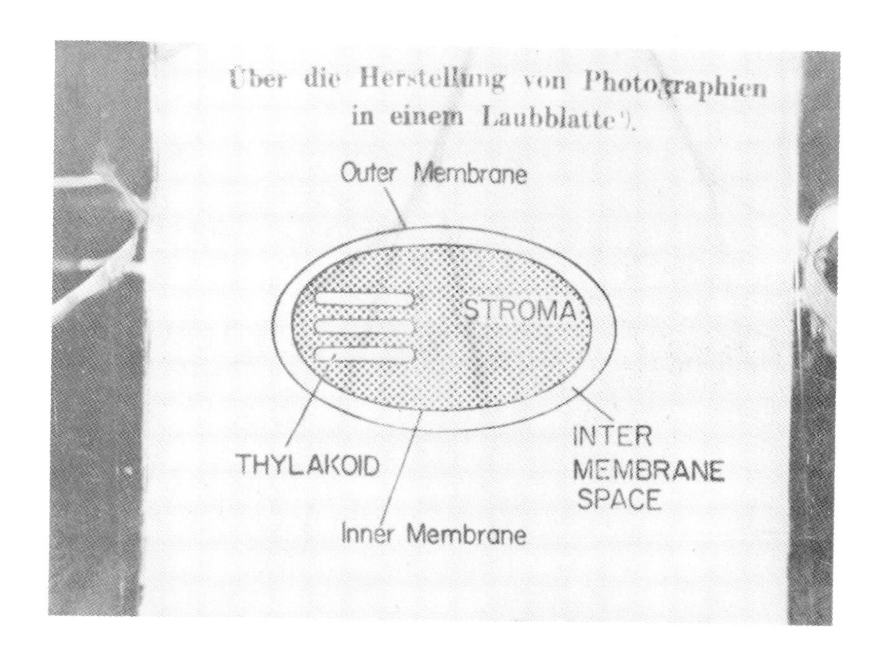

**Plate 8.1 c & d** are successive enlargements of this starch picture.

located by visual examination (Plates 8.1a, b, c, d). If the chlorophyll is extracted from a leaf in hot ethanol and the white tissue then stained with iodine, starch is readily seen, usually as blue/black grains, resulting from the amylose–iodine reaction. In the leaf these grains are restricted to the plastids]. Rapid fractionation in aqueous solvents is of more limited use but again it may suggest the in-vivo location of metabolites which are not immediately leaked from ruptured organelles. [Refs. 10, 11, **42, 50, 51, 62, 64**].

## 8.3 The experimental basis of transport studies *in vitro*

An alternative approach to the fractionation of whole tissues is to isolate functional organelles, to see what they produce *in vitro* and what they export to the surrounding medium.

The isolation of functional chloroplasts is not in itself an easy task (Appendix A). For example, nearly 30 years were to elapse before $CO_2$-dependent oxygen evolution

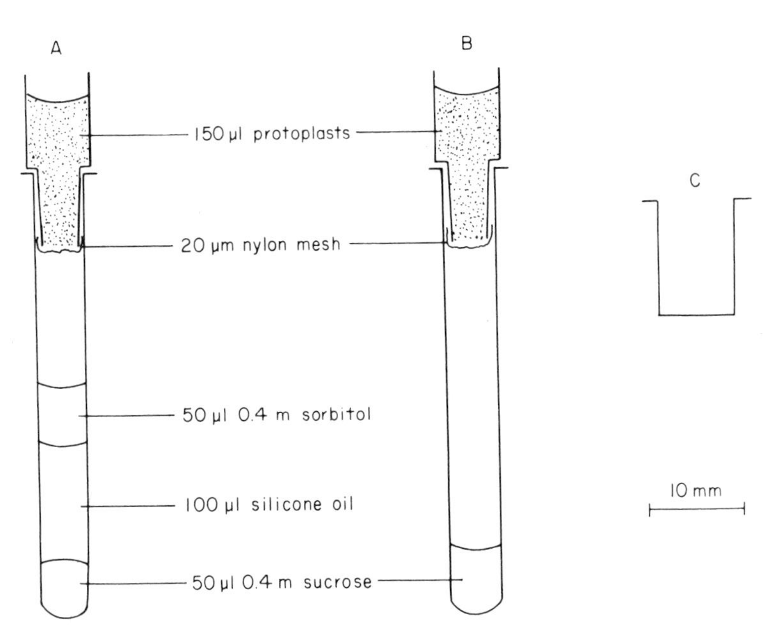


**Fig. 8.3.** A new technique devised by Robinson for the rapid separation of cytoplasm and chloroplasts from isolated protoplasts. The latter are suspended in the top container where they may be illuminated or otherwise pre-treated. At zero time they are ruptured by centrifugation though a 20 $\mu$m net. The cytoplasm is held above the silicone oil, the chloroplasts filtered through it into a denaturing solution. This technique will therefore allow rapid examination of the partitioning of metabolites between these two fractions during photosynthesis or respiration.

by isolated chloroplasts was demonstrated at rates comparable to the release of $O_2$ observed in the presence of artificial hydrogen acceptors by Robert Hill in the 1930s. Moreover chloroplasts are fragile organelles and the presence of chloroplasts with ruptured envelopes and the possibility of continuing rupture during assay inevitably adds to the difficulties of experiment and interpretation. Recently the availability of isolated protoplasts (Appendix A) and the development by Robinson of a neat technique for the rapid separation of chloroplasts from other cellular fractions has greatly extended the usefulness of in-vitro studies (Fig. 8.3). The direct methods used in such studies may be summarized as follows (Sections 8.4, 8.5).
[Refs. **34, 35, 45, 48, 49, 63**]

## 8.4 Chromatographic analysis

The RPP pathway was itself based on chromatographic analysis of algae following feeding with $^{14}CO_2$ (Section 6.7). Bassham and his colleagues in Berkeley, California (particularly Martha Kirk, who also played an important role in the earlier work) have applied virtually the same approach to isolated chloroplasts (Plate 8.2). Like many other procedures applied to isolated chloroplasts its principal disadvantage (apart from its laborious nature) is that it does not readily distinguish between events which occur within the chloroplast and those which occur in the external medium as a consequence of damage to the plastid envelope.
[Refs. **16, 17**]

## 8.5 Centrifugal filtration

This procedure was first devised by Werkheiser and Bartley and used extensively with mitochondria by Klingenberg and his colleagues. It has been applied with great success to chloroplasts by Heldt (Section 8.23). Usually it involves loading chloroplasts with a radioactive metabolite and observing the redistribution of label between the plastid and a bathing solution after the chloroplasts are pulled, by centrifugal force, through a filtering layer of silicone oil into a denaturing agent such as perchloric acid (Fig. 8.4). The addition of various unlabelled metabolites also allows the extent of back-exchange to be studied (e.g. $^{32}Pi$ can be exchanged out of the chloroplasts by triose phosphate on the phosphate transporter). In cases where the labelled compound may be metabolized (e.g. feeding $^{32}Pi$ to chloroplasts in the light at ambient temperature), column chromatography can be used to determine the distribution of products. The small quantities of medium which adhere to the chloroplasts despite passage through the silicone oil may be corrected for by

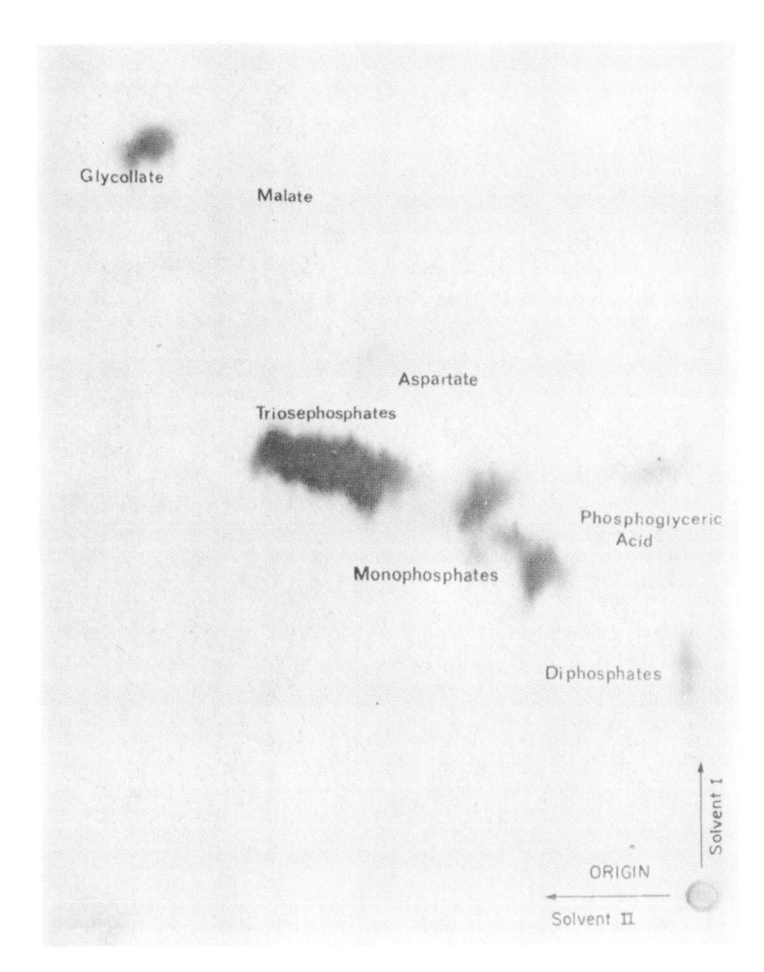


**Plate 8.2.** Chromatography
Two dimensional radioautograph of products of short-term $^{14}CO_2$ fixation by isolated pea chloroplasts. The reaction mixture is acidified after illumination to stop the reaction and liberate $^{14}CO_2$ from bicarbonate. A sample is spotted on to the origin and a mixture of phenol/water is allowed to flow through the paper by capillary action. The paper is then removed from the chromatography tank, dried, turned through 90° and run in butanol/acetic acid in the second direction. This distributes the compounds across the paper, each running to a finite area determined by its characteristics and concentration. Precise location is achieved by co-chromatography with known standards. The position of each compound can be determined by placing the chromatography paper against a piece of X-ray film in a dark room. The radiation 'exposes' the film which shows dark patches (as above) when it is developed. Radiation in spots can be 'counted' directly by applying a thin end-window Geiger counter to the paper or, more accurately, by elution and/or scintillation counting. Two-dimensional paper chromatography of this sort was first devised by Martin & Synge and, with the advent of $^{14}C$, made possible the elucidation of the RPP pathway by Calvin, Benson, Bassham and others (Chapter 6).

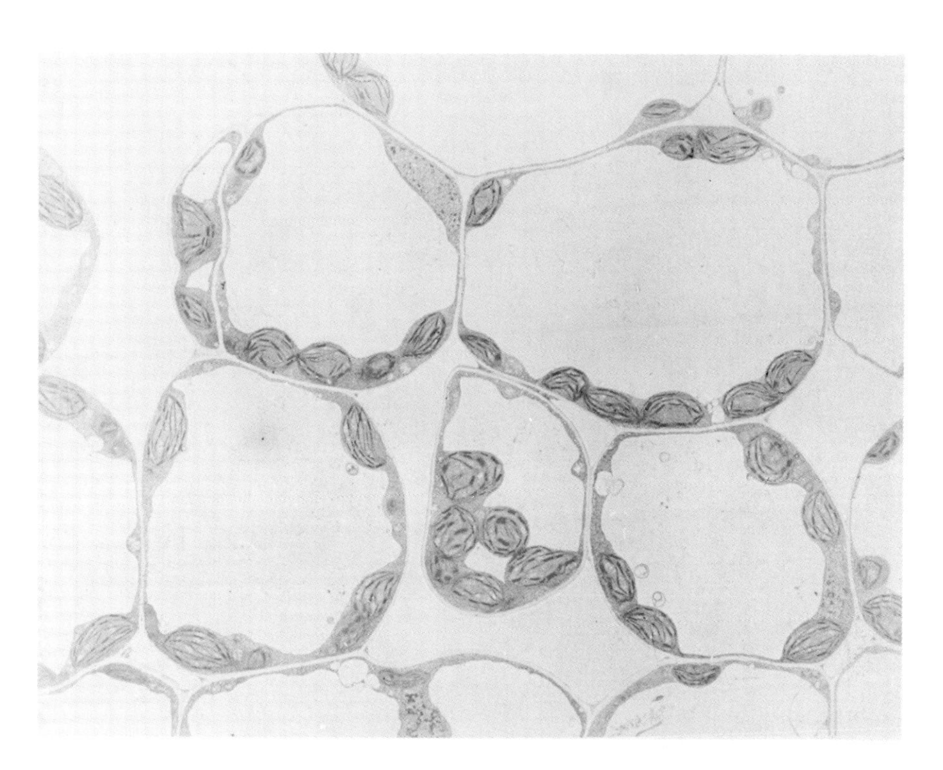

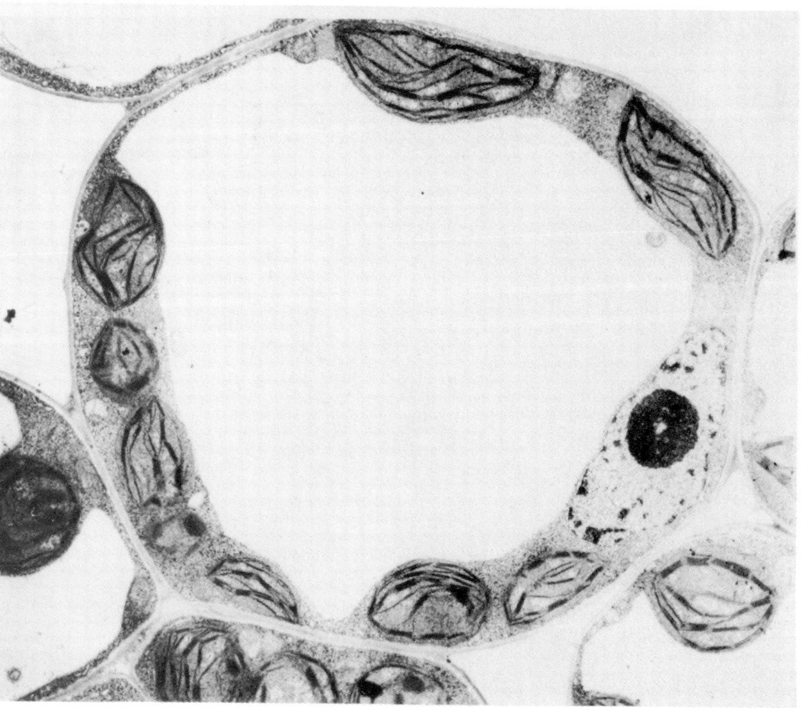

**Plate 8.3.** The chloroplast transports metabolites to its immediate cellular environment. These electron micrographs of spinach mesophyll cells vividly illustrate the relative volumes of chloroplast, cytosol and the large vacuole. (Mag x 6400) (Courtesy of A.D. Greenwood, Imperial College, London.)

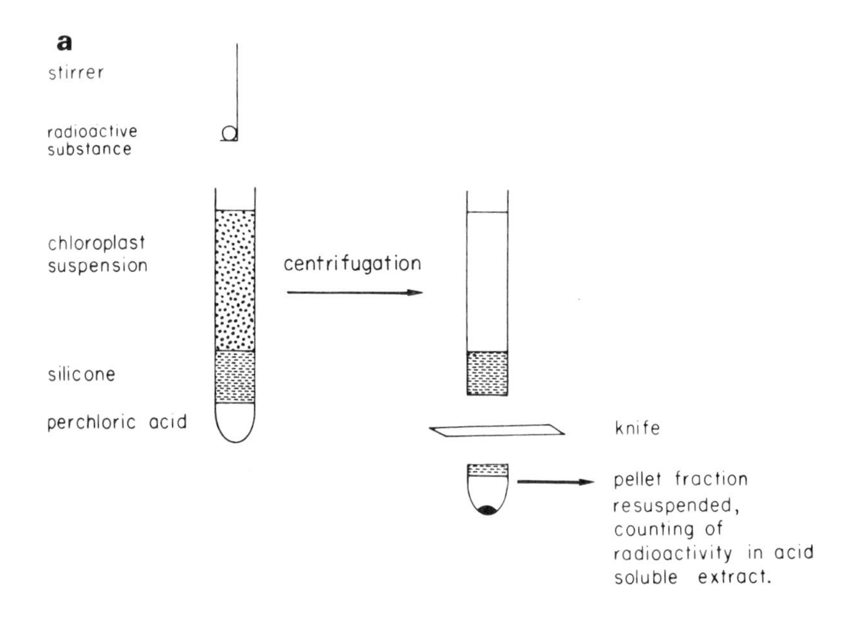


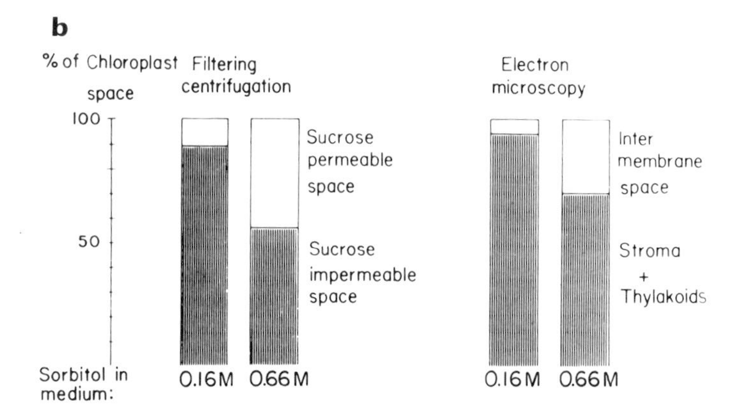


**Fig. 8.4.** Centrifugal Filtration (from Heldt, 1976)
**a** Centrifugation Procedure
**b** Comparison of inter-membrane space derived by feeding labelled sucrose (left) and planimetry of electron micrographs of isolated chloroplasts (right). The uptake of labelled sucrose is much greater than that of dextrans and varies with the osmotic pressure exerted by the external medium. The extent of non-specific penetration by sucrose (and other compounds) indicates that this molecule permeates the outer-envelope but is unable to cross the inner-envelope.

undertaking parallel experiments using labelled compounds, such as dextrans, which do not penetrate membranes (see Section 8.22).
[Refs. 6, **32, 37, 66**]

## 8.6 Indirect methods of following metabolite movement

These are based on attempts to influence the behaviour or function of chloroplasts by adding reagents to the external medium. For example, if a compound is seen to

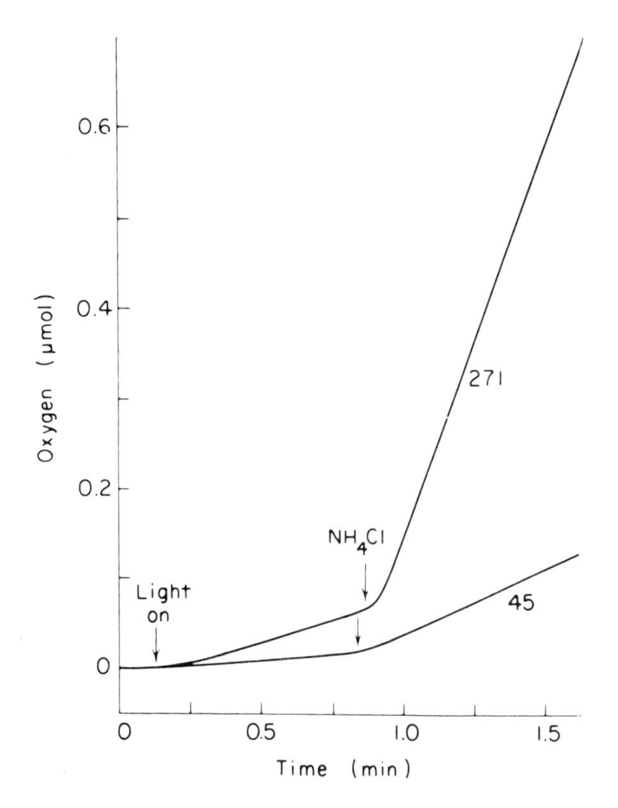


**Fig. 8.5.** Ferricyanide-dependent $O_2$ evolution by 'intact' and envelope-free chloroplasts. 'Intact' chloroplasts were added to both reaction mixtures but the addition of other reagents was so ordered that the chloroplasts in the mixture which gave the upper trace were briefly subjected to osmotic shock. At the onset of illumination therefore the only difference between the two mixtures is that all of the chloroplasts in one have lost their limiting envelopes. This is used as a measure of chloroplast integrity (Appendix A) but, in this context, shows that the reduction of ferricyanide is limited in relatively intact chloroplasts because it does not cross the envelopes. Conversely $NH_3$ crosses readily and brings about an acceleration by uncoupling electron transport (from Lilley *et al*, 1971).

accelerate or inhibit photosynthesis there is a strong implication that it must penetrate, particularly if its action can be related to events which are known to occur within the stromal or thylakoid compartments (Fig. 8.5). The disadvantages of this approach relate to the presumptive nature of interpretation. Clearly some additives may influence photosynthesis without penetrating, others may be changed prior to penetration.
[Ref. **41**]

## 8.7 Shortening of induction, reversal of orthophosphate inhibition

As already noted (Chapter 7), photosynthesis does not start immediately upon illumination but only after a lag which can be extended by the presence of above optimal Pi. Compounds such as PGA and triose phosphates which shorten

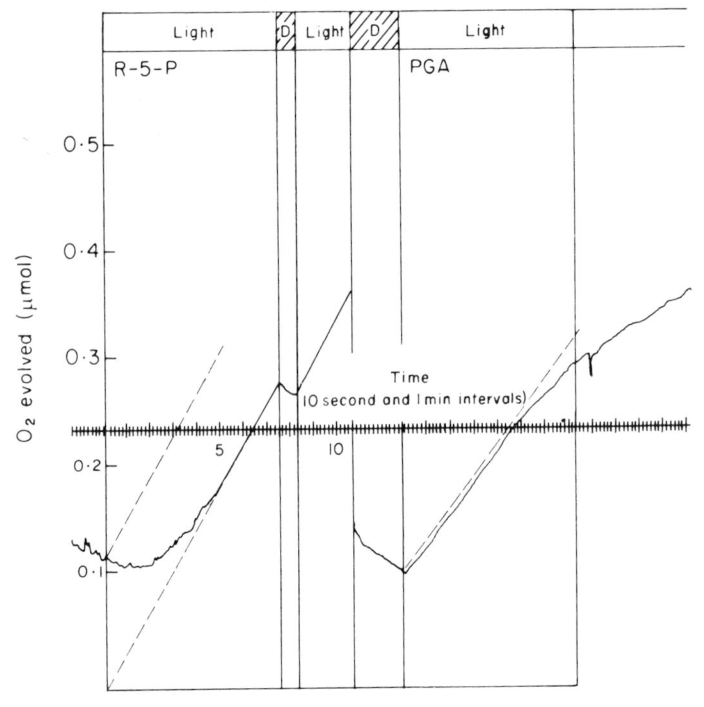


**Fig. 8.6.** The first recording of lag-shortening by the addition of 3-phosphoglycerate to isolated chloroplasts. *Left:* Time-course of $CO_2$-dependent $O_2$ evolution by isolated chloroplasts. *Right.* Time-course of a similar mixture containing PGA. It was assumed that PGA produced this result by internal reduction. Ribulose bisphosphate, which does not cross the envelope, has no corresponding effect on $CO_2$-fixation, nor will it support $CO_2$-fixation by chloroplasts in the dark unless the limiting envelopes are ruptured. (see Walker & Hill, 1967)

induction (see Figs. 8.6 & 7.15) or reverse Pi inhibition are therefore thought to cross the chloroplast envelope.
[Ref. **59**]

## 8.8 Addition of inhibitors

Unless an inhibitor affects photosynthesis by interfering with metabolite transfer it may be assumed to enter the chloroplast. On this basis the simple sugar DL-glyceraldehyde, which is a potent inhibitor of carbon assimilation, is believed to cross the envelope very readily (see also Fig. 8.14).
[Ref. **53**]

## 8.9 Catalysis by intact and ruptured chloroplasts

Stromal enzymes are 'cryptic' in the sense that they are separated from the surrounding medium by the chloroplast envelopes. The failure of a substrate to react (such as the lack of carboxylation in the presence of ribulose bisphosphate) implies lack of penetration. Nevertheless, it must be noted again that other factors may be involved. Thus a reconstituted chloroplast system (essentially ruptured chloroplasts plus additives such as ferredoxin) will evolve $O_2$ readily in the presence of PGA and 1 mM $MgCl_2$ indicating that all of the necessary photochemical events and the enzymes concerned in the reduction of PGA to triose phosphate work effectively under these conditions (Fig. 9.4). With RBP and $CO_2$ as substrates, however, there is virtually no $O_2$ evolved until the $MgCl_2$ concentration is raised to 2 mM. In this instance, envelope rupture makes external RBP accessible to the carboxylase but it also dilutes the stromal $Mg^{2+}$ to the extent that it is no longer able to bring about much activation of RBP carboxylase under these experimental conditions. [Intact photosynthesising chloroplasts are also inhibited by an ionophore which renders the envelope permeable to $Mg^{2+}$ and activity can then be restored by increasing the $Mg^{2+}$ in the external solution].
[Refs. **41a, 45, 58**]

## 8.10 Osmotic volume changes

In some circumstances chloroplasts behave as almost perfect osmometers. In theory, if a space is enclosed by a truly semi-permeable membrane (one which will permit the passage of water molecules but not the transfer of substances in solution) then it should undergo shrinking and swelling according to the relative concentration of the internal and bathing solutions. Thus if the internal solute concentration is high and the external concentration is low, then water will tend to move inwards (endosmosis) so that it dilutes the more concentrated solution. This influx of water causes swelling. Conversely, exposure to a hypertonic solution (an external solution which is more concentrated and therefore exerts a greater osmotic pressure) will cause exosmosis of water and shrinkage (Section 4.1). The fact that chloroplasts may be successfully isolated in solutions containing sucrose or sugar alcohols, such as sorbitol, as osmotica implies that these compounds do not themselves readily penetrate the chloroplast envelope. (Osmotica are supposedly relatively unreactive compounds which are added with the express intention of preventing osmosis. A compound which entered the chloroplast under its own diffusion gradient would increase the internal osmotic pressure inducing endosmosis and swelling).

Osmotic swelling or shrinkage gives rise to changes in apparent absorbance at 535 nm and may therefore be used as a measure of penetration. Thus hypertonic sorbitol causes an increase in absorbance indicating exosmosis of water. Ribose causes a similar increase but this is followed by fairly rapid recovery suggesting that the initial efflux of water is followed by re-entry as the ribose penetrates the chloroplast envelope (Fig. 8.7). Recently, Heber and his colleagues have used an ingenious variant of this approach which has allowed them to draw conclusions concerning single ions (Section 8.21).
[Refs. **24, 30**]

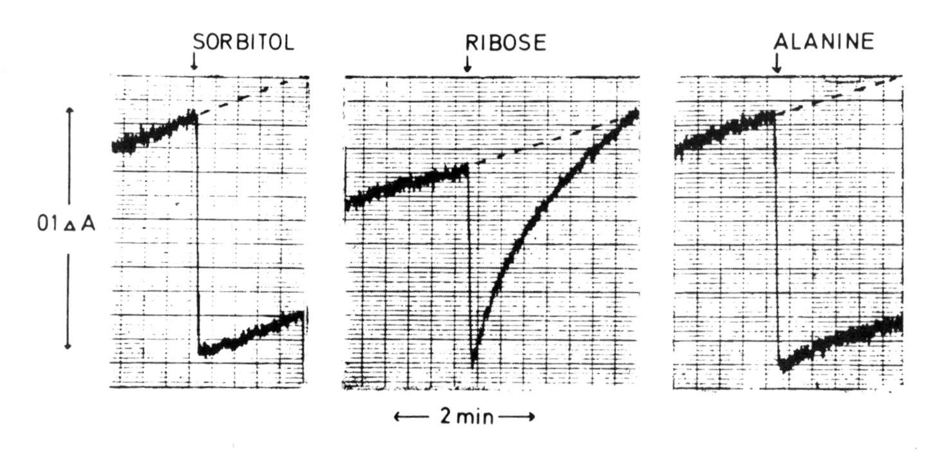


**Fig. 8.7.** Changes in absorbance as an indication of penetration. At appropriate concentration sorbitol, ribose and alanine all cause chloroplasts to shrink as a result of exosmosis of water and this volume change brings about a rapid increase in absorbance. In the presence of ribose, however, there is a fairly rapid recovery indicating that, in this case, exosmosis is followed by endodiffusion of ribose and consequent increase in turgor. (Gimmler *et al*, 1974)

## 8.11 The movement of metabolites

As indicated above (Section 8.1), the principal carbon traffic between the $C_3$ chloroplast and its cellular environment is now believed to be based on $CO_2$ import and triose phosphate export. [Transport of metabolites in $C_4$ photosynthesis is considered separately in Chapter 12].

## 8.12 Carbon dioxide

Carbon dioxide influx is self-evident in the sense that if photosynthesis is accurately represented by Equation 1.1 then $CO_2$ must reach the site of carbon assimilation at the rate needed to maintain this process and that if the choloroplast stroma is indeed the site of carboxylation then $CO_2$ must pass freely across the envelope (Fig. 8.8) in one form or another. The concept of the chloroplast as a self-sufficient photosynthesizing organelle is one which has been in dispute for many years. In the last century the chloroplast came to be regarded as the site of photosynthesis because starch was regarded as the end-product of this process. Sachs was able to demonstrate that starch formation in green leaves was restricted to chloroplasts in those parts of a leaf which were illuminated (cf. Plate 8.1). When Hill isolated chloroplasts, in the 1930s, which were able to evolve $O_2$ but which were not apparently able to fix $CO_2$ the whole question of chloroplast self-sufficiency was reopened and it was not until the 1950s when Arnon and his colleagues demonstrated $^{14}CO_2$ fixation by isolated chloroplasts that the matter appeared to be finally resolved. More recently it has been suggested that the chloroplast can only photosynthesize at maximum rates when it is incubated in a reaction mixture which simulates the action of the cytoplasm. [According to this view, rapid exchange of internal triose phosphate with external Pi is necessary if full photosynthesis is to be achieved and, in this regard at least, the chloroplast is not fully self-sufficient]. Despite these doubts and arguments it seems quite clear that RBP carboxylase is located within the stroma and that by far the greatest proportion of all $CO_2$ fixed by $C_3$ plants in the light is incorporated at this site.

Dissolved $CO_2$ undergoes hydration to carbonic acid

$$CO_2 + H_2O \rightleftarrows H_2CO_3 \rightleftarrows H^+ + HCO_3^- \qquad \text{Eqn. 8.1}$$

in a reaction catalysed by various inorganic ions such as Pi and by the enzyme carbonic anhydrase. For this reason it is not easy to decide which particular molecular species of $CO_2$ is actually used. However, Cooper *et al* have carried out experiments with RBP carboxylase in which fixation has been compared in mixtures containing either $H^{14}CO_3^- + {}^{12}CO_2$ or ${}^{14}CO_2 + H^{12}CO_3^-$ and on this basis have

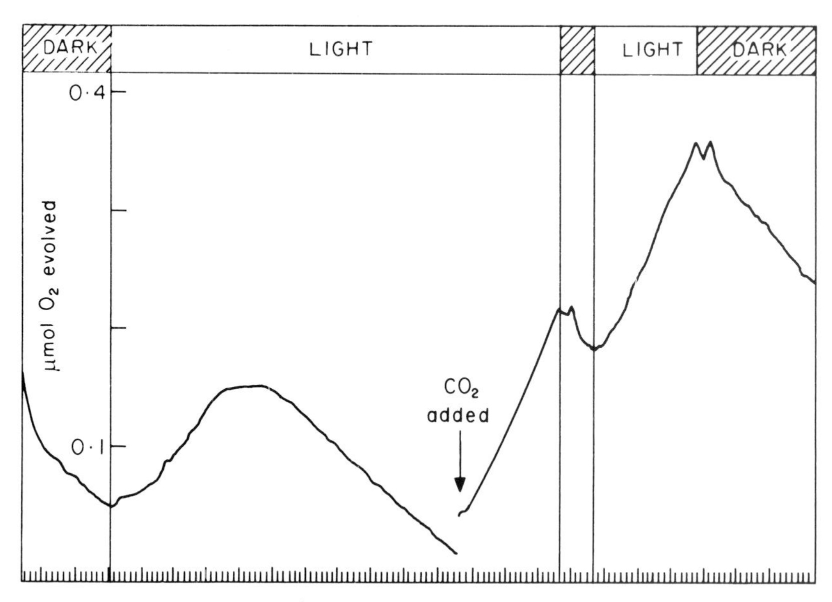


**Fig. 8.8a.** First demonstration of $CO_2$ dependence in photosynthetic carbon assimilation by isolated chloroplasts (cf. Fig. 7.13). The $O_2$ uptake in the dark was largely artefactual and so were some of the peaks in the trace. Nevertheless it can be seen that $O_2$ evolution soon ceased in the absence of $CO_2$ (initially omitted from the reaction mixture) and recommenced immediately after its addition. Light dependence is also shown. (see Walker & Hill, 1967)

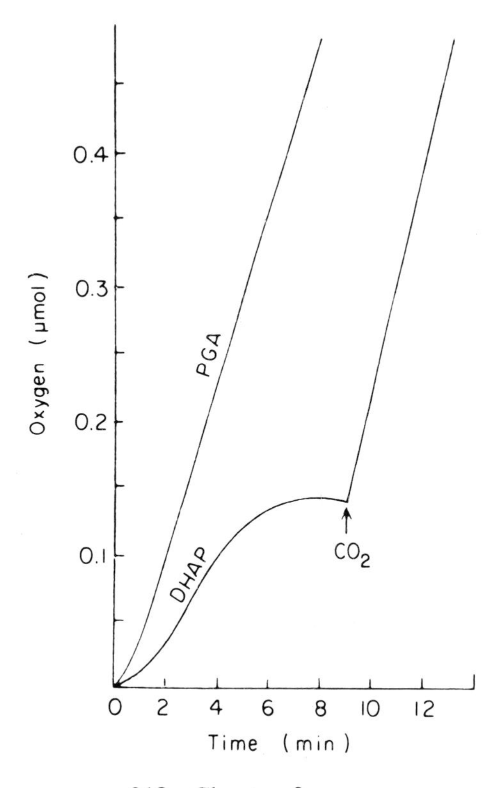


**Fig. 8.8b**

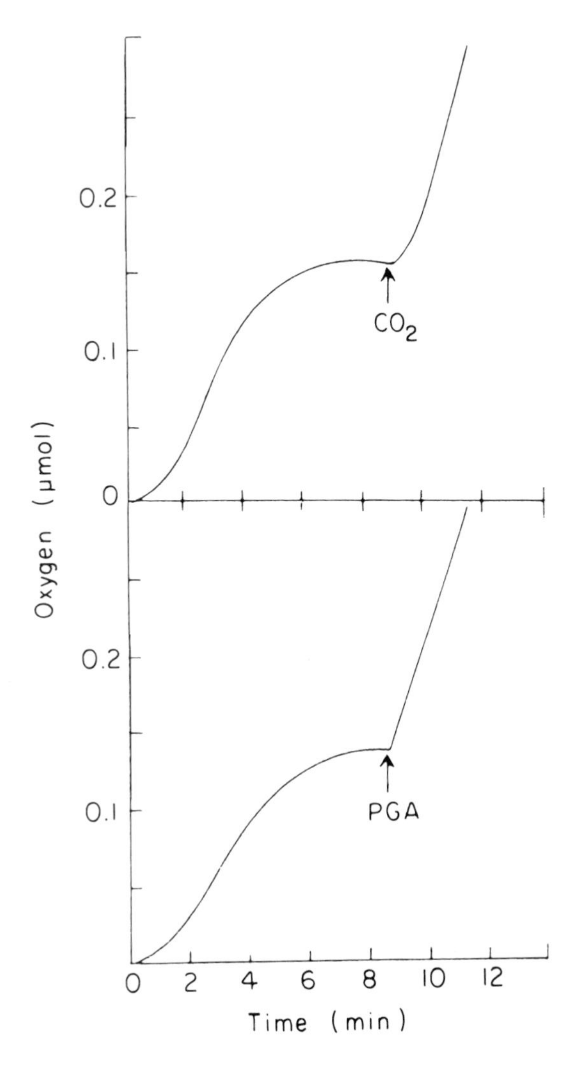


**Fig. 8.8b & c.** Similar experiments, using improved techniques, but also showing that $O_2$ evolution can be started by PGA in the absence of $CO_2$ (from Walker *et al*, 1967).

concluded that it is $CO_2$ rather than bicarbonate which is utilized in this reaction. [The converse is probably true of the reaction catalysed by PEP carboxylase—see Section 14.4]. Poincelot used the same approach with intact chloroplasts and concluded that $HCO_3^-$ was taken up much more rapidly than $CO_2$. On the other hand Werdan and Heldt found that the accumulation of bicarbonate within the chloroplast is in accord with the pH gradient across the chloroplast envelope and the Henderson–Hasselbach equation (an equation which relates the distribution of $CO_2$ between its various ionic forms to the prevailing pH). They concluded that this behaviour was consistent with free and rapid movement of $CO_2$ (but only limited movement of bicarbonate) as follows:

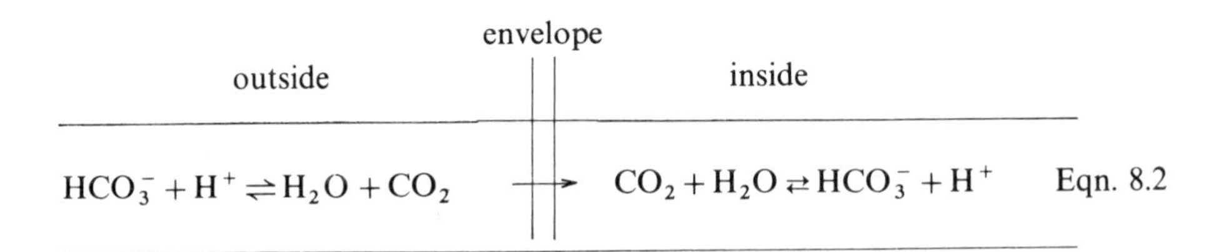


Similarly, Heber and Purczeld showed that bicarbonate caused an irreversible osmotic shrinkage of chloroplasts (implying lack of ready penetration). Moreover they were able to confirm the observation of Werden *et al* that carbonic anhydrase behaves as a cryptic enzyme (the rate of hydration of $CO_2$ was not greatly accelerated by intact chloroplasts but was rapid in mixtures containing chloroplasts with ruptured envelopes) (see Figs. 8.8b & c).

The role of carbonic anhydrase in the chloroplast stroma remains obscure. It is possible that it facilitates $CO_2$ transport *within* this compartment by building up a large pool of bicarbonate (98% of the total $CO_2$ is present as bicarbonate at pH 8.0; the pH of the stromal compartment in the illuminated chloroplast (Section 5.9).

$$\begin{array}{ccc} CO_2\,(2\%) & \xrightarrow[\text{transport}]{\text{slow}} & CO_2\,(2\%) \\ \downarrow\uparrow & & \uparrow\downarrow \\ HCO_3^-\,(98\%) & \xrightarrow[\text{transport}]{\text{fast}} & HCO_3^-\,(98\%) \end{array} \qquad \text{Eqn. 8.3}$$

Both $CO_2$ species would diffuse at rates proportional to their concentrations and advantage would therefore derive from the large bicarbonate pool even though RBP carboxylase utilizes $CO_2$.

[Refs. **13, 14, 21, 30, 34, 44, 51, 54, 61, 65**]

## 8.13 Triose phosphates and 3-phosphoglycerate

Intact chloroplasts will evolve $O_2$ at rapid rates when PGA is added in the absence of $CO_2$ (Fig. 8.8 b, c). This indicates that PGA can act as a Hill oxidant in the overall reaction

$$\begin{array}{ccccc} H_2O & & NADP & \leftarrow & G3P + ADP + Pi \\ \downarrow & & \uparrow\downarrow & & \uparrow \\ \frac{1}{2}O_2 & & NADPH_2 & \rightarrow & PGA + ATP \end{array} \quad \text{Eqn. 8.4}$$

and that in order to do this it must enter the chloroplast at rates commensurate with the observed $O_2$ evolution. Similarly both PGA and the two triose phosphates (glyceraldehyde 3-phosphate and dihydroxyacetonephosphate) will shorten induction and reverse Pi inhibition (Sections 7.11 & 7.12). The latter observation led Heldt and his colleagues to undertake the direct measurements (Section 8.5) which permitted the formulation of the Pi translocator in its present form (see Sections 7.12 & 8.23). All three compounds move, via the translocator at rates which may equal (or exceed) the rate of photosynthesis at its fastest.
[Refs. 6, **32, 55, 56**]

## 8.14 Orthophosphate and inorganic pyrophosphate

Exogenous Pi readily restores photosynthesis by isolated chloroplasts when this has declined to a low level as a consequence of Pi depletion. When this occurs, as little as 0.05 $\mu$mol in 2 ml (Fig. 8.9) restores $CO_2$-dependent $O_2$ evolution to a relatively high level. The stoichiometry of the response is then approximately 3 $O_2$ evolved for every Pi added, a ratio which is consistent with the formation of one triose phosphate from 3 molecules of $CO_2$ and one molecule of Pi. When this response was first studied in spinach chloroplasts, inorganic pyrophosphate (PPi) was found to be equally effective although there was a slight but significant lag (of about 15 s) before the response was seen. The stoichiometry was then approximately 6 $O_2$ to 1 PPi, implying that the PPi probably underwent complete hydrolysis. This was a very puzzling response in view of the known inhibition of photosynthesis by Pi because as the PPi concentration was increased no corresponding inhibition was observed. Indeed, not only did PPi fail to inhibit, but at high concentrations (say 5 mM) it also offered substantial protection against Pi inhibition (Fig. 8.10). Partial resolution of these seemingly contradictory results came with the observation that with some chloroplast preparations the response to PPi was delayed, or even entirely absent

(Fig. 8.11). In such circumstances it could be restored by the inclusion in the reaction mixture of a small quantity of osmotically shocked chloroplasts or by the addition of inorganic pyrophosphatase prepared from chloroplast stroma. Similarly, there was no response to added PPi in the absence of external $Mg^{2+}$ but when $Mg^{2+}$ was added subsequently the response was more or less immediate, particularly in the presence of added PPiase. By this time it had been established that the substrate for PPiase was Mg-PPi and that anionic PPi was not only ineffective as a substrate but even inhibited hydrolysis of Mg-PPi. Thus it seemed evident that PPi could not



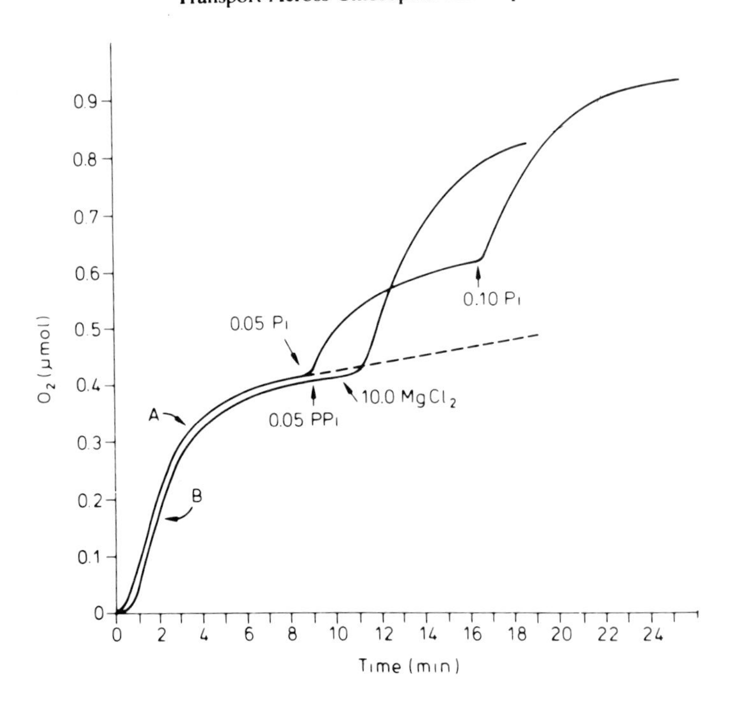


**Fig. 8.9.** Response of isolated spinach chloroplasts to exogenous phosphate. Reaction mixtures initially contained no added Pi but endogenous Pi was sufficiently high to permit detectable photosynthetic $O_2$ evolution which fell to a level commensurate with starch synthesis as the stromal Pi was consumed. Photosynthesis was then restored by Pi with a stoichiometry of approximately 3 molecules $O_2$ evolved per molecule of Pi added. [This is in accord with the overall equation $Pi + 3CO_2 + 2H_2O \rightarrow$ triose phosphate $+ 3O_2$ and measurements with $^{14}C$ and $^{32}P$ which identify triose phosphates as major products of short-term photosynthesis by intact chloroplasts]. PPi is ineffective in the absence of Mg but, in its presence, undergoes external hydrolysis by a stromal pyrophosphatase released by ruptured chloroplasts (See Fig. 8.11). (Walker & Herold, 1977)

penetrate the chloroplast envelope at an appreciable rate but had to undergo external hydrolysis by PPiase released from ruptured chloroplasts in order to substitute for Pi. Since the quantities of PPiase in the medium were normally low, and since free Pi was inhibitory, the rate of hydrolysis would not be substantially increased in the presence of high [PPi]. Subsequently, direct measurements confirmed that PPi penetration was slow and showed that PPi interfered with the action of the Pi translocator (Section 8.23) thus offering a degree of protection against Pi inhibition. It is for this reason that optimal rates of photosynthesis by isolated chloroplasts are most easily observed in the presence of 5 mM PPi. In the

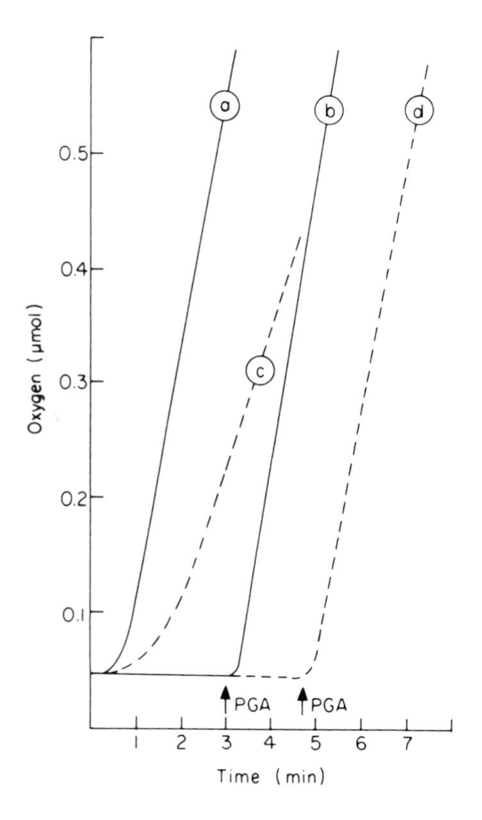


**Fig. 8.10.** Protection by PPi against Pi inhibition of photosynthesis. Pi inhibits photosynthesis in isolated chloroplasts by enforcing the export of intermediates of the RPP pathway via the Pi translocator (Section 8.23). Curves 'b' & 'd' show complete inhibition by Pi and reversal of this inhibition by PGA as indicated. A large excess of PPi (curve 'a') did not cause inhibition. When both PPi and Pi were present (curve 'c') the PPi exerted a large measure of protection against Pi inhibition (Ref. 12)

presence of external $Mg^{2+}$ the PPi acts as 'Pi-stat', releasing Pi by hydrolysis at about the rate that it is consumed by the chloroplast and at the same time avoiding the Pi inhibition which is readily observed if the [Pi] is increased above about 0.5 mM (Section 8.7). In addition PPi may also protect the chloroplast by chelating divalent cations (although the precise nature of this deleterious effect of cations has yet to be established).

In pea chloroplasts (or possibly in chloroplasts isolated from immature tissue) response to PPi is further complicated by a multiplicity of effects. Thus PPi may exchange with internal ATP via the adenylate translocator (Section 8.25) thereby producing an inhibition (Fig. 8.12). Robinson and Wiskich suggest that PPi may also enter via the Pi translocator (Section 8.23) and, after internal hydrolysis, diminish loss of intermediates by competing for efflux with accumulated PGA. This would

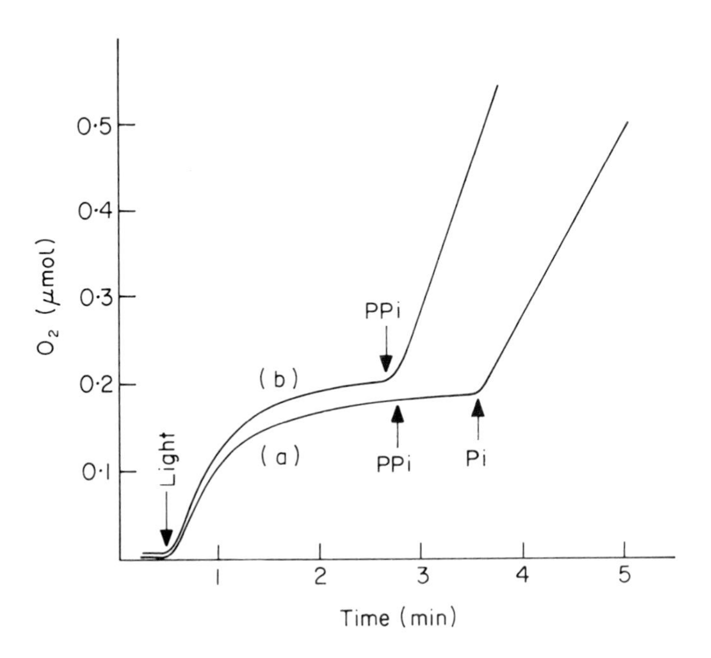


**Fig. 8.11.** Failure of PPi to cross the chloroplast envelope. If chloroplasts are prepared which show a high percentage intactness then Pi, but not PPi, will correct a Pi deficiency (Curve a). The inclusion of a small fraction (5%) of ruptured chloroplasts (Curve b) will release sufficient pyrophosphatase to hydrolysize PPi and allow it to function like two molecules of Pi (See Fig. 8.9). (Ref. 12)

help to account for the fact that while ADP also inhibits in peas (either by inhibiting the conversion of PGA to triose phosphate or producing a Pi deficiency by virtue of ATP/ADP exchange) it actually stimulates when added together with (otherwise inhibiting) PPi. It is argued that together, ADP and PPi would interfere with each other's entry and that PPi could then produce beneficial effects as indicated above or by offsetting the internal depletions caused by ADP-induced export of ATP. [Refs. **18, 19, 20, 33, 40, 46, 47, 60, 61**]

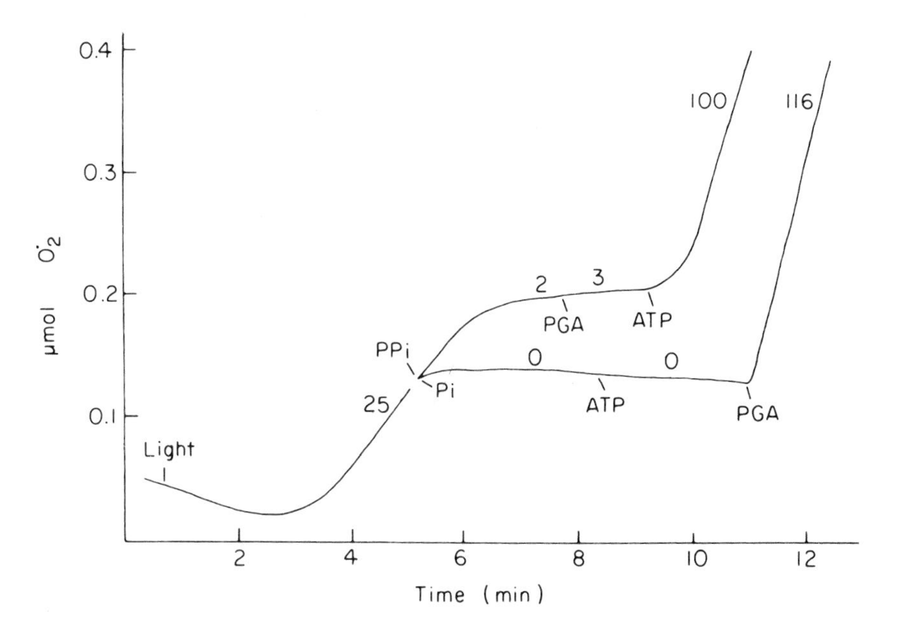


**Fig. 8.12.** Inhibition of photosynthesis in pea chloroplasts by PPi. In contrast to spinach, chloroplasts from young peas are more permeable to PPi which inhibits and to ATP which reverses this inhibition (upper curve). Lower curve shows that 'normal' Pi inhibition is reversed by PGA as in spinach but not by ATP. (Robinson & Wiskich, 1977. Ref. 46)

## 8.15 Pentose and hexose monophosphates

All three pentose monophosphates of the RPP pathway will reverse Pi inhibition after a short lag (see Fig. 7.22) and it seems probable that this is a measure of slow penetration. Similarly Heldt *et al* have observed penetration by direct measurement of R5P uptake which is inhibited by Pi implying slow uptake (at about 1–2 $\mu$mol mg$^{-1}$ Chl h$^{-1}$) via the Pi translocator (Table 8.2).

**Table 8.2.** Uptake of phosphorylated metabolites into the stroma space of spinach chloroplasts (from Fliege *et al*, 1978)
Temperature 20°C, darkness

| Metabolite (2mM) | Rate of uptake $\mu mol\ mg^{-1}\ Chl\ h^{-1}$ | | |
|---|---|---|---|
| | – | + Inorganic phosphate (0.2mM) | + D-glucose (20mM) |
| Inorganic pyrophosphate | 1.3 | 0.4 | n.d |
| Ribose 5-phosphate | 1.7 | 1.1 | 1.9 |
| Glucose 1-phosphate | 0.07 | 0.04 | 0.08 |
| Glucose 6-phophate | 0.43 | 0.44 | 0.24 |
| Fructose 6-phosphate | 0.32 | 0.20 | 0.34 |
| Fructose 1,6-bisphosphate | 0.85 | 0.53 | 0.86 |
| Gluconate 6-phosphate | 0.23 | 0.22 | 0.28 |

n.d. not determined

Direct measurement also shows penetration by F6P and G6P at rates of about a quarter of those exhibited by R5P. Uptake of F6P was inhibited by Pi and uptake of G6P by glucose, suggesting translocation via the Pi translocator (Section 8.23) and glucose translocator (Section 8.17) respectively. F6P has been reported to reverse Pi inhibition but this effect is not readily reproducible. Both F6P and G6P appear fairly quickly in the cytoplasm of leaves which are illuminated in the presence of $^{14}CO_2$ but this could be a consequence of external formation from triose phosphate. [Refs. **18, 22**]

### 8.16 Sugar bisphosphates

RBP does not appear in the cytoplasm of illuminated leaves or in a medium in which chloroplasts are illuminated. It does not reverse Pi inhibition or shorten induction. When added in the dark to ruptured chloroplasts it supports much more $^{14}CO_2$ fixation than when the chloroplasts are largely intact. In short, the chloroplast envelope seems to be largely impermeable to this compound. It is supposed that SBP and FBP may behave like RBP. It is, however, more or less impossible to obtain unequivocal results with FBP because of the ease with which this compound is formed from triose phosphates in the presence of aldolase.
[Refs. **3, 10, 11**]

## 8.17 Free sugars

Chloroplasts may be isolated in solutions containing either sucrose, fructose or glucose as osmotica (as well as the sugar alcohols such as sorbitol or mannitol which are more normally used for this purpose). This, in itself, indicates lack of rapid penetration (Section 8.10). Direct measurement shows that neither sucrose nor sorbitol rapidly penetrate the inner chloroplast envelope (Section 8.22). Although there was some indirect evidence in favour of fructose penetration, the chloroplast was, until recently, thought to be largely impermeable to glucose. However, when specifically labelled glucose was fed to tobacco leaf discs, starch was formed in which the labelling pattern was largely retained. (Entry as triose phosphate followed by internal resynthesis of hexose would lead to randomization of label). Further investigation by Schäfer, Heber and Heldt using centrifugal filtration, then showed that glucose and several other free sugars were, in fact, transported at low rates (Table 8.3) and that the behaviour of the chloroplasts was consistent with carrier mediated diffusion (Fig. 8.13). Ribose has also been shown to behave as a penetrating compound in experiments based on osmotic volume changes (Fig. 8.7) and, in some circumstances, fructose and other free sugars have been found to affect induction. It is of interest that the transport of D-glucose is inhibited by a high concentration of D-glucose thus explaining the fact that this sugar is as effective as an osmoticum as

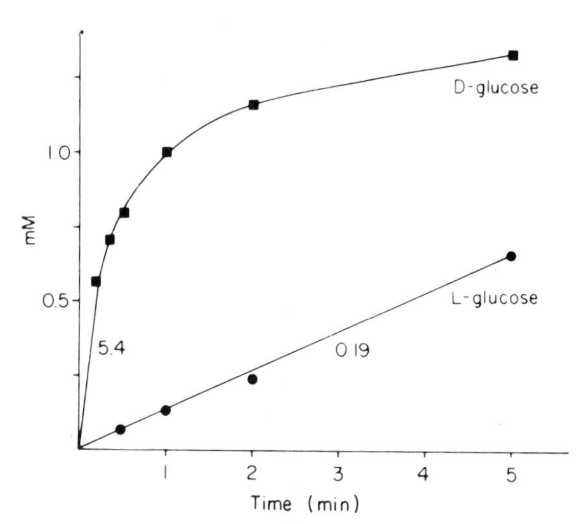


**Fig. 8.13.** Uptake of stereoisomers of glucose by intact spinach chloroplasts. The numbers are rates of uptake in $\mu$mol mg$^{-1}$ Chl h$^{-1}$ (from Schäfer *et al*, 1977).

**Table 8.3** Rate of sugar uptake into chloroplasts
The measuring time was 20s. (from Schafer *et al* 1977)

| Sugar 5mM | Rate of uptake |
|---|---|
| | $\mu$mol mg$^{-1}$ Chl h$^{-1}$ |
| D-glucose | 7.4 |
| L-glucose | 0.3 |
| D-mannose | 8.4 |
| D-fructose | 2.8 |
| D-xylose | 9.3 |
| D-ribose | 6.0 |
| L-arabinose | 7.7 |
| D-arabinose | 1.6 |

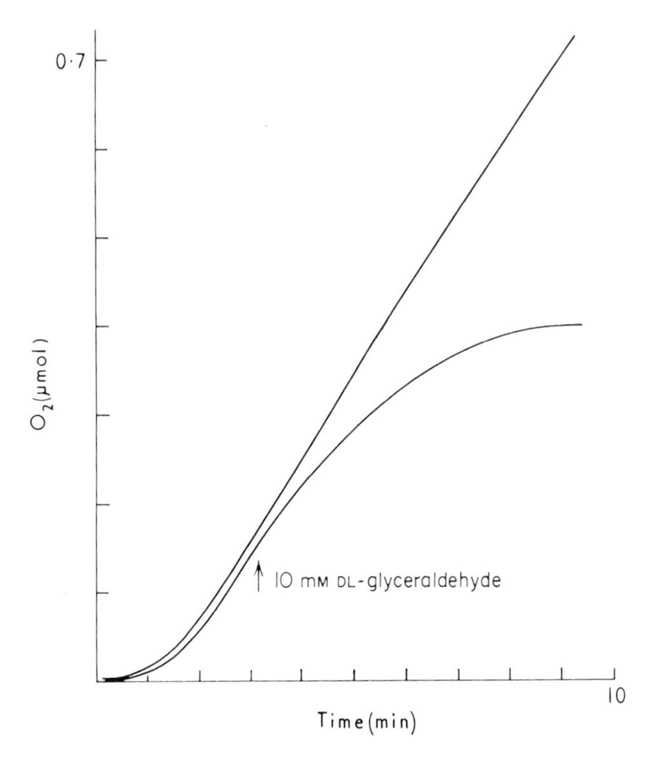


**Fig. 8.14.** Inhibition of $CO_2$-dependent $O_2$ evolution by DL glyceraldehyde. This compound inhibits photosynthetic carbon assimilation, probably by interfering with the transketolase reaction. The fact that it completely inhibits photosynthesis by intact chloroplasts in a matter of minutes indicates reasonably rapid penetration. (Ref. 53)

sucrose or sorbitol. The triose, DL-glyceraldehyde is a potent inhibitor of photosynthetic carbon assimilation (Section 8.8) and the response to its addition implies more rapid penetration than that shown by sugars of longer chain length (Fig. 8.14).
[Refs. 10, 11, **22, 52, 53**]

## 8.18 Carboxylic acids

Dicarboxylic acids such as malate, oxaloacetate, succinate, fumarate and 2-oxoglutarate are all transported at relatively fast rates by the dicarboxylate translocator. Glycerate leads to fairly rapid $O_2$ evolution by intact chloroplasts (Fig. 8.15) implying equally rapid uptake. Comparative biochemistry and the role of glycolate and phosphoglycolate in photorespiration (Chapter 14) implies fairly free movement of one or both of these molecules but definitive evidence is surprisingly lacking.
[Refs. **29, 39**]

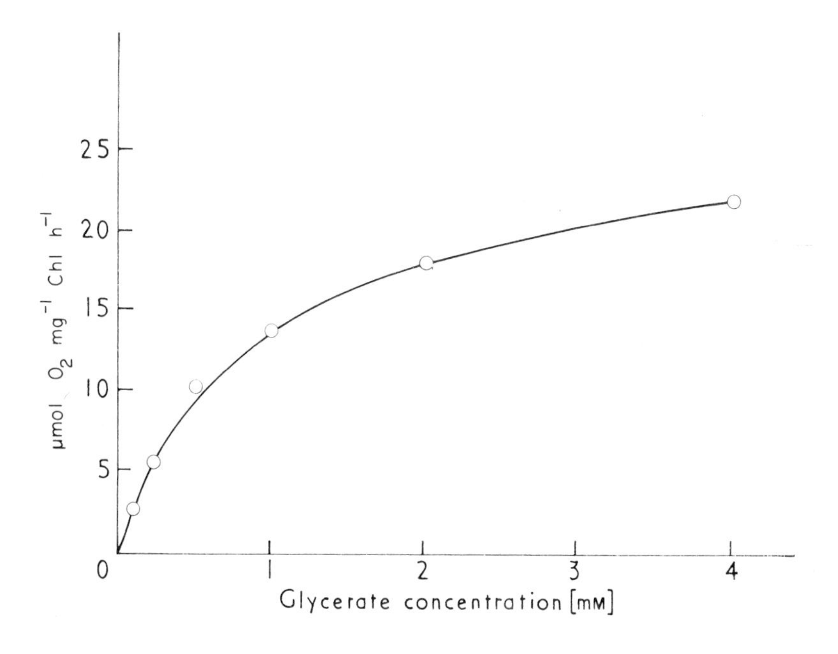


**Fig. 8.15.** Oxygen evolution by intact spinach chloroplasts as a function of glycerate concentration in the medium. The reaction depends on glycerate penetration followed by conversion to PGA which is then reduced at the expense of photosynthetically generated ATP and NADPH (after Heber *et al*, 1974).

## 8.19 Amino acids

Aspartate and glutamate are transported about as well as malate by the dicarboxylate translocator (and asparagine and glutamine somewhat slower) but present evidence does not favour movement of neutral amino acids at rates faster than that which would be needed to maintain protein turnover (about 0.1 $\mu$mol $mg^{-1}$ Chl $h^{-1}$). The evidence relating to transfer of glycine and serine does not permit the conclusion that these compounds could be directly transported at rates which would allow them to play a significant role in carbon export from the chloroplast.
[Refs. 3, 11, **24**]

## 8.20 ATP and NADP

NADP will act as a Hill oxidant in the presence of osmotically shocked chloroplasts and ferredoxin but there is no appreciable reaction with intact chloroplasts and it is generally accepted that direct transport of nicotinamide nucleotides is negligible. *Direct* movement of adenine nucleotides (other than at the slow rates brought about by the adenylate translocator) has been a much more controversial matter but it seems clear that osmotically shocked chloroplasts phosphorylate exogenous ADP much more rapidly than intact chloroplasts. Moreover, the inhibition of photosynthesis by uncouplers (Section 5.11) such as nigericin or ammonia or by phloridzin (which is believed to inhibit a terminal step in ATP synthesis) is not reversed by ATP, as would be expected if all of these compounds did nothing more than prevent ATP formation and if ATP moved freely across the chloroplast envelope (Fig. 8.16). On balance, therefore, there seems no real doubt that adenylates cannot move freely across the mature spinach chloroplast envelope (Fig. 8.17) but recent work suggests that the adenylate translocator (Section 8.25) may be much more active in pea chloroplasts (Fig. 8.12). Chloroplasts are normally prepared from very young pea shoots (10–15 days) and this, together with experiments on somewhat older material, suggests that more active exchange of adenylates which was observed might be a feature of young chloroplasts rather than simply a difference between species. If the permeability of the chloroplast envelope does change as it matures this could obviously reflect a changing role for the chloroplast in the general metabolic economy of the plant.
[Refs. 3, 4, 5, 11]

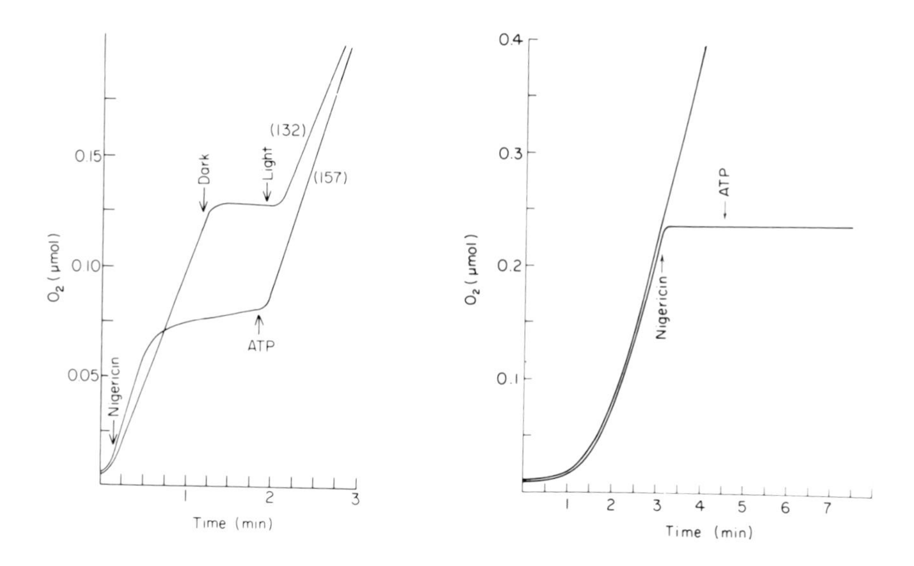


**Fig. 8.16.** ATP reverses nigericin inhibition of $O_2$ evolution by reconstituted chloroplasts (left, lower trace) but not of intact chloroplasts (right). Nigericin uncouples ATP formation from electron transport and this experiment shows that this inhibition can be reversed by the addition of ATP to the reconstituted system but not by addition to intact chloroplasts because of the permeability barrier to ATP imposed by the chloroplast envelope. (Ref. 12)

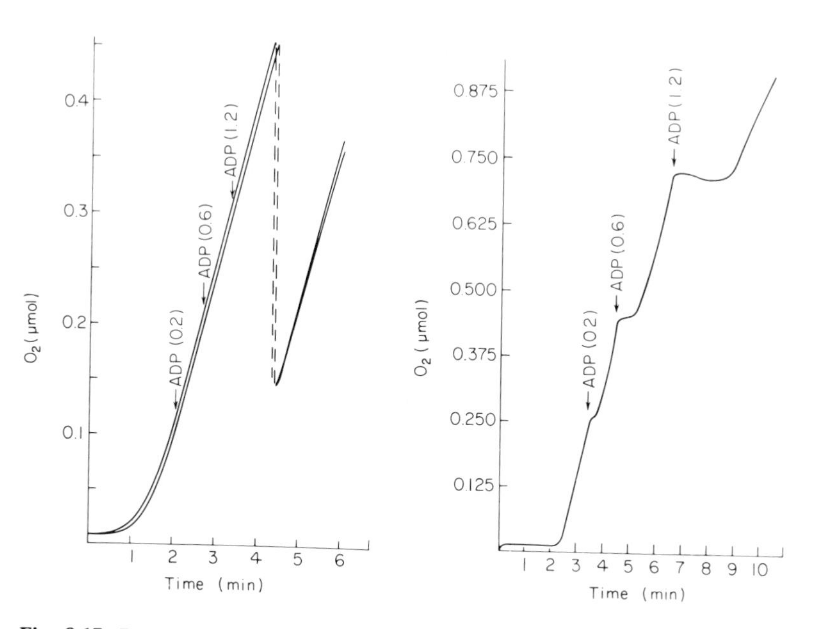


**Fig. 8.17.** Responses to ADP by intact spinach chloroplasts (left) and reconstituted spinach chloroplasts (right). The presence of the envelope (left) slows ADP uptake and prevents the marked inhibition seen in its absence (right). (Ref. 12)

## 8.21 Ion fluxes

When osmotically-shocked chloroplasts are illuminated, protons are taken up from the medium and the pH of the thylakoid compartment falls by about 3 pH units (Section 5.14). When the chloroplasts are intact it is the stromal compartment which becomes more alkaline rather than the external medium. The fact that the stromal compartment is maintained at a relatively alkaline pH (Chapter 5.9) shows that the envelope is not readily permeable to $OH^-$ or $H^+$.

When $KNO_2$, $NaNO_2$ and $NH_4Cl$ are added to suspensions of intact chloroplasts, osmotic shrinkage occurs but this is reversed upon the addition of valinomycin to mixtures containing the $K^+$ and $NH_4{}^+$ salts (valinomycin renders the envelope permeable to $K^+$ and $NH_4{}^+$). It is argued that $NO_2^-$ and $Cl^-$ must permeate because recovery would not follow in the presence of facilitated cation transport alone because of the inhibitory charge imbalance which would be created. No recovery is seen with $NaNO_2$, supporting this interpretation and pointing to the fact that the envelope is impermeable to $Na^+$. Impermeability to $NH_4{}^+$ is also indicated by the fact that the external medium becomes acid when intact chloroplasts are illuminated in the presence of $NH_4Cl$. This is attributed to dissociation:

$$NH_4^+ \rightarrow NH_3 + H^+ \qquad \text{Eqn. 8.5}$$

followed by inward movement of $NH_3$. Addition of $(NH_4)_2SO_4$ to chloroplasts suspended in $KNO_2$ led to rapid osmotic swelling which was explained by:

$$NH_4{}^+ + NO_2{}^- \rightarrow NH_3 + HNO_2 \qquad \text{Eqn. 8.6}$$

and penetration of both $NH_3$ + $HNO_2$. Valinomycin had little effect on the osmotic response of chloroplasts in $K_2SO_4$ or $(NH_4)_2SO_4$ implying that the sulphate ion does not readily penetrate the envelope. On the other hand it has been known since 1968 that sulphate inhibits photosynthesis by intact chloroplasts (probably by substituting for Pi in photophosphorylation) and that this inhibition is readily reversed by Pi. Similarly it is widely accepted that assimilatory sulphate reduction occurs within the chloroplast. On the basis of centrifugal filtration studies Mourioux and Douce have recently proposed a sulphate translocator, which exchanges sulphate with Pi at rates of about $20\,\mu mol\,mg^{-1}\,Chl\,h^{-1}$. Many of these inter-relationships, including the reduction of nitrite to ammonia (Section 12.9) are summarized in Fig. 8.18.

Valinomycin does not reverse osmotic shrinkage in the presence of potassium phosphate, indicating that phosphate uptake cannot occur except by counter exchange of triose phosphate or PGA via the Pi-translocator.

[Refs. **15, 21a, 25, 28, 30, 43**]

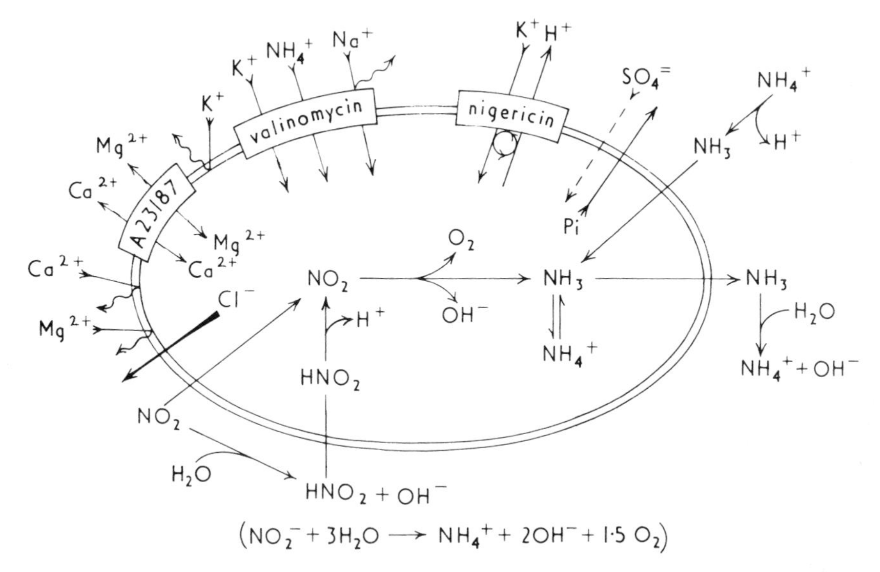


**Fig. 8.18.** Ion-movements across the chloroplast envelope. Generally the envelope is relatively impermeable to divalent cations such as $Mg^{2+}$ and monovalent cations such as $Na^+$ and $H^+$ whereas monovalent anions such as $Cl^-$ and $NO_2{}^-$ pass reasonably freely. The permeability properties may be modified in the presence of certain ionophores etc. as indicated. The reduction of $NO_2^-$ is also summarized (see Section 12.9)

## 8.22 Specific permeability of the inner envelope

Chloroplasts are not penetrated by labelled dextran but centrifugal filtration indicates rapid penetration by all compounds of relatively low molecular weight which have been tested. Work with labelled sucrose indicates a correlation between the extent of this penetration and the space between the two envelopes which is increased in hypertonic medium (Section 8.10). The 'sucrose-permeable' space (Fig. 8.4b) is therefore regarded as the intermembrane space and it is concluded that it is the inner of the two envelopes which exhibits specific permeability. The degree of selective permeability exhibited is quite remarkable when it is considered that Pi is transported very rapidly into spinach chloroplasts whereas PPi scarcely moves at all (Figs. 8.9, 8.11, 8.15) and that the entry of D-glucose is 25 times as fast as that of L-glucose.

## 8.23 The phosphate translocator

As previously noted (Section 7.12) indirect measurements led to the notion that Pi might exchange with sugar phosphates. Work by Heldt and his colleagues established such exchange (in relation to Pi and triose phosphates) by direct measurement (Section 8.5) and the Pi translocator (Fig. 8.19) became a reality. The amount of Pi released from chloroplasts preincubated with $^{32}$Pi is equivalent to the quantity of $^{14}$C labelled PGA which is taken up indicating a strict coupling of inward and outward transport (Fig. 8.20). Pi does not leak from chloroplasts but only undergoes export at a rapid rate in the presence of exogenous PGA or triose phosphates. In addition, the translocators will accept arsenate and, to some extent, substances like 1-glycerophosphate. Ribose 5-phosphate is translocated slowly and compounds like fructose 6-phosphate are virtually excluded. The movement of Pi, PGA and triose phosphate is very rapid (in the range of 200–300 $\mu$mol mg$^{-1}$ Chl h$^{-1}$ at 20° C). After as little as 240 s the concentration of Pi or PGA in the stroma has been found to be 20 times higher than that in the medium. The kinetic behaviour of the translocator is similar to that of an enzyme so that when 1/v is plotted against 1/[S] a straight line is obtained (Fig. 8.21). The four principal compounds translocated compete with one another and their *K*i values (concentration of one needed to halve the rate of transport of another) are similar to their *K*m values (concentration needed to give half maximal rate of transport). *K*i and *K*m values etc. are given in Tables 8.4 and 8.5.

Simultaneous measurements of the proton concentration in the medium and of transport into the chloroplast indicates that transfer of PGA is accompanied by transport of a proton (probably as a consequence of $PGA^{3-} + H^+ \rightarrow PGA^{2-}$).

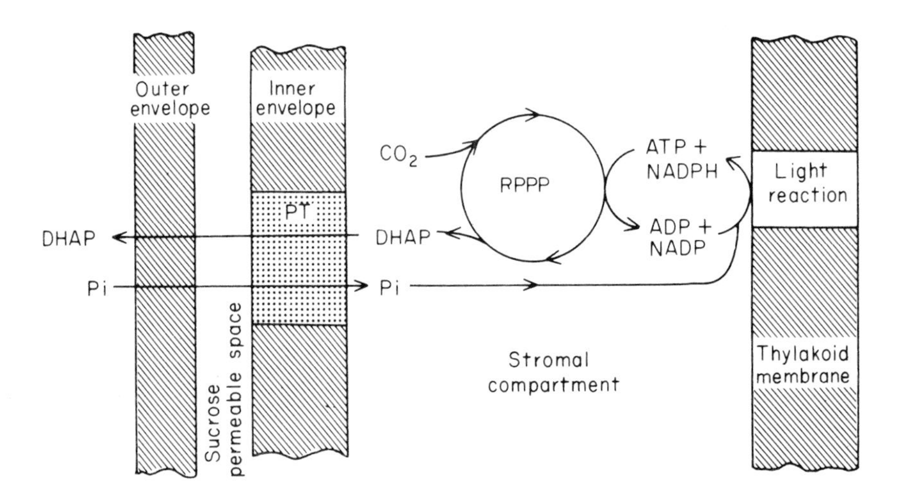


**Fig. 8.19.** The Pi translocator. This is located in the inner envelope and permits a strict stoichiometric exchange of external Pi with internal DHAP. Other compounds such as PGA will also exchange with Pi or DHAP. Movements may occur in either direction according to circumstance (after Heldt. Refs. 11 & 12).

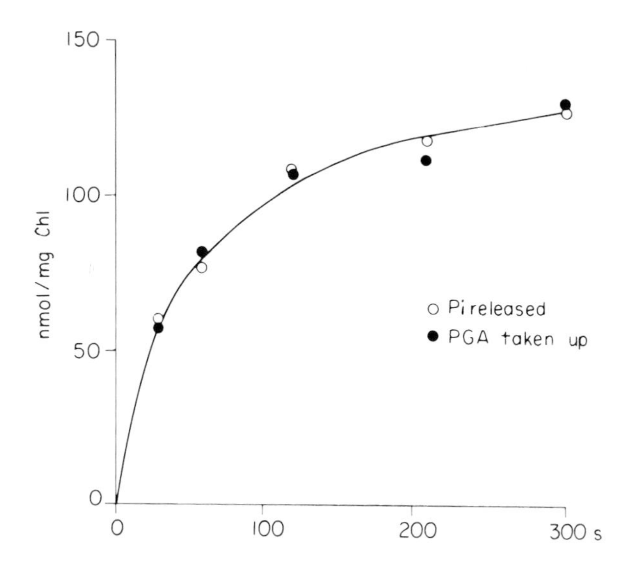


**Fig. 8.20a.** Stoichiometry of Pi release and PGA uptake via the Pi-translocator (after Heldt, 1976).

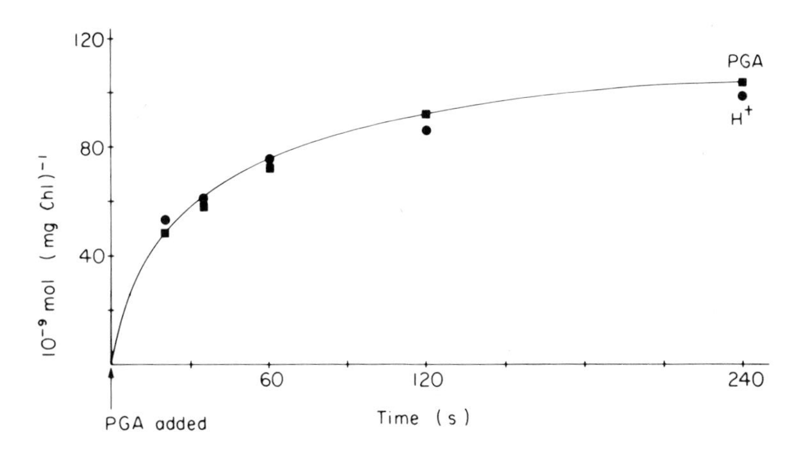


**Fig. 8.20b.** Simultaneous measurement of the uptake of 3-phosphoglycerate into the stroma and uptake of protons in the external medium (from Fliege *et al*, 1978).

Similarly, the relationship between pH and transport suggests that the other three compounds are also transferred as divalent anions. That PGA is transported as the divalent anion may help to account for the fact that in steady-state photosynthesis there is about 10 times as much PGA as triose phosphate in the chloroplast stroma.

**Table 8.4.** Kinetic constants of phosphate transport into the chloroplast stroma (from Fliege *et al*, 1978). Values obtained by back exchange. Temperature 4° C, darkness

| | *K*m (mM) | *V*max μmol mg$^{-1}$ Chl h$^{-1}$ |
|---|---|---|
| Inorganic phosphate | 0.30 (0.15–0.50) | 57 (40–91) |
| Dihydroxyacetone phosphate | 0.13 (0.06–0.22) | 51 (36–74) |
| D-Glyceraldehyde phosphate | 0.08 (0.03–0.17) | 41 (37–44) |
| 3-Phosphoglycerate | 0.14 (0.09–0.19) | 36 (29–45) |
| 2-Phosphoglycerate | 2.8 (1.8–3.7) | 24 (20–26) |
| Glycerol 1-phosphate | 1.1 (0.7–1.4) | 59 (38–71) |

**Table 8.5.** Competitive inhibition of the transport of inorganic phosphate (*K*m = 0.20 mM) into the chloroplast stroma. Temperature 4° C, darkness (From Fliege *et al*, 1978).

| Inhibitor (mM) | *K*i (mM) |
|---|---|
| Dihydroxyacetone phosphate | 0.13 |
| 3-Phosphoglycerate | 0.15 |
| 2-Phosphoglycerate | 6.5 |
| 2,3-Diphosphoglycerate | 9.1 |
| Phosphoenolpyruvate | 4.7 |
| Erythrose 4-phosphate | 2.3 |
| Ribose 5-phosphate | 10.0 |
| Glucose 6-phosphate | 40.0 |
| Fructose 6-phosphate | 13.0 |
| Fructose 1,6-bisphosphate | 8.5 |
| 6-Phosphogluconate | 20.0 |
| Inorganic pyrophosphate | 1.8 |

The stromal pH is more alkaline during illumination and, if it needed to acquire a proton, PGA would escape less readily from an alkaline compartment than would a compound which did not have this requirement.

The translocator is inhibited by p-chloromercuriphenyl sulfonate, pyridoxal-5-phosphate (Table 8.6) and trinitrobenzene sulfonate. Inorganic pyrophosphate binds to the carrier with a *K*i about 10 times higher than the *K*m for Pi but is translocated at only 1/100 of the rate. Accordingly, at high concentration it is an extremely good inhibitor and will protect isolated chloroplasts from inhibition by excess Pi (see Fig. 8.22). There seems little doubt that the principal function of the Pi-

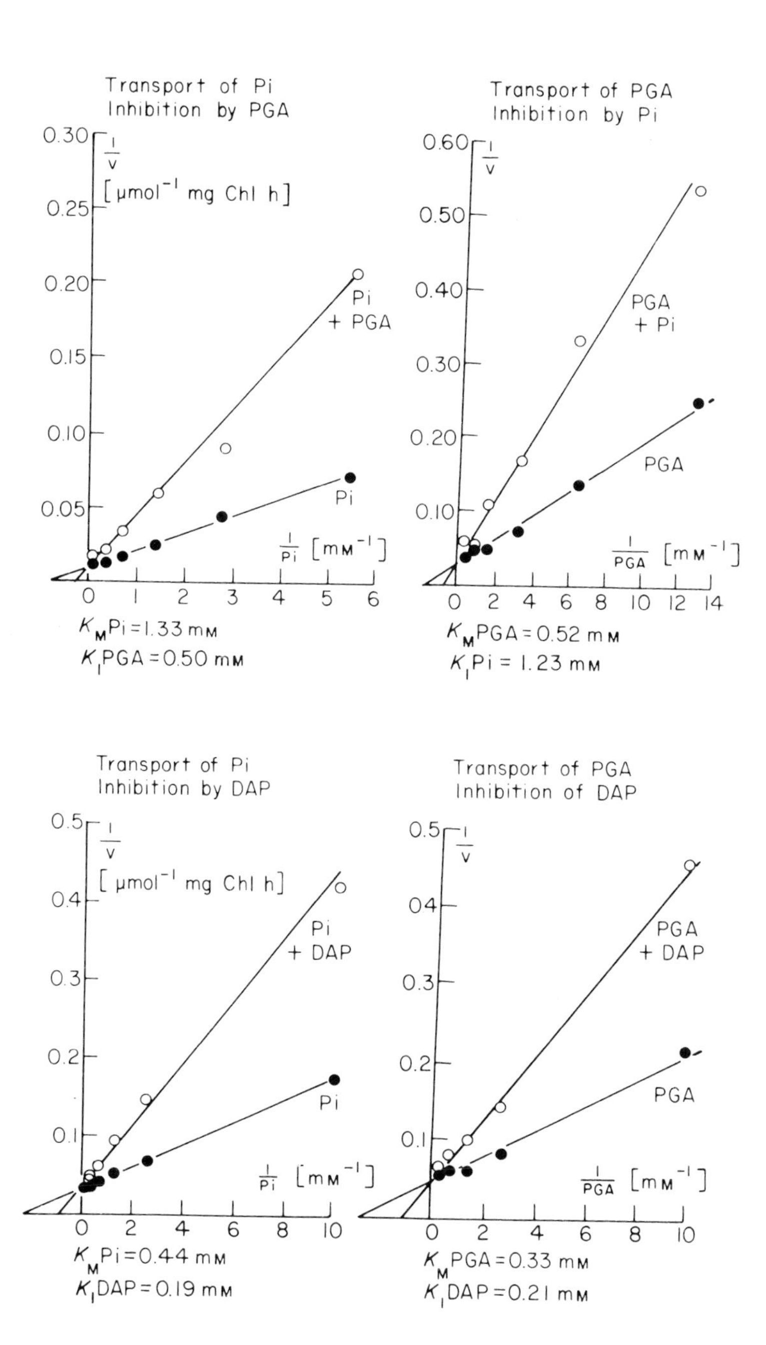


**Fig. 8.21.** The Pi translocator. Each metabolic translocated can interfere with the transport of other transported compounds in a competitive manner. (Werdan & Heldt, 1971).

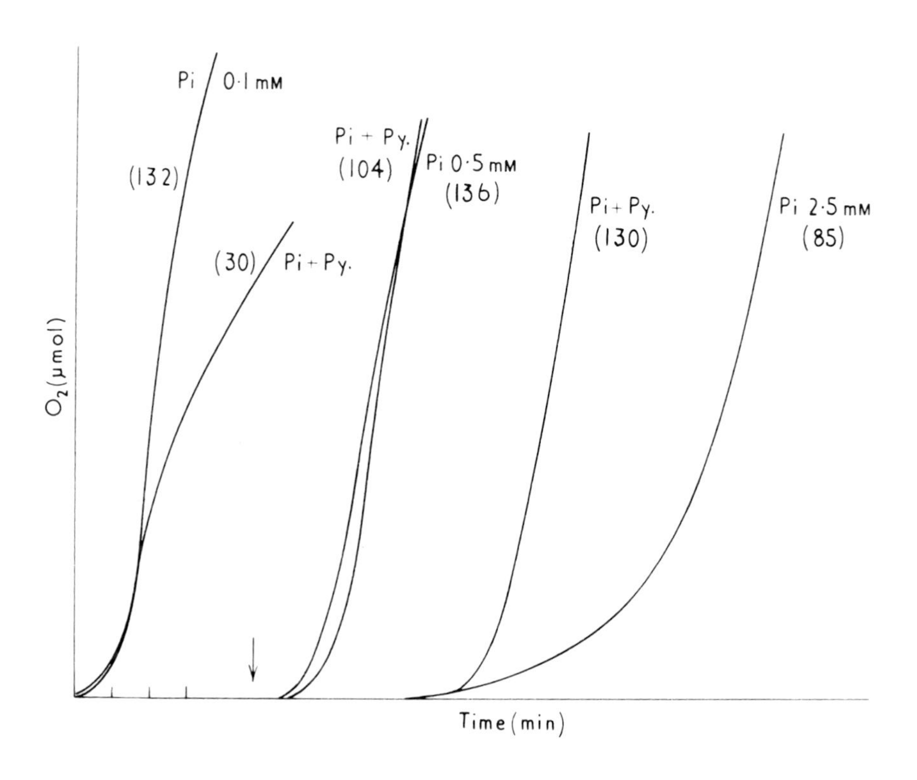


**Fig. 8.22.** Inhibition by pyridoxal phosphate of the Pi translocator and its effect on photosynthetic carbon assimilation (see Table 8.3). Like PPi, pyridoxal phosphate inhibits the Pi translocator but its effect is less ambiguous because, unlike PPi, it does not also act as a potential source of Pi. Photosynthesis at three concentrations of Pi is depicted. At left, 0.1 mM is sub-optimal; in centre, 0.5 mM is near optimal and, on right, 2.0 mM is inhibitory due to enforced export of intermediate of the RPP-pathway. Accordingly, pyridoxal phosphate (Py) inhibits at left because it depresses entry of Pi still further below the optimal rate. At centre, pyridoxal phosphate inhibits less strongly at the Pi optimum. At right, where Pi is itself inhibitory, pyridoxal phosphate protects as would PPi (see Fig. 8.10). (Flügge *et al*, 1980)

translocator is to allow the chloroplast to import Pi and export DHAP at fast rates. The latter is almost certainly the major metabolite exported to the cytoplasm and the precursor of cytoplasmic sucrose synthesis or the substrate for glycolytic degradation. DHAP may also contribute to the export of assimilatory power from the chloroplast by its role in shuttle mechanisms (Section 8.26).
[Refs. 6, **22**, **23**, **32**]

**Table 8.6.** Loss of stromal metabolites induced by high external Pi and protection against loss by inhibition of the Pi-translocator by pyridoxal phosphate (from Flügge *et al*, 1979).

| | | | |
|---|---|---|---|
| Pi (mM) | 0.5 | 2.0 | 2.0 |
| pyridoxal 5′-phosphate (mM) | 0 | 0 | 1.0 |
| | Metabolite conc. in stroma (mM) | | |
| Triose phosphates | 0.14 | trace | 0.18 |
| 3-phosphoglycerate | 4.44 | trace | 3.69 |
| Hexose and heptose monophosphates | 1.96 | 0.10 | 1.85 |
| Fructose bisphosphate | 0.29 | trace | 0.37 |
| Sedoheptulose bisphosphate | 0.11 | trace | 0.10 |
| Pentose monophosphate | 0.08 | trace | 0.02 |
| Ribulose bisphosphate | 0.46 | 0.06 | 0.30 |
| ATP/ADP | 0.42 | 1.75 | 0.98 |

## 8.24 The dicarboxylate translocator

This is similar to the Pi translocator, in principle, except that it transports dicarboxylic acids and counter exchange is not as strictly coupled. The *K*i for inhibition by fumarate of malate transport is close to that of the *K*m for fumarate transport and competitive inhibition by other dicarboxylic acids (Fig. 8.23) has been demonstrated. At 4°C *V*max values of 19 $\mu$mol mg$^{-1}$ Chl h$^{-1}$ have been recorded for malate and 31 for aspartate (Table 8.7). (These are about 1/2–1/3 of the *V*max values at 4°C for the Pi-translocator).

**Table 8.7.** Kinetic constants of dicarboxylate transport into the sorbitol-impermeable space of spinach chloroplasts (from Lehner & Heldt, 1978). Temperature 4°C.

| | *V*max $\mu$mol mg$^{-1}$ Chl h$^{-1}$ | *K*m (mM) |
|---|---|---|
| L-malate | 18.6 | 0.39 |
| Succinate | 14.0 | 0.26 |
| Fumarate | 18.6 | 0.21 |
| L-aspartate | 31.1 | 0.72 |
| α-ketoglutarate | 26.4 | 0.19 |
| L-glutamate | 7.9 | 1.17 |

Values for *V*max exceeding the above by a factor of 5 have been obtained for spinach grown in water culture.

The dicarboxylate translocator is thought to play a more important role in shuttle mechanisms (Section 8.26) than in carbon traffic.
[Refs. 6, **23, 26, 32, 39**].

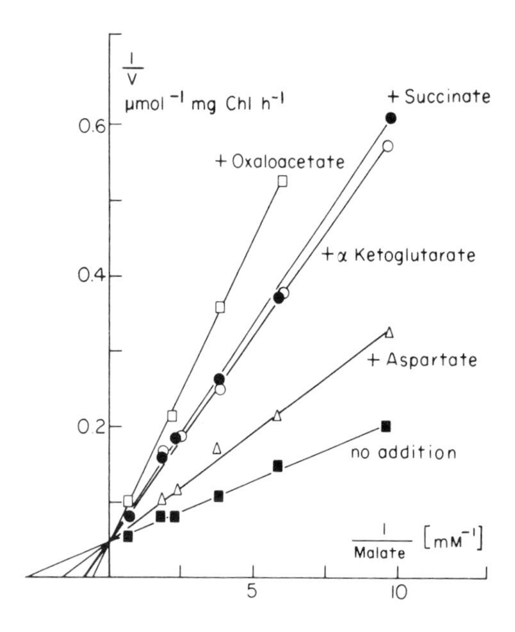


**Fig. 8.23.** Malate transport via the dicarboxylate translocator as a function of concentration and its inhibition by other dicarboxylic acids. Spinach chloroplasts at 4°C (from Heldt, 1976).

## 8.25 The adenylate translocator

This facilitates a counter exchange of adenine nucleotides. In spinach chloroplasts the fastest rates (about 5 $\mu$mol mg$^{-1}$ Chl h$^{-1}$ at 20° C) are observed in the presence of external ATP. In pea chloroplasts (or perhaps in chloroplasts from immature leaves) the rates of adenylate transport are almost certainly higher (by a factor of 5 or 10) and PPi also appears to be transported at significant rates producing an inhibition which is reversed by ADP or ATP (Section 8.14). Whatever the role of the adenylate translocator, its low rate of function makes it seem unlikely that it can make an important contribution to the export of assimilatory power in mature chloroplasts although it may be more important in this regard in young tissues.
[Ref. 6, **31, 32, 39**].

## 8.26 Shuttles

When it became evident that assimilatory power could not be rapidly exported from the chloroplast by direct movement it was proposed, on purely theoretical grounds, that the same effect could be achieved by export of triose phosphate followed by cytoplasmic oxidation of the metabolite to PGA. Accordingly ATP and NADH would be generated externally (Fig. 8.24) and, if the PGA re-entered the chloroplast

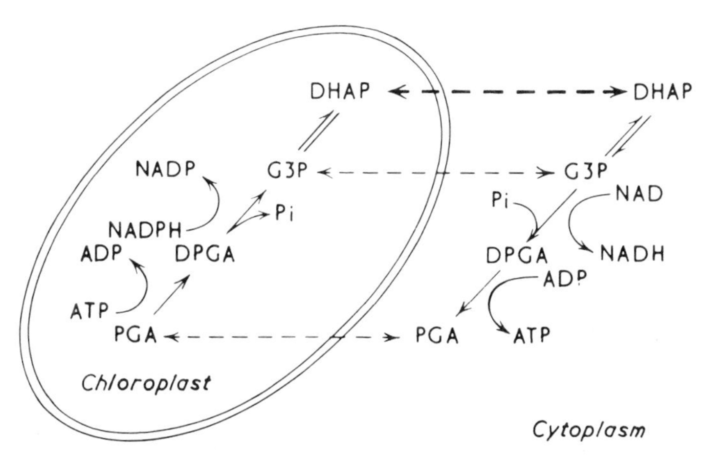


**Fig. 8.24.** PGA/DHAP shuttle. Inside chloroplast PGA is reduced to TP at expense of NADPH and ATP. Externally NADH and ATP are formed from TP released from chloroplast. PGA can re-enter chloroplast to continue shuttle. (After Heber, 1974; Walker, 1974).

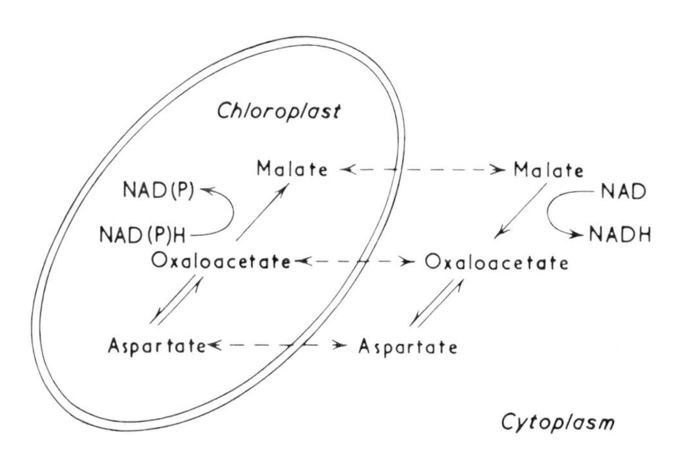


**Fig. 8.25.** Malate/oxaloacetate shuttle. Inside chloroplast, oxaloacetate is reduced to malate at expense of NADPH. Externally it is reoxidized by cytoplasmic malate dehydrogenase. Reducing equivalents are thus exported. (After Heber, 1974; Walker 1974).

and underwent internal reduction, this process might continue indefinitely. Subsequently such transport was actually measured by Stocking and Larson. Heber and Krause have observed rates as high as 50 $\mu$mol mg$^{-1}$ Chl h$^{-1}$. External oxidation of triose phosphate may also be brought about by the irreversible NADP specific triose phosphate dehydrogenase system of Kelly and Gibbs. This reaction is not linked to ATP synthesis and would export NADPH alone. Reducing equivalents may also be transported across the chloroplast envelope by a dicarboxylate shuttle (Fig. 8.25). Although these shuttles were originally visualized as aids to exports they could contribute, for example, to ATP generation within the chloroplast at night. (e.g. The triose phosphate-PGA shuttle could (in effect) transfer ATP and reducing power to chloroplasts, and the dicarboxylate shuttle could transfer reducing power from the chloroplasts resulting in ATP generation in the chloroplast). [Refs. **26, 27, 36, 38, 57**].

## General Reading


1 DOUCE R. & JOYARD J. (1979) Structure and function of the plastid envelope. In *Adv. Bot. Res.*, **7**, 1–116.

2 GIBBS M. (1971) Carbohydrate metabolism by chloroplasts. In *Structure and Function of Chloroplasts*, (ed. M. Gibbs) pp. 169–214. Berlin: Springer, Verlag.

3 HEBER U., (1970) Flow of metabolites and compartmentation phenomena in chloroplasts. In *Transport and Distribution of Matter in Cells of Higher Plants*, (eds. K. Mothes, E. Muller, A. Nelles, D. Neumann), pp. 152–84. Berlin: Akademie-Verlag.

4 HEBER U., (1974) Metabolite exchange between chloroplasts and cytoplasm. *Ann. Rev. Plant Physiol.* **25**, 393–421.

5 HEBER U. & KRAUSE G. H., (1971) Transfer of carbon, phosphate energy and reducing equivalents across the chloroplast envelope. In Photosynthesis and Photorespiration, (eds. M. D. Hatch, C. B. Osmond, R. O. Slatyer), pp. 218–25. (*Proc. Conf. Australian Nat. Univ. Canberra, 1970*). New York: Wiley Interscience.

5a HEBER U & HELDT H. W. (1981) The chloroplast envelope: structure, function and role in leaf metabolism. *Ann. Rev. Plant Physiol.* **32**, 139–68.

6 HEBER U., & WALKER D. A. (1979) The Chloroplast Envelope, Barrier or Bridge? *T. I. B. S.* **4**, 11.

7 HELDT H. W. (1976) Metabolite transport in intact spinach chloroplasts. In *Topics in Photosynthesis*, Vol. 1: The Intact Chloroplast (ed. J. Barber), pp. 215–34. Amsterdam: Elsevier, 1976.

8 HEROLD A. & WALKER D. A., Transport across chloroplast envelopes—the role of phosphate. In: *Membrane Transport in Biology*, (eds. G. Giebisch, D. C. Tosteson & J. J. Ussing) Springer-Verlag, Heidelberg, New York, Vol. II, 412–39 (1979).

9 KELLY G. J., LATZKO E. & GIBBS M. (1976) Regulatory aspects of photosynthetic carbon metabolism. *Ann. Rev. Plant Physiol.* **27**, 181–205.

10 KELLY G. J. & LATZKO E., (1979) The Cytosol. In *The Biochemistry of Plants: A Comprehensive Treatise*. (P. K. Stumpf and E. E. Conn, eds.) Volume 1, *The Plant Cell* (ed. N. E. Tolbert), Chapter 4. New York: Academic Press.

10a KRAUSE G. H. & HEBER U. (1976) Energetics of intact chloroplasts. In *The Intact Chloroplast*. (ed. J. Barber) Chapter 5, 171–214. Amsterdam: Elsevier.

11 WALKER D. A., (1974) Chloroplast and Cell. Concerning the movement of certain key metabolites

etc. across the chloroplast envelope. In: *Med. Tech. Publ. Int. Rev. Sci. Biochem. Ser. I* (ed. D. H. Northcote), Vol. **XI**, pp. 1–49. London: Butterworths.
12 WALKER D. A. (1976) Plastids and Intracellular Transport. In *Encyclopedia of Plant Physiology. Transport in Plants III* (eds. C. R. Stocking & U. Herber) New Series, Volume 3, pp. 85–136 Berlin: Springer-Verlag.

## Specific Citations

13 ALLEN M. B., ARNON J. B., CAPINDALE J. B., WHATLEY F. R. & DURHAM L. J. (1955) Photosynthesis by isolated chloroplasts. III. Evidence for complete photosynthesis. *J. Am. Chem. Soc.* **77**, 4149–55.
14 ARNON D. I., ALLEN M. B., & WHATLEY F. R. (1954) Photosynthesis by isolated chloroplasts. *Nature* **174**, 394–6.
15 BALDRY C. W., COCKBURN W. & WALKER D. A. (1968) Inhibition by sulphate of the oxygen evolution associated with photosynthetic carbon assimilation. *Biochim. Biophys. Acta.* **153**, 476–83.
16 BASSHAM J. A. & CALVIN M. (1957) *The Path of Carbon in Photosynthesis.* pp. 1–104, Englewood Cliffs, N. J: Prentice–Hall Inc.
17 BASSHAM J. A. & JENSEN R. G. (1967) Photosynthesis of carbon compounds. In: *Harvesting the Sun* (eds. A. San Pietro, F. A. Greer, T. J. Army), pp. 79–110. New York: Academic Press.
18 COCKBURN W., BALDRY C. W. & WALKER D. A. (1967) Oxygen evolution by isolated chloroplasts with carbon dioxide as the hydrogen acceptor. A requirement for orthophosphate or pyrophosphate. *Biochim. Biophys. Acta*, **131**, 594–6.
19 COCKBURN W., BALDRY C. W. & WALKER D. A. (1967) Some effects of inorganic phosphate on $O_2$ evolution by isolated chloroplasts. *Biochim. Biophys. Acta*, **143**, 614–24.
20 COCKBURN W., WALKER D. A. & BALDRY C. W. (1968) Photosynthesis by isolated chloroplasts. Reversal of orthophosphate inhibition by Calvin cycle intermediates. *Biochem. J.* **107**, 89–95.
21 COOPER T. G., FILMER D., WISHNICK M. & LANE M.D. (1969) The active species of '$CO_2$' utilized by ribulose diphosphate carboxylase. *J. Biol. Chem*, **244**, 1081–3.
21a ENSER U. & HEBER H. (1980) Metabolic regulation by pH gradients. Inhibition of photosynthesis by indirect proton transport across the chloroplast envelope. *Biochim. Biophy. Acta* **592**, 577–91.
22 FLIEGE R., FLÜGGE U. I., WERDAN K & HELDT H. W., (1978) Specific transport of inorganic phosphate, 3-phosphoglycerate and triosephosphates across the inner membrane of the envelope in spinach chloroplasts. *Biochim. Biophys. Acta*, **502**, 232–47.
23 FLÜGGE U. I., FREISL M. & HELDT H. W. (1980) Balance between metabolite accumulation and transport in relation to photosynthesis by isolated spinach chloroplasts *Plant Physiol.* **65**, 574–7
24 GIMMLER H., SCHAFER G., KRAMINER A. & HEBER U. (1974) Amino acid permeability of the chloroplast envelope as measured by light scattering, volumetry and amino acid uptake. *Planta*, **120**, 47–61.
25 HAMPP R. & ZIEGLER I. (1977) Sulfate and sulfite translocation via the phosphate translocator of the inner envelope membrane of chloroplasts. *Planta* **137**, 309–12.
26 HEBER U. & SANTARIUS K. A. (1970) Direct and indirect transport of ATP and ADP across the chloroplast envelope. *Z. Naturforsch*, **25b**, 718–78.
27 HEBER U. & KRAUSE G. H. (1971) Transfer of carbon, phosphate energy and reducing equivalents across the chloroplast envelope. In Photosynthesis and Photorespiration, (eds. M. D. Hatch, C. B. Osmond, R. O. Slatyer), pp. 218–25. *Proc. Conf. Australian Nat. Univ. Canberra, 1970.* Wiley Interscience, New York.
28 HEBER U. & KRAUSE G. H. (1972) Hydrogen and proton transfer across the chloroplast envelope. In: Progress in Photosynthesis, (eds. G. Forti, M. Avron, A. Melandri), Vol. II, pp. 1023–1033 (*Proc. 2nd Intern. Congr. Photosyn. Res. 1971*) The Hague: N. V. Junk.
29 HEBER U., KIRK M. R., GIMMLER H. & SCHÄFER F. (1974) Uptake and reduction of glycerate by isolated chloroplasts. *Planta* **120**, 31–46.

30 HEBER U. & PURCZELD P. (1978) Substrate and product fluxes across the chloroplast envelope during bicarbonate and nitrate reduction. *Proceedings of the Fourth International Congress on Photosynthesis 1977.* (eds. D. O. Hall, J. Coombs, T. W. Goodwin), pp. 107–18. London: The Biochemical Society.
31 HELDT H. W. (1969) Adenine nucleotide translocation in spinach chloroplasts. *FEBS Letters*, **5**, 11–4
32 HELDT H. W. (1976) Metabolite carriers of chloroplasts. *Encyclopedia of Plant Physiology* New Series Vol. 3. *Transport in Plants III.* (eds. C. R. Stocking and U. Heber) pp. 137–43. Berlin-Heidelberg-New York:Springer-Verlag.
33 HEROLD A. & WALKER D. A. (1979) Transport across chloroplast envelopes—the role of phosphate. In *Membrane Transport in Biology*, Vol. II (eds. G. Giebisch, D. C. Tosteson and J. J. Ussing), Springer-Verlag, Heidelberg, New York, Vol. II, pp. 412–39.
34 HILL R. (1937) Oxygen evolved by isolated chloroplasts. *Nature*, **139**, 881–2.
35 HILL R. (1965) The biochemists' green mansions. The photosynthetic electron-transport chain in plants. *Essays Biochem.* **1**, 121–51.
36 KELLY G. J. & GIBBS M. (1973) Nonreversible D-glyceraldehyde 3-phosphate dehydrogenase in plant tissue. *Plant Physiol*, **52**, 111–18.
37 KLINGENBERG M. & PFAFF E. (1967) Means of terminating reactions. *Methods Enzymol.* **10**, 680–4.
38 KRAUSE G. H. (1971) Indirekter ATP-Transport zwischen Chloroplasten und Zytoplasma während der Photosynthese. *Z. Pflanzenphysiol*, **65**, 13–23.
39 LEHNER K. & HELDT H. W. (1978) Dicarboxylate transport across the inner membrane of the chloroplast envelope. *Biochim. Biophys. Acta.* **501**, 531–44.
40 LILLEY R. MCC., FITZGERALD M. P., RIENITS K. G. & WALKER D. A. (1975) Criteria of intactness and the photosynthetic activity of spinach chloroplast preparations. *New Phytol*, **75**, 1–10.
41 LILLEY R. MCC., SCHWENN J. & WALKER D. A. (1973) Inorganic pyrophosphatase and photosynthesis by isolated chloroplasts. II. The controlling influence of orthophosphate. *Biochem. Biophys. Acta*, **325**, 596–604.
41a LILLEY R. MCC. & WALKER D. A. (1979) Studies with the reconstituted chloroplast system. *Encyclopedia of Plant Physiology (New Series)—Photosynthesis Vol. II.* pp. 41–53. Regulations of photosynthetic carbon metabolism and related processes. (eds. M. Gibbs and E. Latzko). Springer-Verlag.
42 MOLISCH H. (1922) *Populare Biologische-Vortrage*, pp. 1–306. Jena: G. Fischer.
43 MOURIOUX, G. & DOUCE, R. (1979) Transport du sulfate à travers la double membrane tumitante, ou enveloppe, des chloroplasts d'épinard, *Biochimie*, **61**, 1283–92.
44 POINCELOT R. P. (1974) Uptake of bicarbonate ion in darkness by isolated chloroplast envelope membrane and intact chloroplast of spinach. *Plant Physiol.* **54**, 520–6.
45 PORTIS A. R. Jr., & HELDT H. W. (1976) Light-dependent changes of the $Mg^{2+}$ concentration in the stroma in relation to the $Mg^{2+}$ dependency of $CO_2$ fixation in intact chloroplasts. *Biochim. Biophys. Acta.* **449**, 434–46.
46 ROBINSON S. P. & WISKICH J. T. (1977) Pyrophosphate inhibition of carbon dioxide fixation in isolated pea chloroplasts by uptake in exchange for endogenous adenine nucleotides. *Plant Physiol.* **59**, 422–7.
47 ROBINSON S. P. & WISKICH J. T. (1977) Inhibition of $CO_2$ fixation by adenosine 5′-diphosphate and the role of phosphate transport in isolated pea chloroplasts. *Archs. Biochem. Biophys.* **184**, 546–54.
48 ROBINSON S. P., EDWARDS G. E. & WALKER D. A. (1979) Established methods for the isolation of intact chloroplasts. In *Methodological Surveys in Biochemistry*, Ch. 9, pp. 13–24 (ed. E. Reid), Chichester: Ellis Horwood.
49 ROBINSON S. P. & WALKER D.A. (1979) Rapid separation of the chloroplast and cytoplasmic fractions from intact leaf protoplasts. *Arch. Biochem.* **196**, 319–23.
50 SACHS J. (1862) Uber den Einflusz des Lichtes auf die Bildung des Amylums in den Chlorophyllkornern. *Botan. Z.* **20**, 365–73.

51 SACHS J. (1887) Lectures on the Physiology of Plants. Translated by H. M. WARD, p. 309 seq. Oxford: Clarendon Press.
52 SCHÄFER G., HEBER U. & HELDT H. W. (1977) Glucose transport into spinach chloroplasts. *Plant Physiol*, **60**, 286–9 (1977).
53 STOKES D. M. & WALKER D. A. (1972) Photosynthesis by isolated chloroplasts. Inhibition by DL-glyceraldehyde of carbon dioxide assimilation. *Biochem. J.* **128**, 1147–57.
54 UMBREIT W. W., BURRIS R. H. & STAUFFER J. F. (1945) *Manometric Techniques and Related Methods for the Study of Tissue Metabolism*, pp. 1–387. Minneapolis, Minnessota, Burgess.
55 WALKER D. A., COCKBURN W. & BALDRY C. W. (1967) Photosynthetic oxygen evolution by isolated chloroplasts in the presence of carbon cycle intermediates. *Nature,* **216**, 597–9.
56 WALKER D. A. & HILL R. (1967) The relation of oxygen evolution to carbon assimilation with isolated chloroplasts. *Biochem. Biophys. Acta,* **131**, 330–8.
57 WALKER D. A. & CROFTS A. R. (1970) Photosynthesis. *Ann. Rev. Biochem.* **39**, 389–428.
58 WALKER D. A. (1973) Photosynthetic induction phenomena and the light activation of ribulose diphosphate carboxylase. *New Phytol.* **72**, 209–35.
59 WALKER D. A. (1976) Photosynthetic induction and its relation to transport phenomena and control mechanisms in chloroplasts. In: *The Intact Chloroplast.* (ed. J. BARBER) Ch 7, 235–78. Amsterdam: Elsevier.
60 WALKER D. A. & STANKOVIC Z. S. (1977) Photosynthesis by isolated chloroplasts. Some effects of adenylates and inorganic pyrophosphate. *Plant Physiol.* **59**, 428–32.
61 WALKER D. A. & HEROLD A. (1977) Can the chloroplast support photosynthesis unaided? In Photosynthetic Organelles: Structure and Function. (Eds. Y. Fujita, S. Katoh, K. Shinata and S. Miyachi). *Special Issue of Plant and Cell Physiol.* 295–310.
62 WALKER D. A. & ROBINSON S. P. (1978) Chloroplast and Cell. A contemporary view of photosynthetic carbon assimilation. Vorgetragen auf der Botaniker—Tagung in Marburg am 13 September 1978). *Ber. Deutsch. Bot. Ges. Bd.* **91**, 513–26.
63 WALKER D. A. (1979) Preparation of higher plant chloroplasts. Photosynthesis and nitrogen fixation. *Methods in Enzymology*, pp. 94–104. New York: Academic Press.
64 WALKER D. A. (1978) *Energy, Plants and Man.* (A booklet for first year students in Environmental Sciences etc.). Packard Publishing Ltd., (Chichester) 1978.
65 WERDAN K., HELDT H. W. & GELLER G. Accumulation of bicarbonate in intact chloroplasts following a pH gradient. *Biochim. Biophys. Acta,* **283**, 430–41 (1972).
65a WERDAN K. & HELDT H. W. (1971). The phosphate translocator of spinach chloroplasts. In Progress in Photosynthesis, (eds. G. Forti, M. Avron, A. Melandri), Vol II pp. 1337–44 (*Proc. 2nd. Intern. Congr. Photosynthesis Res* 1971) The Hague: N. V. Junk.
66 WERKHEISER, W. C. & BARTLEY W. (1957) The study of steady-state concentrations of internal solutes of mitochondria by rapid centrifugal transfer to a fixation medium. *Biochem. J.* **66**, 79–91.

# Chapter 9
# The Regulation of Photosynthetic Carbon Assimilation



## SUMMARY

Like other aspects of metabolism, photosynthetic carbon assimilation is a controlled or regulated process although neither the precise mechanisms nor the full effects of regulation are clearly understood. Regulation is based on catalysis (modulation of enzyme activity) mass action (influence of metabolite concentrations, etc.,) and transport (the movement of metabolites between the chloroplast and its cellular environment). Typically, the regulated enzymes are those which catalyze 'irreversible' reactions (those with a high 'forward to back' reaction rate such as: Ru5P → RBP → PGA and FBP → F6P). The sequence PGA → DPGA → triose phosphate is very dependent on mass action and the first stage is strongly influenced by the concentrations of ADP, ATP and PGA. Transport across the chloroplast envelope is governed by the availability of Pi and this is a major factor in determining the distribution of photosynthetic product between internal utilization, internal storage (as starch), and export.

## 9.1 General principles

If we think of a factory, building complicated machines like aircraft, we are immediately conscious of a high degree of organization. Clearly, if it is to function effectively, it is essential that the various raw materials and parts which are to be assembled arrive at the right time and in the right quantities. Within the factory an excess of landing-wheels will not compensate for a shortage of radar components. The living organism is faced with similar problems. Its metabolism must be regulated to a very high degree. This is particularly true of the higher plant which, unlike some organisms, cannot come in out of the rain and, by and large, has a very limited capacity for acquiring the compounds that it needs and disposing of those that it does not. Like animals, plants regulate various aspects of their growth, etc., by hormonal action but, in addition, there are more direct controls. Broadly speaking, these may be divided into three main categories—regulation of catalysis, regulation by mass action, and regulation by transport.

## 9.2 Regulation of catalysis

The extent to which a reaction will proceed is governed by its equilibrium constant, by its free energy change, and where appropriate, by its redox potential. As already noted (Section 2.5, 2.13), these are all interrelated, so that, e.g., a large negative $\Delta F'$ is equivalent to a large value for $K$ and both indicate that a reaction will go to virtual completion if other factors allow it to proceed at all. The most important of these factors, in a biological context, is the presence of an enzyme which catalyzes the reaction. It is then evident that if a reaction A → B has a large negative $\Delta F'$ and is catalyzed by an effective enzyme, then A will be converted into B as fast as it is formed unless the catalytic process is regulated. In the RPP pathway the rate of the reaction catalysed by fructose bisphosphatase is increased at high pH and high [$Mg^{2+}$] and this reaction therefore provides an excellent example of the way in which changes in the stromal environment can modulate enzyme action. In the dark, the more acid conditions in the stroma do not favour catalysis and for this reason and because the enzyme itself is in a largely inactive state (Section 9.3), fructose

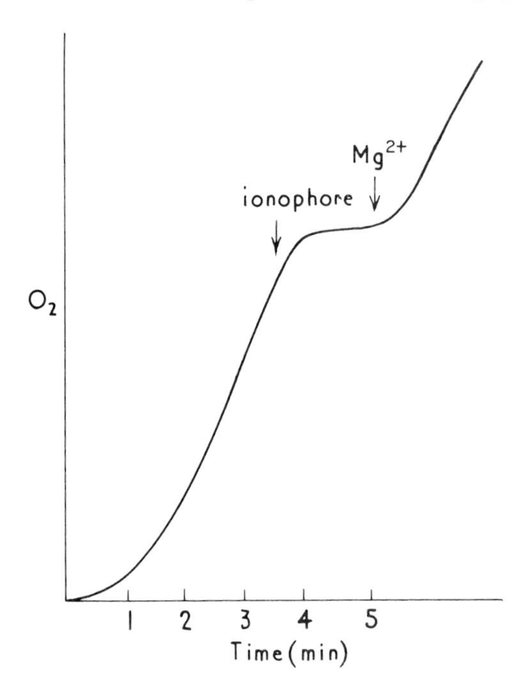


**Fig. 9.1** Inhibition of photosynthesis by $Mg^{2+}$ ionophore. Oxygen evolution by isolated chloroplasts is rapidly inhibited by addition of ionophore which permits escape of $Mg^{2+}$ from the stroma. Restoration follows addition of $Mg^{2+}$ to the medium. See also Fig. 9.4 (After Portis and Heldt, 1976).

bisphosphate hydrolysis is unlikely to proceed at an appreciable rate. Upon illumination, protons move rapidly from the stromal compartment into the thylakoid compartment and the stromal pH increases from about 7.0 to about 8.0. The efflux of $H^+$ from the stroma is countered by an influx of $Mg^{2+}$ from the thylakoid and it has been shown by Baier and Latzko that changes of this nature and magnitude are sufficient to bring about marked activation of the bisphosphatase. At York, in 1979, Ulrike Enser reported that, at high concentration, the salts of weak acids (such as acetate and propionate) reverse the $\Delta pH$ across the chloroplast envelope causing the stroma to become more acid than the medium. In such circumstances FBP accumulated and hexose monophosphate declined, indicating inhibition of FBPase. Similarly, in experiments with the reconstituted chloroplast system there is little or no photosynthesis at neutral pH and low $[Mg^{2+}]$ whereas at pH 7.8 and 5 mM $Mg^{2+}$, the rates of $O_2$ evolution with FBP as substrate are as high as those with PGA. Moreover, if an ionophore which facilitates the passage of $Mg^{2+}$ across the chloroplast envelope is added to intact chloroplasts, photosynthesis ceases but may be restored if the external $[Mg^{2+}]$ is increased (Fig. 9.1). All of these experiments point to a requirement for $Mg^{2+}$ and high pH and the realization of these requirements in the illuminated stroma.

## 9.3 Activation of catalysts

### (a) Light activation or deactivation

In Section 9.2 it is implied that FBPase 'works well' at high pH and in high $[Mg^{2+}]$. A difference between light activation in that sense (i.e. that light occasions a change in the stromal environment which favours catalysis) and in the sense that light can also bring about a change in the properties of an enzyme is somewhat artificial and certainly not easy to sustain in all circumstances (see below and Section 9.5). It is clear that a change in the reduction status of the sulfhydryl groups of an enzyme is frequently crucial and may often be accomplished *in vitro* by the addition of dithiothreitol. Buchanan *et al* have proposed a role for two proteins in the light-generation of the reducing conditions required for FBPase activation *in vivo*. These are thioredoxin and its reductase. In the light, ferredoxin is reduced in the normal fashion and electrons are then transferred from ferredoxin via the reductase to thioredoxin (Fig. 9.2a). In turn, in the reduced state, thioredoxin activates FBPase, SBPase, and possibly Ru5P kinase and triose phosphate dehydrogenase (Fig. 9.2b). It has also been proposed that the active enzymes could be oxidized (and thereby rendered inactive) by oxidized glutathione (Fig. 9.2c).

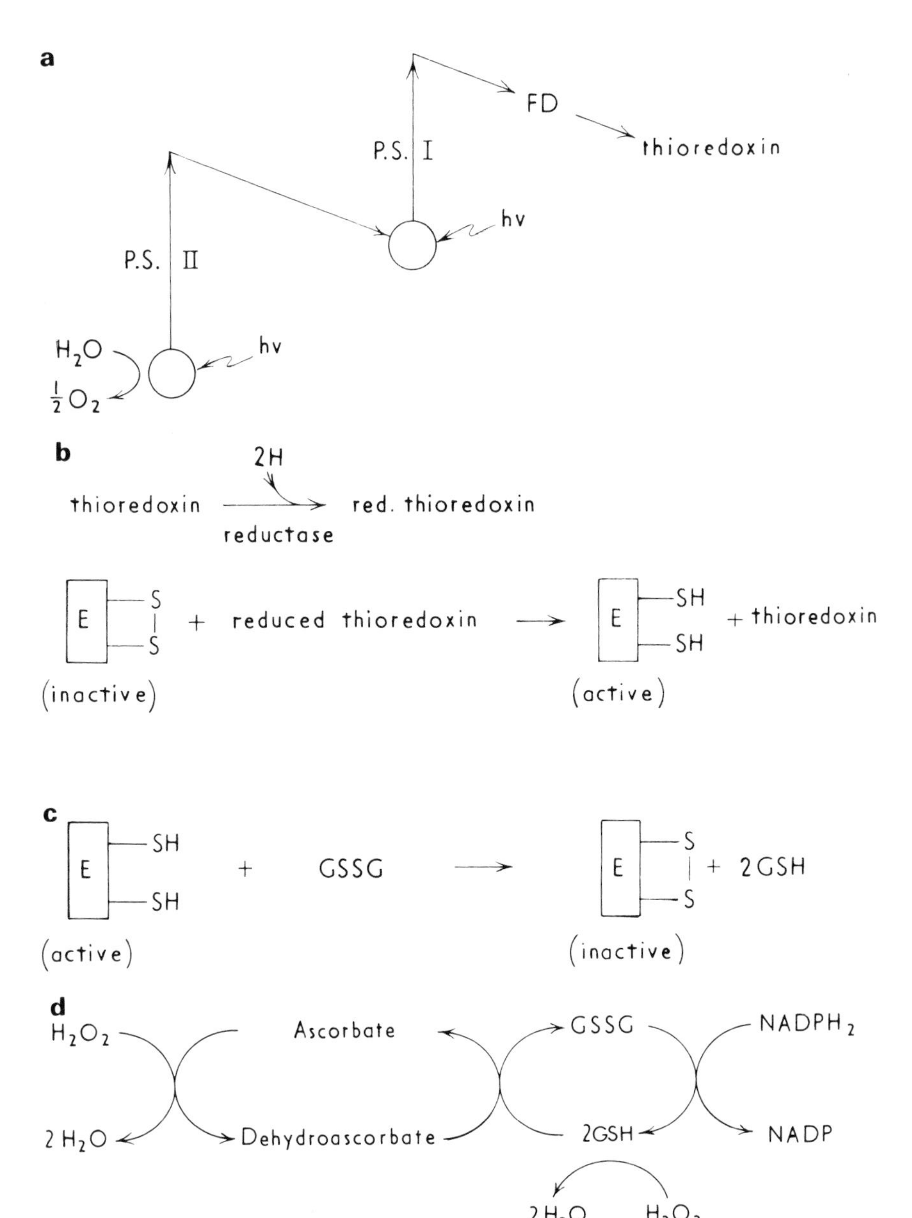


**Fig. 9.2** Activation by thioredoxin and deactivation by oxidized glutathione (GSSG). (a) electron transfer from water to thioredoxin (b) reduction by thioredoxin of disulfide bonds (c) reoxidation of SH groups by GSSG (d) oxidation and reduction of glutathione by $H_2O_2$ and $NADPH_2$.

More recently Buchanan *et al* have found evidence that two thioredoxins may be involved. The larger of the two is believed to activate fructose bisphosphatase, ribulose kinase, and triose phosphate dehydrogenase. The smaller only activates $NADP^+$-specific malic dehydrogenase. Both are reduced either by ferredoxin/ ferredoxin-thioredoxin reductase or by dithiothreitol.

Louise Anderson and her colleagues at Chicago have defined two light-effect mediators (LEMs) which are firmly bound to the thylakoid membranes. Both accept electrons on the reducing side of photosystem I (Fig. 9.3a), one prior to ferredoxin, the other after. Again, these are thought to activate enzymes of the RPP pathway (or to deactivate glucose phosphate dehydrogenase) by reducing disulfide bonds and

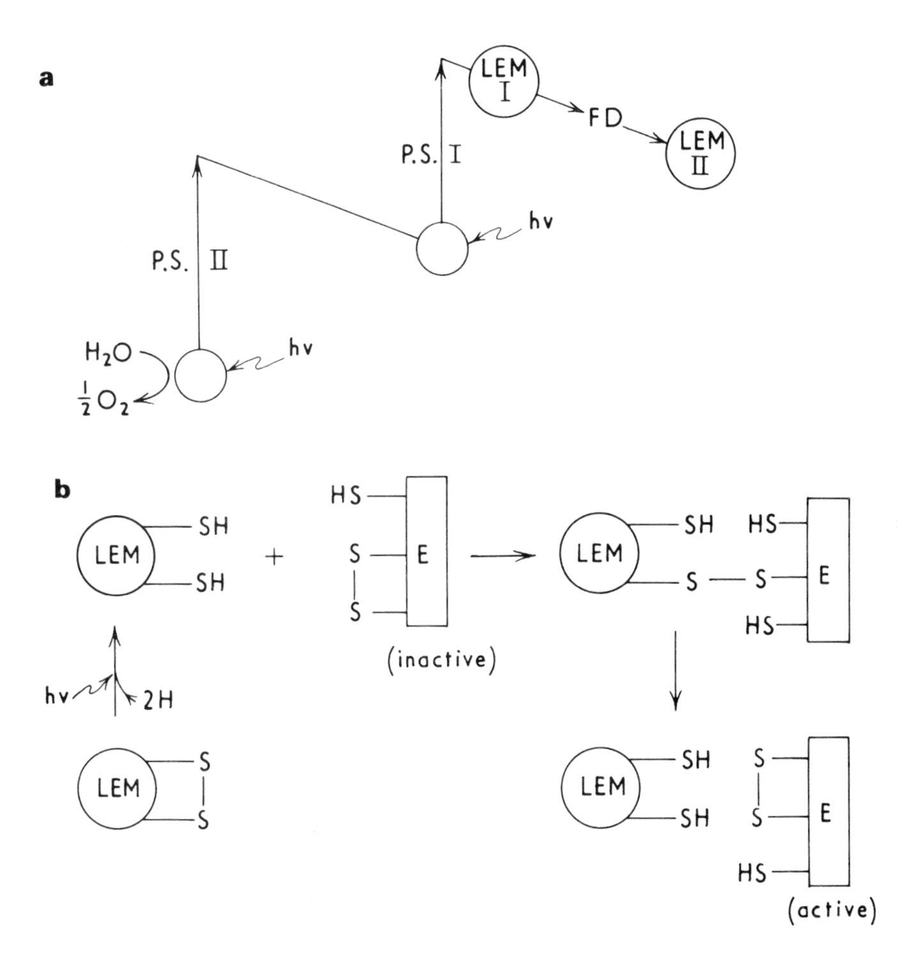


**Fig. 9.3** Activation by LEM, involving thiol disulfide exchange. (After Anderson *et al.* 1978)

thereby changing the status of potentially active sites on the enzymes concerned. Their experiments suggest, however, that the mechanism involved is not one of direct reduction but rather thiol disulfide exchange (Fig. 9.3b). It should be noted that despite some apparent similarities, the LEMs differ from the thioredoxin system in the fact that they are thylakoid-bound rather than freely soluble compounds and the relationship (if any) between the two systems remains to be established.

### (b) ACTIVATION BY $Mg^{2+}$

RBP carboxylase is also activated by $Mg^{2+}$ (Section 6.11). For many years this enzyme was believed to be inadequate for its postulated role *in vivo* because it did not seem to afford sufficient catalytic potential. The inadequacy was largely centred on its affinity for $CO_2$ which seemed too low to be acceptable. [Allowing for different assumptions in calculation, it seemed that it might require as much as 8 % $CO_2$ in the gas phase to achieve half maximal rate whereas the concentration of $CO_2$ in the stroma of the $C_3$ chloroplast is of the order of 0.01 %.] Eventually, it turned out that most measurements must have been made with the enzyme in a more or less inactive state. Lorimer *et al* have shown that the carboxylase forms a complex with $CO_2$ which is then rapidly activated at alkaline pH by $Mg^{2+}$. Its affinity for $CO_2$ is then in the region of 10 $\mu$M $CO_2$ (equivalent, at pH 8.0 and 20°, to about 0.03 % $CO_2$). It is possible that the carboxylase is activated by changes in $[Mg^{2+}]$ and $[H^+]$ *in vivo* in the same way as the FBPase. Activation in this manner has been simulated in a reconstituted chloroplast system in which 1 mM $MgCl_2$ is sufficient to provide the $Mg^{2+}$ requirement to maintain 90 % of maximal electron transport from $H_2O$ to diphosphoglycerate (Fig. 9.4); but $CO_2$ fixation is minimal until the $[Mg^{2+}]$ is raised to about 2.5 mM. (These changes in $[Mg^{2+}]$ are of about the same magnitude as those which are thought to occur within the stroma upon illumination. However, at a meeting held in York in April 1979 Krause and Ben-Hayyim presented evidence, based on studies with the metalochromic indicator Eriochrome Blue S.E. [*Biochim. Biophys. Acta* **460**, 500–10, 1977], that low concentration of $NH_4Cl$ abolished the light induced increase in stromal $[Mg^{2+}]$ in isolated chloroplasts without significantly affecting $CO_2$ fixation. They concluded that the dark levels of $Mg^{2+}$ in the stroma might be sufficient for high activities of 'regulated enzymes'.) Bahr and Jensen have examined the activity of carboxylase within illuminated chloroplasts and find it to be 40–60 % of full activation. Lorimer and Heldt have found a role for Pi during activation and have reported complete activation of the enzyme during illumination of intact chloroplasts. Some metabolites and coenzymes such as 6-phosphogluconate, fructose 1,6-bisphosphate, PGA and NADPH may also influence the activation of the carboxylase at low [$CO_2$]. As indicated above, there is no doubt that [RBP] carboxylase can lose activity when it is released from the chloroplast (particularly if it is subjected to purification procedures), and that this activity can be restored by incubation with $Mg^{2+}$ and $CO_2$ at slightly alkaline pH. There is now, however, real doubt that anything approaching full deactivation occurs in the dark. Originally, the idea of dark deactivation derived from observations that RBP did not decline in the dark as fast or as far as might have been predicted from the rate of $CO_2$ fixation in the light and the known favourable equilibrium position of the reaction. Apart from the difficulties involved in accurately measuring small quantities of RBP, however, some of these results might be explained by declining catalysis and binding of RBP at sites which do not lead to carboxylation. Whatever the explanation a large number of recent observations at Sheffield point to very high levels of carboxylase activity in chloroplasts and protoplasts separated from dark-stored tissue. Where this is in excess of the rate of photosynthesis, it is clearly difficult to understand how 'dark deactivation' might afford some metabolic advantage. The possibility must remain that some deactivation is an artefact of isolation and that it is simply *catalysis* which proceeds more effectively in the conditions which obtain in the illuminated stroma (Section 9.5). Indeed, recent work at Sheffield has failed to detect appreciable dark deactivation of several enzymes of the RPP pathway (with FBPase a notable exception—Table 9.1). This suggests that 'light effect mediators' or the ferredoxin/thioredoxin systems (Section 9.3) may sometimes function as repair mechanisms, which repair damage caused by super-oxide radicals and the like (Section 5.8), rather than regulators as such.

[It should be emphasized that RBP carboxylase has proved to be a notoriously difficult enzyme to assay in the past (Section 6.11), partly because of the requirement for activation and partly because of the inhibition by degradation products of RBP which have been demonstrated by Tolbert *et al.* For this reason, much of the work

**Table 9.1.** Light activation of RPP pathway enzymes

| Enzyme | (Fold) Increase | Final Rate ($\mu$mol mg$^{-1}$ Chl h$^{-1}$) |
|---|---|---|
| FBPase | 5.0 | 120 |
| RBP carboxylase | 1.1 | 200 |
| NADP-G3P dehydrogenase | 1.4 | 1000 |
| Ru5P kinase | 1.5 | 1100 |
| Phosphoglycerate kinase | 1.0 | 2200 |

(Courtesy of R. Leegood, unpublished)

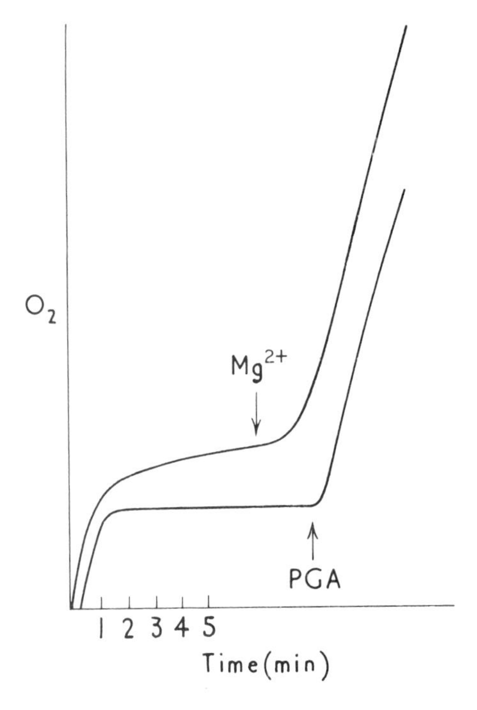


**Fig. 9.4** Activation of RBP-dependent $O_2$ evolution by $Mg^{2+}$ in a reconstituted chloroplast system. In lower trace, the [$Mg^{2+}$] is sufficient to support electron transport to $NADP^+$ (first evolution of $O_2$, at left) or to triose phosphate, which requires ATP formation (second evolution, on right, following PGA addition). In upper trace RBP is present from outset but $O_2$ evolution falls to a low rate after initial reduction of $NADP^+$ during the first minute. Subsequent addition of $Mg^{2+}$ permits conversion of RBP → PGA and resumption of rapid oxygen evolution. (After Lilley *et al.*) [Ref. 29a]

prior to 1975 must be regarded with added caution in regard to such aspects as maximal rate, affinity for $CO_2$, etc.] In recent years there has been a great deal of speculation about the role of ascorbate and other reducing agents in chloroplasts. Ascorbate can be present in very large amounts, particularly in the summer and has often been used as a protective agent or electron donor in work with isolated chloroplasts. No function other than an antioxidant has been apparent but Halliwell and Foyer have defined this more closely in their proposal that it may help to dissipate $H_2O_2$. The chloroplast does not contain catalase and yet is capable of generating $H_2O_2$ by donating electrons to $O_2$ from carriers such as ferredoxin. Reduced glutathione could be oxidized by $H_2O_2$ either directly via glutathione peroxidase or indirectly by hydroascorbate. Subsequent reduction of GSSG by

NADPH would maintain the *status quo* and the oxidation status of enzymes otherwise subject to inactivation by $H_2O_2$(Fig. 9.2d). [Refs. **16–23, 25, 30–34, 39, 41–45**]

## 9.4 Regulation of catalysis and activation of catalysts—a distinction

In Sections 9.2 and 9.3 some distinction is made between the effects of light on catalysts and *catalysis*. In many respects this is a purely academic separation (because it seeks to distinguish between effects on an enzyme and effects on the reaction which it accelerates) but it should not be entirely dismissed even when it is granted that some parameters, such as pH, may affect both simultaneously. For example, the reader of a text-book on enzymology written 20 years ago would almost certainly have found no mention of 'regulation'. On the other hand the fact that most enzymes were active over a limited range of pH was already well established and time-honoured (enzymes are protein and, as such, are multivalent electrolytes). Accordingly the ionization state of their constituent amino acids and therefore their catalytic ability is influenced by hydrogen ion concentration. Similarly carboxylate ions of amino acids will interact with cations such as $Mg^+$. Such effects of pH were mostly related to changes in *V*max or *K*m or to the stability of the enzyme itself although it may also be noted that in reactions directly involving protons the effect of hydrogen ion concentration *on the reaction itself* may be of considerable importance [For example the $\Delta F'$ value of the reaction catalysed by malic dehydrogenase (Section 6.12) is so unfavourable in the direction of malate oxidation that it cannot be readily made to go in that direction unless the equilibrium is shifted by the provision of a sink for the end-products, cyanide (by binding oxaloacetate) and high pH (by binding $H^+$) will favour malate oxidation. Thus it has been commonplace for many years to assay enzymes at optimal pH and, very often, (as appropriate) to add cysteine or reduced glutathione to protect sulfhydryl groups. The fact that an enzyme will 'work' at pH 8.0 (as in the illuminated chloroplast) and not at pH 7.0 (as in the darkened chloroplast) is not, therefore necessarily an indication that the enzyme itself has undergone a sufficiently large physical change to render it 'inactive'. On the other hand, like RBP carboxylase, an isolated enzyme may be readily inactivated *in vitro* and require alkaline pH in order to complete a re-activation process. Similarly $Mg^{2+}$ ions may be required for *activation* (RBP carboxylase is again a case in point) but not for *catalysis* as they are in kinase reactions. Similarly an enzyme like FBPase which is evidently very responsive to its reduction status and is largely inactive in the dark will nevertheless display some activity if it is assayed in a mixture containing 10–20 mM $MgCl_2$ at pH 8.0. It may be reasonably assumed that it is the *combination* of the oxidation deactivated process *and* the unfavourable environment offered by the darkened stroma which ensures that the reaction catalysed by this enzyme virtually ceases in the dark.

## 9.5 Dark deactivation

It has been stated (Section 9.3) that light activating systems may function as repair mechanisms. It seems equally probable that they can also function in regulation i.e., that they constitute one half of a system which 'switches on' an enzyme in an illuminated chloroplast and 'switches it off' in the dark. The difference between repair and regulation, if it is real, could be a question of degree. It would manifest itself not in the extent to which an enzyme *could* be light-activated but in the extent to which such light activation appeared to be necessary, i.e. in the degree of dark deactivation which must be reversed in order to permit normal function. For example if an enzyme can be shown to undergo a ten-fold increase in activity when isolated chloroplasts are illuminated in an appropriate mixture then it is clear that it can be light-activated. If, however, when it is taken directly from the darkened chloroplast it still displays most of its activity then there is no apparent requirement for light-activation and the fact that such a potential exists suggests that it might serve to make-good normal 'wear and tear'. Conversely, if an enzyme displays very little activity when taken directly from the dark there is a clear need for it to be 'switched on'. Apparent differences of this sort are recorded in Table 9.1. As already noted, it is possible that they reflect artefacts in experimental procedures but it is evident that when treated in this particular way only one (FBPase) is really switched off. [Even in this case there was some residual activity in the dark but it must be borne in mind that in experiments of this sort the enzymes were assayed under otherwise optimal conditions of pH etc. and these would also change for the worse in the darkened stroma, leading to complete inactivation. It is also suggestive that FBPase was present in relatively low activity, when compared with Ru5P kinase. This is consistent with a regulatory role but, on the other hand the actual rate which is sustained *in vivo* will depend on the affinity of the enzyme for its substrate (s) and their steady state concentration. In this regard, [Ru5P] is almost certainly very low and therefore it could be misleading to read too much into rates achieved under more favourable conditions].

If it is accepted that different enzymes show different degrees of deactivation in the dark, it is of interest to ask how this is accomplished. It is possible that an enzyme may require constant reduction for activity and that it then reverts to an inactive oxidized state as soon as the reductant is removed. On the other hand an entirely new mechanism may come into play. It has been suggested, for example, that glutathione is oxidized in the dark by the action of light-generated $H_2O_2$ and glutathione peroxidase and that GSSG then oxidizes the sulfhydryl groups of susceptible enzymes. This possibility has not been strengthened by the observations of Halliwell and Foyer (who found that glutathione was primarily in the reduced state in the dark as well as in the light) although this is not, in itself, conclusive (see Fig. 9.2 and also Section 9.3).

Work by Leegood favours the former possibility (constant reduction) and indicates a requirement for molecular $O_2$ or an electron acceptor. He has found that FBPase is not only remarkably responsive to electron flow but also to the availability, or otherwise, of a number of electron acceptors. For example carbon dioxide *limits* full activation, presumably by functioning as an effective, alternative electron sink. Similarly light activation is not reversed in the dark if oxygen is excluded from the mixture but the enzyme is deactivated, both in light and dark, and in the absence of $O_2$, by oxaloacetate (which can function, like $CO_2$ or $O_2$ as an electron sink). Antimycin A (which inhibits cyclic electron transport) had little effect on FBPase activity when electron flow was rapid but brought about detectable additional activation when electron flow was slowed by the addition of DCMU (at a concentration sufficient to inhibit electron flow by 50%) or by decreased light intensity. Conversely, pyocyanine (which promotes cyclic electron flow) diminishes FBPase activity. These relations are summarized in Fig. 9.5. Broadly it is seen that anything which deflects electron transport away from the reduction of FBPase will diminish its activity in the light and *vice versa*. In the dark, some part of this electron

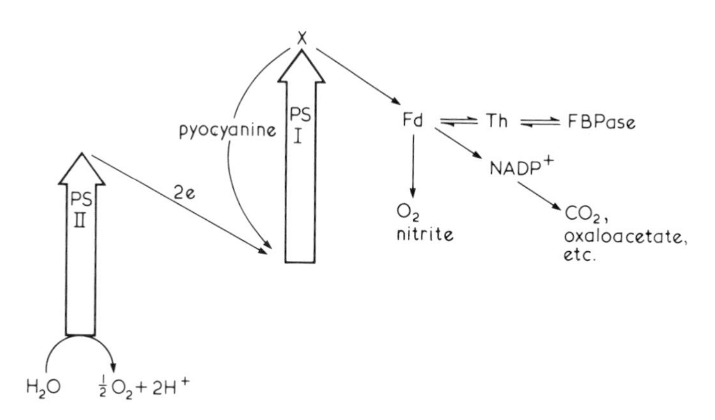


**Fig. 9.5** Light activation and dark deactivation of FBPase. Electrons reaching FBPase via thioredoxin, ferredoxin and the photosystems reduce dithiol groups and increase activity. Other electron acceptors ($CO_2$, $O_2$, oxaloacetate, nitrite) and cyclic electron transport divert electrons from FBPase, diminishing activity in the light and deactivating it in the dark (After Leegood & Walker, 1980) [Ref. **28b**]

flow is reversed and oxidation (either by molecular oxygen, some derivative of oxygen, or an appropriate substitute) will bring about oxidation of sulfhydryl groups and consequent deactivation. Whether thioredoxin is directly concerned in the oxidative deactivation, as well as the reductive activation, remains to be established.
[Refs. **28a, b**].

As pointed out, above, the demonstration that an enzyme can undergo light activation is not, in itself, evidence that this process is truly regulatory in nature. No one today would question the fact that RBP carboxylase can be readily inactivated during extraction or readily reactivated by exposure to concentrations of $H^+$ and $Mg^{2+}$ which may well obtain in the illuminated chloroplast. Similarly, present experience indicates that if an illuminated chloroplast is lysed directly into an assay mixture it does not display its full, potential, carboxylase activity (as judged by pre-incubation with high $Mg^{2+}$ at alkaline pH) but that its activity is, nevertheless, somewhat higher than that displayed by similar chloroplasts taken from the dark. What is at issue, is the relative magnitude and importance of these changes. Given a $V$max of 1000 $\mu$mol. mg$^{-1}$ Chl h$^{-1}$ (i.e. a relative excess) a 1.5 fold light activation may be relatively unimportant whereas a 4-fold activation would almost certainly be of crucial significance. However, although 4-fold activation has been observed the spectre of inadvertent inactivation during manipulation must obviously be entertained and the fact that other workers have recorded changes in the range of 1 to 1.5 can not be lightly dismissed.

## 9.6 Mass action

If a reaction proceeds with a large negative $\Delta F'$ (large $K$), then it will be largely uninfluenced by its end-products (cf. Section 2.5) and the only real possibility of regulation lies in some modulation of the enzyme concerned (as in Section 9.2). Conversely, if the $\Delta F'$ is positive or small ($K$ small or near unity), then the extent to which the reaction proceeds will depend very much on accumulation of substrate and removal of product.

The reduction of PGA to triose phosphate (Section 6.11), the only reductive step in photosynthetic carbon assimilation, provides an excellent example. As shown in Fig. 9.6, the overall reaction proceeds in two stages. The first is unfavourable and involves the expenditure of ATP.

$$\text{PGA} + \text{ATP} \rightarrow \text{DPGA} + \text{ADP} \qquad \text{Eqn. 9.1}$$

The second is favourable and involves the utilization of NADPH as an electron donor.

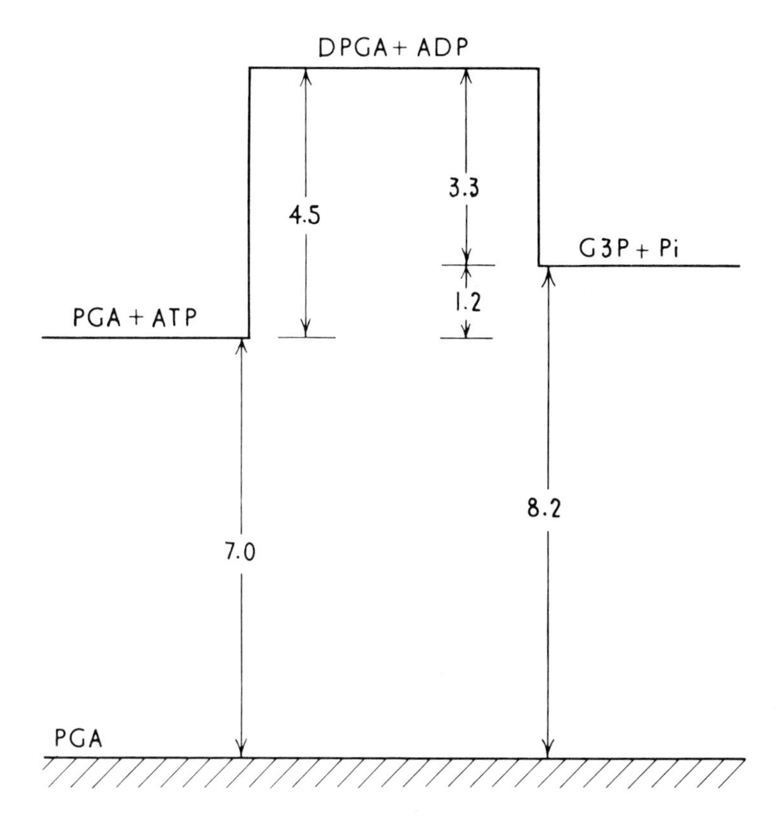


**Fig. 9.6** Approximate free energy relationships in PGA reduction. The direct reduction of PGA to triose phosphate would demand an energy imput of approx. 8 kcal. Instead the utilization of ATP diminishes this formidable gap by about 7 kcal leaving the overall conversion of PGA → G3P as a freely reversible reaction with a $\Delta F'$ of about 1 kcal. Nevertheless, the first stage of this sequence has a $\Delta F'$ of about 4.5 kcal which means that the progress of the reaction PGA + ATP → DPGA + ADP is particularly susceptible to decreases in the ratio of [ATP]/[ADP] and therefore to the rate of photophosphorylation on the one hand and the activity of ATP sinks on the other (see text).

$$\text{DPGA} + \text{NADPH} \rightarrow \text{G3P} + \text{Pi} + \text{NADP}^+ \qquad \text{Eqn. 9.2}$$

Together the two reactions constitute a freely reversible oxidation/reduction which is the same as that in glycolysis except that the coenzyme is $NADP^+$ rather than $NAD^+$. In glycolysis, the oxidation is a substrate linked phosphorylation in which 7 of the 8.2 kcal associated with the oxidation can be considered to be conserved as ATP. In photosynthesis, the reduction of PGA to triose phosphate is made energetically feasible by the expenditure of ATP. Nevertheless, the unfavourable equilibrium position of the first stage makes it particularly susceptible to reversal by

ADP and experiments with the reconstituted chloroplast system have shown that PGA-dependent $O_2$ evolution is readily inhibited by the addition of a small quantity of ADP. (Fig. 8.17).

The reconstituted chloroplast system is comprised of osmotically shocked chloroplasts supplemented with co-enzymes and stromal proteins. Since it lacks the barriers imposed by the intact envelopes, it is immediately open to access and manipulation and, in many ways, may be regarded as one large chloroplast bounded by the walls of the reaction vessel. As indicated above, the reconstituted system will evolve $O_2$ with PGA as substrate and this $O_2$ evolution is temporarily arrested upon the addition of ADP. As $O_2$ evolution ceases, however, photophosphorylation continues (whether this is cyclic, pseudocyclic or both remains to be certainly established) and as the added ADP is converted to ATP, the temporary inhibition is lifted. Accordingly, similar inhibition may be brought about by the addition of ATP sinks such as R5P (Eqn. 9.3)

$$\text{R5P} \longrightarrow \text{Ru5P} \xrightarrow{\text{ATP} \curvearrowright \text{ADP}} \text{RBP} \qquad \text{Eqn. 9.3}$$

or glucose plus hexokinase (Eqn. 9.4)

$$\overset{\text{ATP} \curvearrowright \text{ADP}}{\text{glucose} \rightarrow \text{glucose 6-phosphate}} \qquad \text{Eqn. 9.4}$$

and all may be entirely overcome by the simultaneous inclusion of a powerful ATP generator such as creatine phosphate and its kinase,

$$\text{creatine phosphate} + \text{ADP} \rightarrow \text{creatine} + \text{ATP} \qquad \text{Eqn. 9.5}$$

which immediately converts the added ADP to ATP.

The regulatory nature of this interaction becomes most apparent with FBP as a substrate. Carbon assimilation follows almost immediately upon the addition of this metabolite but $O_2$ evolution commences only after a lag (Fig. 9.7). As already noted (Section 9.3), FBPase is activated by certain reductants, by high pH, and by high $[Mg^{2+}]$. All of these factors extend the lag in $O_2$ evolution and in each case, the lag may be eliminated by creatine phosphate and its kinase. Similarly, the lag may be extended by the addition of extra FBPase or by adding part of the fructose

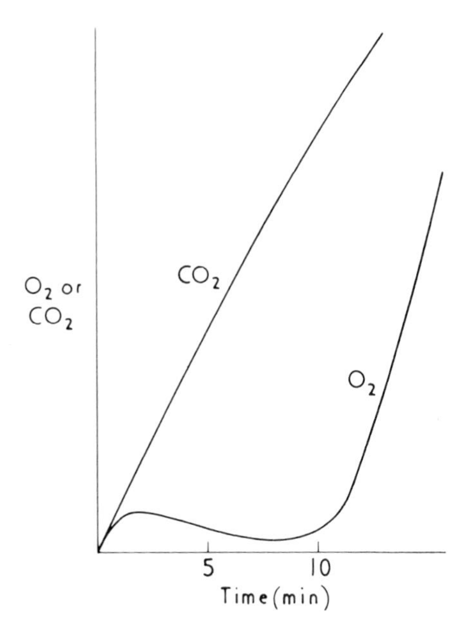


**Fig. 9.7** Delay in $O_2$ evolution in the reconstituted chloroplast system with fructose bisphosphate as substrate. (After Walker *et al*) [Ref. **43**]. For further explanation see Fig. 9.8.

phosphate as F6P rather than FBP (which is equivalent to hydrolyzing part of the FBP prior to addition). Here then is the apparent contradiction that any of a number of steps which accelerate one reaction in the RPP pathway will automatically arrest another. In each instance the causal factor is the same (Fig. 9.8). Thus, increased FBPase leads (via FBP, etc.) to a higher steady-state concentration of Ru5P. This in turn acts as an ATP sink and the consequent lowering of the [ATP]/[ADP] ratio blocks PGA reduction and its associated $O_2$ evolution. In steady-state photosynthesis, stromal [PGA] is high and stromal [pentose monophosphate] is low. The reason is now self-evident because any increase in Ru5P will shut down PGA reduction and hence Ru5P production. In this way, only enough triose phosphate will find its way through the regenerative phase of the RPP pathway to match the current rate of ATP production. Should the light decrease, less triose phosphate will enter this phase and vice versa. Similarly, in the steady state, any tendency for excess triose phosphate to enter the regenerative phase will be matched by a corresponding decrease in PGA reduction and, presumably, by a change in the proportions of PGA

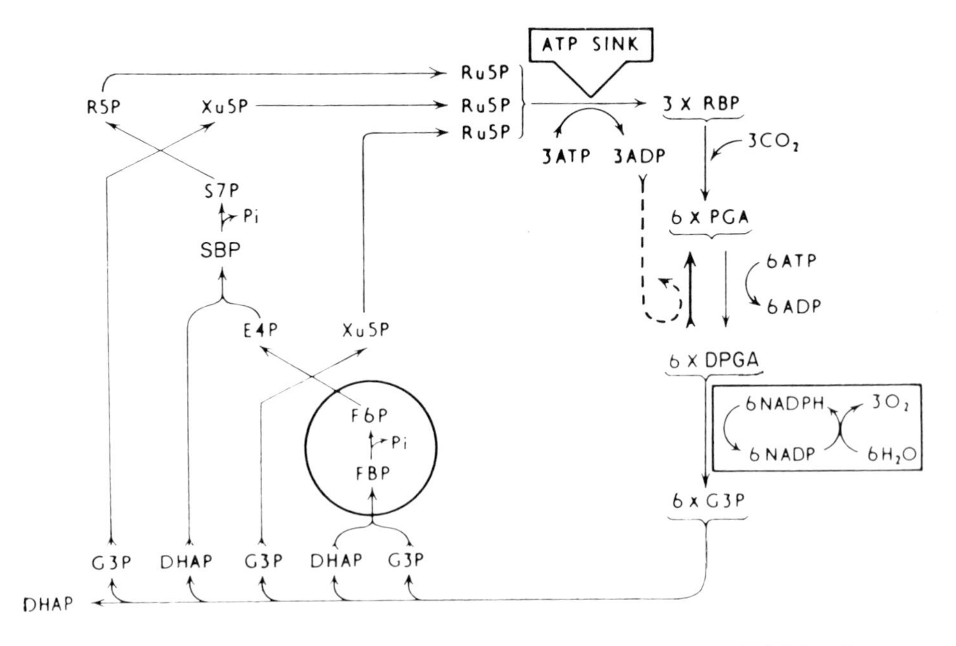


**Fig. 9.8** Explanation of delay in $O_2$ evolution recorded in Fig. 9.7. The events which bring about the lag in $O_2$ evolution with FBP as substrate and influence its extent. The diagram outlines the reactions of the RPP pathway. When FBP is introduced in substrate quantities (circles, bottom left), it initiates a series of reactions. Carbon dioxide fixation starts very quickly indicating that PGA is being formed. $O_2$ evolution (apart from that associated with the reduction of catalytic NADP) is delayed showing that DPGA is not being reduced to G3P. If it were, NADPH would be reoxidized to $NADP^+$ and this would accept electrons from water (via the photosynthetic electron transport sequence) and $O_2$ would be evolved (see box at bottom right). The reason that $O_2$ evolution is delayed despite the availability of PGA, ATP, and NADPH is that Ru5P (top centre) formed from FBP acts as a highly efficient sink for ATP thus regenerating ADP as rapidly as endogenous photosynthetic phosphorylation can form ATP.

The steady state concentration of ADP is therefore high and (heavy arrow) favours the conversion of DPGA to PGA rather than its reduction. Eventually, however, the newly generated Ru5P pool shrinks (as the FBP from which it is derived shrinks), the ADP concentration falls, DPGA is reduced, and $O_2$ is evolved. Factors which favour the conversion of FBP to F6P lead to the formation of a large Ru5P pool and a long lag. Factors which favour the conversion of ADP to ATP shorten the lag. (From Walker *et al*, 1976) [Ref. **43**]

and triose phosphate which are exported to the cytoplasm (see Chapter 7). [Refs. **29, 36, 37, 38, 40**]

## 9.7 Regulation by transport

In order that export (Chapter 8) can continue unchecked, it is necessary for the Pi exported from the stroma as organic phosphate to be returned to the stroma as inorganic phosphate (Fig. 9.9). If the supply of Pi is diminished, the [ATP]/[ADP] ratio will decrease, PGA reduction will fall and again the regenerative phase of the cycle will slow. Maximal photosynthesis will only follow when export is maximal. It may be noted in this regard that triose phosphate is consumed in two reactions. Either it condenses to form hexose bisphosphate or else it enters the transketolase reaction. Entry into the transketolase reaction will, in turn, depend on how much F6P and S7P is available for $C_2$ donation in Equation 6.21.

It has been shown by Preiss *et al* that when stromal [Pi] is low and [PGA] is high (as it will be when export is diminished), the activity of ADP glucose pyrophosphorylase is greatly increased. This enzyme plays a big role in starch synthesis and its allosteric activation by a high [PGA]/[Pi] ratio should ensure that it is most active

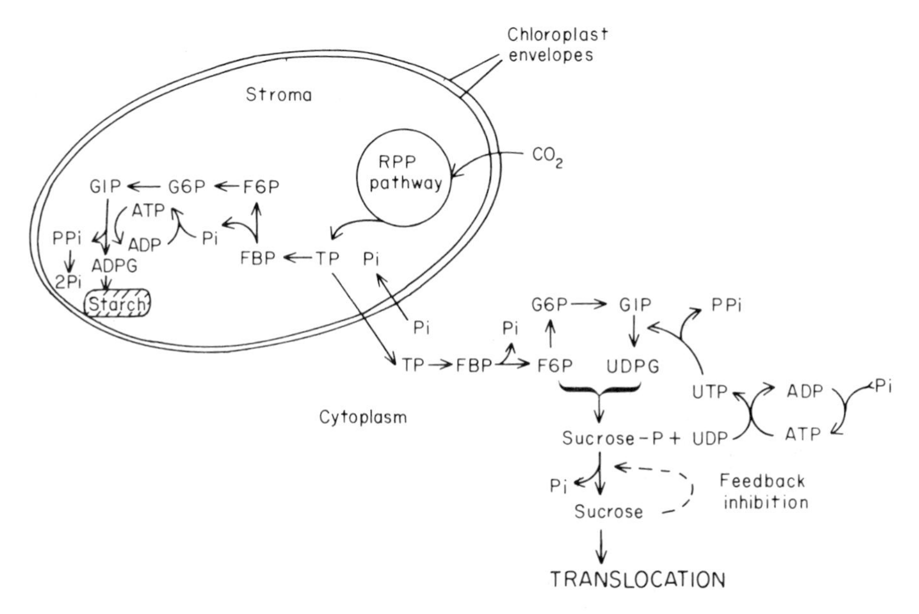


**Fig. 9.9** Regulation by transport. Starch is synthesized in the stroma and sucrose in the cytoplasm. Photosynthesis will be maximal when processes such as sucrose synthesis release Pi from triose phosphates, etc., so that it may be returned to the stroma at a non-limiting rate. When external Pi is limiting, photosynthesis can only proceed as fast as starch synthesis and although low [Pi] and high [PGA] bring about allosteric activation of a key enzyme in this process, accumulation of hexose phosphate in the stroma tends to increase pentose monphosphate formation and this in turn depresses PGA reduction by the mechanism shown in Section 9.6. If translocation of sucrose is slowed for any reason, accumulating sucrose may bring about a feedback inhibition of its own synthesis and, by slowing Pi release, bring about a corresponding decrease in photosynthesis. Much of this is speculative but there is no doubt that low external [Pi] retards photosynthesis (Fig. 9.10) and that reactions which sequester cytoplasmic Pi bring about a similar response (Fig. 9.11). The relationship between [Pi], transport and photosynthetic rate is also explored in Chapter 7.

when conditions within the stroma best favour starch synthesis. It has, in fact, now been clearly demonstrated by a number of workers that this prediction is realized when isolated chloroplasts are deprived of an adequate supply of Pi (Fig. 9.10). As would be expected, export is slowed and the decreasing [ATP]/[ADP] ratio slows PGA reduction. Increased consumption of F6P in starch synthesis will diminish the amount of triose phosphate entering the first transketolase reaction ( Section 6. 11).

It may be noted that the equilibrium position of the aldolase reaction may be important in this regard. This is well over in the direction of hexose bisphosphate

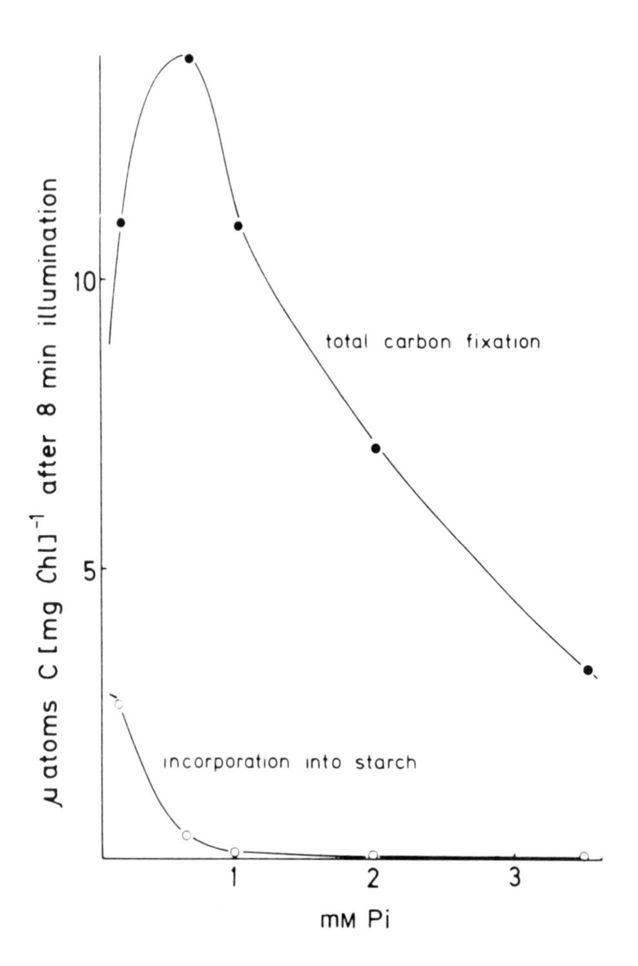


**Fig. 9.10** Starch synthesis as a function of external [Pi] in isolated chloroplasts. Total assimilation first increases and then falls as [Pi] is increased but starch decreases with increasing [Pi] and soon falls to a very low level. (After Heldt *et al*) [Ref. **26**].

formation but because there are two reactants and only one product, any build-up of triose phosphate will bring about an additional shift towards hexose accumulation.

The effect of [Pi] on the rate of photosynthesis and distribution of products between starch and soluble components has also been studied in leaf tissues by taking advantage of a known response to mannose feeding (Fig. 9.11). This sugar is actively phosphorylated by ATP in the presence of hexokinase but in many plants, the mannose phosphate which results does not readily undergo further metabolism. In consequence, cytoplasmic Pi is depleted. In $C_3$ species, the response is exactly what would be predicted from experiments with isolated chloroplasts, i.e., the rate of

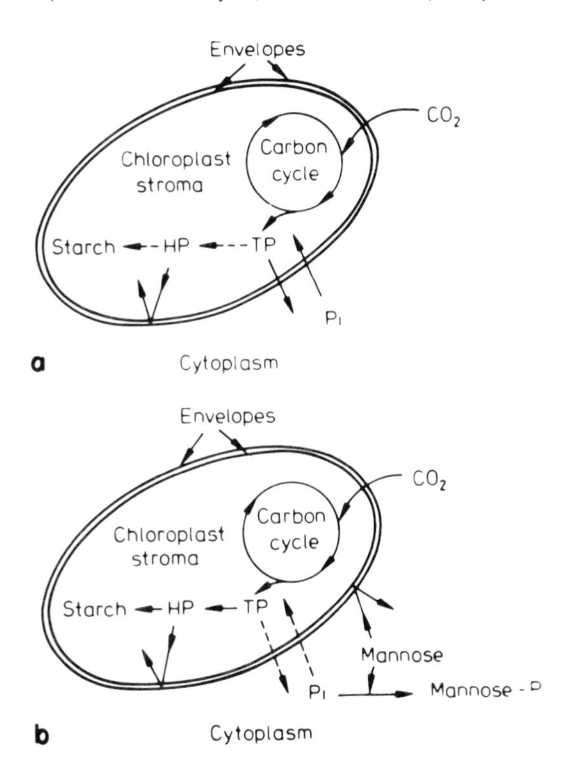


**Fig. 9.11** Stimulation of starch synthesis in $C_3$ species by mannose (cf. Fig. 9.12). When mannose is fed to leaf discs, it is probably unable to penetrate the chloroplast envelope but it is rapidly converted to mannose phosphate in the cytoplasm. Particularly in some $C_3$ species (such as spinach beet, a cultivar of *Beta vulgaris*) which are unable to bring about the further metabolism of mannose phosphate, mannose phosphate formation leads to a sequestration of cytoplasmic Pi. As the [Pi] in the cytoplasm falls, exchange of Pi and triose phosphate across the chloroplast envelop is diminished. This results in decreased photosynthesis but increased starch synthesis.

photosynthesis falls but at the same time, there are increases in starch formation. Under some conditions, these can be very large. (10-fold increases having been recorded). Feeding with labelled mannose shows that there is no direct contribution of this compound to the starch skeleton but that radioactive C and radioactive P both accumulate in a fraction with the chromatographic characteristics of mannose phosphate. In $C_4$ plants, mannose tends to inhibit starch synthesis rather than stimulate it (Fig. 9.12). This has again been attributed to sequestration of

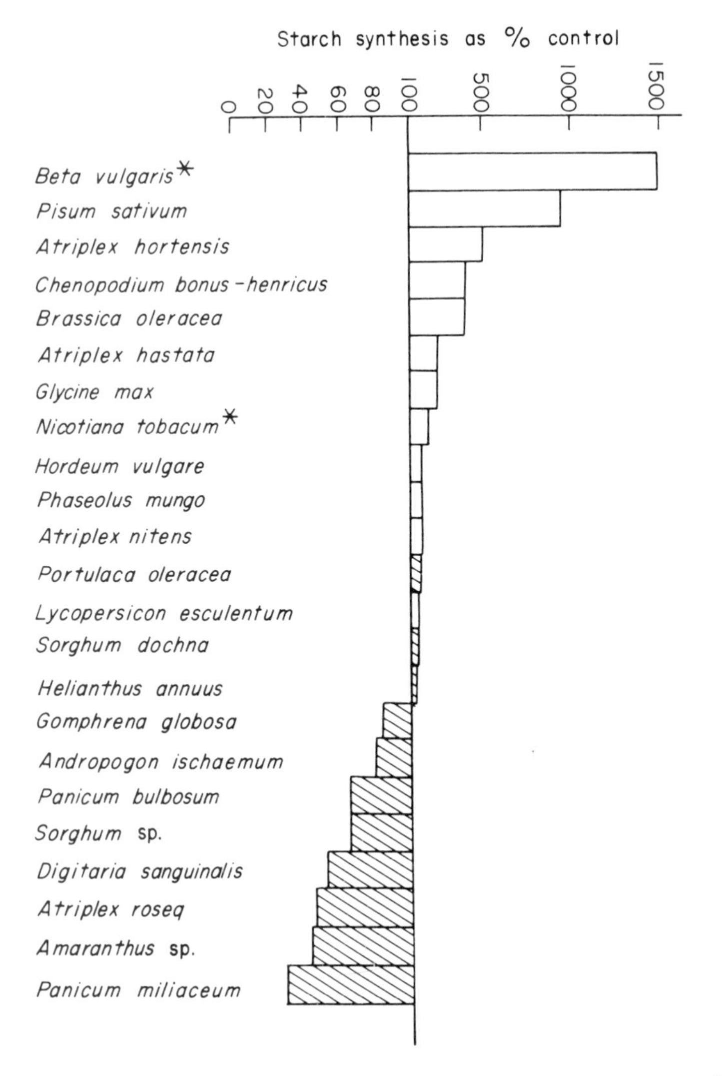


**Fig. 9.12** Differential effect of mannose on $C_3$ and $C_4$ species. Mannose ($10^{-2}$ M) stimulated starch synthesis in $C_3$ species (open bars) and inhibited starch synthesis in most $C_4$ species (hatched bars). Those $C_3$ species (left, open bars) which are most responsive to mannose are those which are unable to metabolize mannose phosphate.

cytoplasmic Pi but it is proposed that interference with the regeneration of PEP in the pyruvate phosphate dikinase reaction inhibits $CO_2$ fixation to such an extent that it outweighs any potential stimulation of starch synthesis (Section 12.10).

It is important to recognize the potential importance of regulation by transport and [Pi] and the fact that if the chloroplast is required to function in a cellular or artificial environment containing low [Pi], its overall rate of photosynthesis will fall whereas both the proportion and the total quantity of carbon entering starch actually increases. In the cell, much of the triose phosphate exported from the chloroplast is converted into sucrose, the principal transport metabolite. The literature concerning the effect of other plant processes on photosynthesis is somewhat confused and doubts have been expressed by Neales and Incoll that depression of photosynthesis by accumulated photosynthetic product has ever been clearly demonstrated. Nevertheless, there seems little doubt that at least in certain circumstances, the manipulation of distant sinks can bring about changes in the rate of photosynthetic carbon assimilation. Regulation by Pi offers an explanation of these results (Fig. 9.9). For example, it has been shown by Hawker that sucrose phosphatase is inhibited by sucrose. This is one step in which Pi exported from the chloroplast will be made available for re-entry into the plastids by cytoplasmic events. Thus, it is conceivable that diminished sink activity could lead to decreased transport of sucrose and that this could, in turn, slow the release of Pi within the cytoplasm and hence retard the export of metabolites from the chloroplast. Similarly, any accumulation of triose phosphate or PGA would tend to decrease export and the effective concentration of these compounds in the cytoplasm would also depend on the Pi concentration. Certainly, the results of mannose feeding (alone) are most readily explained on this basis. Starch accumulation has been observed in Pi deficient plants. Kakie, for example, has shown that starch increases despite diminished photosynthesis in Pi deficient tobacco plants and Alexander has noted starch accumulation and lowered sucrose in Pi deficient sugar cane.
[Refs. **13, 16a, 24, 27, 28, 35**]

Regulation by Pi in steady-state photosynthesis is inseparable from the role of Pi in induction (Sections 7.10–12 & Fig. 7.23) and the principal common features are therefore re-stated here. Firstly it should be borne in mind that when the rate of photosynthesis by isolated chloroplasts is plotted as a function of Pi concentration an extremely sharp optimum can often be observed (Fig. 7.21).

The events which occur on either side of this Pi optimum may be summarized as follows. When Pi is relatively deficient a low [ATP]/[ADP] ratio will slow PGA reduction (Section 9.6 & Fig. 9.8). At the same time the lack of external Pi will diminish the rate of triose phosphate export via the Pi-translocator (Section 8.23). This will divert a larger proportion of a diminished pool of triose phosphate towards starch synthesis—a divergence which will be pulled by increased ADP glucose pyrophosphorylase activity as this enzyme undergoes allosteric regulation induced by a high [PGA]/[Pi] ratio. At the Pi optimum the availability of Pi will ensure that import of Pi is sufficiently high to meet the maximal requirement of 1 Pi to $3CO_2$ in the reaction: $3CO_2 + 2H_2O + Pi \rightarrow$ triose phosphate $+ 3O_2$. This is the basis of spill-over, i.e. at optimal and sub-optimal [Pi] the rate of $CO_2$ fixation can only exceed the rate of supply of external Pi to the extent that it can form starch according to the classic overall equation for photosynthesis (Eqn. 1.1), a process which has no net Pi requirement. Accordingly if the supply of photosynthetically generated triose phosphate exceeds the immediate demands of cytoplasmic metabolism, excess fixation product can be accumulated as starch. If, on the other hand, the turnover of cytoplasmic Pi is high (for example if the cytoplasmic metabolism and distant sink capacity is so high that the [Pi] in the cytosol approaches and exceeds the optimum for photosynthesis) then triose phosphate which might otherwise have become available for starch synthesis will be exported and a still greater part of the Pi-recycling burden will be borne by cytoplasmic events. Under these conditions a lower [PGA]/[Pi] ratio will diminish the pull exerted by ADP glucose pyrophosphorylase and ensure that the regenerative reactions of the RPP pathway will be more favoured than starch synthesis. Eventually, however, triose phosphate export will compete successfully with RBP regeneration and the rate of photosynthesis will be diminished.

## 9.8 Consequences of regulation

It has already been emphasized (Section 9.1) that photosynthetic carbon assimilation, like any other important aspect of metabolism, must be regulated if it is to function effectively. In some instances (e.g., the distribution of product between export, internal consumption and temporary storage as starch; Section 9.7), the picture which emerges is already relatively clear, i.e., some of the likely mechanisms seem reasonably well defined and it is possible for the human observer to see what advantage the plant might derive from the operation of particular control mechanisms. This is equally true of the light deactivation of glucose 6-phosphate dehydrogenase and Bassham *et al* who have studied mechanisms for the activation and deactivation of the enzyme (other than those referred to in Section 9.3) have been quick to point out that this constitutes a potential switch between the RPP pathway in the light and the *oxidative* pentose phosphate pathway in the dark. Similarly, in many leaves, all or most of the starch that accumulates during the day is mobilized in the night and it seems highly probable that synthesis occurs without simultaneous degradation and vice versa. [The regulation of ADP glucose pyrophosphorylase

(Section 9.7) and fructose bisphosphatase (Section 9.3) must be important in this regard. Much remains to be elucidated but there is good evidence that triose phosphate is an important product of starch degradation in the chloroplast. If this occurs via fructose bisphosphate, then it is evident, for example, that (within the same compartment) equally·active fructose 6-phosphate kinase and fructose bisphosphatase would lead to an entirely futile cycle.] (Fig. 9.13). In other circumstances, the purpose of regulation is less readily seen. Tèleology (the tendency of the investigator to look for purpose) is not without its place in science (see Krebs, 1954), but it would be unwise to suppose that every potential regulatory response affords the organism some advantage. It is not surprising that the enzymes of the RPP-pathway work well at the pH, $Mg^{2+}$ concentration and redox levels which occur in the stroma of the illuminated chloroplast. Equally it is possible to suggest advantages which might derive from a decline in their catalytic activity as a result of the changes which ensue in the darkened stroma. It does not follow that some of these responses are essential facets of photosynthesis.

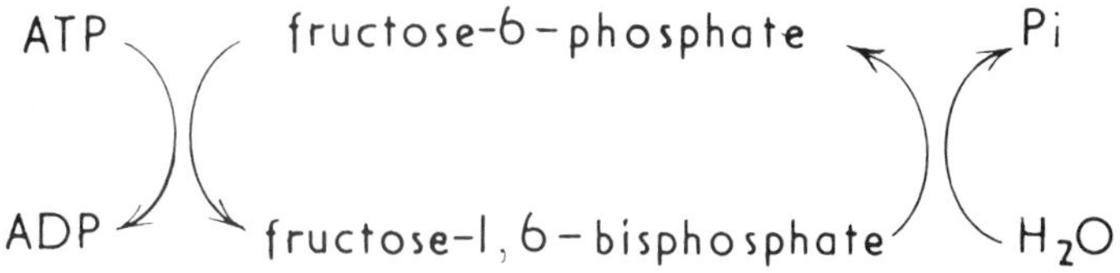

ATP → ADP; fructose-6-phosphate → fructose-1,6-bisphosphate; fructose-1,6-bisphosphate + $H_2O$ → fructose-6-phosphate + Pi

**Fig. 9.13** Fructose 6-phosphate kinase and fructose 1,6-bisphosphatase operating a futile cycle.

Additional aspects of regulation of carbon assimilation in $C_4$ photosynthesis are discussed in Chapter 12.
[Refs. **2, 4, 10, 13, 14**]

### 9.9 Addendum

Importance has been attached (Sections 7.13 & 9.7) to the manner in which chloroplasts 'perceive' that they are in a low [Pi] environment. It has been argued that one or more photosynthetic processes are governed by adenine nucleotide status, perhaps by 'energy charge', mass action or the associated [ATP]/[ADP] ratios. Certainly, if a point is reached at which the total free Pi within the chloroplast becomes appreciably smaller than that required to convert all of the available AMP and ADP to ATP, control of this sort becomes a real possibility. Clearly this notion would gain in credibility if accurate measurements of adenine nucleotide(s) were possible and, at first glance, it would seem that these are not beyond the scope of existing techniques. On reflection, however, it is likely that any changes in adenine nucleotide status that do occur may be extremely subtle. For example a chloroplast which fixes $CO_2$ at a rate of $100\,\mu mol\,mg^{-1}$ $Chl\,h^{-1}$ must synthesize and utilize ATP at a rate of at least $300\,\mu mol\,mg^{-1}$ $Chl\,h^{-1}$ and with a total adenine nucleotide concentration in the region of 0.8 mM the turn over must be accordingly rapid and changes in steady state values could be very small and yet still significant. Moreover the situation is extremely complex when it is borne in mind that the availability of $NADP^+$ and ADP can exert considerable influence on the nature and rate of electron transport. However, the following experiment, recently carried out at Sheffield with the reconstituted chloroplast system (by Hope, Carver & Walker) shows that the adenine nucleotide changes are, in fact, in line with those proposed in Section 9.6. It is seen in Fig. 9.14 that there is an initial and rapid burst of $O_2$ evolution with a stoichiometry consistent with the reduction of the added $NADP^+$. Non-cyclic electron transport then ceases, for want of an oxidant, until $NADP^+$ is regenerated from NADPH by the addition of PGA. Subsequently, PGA reduction and its associated $O_2$ evolution is then stopped by the addition of R5P, which functions as an ATP sink as described in Section 9.6. After a short interval $O_2$ evolution is resumed and is eventually stopped by darkening (D) and re-started by further illumination (L). It will be seen that the adenine nucleotide values are also largely predictable. Some ATP and AMP (the latter not measured in this experiment) are formed in the dark because of the presence of adenylate kinase (myokinase) which catalyses the conversion $2ADP \rightarrow ATP + AMP$. Upon illumination there is a rapid rise in ATP and fall in ADP which soon slows as all of the added NADP is reduced and non-cyclic gives way to cyclic (or pseudo-cyclic) electron transport. As would be expected, the addition of PGA momentarily reverses this trend but despite the fact that 2 molecules of ATP are then required for each molecule of $O_2$ evolved (Eqns. 9.6 & 9.7):

$$2NADP^+ + 2H_2O + 2ADP + 2P \rightarrow 2NADPH_2 + 2ATP + O_2 \qquad \text{Eqn. 9.6}$$

$$2\,PGA + 2ATP + 2NADPH_2 \rightarrow 2G3P + 2Pi + 2ADP + 2NADP^+ \qquad \text{Eqn. 9.7}$$

The ATP required in the second of the reactions is rapidly matched by non-cyclic photophosphorylation in the first and within about a minute the [ATP] and [ADP] have returned to their former steady-state values (Fig. 9.14).

When R5P is added there is a similar, but larger, excursion in adenine nucleotide values. This time, however, the initial depression in the [ATP]/[ADP] ratio stops non-cyclic electron transport and its associated $O_2$ evolution so that a favourable ratio can only be restored by some form of cyclic photophosphorylation. It is of

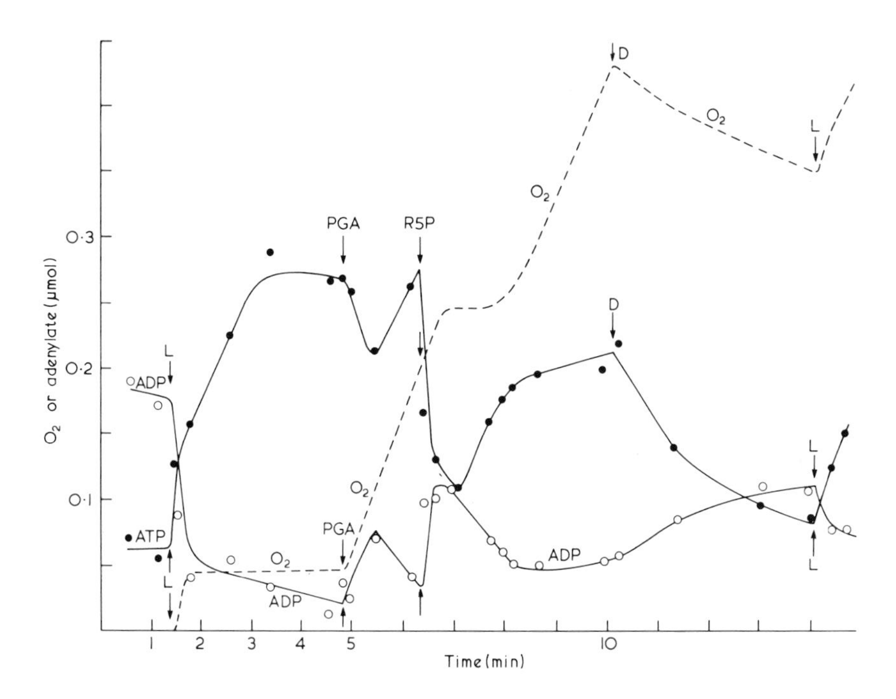


**Fig. 9.14** Time course of $O_2$ evolution and adenine nucleotide changes in a reconstituted chloroplast system made by osmotically shocking chloroplasts (100μg Chl) in a final volume of 1 ml to which was added ferredoxin (55 μg), NADP (0.1 μmol) and ADP (0.3 μmol.) together with DTT, $MgCl_2$, bicarbonate etc. according to previously published procedures (Lilley & Walker, 1979). Ribose 5-phosphate (0.2mM) and 3-phosphoglycerate (0.2mM) were added as indicated. Oxygen (broken line) was measured polarographically. Samples (5 μl) were withdrawn at intervals for assay of ATP (●) and ADP (○) using an LKB luminometer and luciferin-luciferase. ATP was measured directly. ADP was measured following conversion to ATP using phosphoenolpyruvate together with pyruvate kinase. The assay was internally calibrated by adding a known amount of ATP to each assay mixture.

interest that in this particular experiment the steady-state ratio of [ATP]/[ADP] following PGA addition was about 5, that $O_2$ evolution ceased as it approached 1, resumed at something less than 3 and reached a new steady-state (for both $O_2$ and adenine nucleotides) at about 4. This new value was presumably lower than before because of continuing drain on ATP occasioned by R5P and this could also account for the lower rate of $O_2$ evolution now recorded. It must be borne in mind, however, that this reaction mixture is now extremely complex with new PGA being derived from R5P and vice versa.

On darkening, the [ATP]/[ADP] ratio falls and then rises again on subsequent illumination. The first stage of this final rise would presumably reflect the availability of NADP rather than an immediate resumption of PGA reduction.

Results of this sort reinforce the notion that [ATP]/[ADP] ratios and their modulation by [Pi] can be important in the regulation of photosynthesis. They also carry an important warning. This is the fact that large changes in the rate of ATP consumption can clearly occur without altering the steady-state concentration of ATP in anything other than a transitory fashion. It should also be emphasized that it is not the [ATP]/[ADP] ratio alone which is important but rather the $\frac{[ATP]/[PGA]}{[ADP]/[DPGA]}$ ratio.

Thus, photosynthetic $O_2$ evolution occurs in intact chloroplasts in which the ratio of [ATP]/[ADP] is less than unity but this is achieved in the presence of a relatively high concentration of PGA.
[Refs. **36, 37, 40**]

### General Reading


1 AINSWORTH S. (1977) *Steady-state Enzyme Kinetics.* London: Macmillan.
2 ANDERSON L. E. (1979) Interaction between photochemistry and activity of enzymes. *Encyclopedia of Plant Physiology (New Series) Vol VI.* Berlin: Springer-Verlag, p. 271–278.
3 The Intact Chloroplast. *Topics in Photosynthesis, Vol. I,* (ed. J. Barber) Amsterdam: Elsevier, (1976).
4 BASSHAM J. A. (1974) The control of photosynthetic carbon metabolism. *Science,* **172**, 526–534.
5 GIBBS M. & LATZKO E. (1978) Regulation of photosynthetic carbon metabolism and related processes. *Encyclopedia of Plant Physiology (New Series) Photosynthesis Vol, II.* pp. 1–578. Berlin: Springer-Verlag.
6 HALLIWELL B. (1978) The chloroplast at work—a review of modern development on our understanding of chloroplast metabolism. *Prog. Biophys. Molec. Biol.* **33**, 1–54.
7 HEROLD A. & WALKER D. A. (1978) Transport across chloroplast envelopes—the role of phosphate. In: *Handbook on Transport across Biological Membranes* (eds. G. Giebisch, D. C. Tosteson and J. J. Ussing), pp. 412–39. Heidelberg: Springer-Verlag.
8 JENSEN R. G. & BAHR J. T. (1977) Ribulose 1,5 bisphosphate carboxylase oxygenase. *Ann. Rev. Plant Physiol.* **28**, 379–400.
9 KELLY G. K., LATZKO E. & GIBBS M. (1976) Regulatory aspects of photosynthetic carbon metabolism. *Ann. Rev. Plant Physiology.* **27**, 181–205.
10 KREBS H. A. (1954) Excursion into the borderland of biochemistry and philosophy. *Bull. John Hopkins Hosp.* **95**, 45–51.
11 KREBS H. A. (1969) The role of equilibria in the regulation of metabolism. *Curr. Topics in Cellular Reg.* **1**, 45–55.
12 LEHNINGER A. L. (1970) *Biochemistry,* New York: Worth Pub. Inc.
13 PREISS J. & KOSUGE T. (1970) Regulation of enzyme activity in photosynthesis systems. *Ann. Rev. Plant Physiol,* **21**, 433–66.
14 WALKER D. A. (1976) Regulatory mechanisms in photosynthetic carbon metabolism. In *Current Topics in Cellular Regulation,* (eds. B. L. Horecker and E. Stadtman) **11**, 203–41.
15 WALKER D. A. & HEROLD A. (1977) Can the chloroplast support photosynthesis unaided? In Photosynthetic Organelles: Structure and Function. (eds. S. Miyachi, S. Katoh, Y. Fujita, and K. Shibata). *Special Issue of Plant and Cell Physiol.* 295–310.

## Specific Citations


16 Akazawa T. (1979) Ribulose-1,5-bisphosphate carboxylase *New Encyc. Plant Physiol. Vol II* 208–25. (eds. M. Gibbs and E. Latzko) Berlin: Springer-Verlag.

16a Alexander A. G. (1977) *Sugar Cane Physiology*. New York: Elsevier.

17 Anderson L. E. & Advani V. R. (1970) Chloroplast and cytoplasmic enzymes: three distinct enzymes associated with the reductive pentose phosphate cycle. *Plant Physiol.* **45**, 583–5.

18 Anderson L. E., Nehrlich S. C. & Champigny M-L. (1978) Light modulation of enzyme activity. Activation of the light effect mediators by reduction and modulation of enzyme activity by thiol-disulfide exchange? *Plant Physiol.* **61**, 601–5.

19 Bahr J. T. & Jensen R. G. (1978) Activation of ribulose bisphosphate carboxylase in intact chloroplast by $CO_2$ and light. *Archs. Biochem. Biophys.* **185**, 39–48.

19a Buchanan B. B. (1980) Role of light in the regulation of chloroplast enzymes. *Ann. Rev. Plant Physiol.*, **31**, 341–74.

20 Baier D. & Latzko E. (1975) Properties and regulation of C-1-fructose-1,6-diphosphatase from spinach chloroplasts. *Biochim. Biophys. Acta*, **396**, 141–8.

21 Buchanan B. B., Kalberer P. P. & Arnon D. I. (1967) Ferredoxin-activated fructose diphosphatase in isolated chloroplasts. *Biochem. Biophys. Res. Commun.* **29**, 74–9.

22 Buchanan B. B., Schurmann P. & Kalberer P. P. (1971) Ferredoxin-activated fructose diphosphatase of spinach chloroplasts. *J. Biol. Chem.* **246**, 5952–59.

23 Buchanan B. B. & Wolosiuk R. A. (1976) Photosynthetic regulatory protein found in animal and bacterial cells. *Nature*, **264**, 669–70.

24 Chen-she S. H., Lewis D. H. & Walker D. A. (1975) Stimulation of photosynthetic starch formation by sequestration of cytoplasmic orthophosphate. *New Phytol.* **74**, 383–92.

24a Halliwell B. & Foyer C. H. (1978) Properties and physiological function of a glutathione reductase purified from spinach leaves by affinity chromatography. *Planta*, **139**, 9–17.

25 Heldt H. W., Chon C. J. & Lorimer G. H. (1978) Phosphate requirement for the light activation of ribulose-1,5-bisphosphate carboxylase in intact spinach chloroplasts. *FEBS Letters* **92**, 234–40.

26 Heldt H. W., Chon C. J., Maronde D., Herold A., Stankovic Z. S., Walker D. A., Kraminer A., Kirk M. A. & Heber U. (1977) The role of orthophosphate and other factors in the regulation of starch formation in leaves and isolated chloroplasts. *Plant Physiol.* **59**, 1146–55.

27 Herold A., Lewis D. H. & Walker D. A. (1976) Sequestration of cytoplasmic orthophosphate by mannose and its differential effect on photosynthetic starch synthesis in $C_3$ and $C_4$ species. *New Phytol.* **76**, 397–407.

28 Kakie T. (1969) Effects of phosphorous deficiency on the photosynthetic carbon dioxide fixation products in tobacco plants. *Soil Sci. Plant Nutr.* **15**, 245–51.

28a Leegood R. C. & Walker D. A. (1980) Modulation of fructose bisphosphatase activity in intact chloroplasts. *FEBS Letters* **116**, 21–4.

28b Leegood R. C. & Walker D. A. (1980) Regulation of fructose 1,6-bisphosphatase activity in intact chloroplasts. Studies of the mechanism of inactivation. *Biochim. Biophys. Acta*, **593**, 362–70.

29 Lilley R. McC. & Walker D. A. (1974) The reduction of 3-phosphoglycerate by reconstituted chloroplasts and by chloroplast extracts. *Biochim. Biophys. Acta*, **368**, 269–78.

29a Lilley R. McC., Holborow K. & Walker D. A. (1974) Magnesium activation of photosynthetic $CO_2$ fixation in a reconstituted chloroplast system. *New Phytol.* **73**, 657–62.

30 Lilley R. McC. & Walker D. A. (1979) Studies with the reconstituted chloroplast system. *Encyclopedia of Plant Physiology (New Series), Photosynthesis Vol. II.* Regulation of photosynthetic carbon metabolism and related processes, (eds. M. Gibbs and E. Latzko), pp. 41–52. Berlin: Springer-Verlag.

31 Lorimer G. H., Badger M. R. & Andrews T. J. (1976) The activation of ribulose-1,5-bisphosphate carboxylase by carbon dioxide and magnesium ions. Equilibria, kinetics, a suggested mechanism and physiological implication. *Biochemistry* **15**, 529–36.

32 LORIMER G. H., BADGER M. R. & HELDT H. W. (1978) The activation of ribulose 1,5-bisphosphate carboxylase/oxygenase. In Photosynthetic Carbon Assimilation. Ribulose 1,5-bisphosphate Carboxylase/oxygenase. *Brookhaven Symposium in Biology, Vol. 30* (eds. H. W. Siegelman and G. Hind), pp 283–306. New York: Brookhaven Nat. Lab., Upton.
33 NEALES T. F. & INCOLL L. D. (1968) The control of leaf photosynthetic rate by the level of assimilate concentration in the leaf. A review of the hypothesis. *Bot. Rev.* **34**, 107–25.
34 PORTIS A. R. & HELDT H. W. (1976) Light-dependent changes of the $Mg^{2+}$ concentration in the stroma in relation to the $Mg^{2+}$ dependency of $CO_2$ fixation in intact chloroplasts. *Biochim. Biophys. Acta*, **449**, 434–46.
35 PRIESS J. (1969) The regulation of the biosynthesis of α-1, 4-glucans in bacteria and plants. *Curr. Topics in Cell Reg.* **1**, 125–60.
36 ROBINSON S. P. & WALKER D. A. (1979) The control of PGA reduction in isolated chloroplasts by the concentrations of ATP, ADP and PGA. *Biochim. Biophys. Acta*, **545**, 528–36.
37 WALKER D. A. & ROBINSON S. P. (1978) Regulation of photosynthetic carbon assimilation. In Photosynthetic Carbon Assimilation. Ribulose 1,5-bisphosphate Carboxylase/oxygase. *Brookhaven Symposium in Biology, Vol. 30* (eds. H. W. Siegelman and G. Hind) pp. 43–49. New York: Brookhaven Nat. Lab., Upton.
38 ROBINSON S. P., MCNEIL P. H. & WALKER D. A. (1979) Ribulose bisphosphate carboxylase—lack of dark inactivation of the enzyme in experiments with protoplasts. *FEBS Letters* **97**, 296–300.
39 SCHÜRMANN P., WOLOSIUK R. A., BREAZEALE V. D. & BUCHANAN B. B. (1976) Two proteins function in the regulation of photosynthetic $CO_2$ assimilation in chloroplasts. *Nature*, **263**, 257–8.
40 SLABAS A. R. & WALKER D. A. (1976) Localization of inhibition by adenosine diphosphate of phosphoglycerate-dependent oxygen evolution in a reconstituted chloroplast system. *Biochem. J.* **154**, 185–92.
41 PAECH C., MCCURRY S. D., PIERCE J. & TOLBERT N. E. (1978) Active site of ribulose 1,5-bisphosphate carboxylase/oxygenase. In Photosynthetic Carbon Assimilation. Ribulose 1,5-bisphosphate Carboxylase/oxygenase. *Brookhaven Symposium in Biology, Vol.* 30 (eds. H. W. Siegelman and G. Hind), pp. 227–43. New York: Brookhaven Nat. Lab., Upton.
42 WALKER D. A. & LILLEY R. McC. (1974) Ribulose bisphosphate carboxylase—an enigma resolved. *Proceedings of the 50th Anniv. Meeting. SEB Cambridge*, (ed. N. Sunderland) pp. 189–98 Oxford: Pergamon Press.
43 WALKER D. A., SLABAS A. R. & FITZGERALD M. P. (1976) Photosynthesis in a reconstituted chloroplast system from spinach. Some factors affecting $CO_2$-dependent oxygen evolution with fructose-1,6-bisphosphate as substrate. *Biochim. Biophys. Acta*, **440**, 147–62.
44 WOLOSIUK R. & BUCHANAN B. B. (1977) Thioredoxin and glutathione regulate photosynthesis in chloroplasts. *Nature*, **266**, 565–7.
45 WOLOSIUK R. & BUCHANAN B. B. (1978) Activation of chloroplast NADP-linked glyceraldehyde-3-phosphate dehydrogenase by the ferredoxin/thioredoxin system. *Plant Physiol.* **61**, 669–71.

# Chapter 10
# Discovery of the $C_4$ Pathway



SUMMARY


This chapter provides a simple overview of the discovery of $C_4$ photosynthesis and some essential features of the pathway. The early experiments of Hatch and Slack provided a framework for subsequent studies on the biochemistry of photosynthesis in these plants. After more than 10 years of research, it is now clear that the mechanism of photosynthesis in $C_4$ plants involves decarboxylation of $C_4$ acids followed by donation of the $CO_2$ to the RPP pathway, rather than a transcarboxylase as initially envisaged. During this period considerable progress was made in developing techniques for isolating cells, protoplasts and chloroplasts from $C_4$ plants, which has aided biochemical studies. Without exception $C_4$ photosynthesis is strictly associated with two photosynthetic cell types in the angiosperms (the only group of plants clearly shown to contain $C_4$ species). In the light, $C_4$ plants catalyse the transfer of atmospheric $CO_2$ to the RPP pathway via a $C_4$ acid cycle.


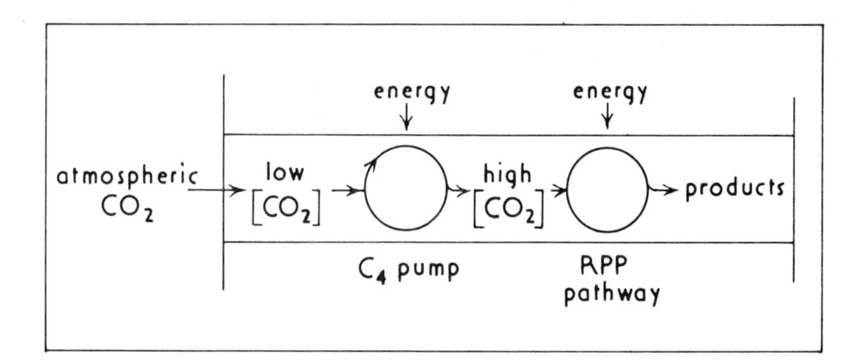


**Fig. 10.1** $CO_2$ pump.
$C_4$ photosynthesis. In order to provide a high, substrate-saturating $[CO_2]$ to the reductive pentose phosphate pathway some plants evolved a metabolic $CO_2$ pump. This is the $C_4$ cycle which is driven by energy (ATP). It functions to trap atmospheric $CO_2$ in an outer layer of photosynthetic cells (mesophyll) and shuttle it to an inner layer (Kranz cells) of the leaf where it is assimilated.

These facts can be used as criteria for determining whether $C_4$ acid metabolism in any photosynthetic tissue, from algae to angiosperms, is analogous to that in $C_4$ plants.

## 10.1 $C_4$ dicarboxylic acids as early products of photosynthesis

The discovery of the $C_4$ pathway of photosynthesis in certain higher plants began with the studies of H. Kortschack and colleagues at the Experimental Station of Hawaiian Sugar Planters' Association in the late 1950s. These researchers noticed that sugarcane leaves, supplied with $^{14}CO_2$ in the light for a few seconds fixed a large percentage of the label (70–80 %) into malate and aspartate. In the dark the amount of $CO_2$ fixed was less than 1 % of that in the light. Malate and aspartate contain 4 carbon atoms and the '$C_4$' was therefore introduced to distinguish this process from the unassisted $C_3$ or RPP pathway (Chapter 6) in which the initial product contains only 3 carbons. At about the same time as the studies in Hawaii, Y. Karpilov and associates in Russia were making similar observations with maize. It was not until the mid-1960s with the publication by Kortschack, *et al.* (1965) that these observations became more widely known. The discovery of a high percentage of label in $C_4$ acids as early photosynthetic products in sugarcane and corn thus represented a significant difference from the initial products of photosynthesis previously observed with algae and some higher plants.

Studies in the 1940s by Calvin and Benson had shown that PGA was the major initial product of photosynthesis in *Chlorella* and *Scenedesmus* and several higher plant species (Chapter 6). Even so malate is a commonly labelled product of photosynthesis and this was recognized in early studies of photosynthesis in algae. With *Chlorella* and *Scenedesmus* exposed to $^{14}CO_2$ for a few seconds, PGA was the major labelled product with a small amount of label appearing in malate and glycolate. This led, around 1950, to proposals by Benson and Calvin, Gaffron and Fager, and by Ochoa for $CO_2$ fixation cycles which involved $C_2$ and $C_4$ intermediates and entailed two separate carboxylations. The possibility of $C_4$ acids such as oxaloacetate, malate or succinate being directly involved in the cycle was considered. (Fig. 10.2) A significant feature of this was the double carboxylation in the cycle, both reactions contributing to net fixation of $CO_2$. As will be discussed later there are two carboxylations in $C_4$ plants but only the carboxylation through RBP leads to net fixation of $CO_2$. In early studies with *Scenedesmus*, Bassham *et al.* (1950) found that poisoning with malonate caused a strong inhibition of malate formation but had little influence on $^{14}CO_2$ incorporation into PGA. Thus malate was considered a secondary product of $CO_2$ fixation and not an intermediate of the primary pathway of carbon assimilation. It was not shown how malonate inhibited,

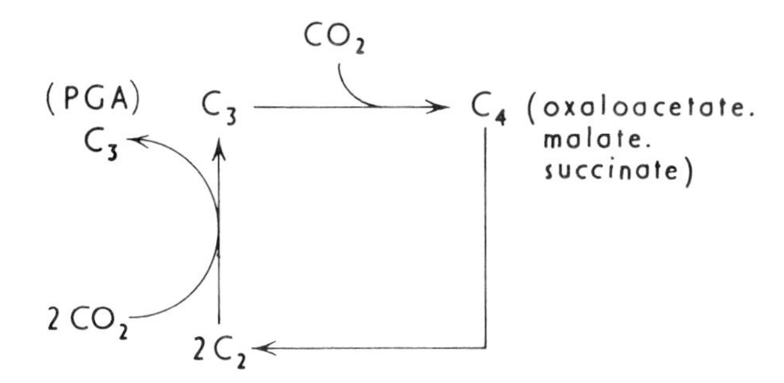


**Fig. 10.2.** Hypothetical double carboxylation of the 1950s. (After Ref. 3, also see Bassham *et al*, 1950)

but it was a known inhibitor of succinate dehydrogenase and more recently has been shown to inhibit PEP carboxylase. In contrast to these results with the alga, Kortschack *et al*, found that malonate strongly inhibited photosynthesis, malate and PGA synthesis in sugar-cane, suggesting that malate formation might be required for synthesis of PGA.

Calvin *et al*, also found, with *Scenedesmus*, that malate was a major labelled product of $CO_2$ fixation under very low light (45 ft. candles), and sucrose was synthesized as a major secondary product. Evidence was already available that malate was a primary product of *dark* $CO_2$ fixation. Thus at very low light the $CO_2$ fixed by the algae could represent a balance between light and dark fixation. Fagar and Gaffron noted at the time this may be an entirely secondary process unrelated to the mechanism of photosynthesis.

In $C_3$ plants the primary carboxylation of atmospheric $CO_2$ occurs via RBP carboxylase (Chapter 6). In leaves of $C_3$ plants a small percentage of the $CO_2$ fixed (less than 5%) is incorporated into $C_4$ acids. Formation of malate may occur via a secondary carboxylation following RBP carboxylase (Eqns. 10.1–10.5) or through PEP derived from the glycolytic pathway.

$\text{RBP} + CO_2 \rightarrow 2(\text{3-PGA})$ (RBP carboxylase) Eqn. 10.1

$\text{3-PGA} \rightarrow \text{2-PGA}$ (3-phosphoglyceromutase) Eqn. 10.2

$\text{2-PGA} \rightarrow \text{PEP}$ (enolase) Eqn. 10.3

$\text{PEP} + HCO_3^- \rightarrow \text{oxaloacetate} + \text{Pi}$ (PEP carboxylase) Eqn. 10.4

$\text{oxaloacetate} + \text{NAD(P)H} + H^+ \rightarrow \text{malate} + \text{NAD(P)}^+$
(NAD(P)H malate dehydrogenase) Eqn. 10.5

[Refs. 4, **12**, **25**, **36**]

## 10.2 Contribution by Hatch and Slack

Intrigued by the large percentage of labelled $^{14}CO_2$ incorporated into malate and aspartate by sugar-cane leaves, Kortschack persuaded Hatch and Slack in Australia in the 1960s to consider the biochemistry involved. Over a 4-year period, Hatch, Slack and Johnson made several important contributions:

(a) They performed pulse-chase experiments with sugar-cane.
(b) They identified several enzymes of the $C_4$ pathway.
(c) They provided preliminary evidence for differential localization of certain enzymes between mesophyll and bundle sheath chloroplasts of maize, and
(d) They identified several other species with a labelling pattern similar to sugar-cane.

In pulse-chase experiments plants are illuminated to give steady-state photosynthesis under atmospheric $CO_2$ ($^{12}C$). A pulse of radioactive $CO_2$ ($^{14}C$) is then provided for a given time (in this context a few seconds) after which the plant is immediately transferred to atmospheric $CO_2$. Samples of leaves are taken at intervals after exposure to $^{14}CO_2$ and during the chase in $^{12}CO_2$. The leaves are killed (e.g. with boiling ethanol or liquid nitrogen) and the $^{14}C$ products determined by chromatography. This type of experiment, in which the total radioactivity does not change during the chase period, reveals the first primary products of photosynthesis and their subsequent metabolism. As illustrated in Fig. 10.3 the early products of photosynthesis in a $C_4$ plant are aspartate and malate. Upon transfer to $^{12}CO_2$ the label in aspartate and malate decreases and the label in PGA, sucrose and starch increases. Most of the label is in C-4 of malate, and during the chase, C-1 of PGA is initially most heavily labelled. This type of experiment by Hatch and Slack provided the first evidence that, in $C_4$ plants, carbon from $C_4$ acids was being donated or metabolized rapidly to other products and that the $C_4$ acids were not end-products.

In contrast to $C_4$ plants, where malate and aspartate are the primary products of photosynthesis, PGA is the initial product of $CO_2$ fixation in $C_3$ plants. In a pulse-chase experiment as illustrated in Fig. 10.3, labelling in PGA decreases during the chase and labelling in sucrose and starch increases. This pattern is consistent with that seen in earlier studies on photosynthetic carbon metabolism and now typifies that of a $C_3$ plant.

To ensure that a valid indication is obtained from the metabolic conversions which occur during in-vivo photosynthesis, the methodology in this type of experiment is important. This includes using adequate light, adding $^{14}CO_2$ under steady-state photosynthesis, and maintaining a relatively low total $CO_2$ concentration during the pulse (i.e. of 0.04–0.05% after adding a high specific activity of

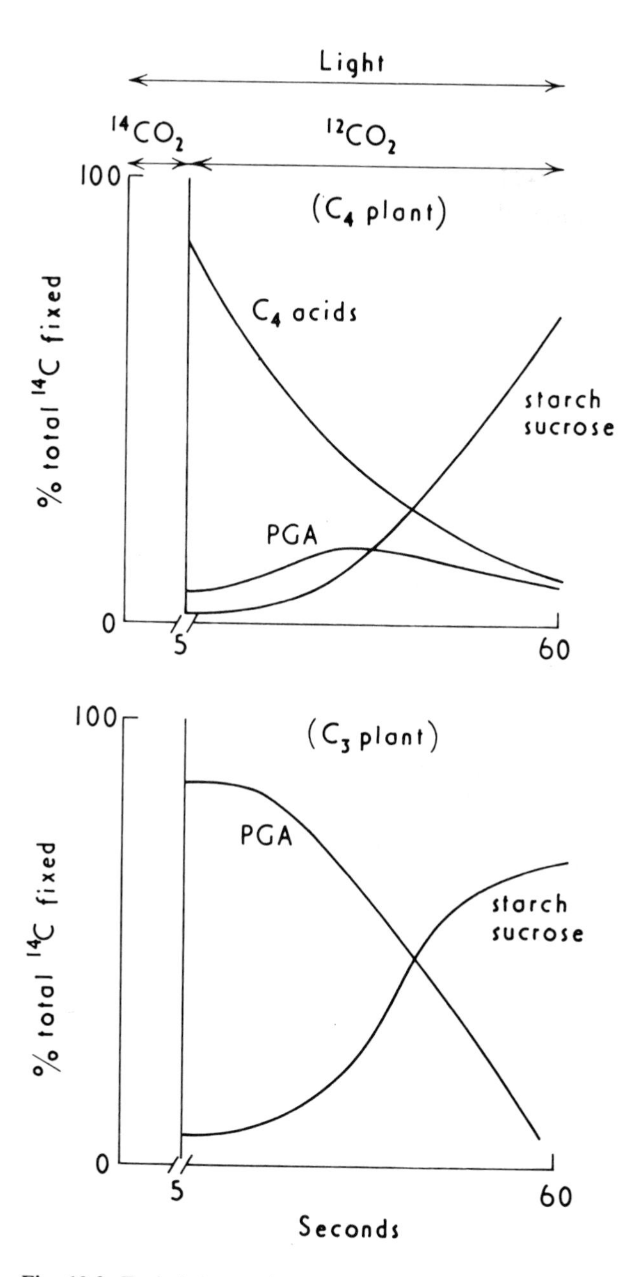


**Fig. 10.3.** Typical changes in percentage label in $C_4$ acids (malate + aspartate), PGA, and starch + sucrose with $C_4$ versus $C_3$ plants during a pulse-chase experiment. Other products such as triose- and hexose-phosphates are not shown. (e.g. Chen *et al*, 1971; Galmiche, 1973; Hatch and Osmond, 1976).

labelled $CO_2$ to atmospheric $CO_2$ of about 0.03 %). Variations on this general procedure may be of interest, but results should be interpreted using normal environmental conditions as a point of reference.

Both malate and aspartate are generally accepted as early products of $C_4$ photosynthesis. Oxaloacetate is considered the first but not the main product even in the shortest labelling experiments conducted (about 1 s). It is difficult to detect the initial product of a sequence of reactions if the pool size of this product is relatively small in comparison to other products of the pathway, as is apparently the case with oxaloacetate in $C_4$ plants.

Consider $A + CO_2 \rightarrow B \rightarrow C$

If the active pool of B is very small relative to the active pool of C, then a relatively small percentage of the total counts fixed will appear in B (total counts incorporated = specific activity × total active pool). However, at the shortest exposure time to $^{14}CO_2$ the specific activity (cpm/$\mu$ mol in the active pool) will be highest in the initial product. Exposing leaves of $C_4$ plants to $^{14}CO_2$ for less than a second and immediately killing results in very low label in oxaloacetate and high label in aspartate and malate. This indicates the pool size of oxaloacetate is very small and the turnover very fast.

If, during steady-state photosynthesis at near atmospheric levels of $CO_2$ (0.03 %), $C_4$ plants are suddenly given a pulse of labelled $^{14}CO_2$ roughly 10 fold higher and the initial products determined after a few seconds, 50–60 % of the label appears in oxaloacetate (Huber & Edwards, 1975). A sudden increase of the level of a substrate under steady-state conditions may cause a transient increase in the labelling of the product, even though its pool size may normally be small.

In addition oxaloacetate is unstable and decarboxylates nonenzymatically to pyruvate and $CO_2$, particularly in the presence of divalent cations and at elevated temperature. This may account for the relatively low pool sizes of oxaloacetate in biological systems. In the killing procedures following the feeding of $^{14}CO_2$, it is necessary to stabilize oxaloacetate, for example by reacting it with dinitrophenylhydrazine to form a phenlhydrazone derivative. Otherwise, if oxaloacetate is labelled in carbon 4, nonenzymatic decarboxylation could lead to complete loss of $^{14}C$ from this compound.

| | | | |
|---|---|---|---|
| C—1 | COOH | | COOH |
| C—2 | C=O | → | C=O + $^{14}CO_2$ |
| C—3 | $CH_2$ | | $CH_3$ |
| C—4 | $^{14}COOH$ | | |

Incorporation of label from $^{14}CO_2$ into malate and aspartate by $C_4$ plants does not in itself indicate the relative rate of asparatate and malate synthesis. For example, aspartate could be rapidly labelled by equilibration with a pool of labelled oxaloacetate in the leaf (exchange transamination is faster than normal transamination), while the aspartate pool remained constant. The net effect is; $^{14}$C-oxaloacetate + aspartate → oxaloacetate + $^{14}$C-aspartate through aspartate aminotransferase (Fig. 10.4)

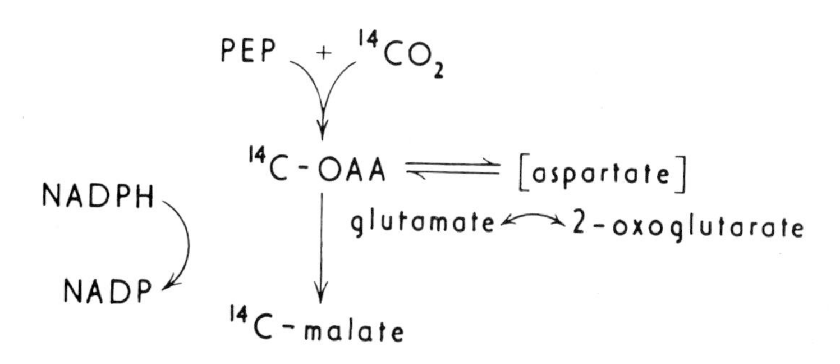


**Fig. 10.4** $^{14}$C labelling of aspartate without net synthesis.

During the chase with $^{12}CO_2$, the label in aspartate would again decrease by equilibration with unlabelled oxaloacetate but the primary flow of carbon may be through malate. Although malate and aspartate may both directly donate carbon into other products, it is obvious that the relative labelling in these acids after a few seconds in the light does not show their degree of direct participation in the $C_4$ cycle.

$C_4$ acids are not end-products of photosynthesis in $C_4$ plants, and synthesis of hexose phosphates, sucrose and starch requires three carbon and six carbon precursors. Therefore it was necessary to consider the mechanism of carbon transfer from $C_4$ acids into three carbon or six carbon compounds as well as the continuous regeneration of the $CO_2$ acceptor, i.e. PEP for the initial $CO_2$ fixation (Section 10.3). As previously stated, Hatch and Slack identified several enzymes of the $C_4$ pathway, initially in studies with sugarcane. Of these enzymes pyruvate,Pi dikinase, PEP carboxylase, and NADP-malate dehydrogenase are involved in the carboxylative phase of the $C_4$ pathway and are found in the mesophyll cells of $C_4$ plants. This phase of the pathway converts pyruvate plus $CO_2$ to malate (see Section 11.1 for details).

Pyruvate + Pi + ATP → PEP + AMP + PPi (Pyruvate,Pi dikinase) Eqn. 10.6

$PEP + HCO_3^- \rightarrow$ oxaloacetate + Pi (PEP carboxylase) Eqn. 10.4

oxaloacetate + NADPH + $H^+ \rightarrow$ malate + $NADP^+$

(NADP-malate dehydrogenase) Eqn. 10.7

Pyruvate,Pi dikinase is generally not detectable in leaves of $C_3$ plants, PEP carboxylase is some 50-fold higher in $C_4$ leaves than in $C_3$ leaves while substantial levels of NADP-malate dehydrogenase are found in both $C_3$ and $C_4$.

Using nonaqueous media of hexane + carbon tetrachloride, (in mixtures which give densities from 1.3 to 1.4 g/ml) Slack *et al*, (1969) provided the first evidence that there may be differential localization of certain photosynthetic enzymes between mesophyll and bundle sheath chloroplasts of $C_4$ plant leaves. The basis for separation of the two types of chloroplasts (from maize in their studies) was that the starch-free mesophyll chloroplasts had a lower density than the starch-containing bundle sheath chloroplasts. Pyruvate,Pi dikinase and NADP-malate dehyrogenase of the $C_4$ pathway were associated with mesophyll chloroplasts. The localization of PEP carboxylase was in doubt due to variability in results obtained by the nonaqueous method. Both the limitations of this technique for chloroplast separation (low yields of bundle sheath chloroplasts, lack of complete purity of the chloroplasts and possible contamination of chloroplasts by cytosolic enzymes) and uncertainties about the enzymology of $C_4$ photosynthesis at the time restricted the conclusions. Clear evidence for enzyme localization was provided by subsequent studies using isolated cells and protoplasts (Chapter 11).
[Refs. 8, 9, **28, 40**]

## 10.3 Metabolism of $C_4$ dicarboxylic acids

Research in a number of laboratories beginning around 1970 determined the mechanism of donation of carbon from $C_4$ acids to PGA and other products in $C_4$ plants. Initially Slack and Hatch (1967) found very low levels of RBP carboxylase in leaves of $C_4$ plants, of the order of 20–35 $\mu$ mol $mg^{-1}$ Chl $h^{-1}$, while $C_4$ plants photosynthesize at rates of about 200 $\mu$mol $CO_2$ $mg^{-1}$ Chl $h^{-1}$. The $C_4$ pathway was then considered as an alternative to the RPP pathway. They suggested a transcarboxylase as a logical mechanism to account for the observations made during the pulse-chase studies with whole leaves. The transcarboxylase would catalyze the direct transfer of a carboxyl group from a $C_4$ acid to either a $C_2$ or $C_5$ compound, resulting in subsequent formation of PGA. Since autocatalysis is required for net fixation of $CO_2$ (Chapter 6) the transcarboxylation proposal would require, in addition to the transcarboxylase, enzymes for regenerating the $C_2$ or $C_5$ acceptor.

Glycoaldehyde phosphate was suggested as a possible $C_2$ acceptor which would be converted to phosphohydroxypyruvate during transcarboxylation. With reductive transcarboxylation of glycoaldehyde phosphate, PGA would be the product. If glycoaldehyde were the $C_2$ acceptor the product of transcarboxylation would be hydroxypyruvate, which could be metabolized to glycerate and then PGA. Transcarboxylation with RBP as the possible acceptor was also suggested. Such an enzyme would directly transfer a carboxyl from the $C_4$ acids to a five-carbon acceptor, which would eliminate a requirement for RBP carboxylase. (Fig. 10.5). The idea of decarboxylation of $C_4$ acids and recarboxylation through RBP carboxylase was initially not appealing as a hypothesis for several reasons:

(a) The measured levels of RBP carboxylase and NADP-malic enzyme in leaf extracts were very low, about 10% of the rates of leaf photosynthesis.

(b) When leaves were labelled from $^{14}CO_2$ fixation and transferred to a $CO_2$-free atmosphere there was no loss of label from the leaves as might have been expected if $C_4$ acid decarboxylation was taking place. Also, chasing with 5% $^{12}CO_2$ has no effect on the transfer of label from $C_4$ acids to triose- and hexose-phosphates, whereas massive dilution of label would have been expected if small quantities of $^{14}CO_2$ were released into atmospheres containing 5% $^{12}CO_2$. Carboxylation, leading to $C_4$ acid formation, and decarboxylation of the $C_4$ acids are now known to occur in separate compartments in the leaf. Decarboxylation of the $C_4$ acids in

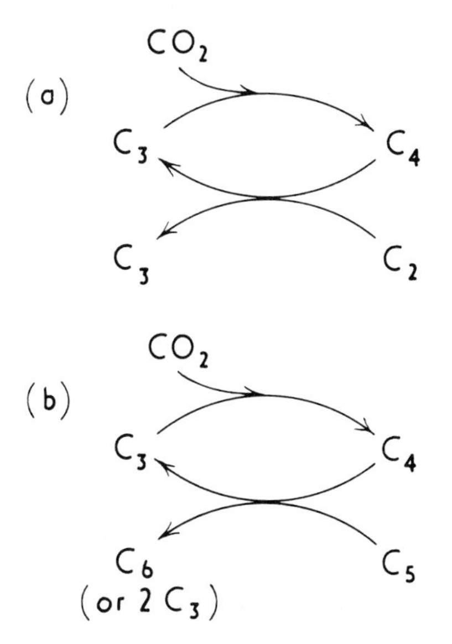


**Fig. 10.5** Hypothetical transcarboxylation reactions following $CO_2$ fixation through PEP carboxylase. (Hatch and Slack, 1966; 1970)

bundle sheath cells apparently restricts loss of $CO_2$ and prevents free equilibration of released $CO_2$ with that in the external atmosphere.

(c) With the limited information at that time, a transcarboxylation reaction seemed the simplest explanation. Perhaps the hypothesis arose because Hatch had previously worked with Stumpf on a transcarboxylase (Hatch and Stumpf, 1961).

Hatch and Slack were reluctant to give up the transcarboxylation hypothesis until considerable evidence was obtained for $C_4$ acid decarboxylation and refixation of $CO_2$ through RBP carboxylase. In 1969, Björkman and Gauhl reported substantial levels of RBP carboxylase in leaf extracts from several $C_4$ species. The failure of Hatch and Slack to detect similarly high levels of RBP carboxylase was apparently a technical one. The enzyme is localized in bundle sheath cells of $C_4$ plants which are very resistant to breakage because of a very thick cell wall. Such preferential breakage of mesophyll cells, as may have occurred in the initial studies, would give an apparently low level of this enzyme on a per unit Chl basis. Also, less than optimal conditions may have been used to assay the enzyme. (Section 6.11) The best assay conditions for RBP carboxylase require preactivation of the enzyme with magnesium and bicarbonate. Even in many present studies the level of RBP carboxylase, as currently extracted and assayed with various $C_4$ plants, does not match the rates of whole leaf photosynthesis. Table 10.1 shows typical values for leaf photosynthesis, and levels of PEP carboxylase and RBP carboxylase from leaf extracts of $C_3$ and $C_4$ species.

**Table 10.1.** Average rates of leaf photosynthesis and activity of carboxylases in leaf extracts of $C_3$ and $C_4$ plants

| Group | Photosynthetic rate: atmospheric $CO_2$ | Photosynthetic rate: $CO_2$-saturated air | Enzyme activity: PEP carboxylase | Enzyme activity: RBP carboxylase |
|---|---|---|---|---|
| | $\mu$mol mg$^{-1}$ Chl h$^{-1}$ | | | |
| $C_3$ plants | 100–200 | 200–400 | 20–40 | 150–500 |
| $C_4$ plants | 200–400 | 200–400 | 800–1200 | 100–200 |

The mechanism of the $C_4$ cycle became clearer in the 1970s with the discovery of substantial levels of $C_4$ acid decarboxylases. These are:

$$\text{malate} + \text{NADP}^+ \rightarrow \text{pyruvate} + CO_2 + \text{NADPH} + H^+ \quad \text{(NADP-malic enzyme)} \quad \text{Eqn. 10.8}$$

$$\text{malate} + \text{NAD}^+ \rightarrow \text{pyruvate} + CO_2 + \text{NADH} + H^+ \quad \text{(NAD-malic enzyme)} \quad \text{Eqn. 10.9}$$

$$\text{oxaloacetate} + \text{ATP} \rightarrow \text{PEP} + CO_2 + \text{ADP} \quad \text{(PEP carboxykinase)} \quad \text{Eqn. 10.10}$$

$C_4$ plants are now divided into three major subgroups based on differences in their primary decarboxylation mechanism (Sections 11.2, 11.8).

## 10.4 Anatomy of photosynthetic tissue of $C_4$ plants

$C_4$ plants have Kranz anatomy. In the late 1800s a layer of chlorophyllous cells which was distinctly different from the other chloroplast-containing cells was first recognized in leaves of certain species of Cyperaceae (Haberlandt, 1884). Haberlandt called this layer of cells Kranz (wreath) as they formed a distinct wreath around the vascular tissue. Considerable variation is now recognized in Kranz anatomy and in the evolution of Kranz cells in different species. In the families Gramineae and Cyperaceae, many species (Kranz and non-Kranz) have two sheaths around the vascular tissue, the mesotome sheath which is surrounded by the parenchyma sheath. In the Gramineae, Kranz cells originated from either the parenchyma sheath (e.g. subfamily Eragrostoideae) or the mesotome sheath (e.g. most genera of the subfamily Panicoideae in which case the parenchyma sheath is lacking). In Cyperaceae the Kranz cells developed just internal to the mesotome sheath. In some species of Cyperaceae, Kranz cells and mesophyll cells are separated by both a mesotome and parenchyma sheath (*Fimbristylis* type) or only a mesotome sheath (*Cyperus* type, where the parenchyma sheath is lacking). Dicots usually only have a parenchyma sheath and this is often where Kranz cells developed. In numerous $C_4$ species of the Chenopodiaceae, Polygonaceae and Compositae having cylindrical leaves or photosynthesizing branches, Kranz cells form a continuous cylindrical layer just internal to the mesophyll at the periphery of the tissue. A large part of the vascular tissue is separated from the Kranz cells by several layers of colourless parenchyma tissue. In this case the Kranz cells originated from the parenchyma tissue. It is clear that Kranz cells have originated from different cell types and in some cases the Kranz cells are separated from the mesophyll cells by nonchlorophyllous cells. In species used in early studies on $C_4$ photosynthesis (grasses and some dicots), Kranz had its origin in the bundle sheath cells. While in many $C_4$ species it is still correct to make reference to bundle sheath cells, it is now recognized that Kranz cells often, but not always, originated from the sheath cells adjacent to the vascular tissue. In general reference to $C_4$ plants, it is appropriate to use the term Kranz cells. $C_3$ plants often have a layer of sheath cells around the vascular tissue which occasionally have a few scattered chloroplasts. However, these cells are clearly different from the Kranz cells, the latter being large, thick-walled cells with prominent chloroplasts. (see Fig. 10.6)

Fig. 10.7 illustrates general anatomical differences in leaves of $C_3$ and $C_4$ monocots and dicots. Note the distinctive bundle sheath cells in the $C_4$ monocot

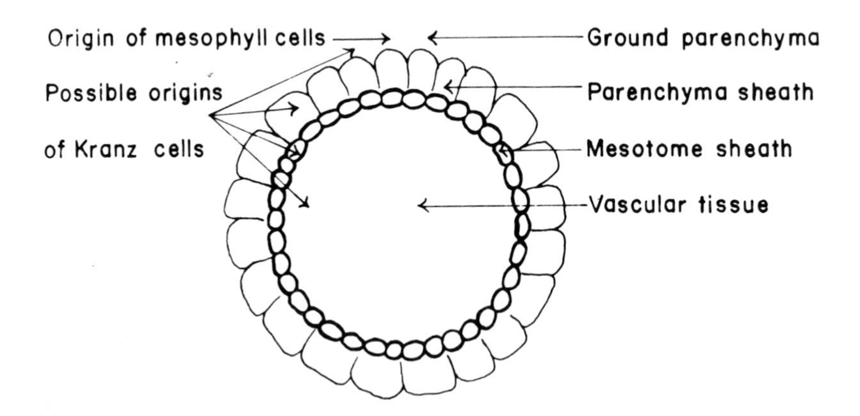


**Fig. 10.6.** Illustration of possible origins of Kranz cells in $C_4$ plants.

(Fig. 10.7a) and the $C_4$ dicot (Fig. 10.7c) and the lack of these cells in the $C_3$ monocot (Fig. 10.7b) and the $C_3$ dicot (Fig. 10.7d). The bundle sheath chloroplasts of maize are located in a centrifugal position (Fig. 10.7a) while those of *G. celosioides* are in a centripetal position (see Section 11.8 for discussion of anatomy of $C_4$ subgroups). Some $C_4$ dicots have palisade and spongy parenchyma surrounding the bundle sheath cells (as in Fig. 10.7c) while others have a common type of mesophyll cell similar to that of $C_4$ monocots. The paradermal sections illustrate the relation of

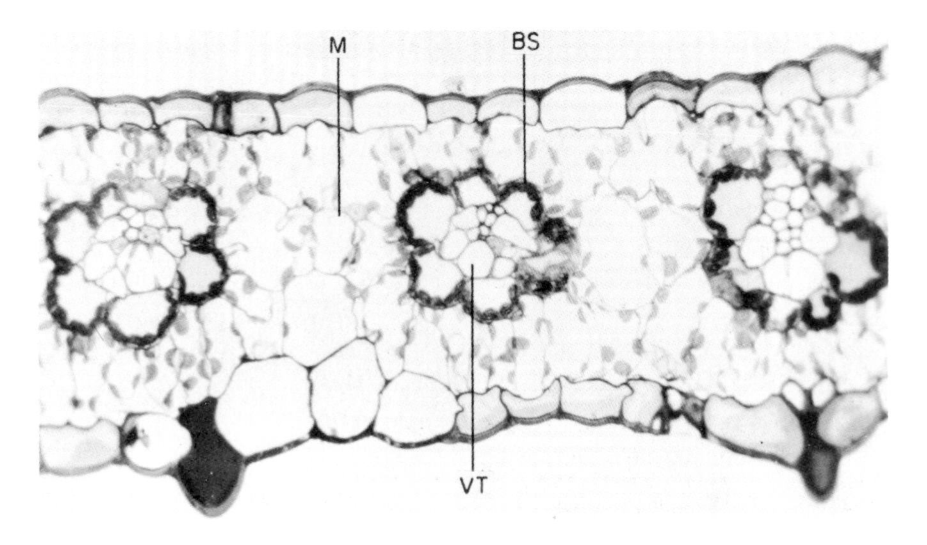


**Fig. 10.7a**

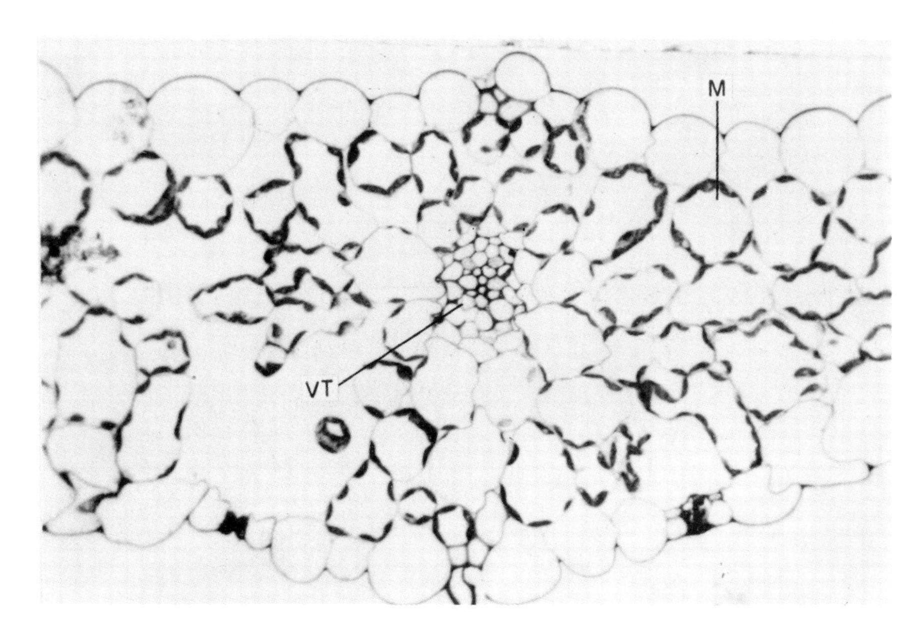


Fig. 10.7b

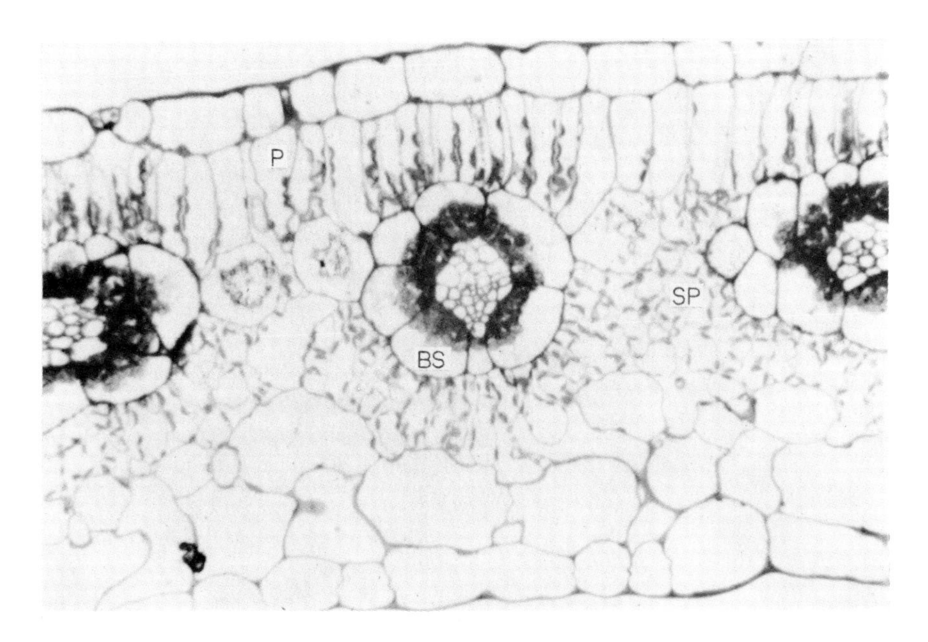


Fig. 10.7c

**Fig. 10.7.** Light micrographs of $C_4$ versus $C_3$ leaf anatomy.
a. Cross section of leaf of $C_4$ monocot (*Zea mays*), × 430.
b. Cross section of leaf of $C_3$ monocot (*Avena sativa*), × 380.
c. Cross section of leaf of $C_4$ dicot (young leaf of *Gomphrena celosioides*), × 1000. Note the mesophyll tissue is composed of palisade parenchyma cells (P) and spongy parenchyma cells (SP) which surround the bundle sheath.

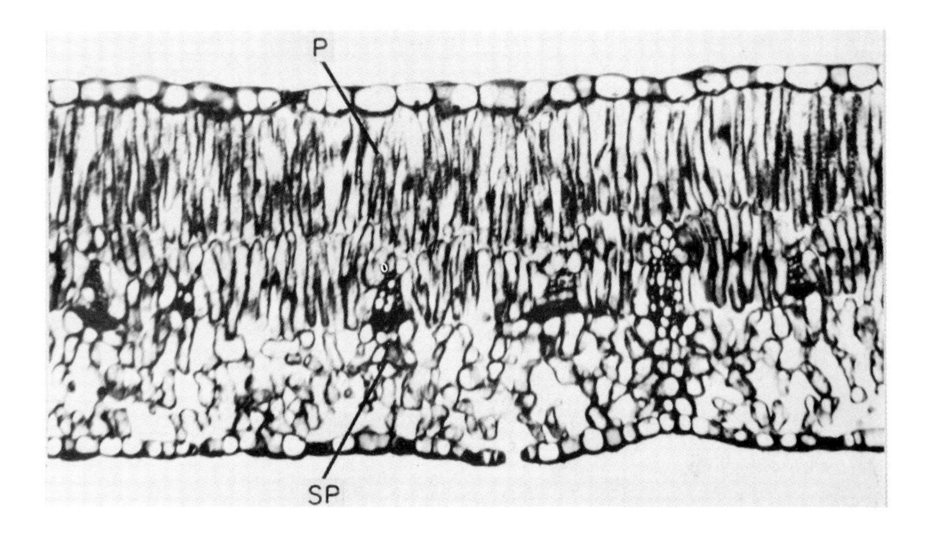


Fig. 10.7d

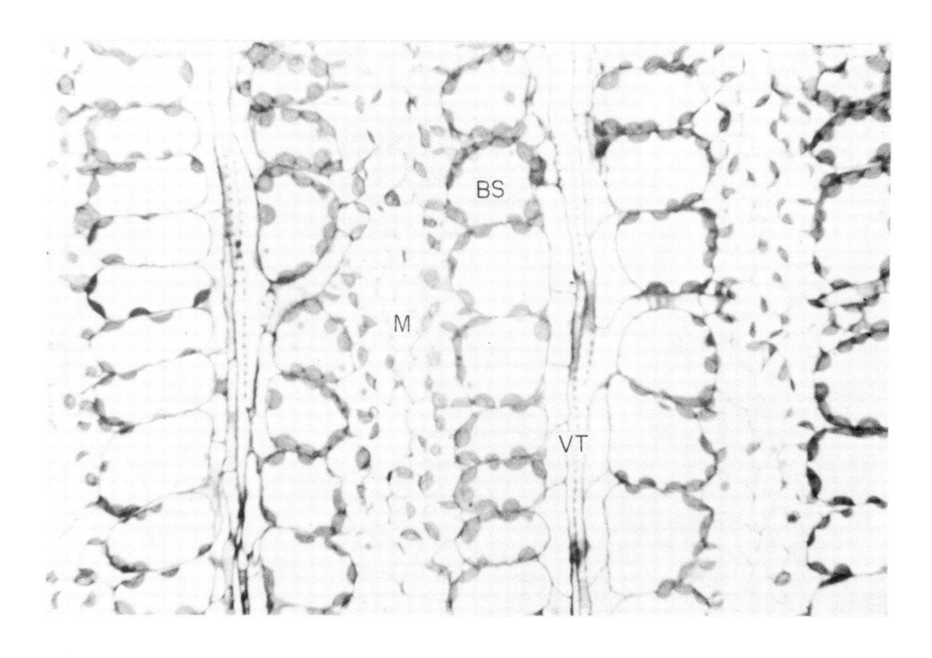


Fig.10.7e

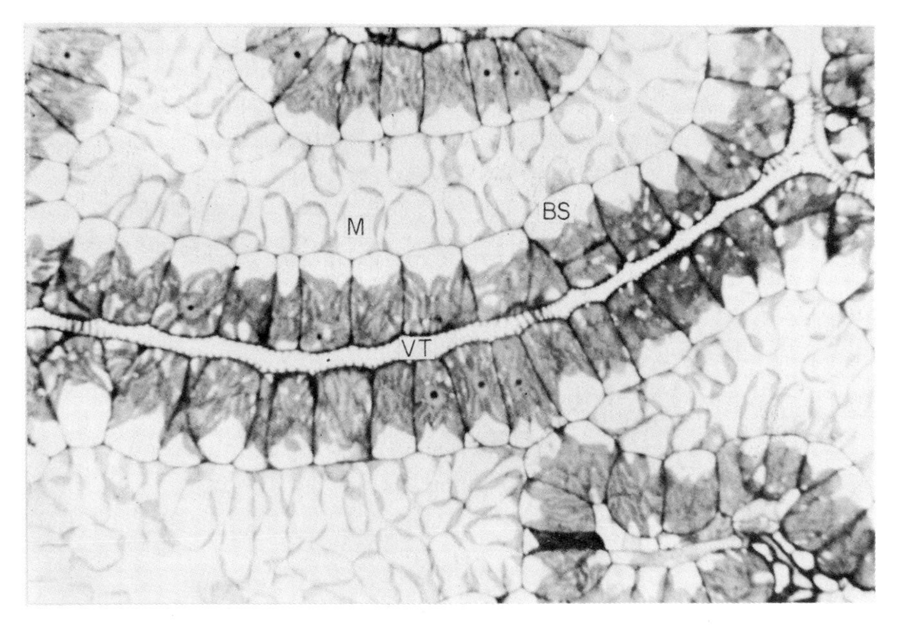


**Fig. 10.7f**

**Fig. 10.7**
d. Cross section of leaf of $C_3$ dicot (*Syringa vulgaris*) illustrating the palisade parenchyma (P) and spongy parenchyma cells (SP), × 137.
e. Paradermal section of $C_4$ monocot (*Chloris gayana*), × 1000.
f. Paradermal section of $C_4$ dicot (young leaf of *Atriplex spongiosa*), × 1000.
BS, bundle sheath cell; M, mesophyll cell; P, palisade parenchyma cell; SP, spongy parenchyma cell; VT, vascular tissue.
Micrographs a and b are by Sue E. Frederick, courtesy of E. H. Newcomb, University of Wisconsin. Micrographs c–f are by E. H. Newcomb, University of Wisconsin.

mesophyll to bundle sheath cells in the $C_4$ monocot which has parallel veins (Fig. 10.7e) and the $C_4$ dicot having the branched veination (Fig. 10.7f). There is a striking size dimorphism between mesophyll and bundle sheath cells and between their chloroplasts in some $C_4$ species.

In the 1940s Rhoades and Carvalho made a light microscopy study of mesophyll and bundle sheath cells of maize. They noted that bundle sheath cells were lighter green in colour than mesophyll cells. Starch accumulated during the day and disappeared in the dark in bundle sheath plastids but not in the mesophyll plastids. Cut leaves fed with sucrose for 78 hours resulted in mesophyll plastids accumulating large amounts of starch. In variegated leaves of maize (containing green and white stripes) they noted that apparently normal chlorophyllous bundle sheath cells did not store starch when they were adjacent to mesophyll cells which were without chlorophyll. They thus concluded: 'Fortunately we were able to show unequivocally that the starch formed in these plastids (bundle sheath) was derived from soluble carbohydrates made in the mesophyll plastids and translocated to the parenchyma sheath cells where the starch synthesis occurred.' Years later, it is now known that high rates of photosynthesis in $C_4$ plants requires the $C_4$ cycle as a $CO_2$-concentrating mechanism which is dependent on a photosynthetically functional mesophyll. $C_4$ mesophyll cells cannot independently synthesize sugars from $CO_2$ although this concept persisted before recent conclusive biochemical studies were made (see Section 12.10 for discussion of sucrose and starch synthesis in $C_4$ plants).

In 1964 Johnson, in a Ph.D. study, made a substantial ultrastructural study of bundle sheath and mesophyll cells of many grasses, including types with Kranz anatomy. Part of this study was published in 1973 after the biochemical significance of Kranz anatomy emerged (Johnson and Brown, 1973). In the mid-1960s Laetsch began electron microscopy studies on sugarcane and subsequently other $C_4$ species. Around 1968 several scientists independently made associations between leaf anatomy and some characteristics of $C_4$ photosynthesis in many species (Downes and Hesketh; Downton and Tregunna; Johnson and Hatch; and Laetsch—see review by Black, 1973).

[Refs. 5, 6, 10, **16, 18, 27, 33, 38, 43**]

## 10.5 Isolation of chloroplasts, protoplasts and cells from $C_4$ plants

The earliest attempts to separate mesophyll and bundle sheath chloroplasts of sugarcane and some other $C_4$ species were with aqueous density gradients, non-aqueous density gradients, and differential grinding procedures (see Black, 1973 for review). Differential grinding procedures rely on mesophyll cells breaking prior to bundle sheath cells, the latter cell type having a thicker cell wall. Mesophyll cells and bundle sheath strands (vascular bundles with the bundle sheath cells attached) were first separated mechanically from *Digitaria sanguinalis* (Edwards *et al*, 1970). The mechanical isolation of mesophyll cells from $C_4$ plants has been limited to the genus *Digitaria*. Mesophyll cells of other species readily break during mechanical grinding of leaf tissue. Ryuzi Kanai first enzymatically separated the cell types from $C_4$ plants (maize, sugarcane and sorghum) in the form of mesophyll protoplasts and bundle sheath strands (Fig. 10.8). This technique has been used with a number of $C_4$ species for various studies on $C_4$ photosynthesis (see Edwards and Huber, 1978 for review).

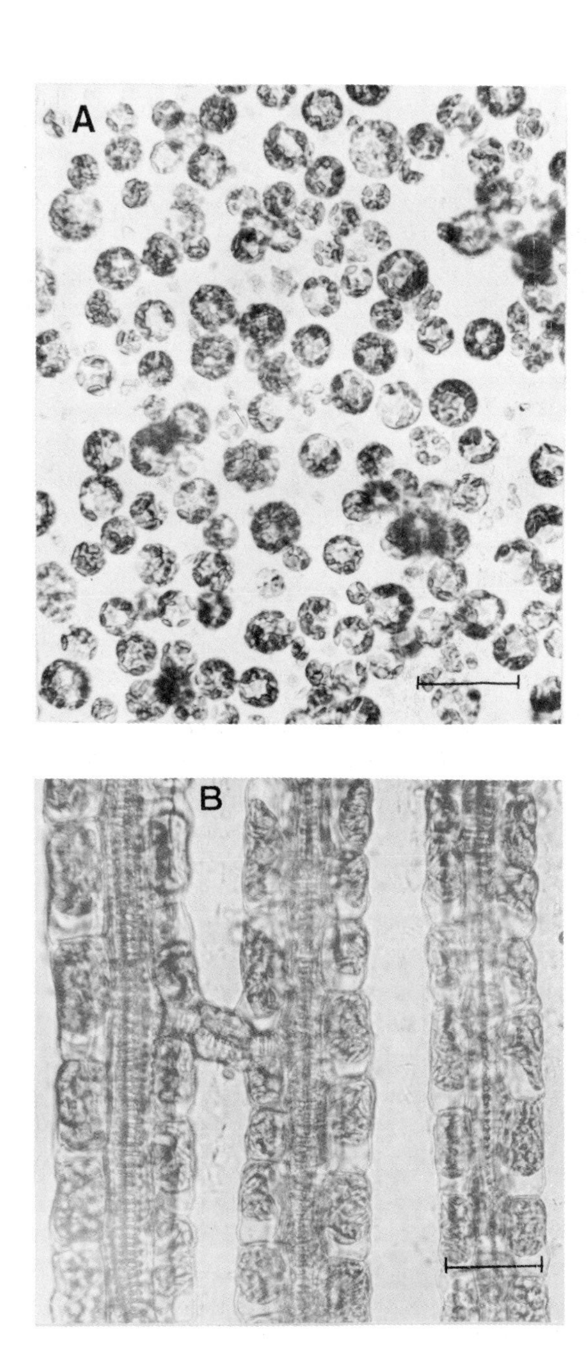


**Fig. 10.8.** Mesophyll protoplasts and bundle sheath strands enzymatically isolated from maize leaves. Bar represents 40 microns. (After Kanai and Edwards, 1973)

See Appendix A for isolation procedures for protoplasts and chloroplasts from protoplasts. A combination enzymatic-mechanical treatment has also been used to isolate cells from some species (Chen *et al*, 1973).

The successful isolation and separation of mesophyll and bundle sheath cells had important consequences. For example such preparations show that PEP carboxylase is confined to mesophyll cells and RBP carboxylase is in bundle sheath cells. $C_4$ mesophyll protoplasts or the chloroplasts isolated from the protoplasts have little or no RBP carboxylase ($< 1\%$ of that in bundle sheath cells) and no fraction I protein or subunits of fraction I protein. Conversely, isolated bundle sheath cells do have RBP carboxylase activity and fraction I protein (Edwards and Huber, 1978). An ingenius method was used by Hattersley *et al* (1977) to study the localization of fraction I protein in intact tissues. Using immunofluorescent labelling of fraction I protein, they found with leaf cross-sections of various $C_4$ species that the protein is localized in bundle sheath cells. Table 10.2 shows the activity of PEP carboxylase and RBP carboxylase in enzymatically isolated mesophyll and bundle sheath preparations of maize.

[Refs. 4, **20, 23, 30, 32, 40**]

**Table 10.2.** Activities of PEP carboxylase and RBP carboxylase in extracts from maize (After Kanai and Edwards, 1973)

| Enzyme | Enzyme extract from mesophyll protoplasts | Enzyme extract from bundle sheath strands |
|---|---|---|
| | $\mu$mol mg$^{-1}$ Chl h$^{-1}$ | |
| RBP carboxylase | 0 | 389 |
| PEP carboxylase | 864 | 10 |

## 10.6 Current simplified scheme of $C_4$ photosynthesis

Extensive studies on levels of photosynthetic enzymes and photosynthetic metabolism of isolated mesophyll and bundle sheath cells and chloroplasts of various species (Chapters 11 and 12) support the simplified scheme of photosynthesis in $C_4$ plants as shown below (Fig. 10.9):

The important features of the scheme are:

(a) The $C_4$ cycle shuttles $CO_2$ from the atmosphere to the RPP pathway in the bundle sheath cells.

(b) This mechanism is thought to provide a relatively high concentration of $CO_2$ to the RPP pathway in bundle sheath cells, above that which would normally exist by

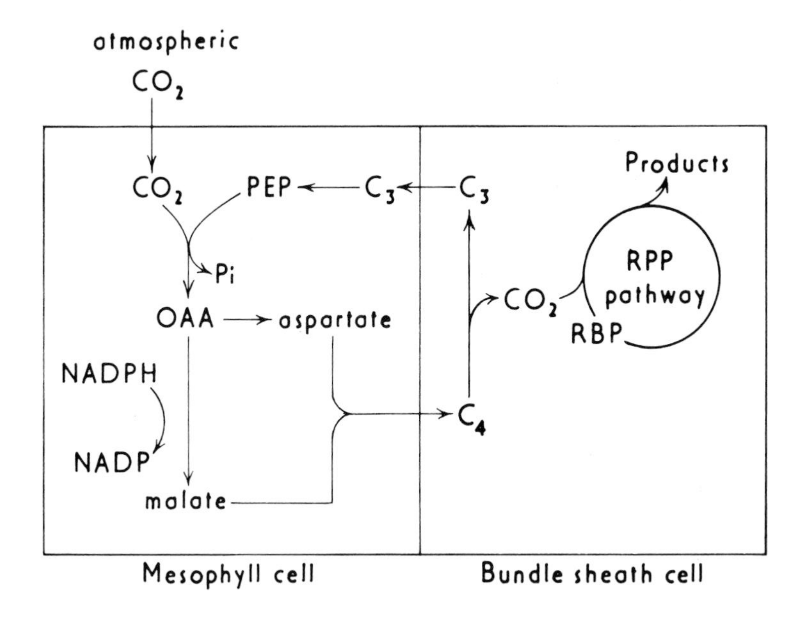


**Fig. 10.9.** $C_4$ photosynthesis.

diffusion of $CO_2$ into the leaf from the low levels of atmospheric $CO_2$ (0.03 %). Thus $CO_2$ is thought to become non-limiting as a substrate for photosynthesis. This results in a repression of photorespiration, the significance of which will be discussed later (Chapter 14).
(c) This mechanism requires rapid transport of certain metabolites between mesophyll and bundle sheath cells.
(d) There is no net fixation of $CO_2$ in mesophyll cells or in the $C_4$ cycle.

## 10.7 When can a species be considered $C_4$?

The terms $C_3$ plant and $C_4$ plant originated out of the recognition of a major difference in the primary initial products of photosynthesis in certain species of higher plants. This terminology was developed as $C_4$ plants were being discovered but before the mechanism of photosynthesis in them was known. $C_4$ plants donate the carbon initially fixed into the $C_4$ acids to the RPP pathway of photosynthesis. When $C_4$ plants are fed $^{14}CO_2$ for a few seconds, 70–80 % of the labelled products are in the $C_4$ acids malate and aspartate. There is only a small percentage (1–5 %) in oxaloacetate (Section 10.2). Since the elucidation of the pathway of photosynthesis in $C_4$ plants, the terminology can be more strictly defined to include these characteristics:

(a) Primary initial products of photosynthesis are the $C_4$ dicarboxylic acids oxaloacetate, malate and aspartate.

(b) Carbon is donated from the $C_4$ acids into the RPP pathway of photosynthesis (C–4 of $C_4$ acids to C–1 of PGA).

(c) These events occur in the light (unlike CAM plants, see Chapter 15).

(d) *Generally* there are two photosynthetic cell types, mesophyll and Kranz cells.

The question has been raised from several studies as to whether Kranz anatomy is required for $C_4$ photosynthesis. To date all species which are identified as $C_4$ plants have Kranz anatomy with two distinct photosynthetic cell types arranged in concentric layers (Section 10.4). In fact, examining the anatomy of a terrestrial plant is one of the simplest and quickest methods of determining if it is $C_4$. In photosynthetic tissue cultures of the $C_4$ species, *Amaranthus retroflexus* (Usuda *et al*, 1971) *Froelichia gracilis* (Laetsch and Kortschack, 1972) and *Portulaca oleracea* (Kennedy, 1976), $C_4$ acids are major products of $^{14}CO_2$ fixation, although they lack Kranz anatomy. In *Portulaca oleracea* cultures photorespiration appears low. This indicates some partial functions of $C_4$ photosynthesis in the tissue culture. However, the extent to which carbon flow in photosynthesis in these cultures is like that in the parent $C_4$ plant is uncertain. For example, with cultures maintained on sucrose, the source of PEP for $CO_2$ fixation could, in part, be through glycolysis, with synthesis of $C_4$ acids coupled to a relatively high inherent PEP carboxylase. Thus, the tissue culture experiments do not unequivocally show that efficient, high rates of $C_4$ photosynthesis can exist without Kranz anatomy.

Occasionally $C_4$ photosynthesis has been suggested from studies on aquatic plants. The results have been inconclusive; in some cases subsequent studies have shown $C_3$ photosynthesis. A tropical marine grass, *Thalassia testudinum*, was reported to have photosynthetic characteristics of $C_4$ plants without Kranz anatomy (Benedict and Scott, 1976). However, analysis of early labelled products and pulse-chase experiments recently showed this and several other marine grasses (*Thalassia* spp. and *Halophila* spp.) to have a typical $C_3$ pattern of photosynthesis (Andrews and Abel, 1979; Benedict *et al*, 1980). Unlike terrestrial $C_3$ species, there is little apparent discrimination against $^{13}CO_2$. Apparently the seagrasses have a high diffusive resistance for $CO_2$ uptake in the aqueous media. This may result in all of the $CO_2$ taken up being fixed such that the carbon isotope composition of $CO_2$ in the seawater and the plant tissue are similar (see Section 15.12 for isotope fractionation in plants). An aquatic angiosperm, *Hydrilla verticillata*, under summer temperature and photoperiod exhibited some characteristics of $C_4$-like photosynthesis without Kranz anatomy (Holaday and Bowes, 1980). The $CO_2$ compensation point was relatively low; aspartate and malate were among the major initial products of $^{14}CO_2$ fixation and there was substantial activity of $C_4$ pathway enzymes pyruvate,Pi dikinase and PEP carboxylase. Rates of $^{14}CO_2$ fixation in the light were relatively low in comparison to rates of dark respiration. In the dark, labelling of $C_4$ acids is known to occur in various plant tissues (even under respiring conditions). This could have a substantial effect on the distribution of label in products of $^{14}CO_2$ fixation under low rates of assimilation, thus making the path of carbon in photosynthesis difficult to evaluate. In addition relatively large internal pools of $CO_2$ may exist in lacunae and make pulse-chase experiments more difficult to perform and interpret. In *Egeria densa*, another aquatic macrophyte, the $CO_2$ is directly fixed in the RPP pathway under conditions found in the natural habitat, whereas rather low $CO_2$ levels used in some studies with aquatic angiosperms may favour fixation into $C_4$ acids (Browse *et al*, 1977). However, the aquatic angiosperm *Potamogeton pectinatus* has $C_3$ photosynthesis as determined by the initial products of $CO_2$ fixation over a range of $CO_2$ concentrations obtained at various pH values (Winter, 1978).

No algae have been identified as $C_4$ plants, although in some cases $C_4$ acids are a significant portion of the initial labelled products. Some blue-green algae in particular are suggested to have $C_4$ photosynthesis (relatively high PEP carboxylase activity, $C_4$ acids as primary products of $^{14}CO_2$ fixation, and malonate inhibition of photosynthesis, see Colman and Coleman, 1978).

Without rigorous study, it is difficult to distinguish between a partial function of $C_4$ photosynthesis, as already defined, versus $C_4$ acid formation by a different mechanism (e.g. secondary carboxylation). Significant labelling of the $C_4$ acids malate and aspartate as early products of photosynthesis in a plant is not sufficient to define that plant as $C_4$. A pulse-chase study similar to those of Hatch and Slack with sugarcane (Section 10.2) can help distinguish between $C_4$ acids as donors of carbon into the RPP pathway (or some other pathway such as the reversed TCA cycle in some algae). It is possible for label in $C_4$ acids to be converted to sucrose without carbon donation to the RPP pathway (see Section 16.6). Thus, in a complete analysis, transfer of label from C–4 of $C_4$ acids to C–1 of PGA must also be demonstrated. Photosynthetic tissue which has the $C_4$ acids malate and aspartate as products of photosynthesis, whether in low or high percentage, are not strictly speaking $C_4$ plants unless carbon from the $C_4$ acid is donated to the RPP pathway. In such cases, a different type of $C_4$ photosynthesis, in itself serving a unique function, may be occurring. Exclusive fixation of atmospheric $CO_2$ into $C_4$ dicarboxylic acids in the light with rapid rates of donation to the RPP pathway has only been found in plants which have a spatial separation of the $C_4$ carboxylation and decarboxylation

reactions between two cell types. A few terrestrial species having characteristics intermediate to those of $C_3$, and $C_4$ plants are discussed later (Section 11.9). Metabolism of $C_4$ acids in plants other than CAM and $C_4$ species is also discussed in Chapter 16.
[Refs. **11, 14, 16, 21, 22, 26, 35, 37, 41, 42**]

## General Reading

### Symposia


1 $CO_2$ Metabolism and Productivity of Plants. (eds. R. H. Burris and C. C. Black.) University Park Press, Maryland, U.S.A. (1976.)
2 Photosynthesis and Photorespiration. (eds. M. D. Hatch, C. B. Osmond and R. O. Slatyer.) Wiley-Interscience, New York (1971).
3 Symposium on Chemical Transformation of Carbon in Photosynthesis. *Fed. Proc.*, **9**, 524–553 (1950).


### Reviews


4 Black C. C. (1973) Photosynthetic carbon fixation in relation to net $CO_2$ uptake. *Ann. Rev. Plant Physiol.*, **24**, 253–86.
5 Brown W. V. (1977) The Kranz syndrome and its subtypes in grass systematics. *Memoirs of Torrey Botanical* Club, **13**, 1–97.
6 Coombs J. (1976) Interaction between chloroplasts and cytoplasm in $C_4$ plants. In *The Intact Chloroplast* (ed. J. Barber). pp. 274–313 The Netherlands: Elsevier.
7 Edwards G. E. & Huber S. C. (1981) The $C_4$ pathway. In *The Biochemistry of Plants.* A Comprehensive Treatise. Vol. 8. Photosynthesis (eds. M. D. Hatch and N. K. Boardman) pp. 237–81 New York: Academic Press
8 Hatch M. D. & Osmond C. B. (1976) Compartmentation and transport in $C_4$ photosynthesis. In *Encyclopedia of Plant Physiology* (New Series) Vol. III. Transport in plants. (eds. C. R. Stocking and U. Heber) pp. 144–84. New York: Springer-Verlag.
9 Hatch M. D. & Slack C. R. (1970) The $C_4$-carboxylic acid pathway of photosynthesis. In *Progress in Photochemistry.* (eds. L. Reinhold and Y. Liwschitz) pp. 35–106. London: Wiley-Interscience.
10 Laetsch W. M. (1974) The $C_4$ syndrome: a structural analysis. *Ann. Rev. Plant Physiol.*, **25**, 27–52.


### Specific citations


11 Andrews T. J. & Abel K. M. (1979) Photosynthetic carbon metabolism in sea grasses. $^{14}C$-labelling evidence for the $C_3$ pathway. *Plant Physiol.*, **63**, 650–6.
12 Bassham J. A., Benson A. A., & Calvin M. (1950) The path of carbon in photosynthesis. VIII. The role of malic acid. *J. Biol. Chem.*, **185**, 781–7.
13 Benedict C. R. & Scott J. R. (1976) Photosynthetic carbon metabolism of a marine grass. *Plant Physiol.* **57**, 876–80.
14 Benedict C. R., Wong W. W. L. & Wong J. H. H. (1980) Fractionation of the stable isotopes of inorganic carbon by seagrasses. *Plant Physiol.*, **65**, 512–7.
15 Björkman O. & Gauhl E. (1969) Carboxydismutase activity in plants with and without $\beta$-carboxylation photosynthesis. *Planta*, **88**, 197–203.
16 Brown W. V. (1975) Variations in anatomy, associations, and origins of Kranz tissue. *Amer. J. Bot.*, **62**, 395–402.

17 BROWSE J. A., DROMGOOLE F. I. & BROWN J. M. A. (1977) Photosynthesis in the aquatic macrophyte *Egeria densa*. I. $^{14}CO_2$ fixation at natural $CO_2$ concentrations. *Aust. J. Plant Physiol.*, **4**, 169–176.
18 CAROLIN R. C., JACOBS S. W. L. & VESK M. (1975) Leaf structure in Chenopodiaceae. *Bot. Jahrb. Syst.*, **95**, 226–55.
19 CHEN T. M., BROWN R. H. & BLACK C. C. (1971) Photosynthetic $CO_2$ fixation products and activities of enzymes related to photosynthesis in bermuda grass and other plants. *Plant Physiol.*, **47**, 199–203.
20 CHEN T. M., CAMPBELL W. H., DITTRICH P. & BLACK C. C. (1973) Distribution of carboxylation and decarboxylation enzymes in isolated mesophyll cells and bundle sheath strands of $C_4$ plants. *Biochem. Biophys. Res. Commun.* **51**, 461–7.
21 COLMAN B. & COLEMAN J. R. (1978) Inhibition of photosynthetic $CO_2$ fixation in blue-green algae by malonate. *Plant Sci. Lett.*, **12**, 101–5.
22 DEGROOTE D. & KENNEDY R. A. (1977) Photosynthesis in *Elodea canadensis* Michx. *Plant Physiol.*, **59**, 1133–5.
23 EDWARDS G. E. & HUBER S. C. (1978) Usefulness of isolated cells and protoplasts for photosynthetic studies. In *Proceedings of the Fourth International Congress on Photosynthesis.* (eds. D. O. Hall, J. Coombs, and D. W. Goodwin) pp. 95–106. London: The Biochemical Society.
24 EDWARDS G. E., LEE S. S., CHEN T. M. & BLACK C. C. (1970) Carboxylation reactions and photosynthesis of carbon compounds in isolated bundle sheath cells of *Digitaria sanguinalis* (L.). Scop. *Biochem. Biophys. Res. Commum.*, **39**, 389–95.
25 GAFFRON H. & FAGER E. W. (1951) The kinetics and chemistry of photosynthesis. *Ann. Rev. of Plant Physiol.*, **2**, 87–114.
26 GALMICHE J. M. (1973) Studies on the mechanism of glycerate 3-phosphate synthesis in tomato and maize leaves. *Plant Physiol.*, **51**, 512–9.
27 HABERLANDT G. (1884) *Physiological Plant Anatomy* (Transl. M. Drummond), London: Macmillan.
28 HATCH M. D. & SLACK C. R. (1966) Photosynthesis by sugarcane leaves. A new carboxylation reaction and the pathway of sugar formation. *Biochem. J.*, **101**, 103–11.
29 HATCH M. D. & STUMPF P. K. (1961) Fat metabolism in higher plants XVI. Acetyl coenzyme A carboxylase and acyl coenzyme A-malonyl coenzyme A transcarboxylase from wheat germ. *J. Biol. Chem.*, **236**, 2879–85.
30 HATTERSLEY P. W., WATSON L. & OSMOND C. B. (1977) *In situ* immunofluorescent labelling of ribulose 1,5-bisphosphate carboxylase in leaves of $C_3$ and $C_4$ plants. *Aust. J. Plant Physiol.*, **4**, 523–39.
31 HOLADAY A. S. & BOWES G. (1980) $C_4$ acid metabolism and dark $CO_2$ fixation in a submersed aquatic macrophyte (*Hydrilla verticillata*). *Plant Physiol.*, **65**, 331–5.
32 HUBER S. C. & EDWARDS G. E. (1975) Regulation of oxaloacetate, aspartate and malate formation in mesophyll protoplast extracts of several $C_4$ plants. *Plant Physiol.*, **56**, 324–31.
33 JOHNSON C. & BROWN W. V. (1973) Grass leaf ultrastructure variations. *Amer. J. Bot.*, **60**, 727–35.
34 KANAI R. & EDWARDS G. E. (1973) Separation of mesophyll protoplasts and bundle sheath cells of maize leaves for photosynthetic studies. *Plant Physiol.*, **51**, 1133–7.
35 KENNEDY R. A. (1976) Photorespiration in $C_3$ and $C_4$ plant tissue cultures. Significance of Kranz anatomy to low photorespiration in $C_4$ plants. *Plant Physiol.*, **58**, 573–5.
36 KORTSCHACK H. P., HARTT C. E. & BURR G. O. (1965) Carbon dioxide fixation in sugarcane leaves. *Plant Physiol.*, **40**, 209–213.
37 LAETSCH W. M. & KORTSCHACK H. P. (1972) Chloroplast structure and function in tissue culture of a $C_4$ plant. *Plant Physiol.*, **49**, 1021–3.
38 RHOADES M. M. & CARVALHO A. (1944) The function and structure of the parenchyma sheath plastids of the maize leaf. *Bull. Torrey Bot. Club* **71**, 335–46.
39 SLACK C. R. & HATCH M. D. (1967) Comparative studies on the activity of carboxylases and other

enzymes in relation to the new pathway of photosynthetic carbon dioxide fixation in tropical grasses. *Biochem. J.*, **103**, 660–5.

40 SLACK C. R., HATCH M. D. & GOODCHILD D. J. (1969) Distribution of enzymes in mesophyll and parenchyma sheath chloroplasts of maize in relation to the $C_4$-dicarboxylic acid pathway of photosynthesis. *Biochem. J.*, **114**, 489–98.

41 USUDA H., KANAI R. & TAKEUCHI M. (1971) Comparison of carbon dioxide fixation and the fine structure in various assimilation tissue of *Amaranthus retroflexus* L. *Plant Cell Physiol.*, **12**, 917–30.

42 WINTER K. (1978) Short-term fixation of $^{14}$carbon by the submerged acquatic angiosperm *Potamogeton pectinatus*. *J. Exp. Bot.*, **29**, 1169–72.

43 WINTER K. & TROUGHTON J. H. (1978) Photosynthetic pathways in plants of coastal and inland habitats of Israel and the Sinai. *Flora*, **16**, 1–34.

# Chapter 11
# Three $C_4$ Subgroups: Biochemistry, Photochemistry and Taxonomy



## SUMMARY


The $C_4$ cycle can be divided into two phases: the carboxylation phase, occurring in mesophyll cells, and the decarboxylation phase occurring in bundle sheath cells. $CO_2$ fixation through PEP carboxylase is common to the carboxylation phase in all $C_4$ species. $C_4$ species can be divided into three subgroups based on differences in the mode of $C_4$ acid decarboxylation: NADP-malic enzyme type, NAD-malic enzyme type and PEP carboxykinase type. The major metabolites of the $C_4$ cycle


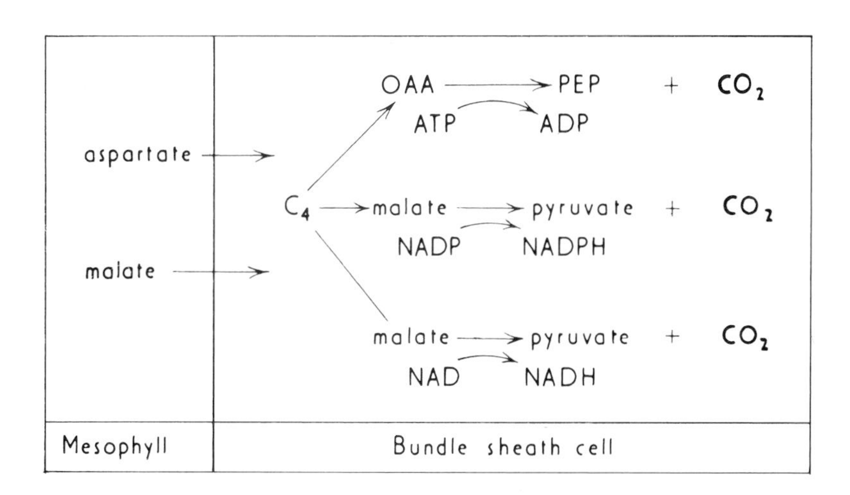


**Fig. 11.1.** Diversity in decarboxylating systems of $C_4$ photosynthesis.
There are different types of '$CO_2$ pumps' in different $C_4$ species. The fixation of atmospheric $CO_2$ into $C_4$ acids in mesophyll cells, is followed by transport of malate and aspartate to bundle sheath cells where $CO_2$ is released by decarboxylation. Some species use PEP carboxykinase, some NADP-malic enzyme and others NAD-malic enzyme as the decarboxylating enzyme. These variations in the $C_4$ cycle are linked to differences in photochemistry, leaf anatomy and taxonomy.

transported between cells are malate pyruvate in NADP-ME species, aspartate alanine in NAD-ME species, and aspartate & PEP in PEP-CK species.

Photochemical differences in the mesophyll and bundle sheath chloroplasts in the three subgroups are consistent with differences in energetics of the $C_4$ cycle. In NADP-ME species, malate transport donates reductive power from mesophyll to bundle sheath cells. The bundle sheath cells of these species are deficient in photosystem II, and may shuttle up to half of the PGA (product of RBP carboxylase activity in the bundle sheath) to mesophyll cells for reduction. Thus, in these species mesophyll cells are the primary site of light-dependent generation of reductive power (NADPH). In PEP-CK and NAD-ME species, only ATP is required for aspartate synthesis in the mesophyll cells. The bundle sheath chloroplasts are the primary site for generating NADPH for reduction of PGA. The energy requirements per $CO_2$ fixed are similar in all three subgroups. The net additional energy beyond that required for the RPP pathway is the ATP needed to drive the $C_4$ cycle.

$C_4$ plants have been found in 16 families. Several major crops and many major weeds are $C_4$. In the Gramineae, a correlation has been found between features of bundle sheath cells (and their chloroplasts) and biochemical differences between the three subgroups.

## 11.1 The carboxylation phase of the $C_4$ pathway

This phase of the $C_4$ pathway occurs in mesophyll cells of $C_4$ plants and results in conversion of $C_3$ precursors and atmospheric $CO_2$ to $C_4$ acids through PEP carboxylase. There are three known alternatives which are common to most $C_4$ species examined, although the proportion of metabolism to malate versus aspartate is variable.

(a) pyruvate $+ CO_2 \rightarrow$ malate — Eqn. 11.1
(b) alanine $+ CO_2 \rightarrow$ aspartate — Eqn. 11.2
(c) PGA $+ CO_2 \rightarrow$ malate $+$ Pi — Eqn. 11.3

In the conversion of pyruvate and $CO_2$ to malate (Eqn. 11.1) five enzymes are utilized as follows:

pyruvate $+$ Pi $+$ ATP $\rightarrow$ PEP $+$ AMP $+$ PPi (pyruvate,Pi dikinase) — Equn. 11.4
PPi $\rightarrow$ 2 Pi (pyrophosphatase) — Eqn. 11.5
AMP $+$ ATP $\rightarrow$ 2 ADP (adenylate kinase) — Eqn. 11.6
PEP $+ CO_2 \rightarrow$ oxaloacetate $+$ Pi (PEP carboxylase) — Eqn. 11.7
oxaloacetate $+$ NADPH $+ H^+ \rightarrow NADP^+ +$ malate (NADP-malate dehydrogenase) — Eqn. 11.8

**Sum** (Eqns. 11.4–11.8):

pyruvate + $CO_2$ + 2ATP + NADPH + $H^+$ → 2 ADP + 2 Pi + $NADP^+$ + malate
Eqn. 11.9

One might ask, what is the role of the chloroplast in the mesophyll cell of $C_4$ plants? As demonstrated in the above sequence of reactions, 2ATP and 1NADPH are required per $CO_2$ fixed in the $C_4$ pathway. Therefore, the $C_4$ pathway is dependent on the mesophyll chloroplast to provide assimilatory power (Chapter 5).

For conversion of alanine and $CO_2$ to aspartate (Eqn. 11.2) six enzymes are required as follows:

alanine + 2-oxoglutarate → pyruvate + glutamate
(alanine aminotransferase) Eqn. 11.10

pyruvate + 2ATP → 2ADP + PEP + Pi Eqns. 11.4–11.6

PEP + $CO_2$ → oxaloacetate + Pi (PEP carboxylase) Eqn. 11.7

oxaloacetate + glutamate → aspartate + 2-oxoglutarate
(aspartate aminotransferase) Eqn. 11.11

---

**Sum:** alanine + $CO_2$ + 2ATP → aspartate + 2ADP + 2 Pi Eqn. 11.12

Note that a continual net synthesis of glutamate and 2-oxoglutarate is not required, as they are recycled through alanine and aspartate aminotransferase. While NADPH is not required, two molecules of ATP are needed to convert alanine and $CO_2$ to aspartate.

For conversion of 3-PGA and $CO_2$ to malate and Pi (Eqn. 11.3) four enzymes are required:

3-PGA → 2-PGA (phosphoglyceromutase) Eqn. 11.13

2-PGA → PEP (enolase) Eqn. 11.14

PEP + $CO_2$ → oxaloacetate + Pi (PEP carboxylase) Eqn. 11.7

oxaloacetate + NADPH + $H^+$ → malate + $NADP^+$
(NADP-malate dehydrogenase) Eqn. 11.8

---

**Sum:** 3-PGA + $CO_2$ + NADPH + $H^+$ → malate + $NADP^+$ + Pi Eqn. 11.15

This conversion requires no ATP but one NADPH is needed per $CO_2$ fixed.

The extractable activity of these enzymes (Eqns. 11.4–11.14) in mesophyll cells of $C_4$ plants is sufficient to support reasonable rates of $CO_2$ fixation in the $C_4$ pathway. However, variable levels of pyruvate,Pi dikinase have been reported, possibly due to different states of activation of the enzyme. There is also evidence from mesophyll preparations (Section 12.3) that all three metabolites, pyruvate, alanine and PGA, can serve as precursors to the carboxylation phase of the $C_4$ pathway.

Some form of assimilatory energy is required in each alternative of the carboxylation phase of the $C_4$ pathway. The balance of ATP and NADPH required per $CO_2$ fixed into $C_4$ acids depends on the relative rates of fixation through the three options. As will be discussed later, variations in carbon flow through the $C_4$ carboxylation phase will influence the distribution of energy requirements between mesophyll and bundle sheath cells, but not the total energy required per $CO_2$ fixed by a $C_4$ plant (Section 11.6).

## 11.2 The decarboxylation phase of the $C_4$ pathway

This phase of the pathway occurs in bundle sheath cells. Again, there are three options based on three $C_4$ acid decarboxylation enzymes. $C_4$ species can be classified into subgroups according to their primary $C_4$-acid decarboxylating mechanism (Table 11.1, Section 11.8).

**Table 11.1.** Examples of species in each of the three subgroups of $C_4$ plants based on difference in the decarboxylating mechanism.

| $C_4$ subgroup | Example of species | |
|---|---|---|
| NADP-malic enzyme | maize | *Zea mays* |
| | sugarcane | *Saccharum officinarum* |
| | sorghum | *Sorghum bicolor* |
| | crabgrass | *Digitaria sanguinalis* |
| NAD-malic enzyme | pigweed | *Amaranthus retroflexus* |
| | purslane | *Portulaca oleracea* |
| | millet | *Panicum miliaceum* |
| PEP-carboxykinase | guineagrass | *Panicum maximum* |
| | rhoadesgrass | *Chloris gayana* |

### (a) NADP-MALIC ENZYME

In this case where malate is the metabolite transported from mesophyll to bundle sheath cells and pyruvate is returned from bundle sheath to mesophyll cells, NADP-malic enzyme is required.

$$\text{malate} + NADP^+ \rightarrow \text{pyruvate} + CO_2 + NADPH + H^+ \qquad \text{Eqn. 11.16}$$

Malate decarboxylation and the RPP pathway are linked through this mechanism. The RPP pathway is dependent on malate decarboxylation to provide $CO_2$; malate decarboxylation is dependent on the RPP pathway for re-oxidation of NADPH.

NADP-malic enzyme is localized in bundle sheath chloroplasts. Typically malate is the primary initial labelled photosynthetic product in NADP-ME species. An enzymatic sequence coupling the conversion of aspartate to $CO_2$ and alanine with NADP-malic enzyme is also possible (Section 12.2). However, the likely transport metabolites between mesophyll and bundle sheath cells during decarboxylation by this enzyme are malate and pyruvate.

### (b) NAD-MALIC ENZYME

Decarboxylation involving NAD-malic enzyme requires four enzymes since in this case aspartate and alanine appear to be the major transport metabolites between mesophyll and bundle sheath cells. NAD-malic enzyme is localized in the mitochondria of bundle sheath cells. Species having NAD-malic enzyme as the primary decarboxylase have abundant and prominent mitochondria in bundle sheath cells (see Fig. 11.7b). These species also have high levels of aspartate- and alanine aminotransferase in both mesophyll and bundle sheath cells, and aspartate is typically the primary initial labelled photosynthetic product.

aspartate + 2-oxoglutarate → oxaloacetate + glutamate — Eqn. 11.11

oxaloacetate + NADH + $H^+$ → malate + $NAD^+$ (NAD-malate dehydrogenase) — Eqn. 11.17

malate + $NAD^+$ → pyruvate + $CO_2$ + NADH + $H^+$ (NAD-malic enzyme) — Eqn. 11.18

pyruvate + glutamate → alanine + 2-oxoglutarate — Eqn. 11.10

---

**Sum:** aspartate → alanine + $CO_2$ — Eqn. 11.19

Note that there is no net utilization or generation of assimilatory power in this sequence of reactions. To a lesser extent malate and pyruvate may be transport metabolites between the cell types with malate decarboxylation through NAD-malic enzyme.

### (c) PEP-CARBOXYKINASE

Aspartate is generally the primary initial photosynthetic product in PEP-CK species and also the main transport metabolite between mesophyll and bundle sheath cells. Thus the decarboxylation phase may utilize two enzymes (Eqn. 11.21).

aspartate + 2-oxoglutarate → oxaloacetate + glutamate — Eqn. 11.11

oxaloacetate + ATP → PEP + $CO_2$ + ADP (PEP-carboxykinase) — Eqn. 11.20

---

**Sum:** aspartate + 2-oxoglutarate + ATP →
PEP + glutamate + $CO_2$ + ADP  Eqn. 11.21

As no evidence has been found for high levels of pyruvate kinase in bundle sheath cells of these species (needed to convert PEP to pyruvate), PEP is believed to move back to mesophyll cells unchanged. This would result in a more complex shuttle of metabolites between the cell types as shown in the summary of the three options (Section 11.3). PEP-carboxykinase is localized in the cytoplasm of bundle sheath cells (Section 12.2). Some differences in cofactor requirements and regulatory properties of the decarboxylases are discussed in Section 12.8.
[Refs. **14, 16, 17, 25**]

## 11.3 Summary of proposed major sequences of carbon flow through three decarboxylating mechanisms (Fig. 11.2)

Where aspartate is a major early product of $C_4$ photosynthesis, pools of the amino acids aspartate and alanine are required in the $C_4$ cycle. Since these pools must be maintained some nitrogen reduction may be required (nitrate reduction to ammonia and amino acid formation). Even so, nitrogen reduction is not directly linked to aspartate synthesis during $CO_2$ fixation in the cycle since the amino group of aspartate is constantly recycled through transamination reactions with aspartate- and alanine aminotransferase. In PEP-CK species either a pyruvate-alanine shuttle

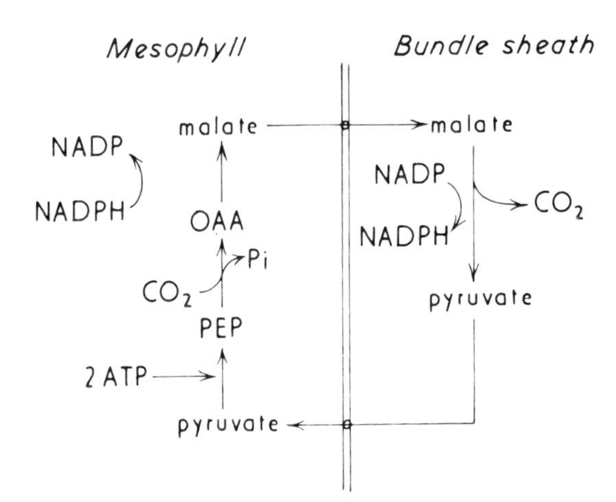


**Fig. 11.2(a)**

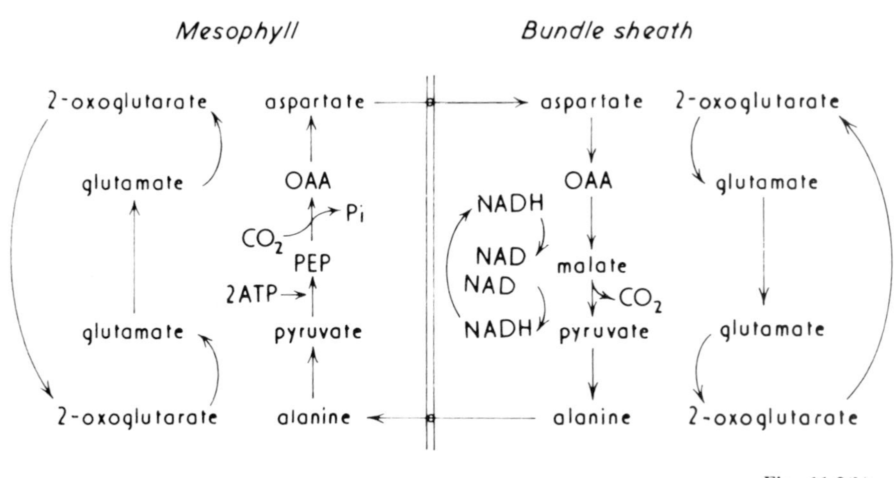


Fig. 11.2(b)

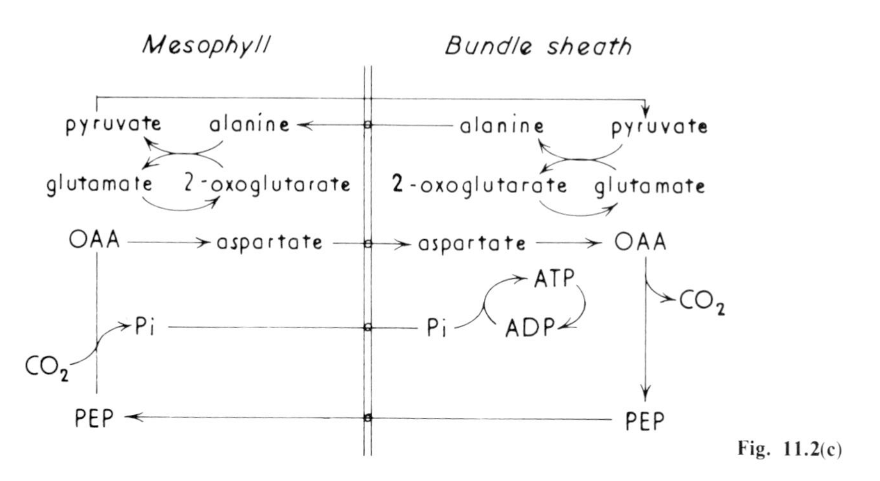


Fig. 11.2(c)

**Fig. 11.2.** Decarboxylating mechanisms.

or 2-oxoglutarate-glutamate shuttle may transfer amino nitrogen from bundle sheath to mesophyll cells to maintain the nitrogen balance. Aspartate decarboxylation by isolated bundle sheath cells of this type is dependent on the addition of pyruvate or 2-oxoglutarate. The pyruvate-alanine shuttle may be favoured as the concentration of 2-oxoglutarate required is an order of magnitude higher than the pyruvate concentration. There is no requirement for pyruvate,Pi dikinase in the cycle. In PEP-CK species some decarboxylation may also occur through NAD-malic enzyme which would utilize pyruvate,Pi dikinase of the mesophyll cells (Section 12.1c).

## 11.4 Charge balance in intercellular transport through the $C_4$ cycle

Carboxylation in the $C_4$ pathway would result in acidification of the mesophyll cells, while decarboxylation of the $C_4$ acids in the bundle sheath cells would result in alkalization. The proposed exchange of metabolites of the $C_4$ cycle between mesophyll and bundle sheath cells would also result in net transfer of negative charges to the bundle sheath cells. Either intercellular transport of protons to bundle sheath cells or hydroxyl ions to mesophyll cells may maintain pH balance and electroneutrality. Alternatively, PGA-DHAP exchange could maintain a charge balance as illustrated for NADP-ME species in Fig. 11.3 (see Section 11.6 for discussion of PGA-DHAP shuttle between the two cells). It is not known whether a pH gradient exists between the cell types during steady-state photosynthesis, and if so what its effect on metabolite transport would be.

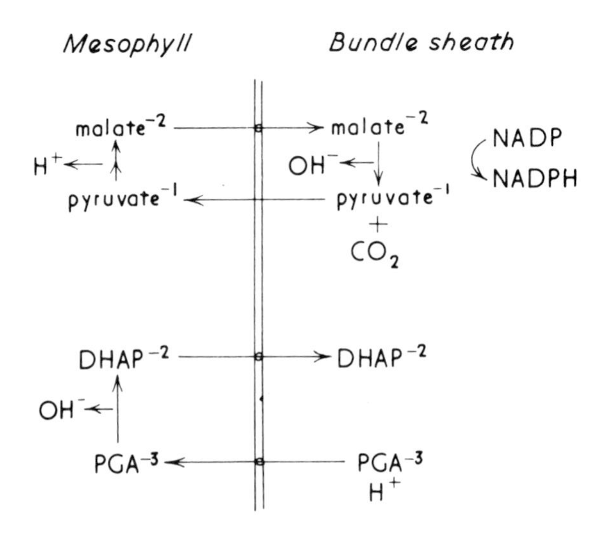


**Fig. 11.3.** Charge balance in intercellular transport.

## 11.5 Intercellular localization of the RPP pathway in $C_4$ plants

As shown in Table 11.2 the regenerative and carboxylative phases of the RPP pathway is only localized in bundle sheath cells while the reductive phase is localized in both cell types. With respect to energetics, this distribution of enzymes suggests that the bundle sheath chloroplasts generate the ATP for the regenerative phase, while both mesophyll and bundle sheath chloroplasts provide energy for the reductive phase of photosynthesis.

**Table 11.2.** Summary of localization of phases in the RPP pathway between mesophyll and bundle sheath cells.

| Phase | Enzyme measured | Localization: Mesophyll | Localization: Bundle sheath |
|---|---|---|---|
| Regeneration of RBP | phosphoribulokinase | − | + |
| Carboxylation | RBP carboxylase | − | + |
| Reduction of PGA to triose-P | phosphoglycerate kinase | + | + |
| | NADP-triose phosphate dehydrogenase | + | + |

## 11.6 Photochemical requirements in $C_4$ photosynthesis

Table 11.3 gives a summary of the photochemical requirements in the $C_4$ cycle and the RPP pathway for the three types of decarboxylating mechanisms. Each $CO_2$

**Table 11.3.** Photochemical requirements in $C_4$ plants per $CO_2$ fixed.

| Group | Energy | Requirements in $C_4$ Cycle* | Additional requirements for RPP pathway | Total per $CO_2$ |
|---|---|---|---|---|
| NADP-malic enzyme | ATP | 2 | 3 | 5 |
| | NADPH | 1 | 1** | 2 |
| NAD-malic enzyme | ATP | 2 | 3 | 5 |
| | NADPH | 0 | 2 | 2 |
| PEP-carboxykinase | ATP | 1 | 3 | 4 |
| | NADPH | 0 | 2 | 2 |

* Based on shuttle in Section 11.3

**There is a net generation of 1 NADPH by decarboxylation through NADP-malic enzyme.

generated by malate decarboxylation provides half of the reductive power required for $CO_2$ fixation in the RPP pathway.

It is more difficult to determine the division of labour between the two cell types in the production of assimilatory power. Presently, some assumptions are made in order to calculate this division.

(a) Assume that the decarboxylation of $C_4$ acids occurs through a single decarboxylating enzyme in a given species (which appears to be generally the case).

(b) Assume that the intercellular transport metabolites are malate and pyruvate in NADP-ME species, aspartate-alanine in NAD-ME species, and aspartate-PEP in PEP-CK species (Section 11.3). This would result in energy requirements in the mesophyll cell and bundle sheath cell for the $C_4$ cycle as shown in Table 11.4.

(c) Assume that there is equal reduction of PGA through the reductive phase of photosynthesis in the two cell types (based on the approximate equal activity of enzymes of the reductive phase in the two cells). This would result in a distribution of photochemical requirements for the RPP pathway as shown in Table 11.4.

Note that the examples in Table 11.4 show a high requirement for NADPH in mesophyll cells of NADP-ME species, a slightly higher requirement for ATP in mesophyll cells of NAD-ME species, and a higher requirement for ATP in bundle sheath cells of PEP-CK species. The photochemical characteristics of mesophyll and bundle sheath chloroplasts support these general differences (Section 11.7).

There are several reasons why the values in Table 11.4 cannot be taken as absolute and strict photosynthetic requirements for energy in the two cell types. A given $C_4$ species probably uses one of these three decarboxylating mechanisms as the primary, although not always exclusive, means of $C_4$ acid decarboxylation. Although the primary means of decarboxylation in PEP-CK species is that implied by their name, NAD-malic enzyme may also be utilized. Pyruvate,Pi dikinase is found in the mesophyll cells of all three groups and in PEP-CK species it might function in a cycle involving NAD-malic enzyme.

In all three groups of $C_4$ plants both aspartate and malate are early labelled products. It is possible, and perhaps likely, that both these $C_4$ acids are transport metabolites from mesophyll to bundle sheath cells in any given species. Obviously, malate transport results in a photochemical requirement for NADPH in mesophyll cells, while aspartate transport requires no photochemical generation of NADPH in the $C_4$ cycle.

Perhaps the greatest flexibility in energy requirements between the cell types is in the site of reduction of PGA following its formation through RBP carboxylase in bundle sheath chloroplasts. PGA may either be reduced in bundle sheath chloroplasts or transported to mesophyll chloroplasts for reduction. In the latter

**Table 11.4.** Calculations of photochemical requirements in mesophyll and bundle sheath cells of $C_4$ plants. Note assumptions made in text, and reasons these cannot be taken strictly as the division of energy for the cell types.

| | | | | $C_3$ pathway | | | | | |
|---|---|---|---|---|---|---|---|---|---|
| | | $C_4$ cycle | | Reductive phase | | Regeneration phase | | Total | |
| Group | Requirement | Mesophyll | Bundle sheath | Mesophyll | Bundle sheath | Mesophyll | Bundle sheath | Mesophyll | Bundle sheath |
| NADP-malic enzyme | ATP | 2 | 0 | 1 | 1 | 0 | 1 | 3 | 2 |
| | NADPH | 1 | * | 1 | * | 0 | 0 | 2 | 0 |
| NAD-malic enzyme | ATP | 2 | 0 | 1 | 1 | 0 | 1 | 3 | 2 |
| | NADPH | 0 | 0 | 1 | 1 | 0 | 0 | 1 | 1 |
| PEP-carboxykinase | ATP | 0 | 1 | 1 | 1 | 0 | 1 | 1 | 3 |
| | NADPH | 0 | 0 | 1 | 1 | 0 | 0 | 1 | 1 |

* See footnote in Table 11.3 (**)

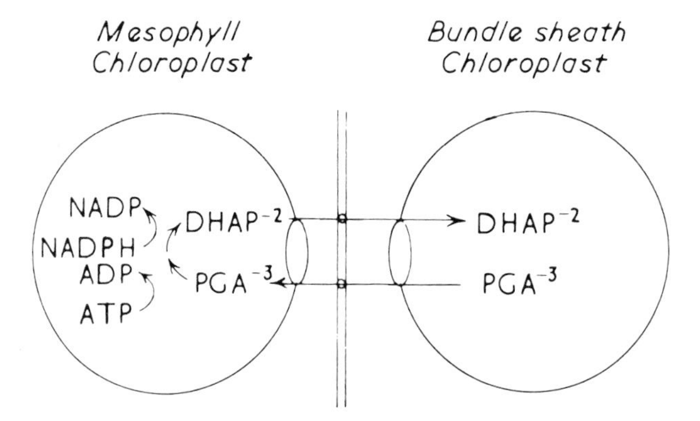


**Fig. 11.4.** Intercellular PGA-DHAP shuttle.

case, there may be a PGA-DHAP shuttle between mesophyll- and bundle sheath chloroplasts (Fig. 11.4).

In mesophyll chloroplasts of spinach (a $C_3$ plant), PGA can be taken up by the chloroplasts in exchange for DHAP on the phosphate translocator (Section 8.23). Similarly, a phosphate translocator in the mesophyll and bundle sheath chloroplasts of $C_4$ plants might catalyze an exchange of PGA-DHAP.

Another option, although more expensive energetically, is transport of PGA to mesophyll cells and conversion to glycerate through PGA phosphatase in the cytosol. Glycerate and Pi might then be taken up by the chloroplasts and converted to PGA through glycerate kinase. This shuttle has been suggested because there is some evidence for PGA phosphatase and glycerate kinase in relatively high levels in $C_4$ mesophyll extracts (Randall & Tolbert, 1971) (see Fig. 11.5).

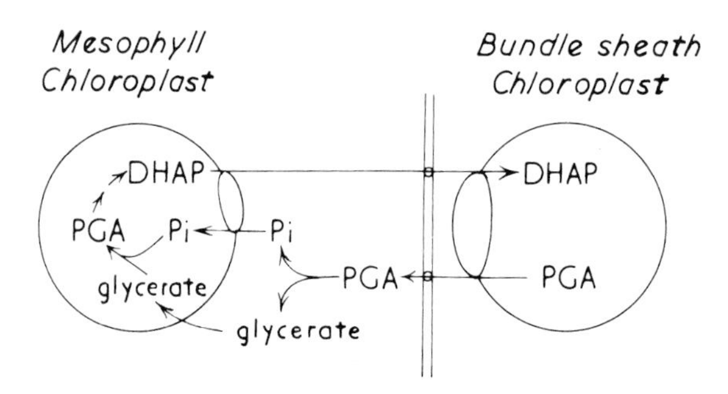


**Fig. 11.5** Alternate shuttle of PGA-DHAP. (Adapted from Randall and Tolbert, 1971).

As much as one half of the PGA formed may be transported from bundle sheath to mesophyll chloroplasts for reduction. In the RPP pathway, 5/6 of the PGA formed through RBP carboxylase is needed to continue regeneration of RBP (see Chapter 6). Thus, if half of the PGA is reduced in mesophyll cells, at least 2/3 of this amount would need to return to the RPP pathway of the bundle sheath chloroplasts.

An alternative to the transport of triose phosphates back to the bundle sheath cells is conversion of DHAP to sucrose in the mesophyll cells (Fig. 11.6). In the few $C_4$ species examined, enzymes involved in sucrose synthesis, such as UDP-glucose pyrophosphorylase and sucrose phosphate synthetase, are found in both mesophyll and bundle sheath cells. Downton and Hawker suggested that the mesophyll cell may be a primary site of sucrose synthesis in maize (Section 12.10).

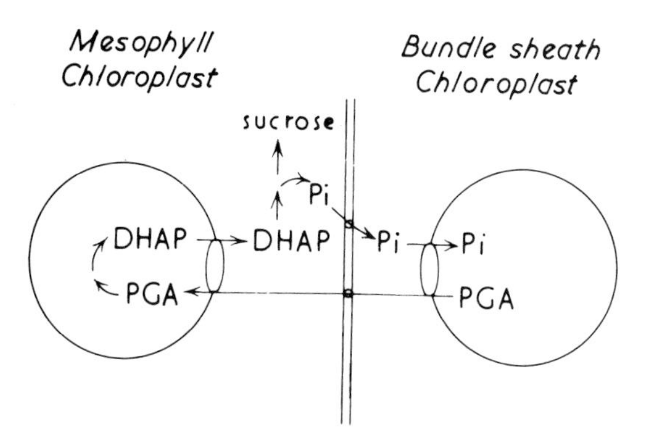


**Fig. 11.6.** Shuttle resulting in sucrose synthesis in mesophyll cells.

Another means of PGA metabolism in mesophyll cells was noted earlier, i.e. that PGA can serve as a precursor to the carboxylation phase of the $C_4$ pathway at substantial rates *in vitro*. This would provide an alternative precursor in the $C_4$ pathway should the products of decarboxylation in the bundle sheath cells (pyruvate or PEP) be partly metabolized in other paths.
[Refs. **10, 14, 16**].

## 11.7 Photochemical differences in chloroplast types

Photochemical differences in the chloroplast types from species of the three $C_4$ groups support the proposed differences in energy requirements during carbon assimilation.

### (a) MEASUREMENTS

Several methods have been used to assess the relative photochemical potential of chloroplasts. The distribution of the leaf Chl between the mesophyll and bundle sheath cells will influence the light-harvesting potential of the chloroplasts. Distribution can be determined by complete separation of the cell types by extensive enzymatic digestion or by derivation from Chl *a/b* ratios of whole leaf (W), mesophyll chloroplasts (M) and bundle sheath chloroplasts (B).

For the latter method the percentage of leaf Chl in mesophyll cells is equal to:

$$\frac{100(M+1)\ (W-B)}{(W+1)\ (M-B)}$$

This method is only reliable if a substantial difference exists in the Chl *a/b* ratio between mesophyll and bundle sheath chloroplasts, as in NADP-ME species.

PSII is required for noncyclic electron flow. This can be assessed by measuring the Hill reaction activity with various oxidants and by measuring delayed light emission (Section 4.13).

The ratio of photosystem I/II activity is indicative of the relative level of cyclic to noncyclic electron flow (i.e. a high photosystem I/II suggests high potential for cyclic electron flow and low potential for noncyclic electron flow and NADP reduction). Chl *a/b* ratio, fluorescence emission under liquid nitrogen and *P700* content together allow estimates of the relative levels of PSI to PSII (Section 4.13).

In $C_3$ chloroplasts, Chl *a* is the principal light harvesting Chl of PSI and Chl *a* and *b* the light harvesting chlorophylls of PSII. This implies that the higher the Chl *a/b* ratio the higher the potential for PSI activity. PSI and PSII are characterized by strong fluorescence emission around 730 and 685 nm respectively. This has been established from studies with $C_3$ chloroplasts where PSI and PSII particles have been separated (Section 4.13).

Since *P700* is the reaction centre Chl for PSI (Section 4.12), the ratio of Chl/ *P700* may vary depending on the relative amount of PSI versus PSII. For example, if a chloroplast has a Chl/*P700* ratio of 400/1 and equal distribution of Chl between the photosystems, the Chl/ *P700* ratio of the PSI particles is then 200 (generalization for $C_3$ chloroplasts, Section 4.13). Chloroplasts having a deficiency in PSII may then have a low Chl/ *P700* ratio, e.g. 200.

A low Chl/ *P700* ratio is not always indicative of low PSII potential. For example, reduction in the light-harvesting Chl of the photosystem but not of the reaction centre Chl may increase the light requirement for saturation of PSII activity without affecting the maximum potential. However, in-vivo this could effectively reduce the PSII activity if the chloroplasts receive limiting light. *P700* content has been routinely measured as the reaction centre Chl of PSI, while the PSII reaction centre Chl has been difficult to measure.

The first measurements of *P700* content with $C_4$ species were with chloroplasts from total leaf extracts. The measurements indicated that $C_4$ chloroplasts had a higher *P700* /Chl content than did $C_3$ chloroplasts (Black and Mayne, 1970). These results are consistent with $C_4$ plants having a higher requirement for ATP than NADPH (5 ATP/2 NADPH compared to 3 ATP/2 NADPH for $C_3$ species) since high *P700* content may be associated with a relatively higher level of cyclic photophosphorylation.

### (b) NADP-ME SPECIES

In species of the NADP-ME group, 70–100% of the capacity for non-cyclic electron flow appears to be in the mesophyll chloroplasts. There is a gradation in structure from bundle sheath chloroplasts with rudimentary grana, as in crabgrass and maize, to ones which are completely agranal (exhibiting no thylakoid stacking) as in sorghum (Fig. 11.7a) and sugarcane. There is a deficiency of PSII activity in these chloroplasts which corresponds to the lack of grana development (Table 11.5). Bundle sheath chloroplasts of NADP-ME species are the only chloroplasts of $C_4$ plants found to have a substantial reduction in grana development.

The full potential for non-cyclic electron transport *in vivo* in these chloroplast types is still uncertain. From measurements on isolated chloroplasts it is difficult to estimate the relative in-vivo level of non-cyclic electron flow in the two cell types. For example, with isolated maize bundle sheath chloroplasts, varying rates of non-cyclic electron flow have been obtained using different Hill oxidants. Some studies imply an equivalent capacity for non-cyclic electron transport in maize mesophyll and bundle sheath chloroplasts (activities with several oxidants of about 75–150 $\mu$mol $O_2$ evolved $mg^{-1}$ Chl $h^{-1}$). Other studies with maize indicate mesophyll chloroplasts as the primary site of non-cyclic electron flow with rates up to 600 $\mu$mol $O_2$ evolved $mg^{-1}$ Chl $h^{-1}$ with p-benzoquinone as the Hill oxidant which is sufficient to accomodate rates of leaf photosynthesis. Discrepancies in results may arise due to difference in age of tissue (bundle sheath chloroplasts lose grana during development), different light intensities during growth, differential loss of activity between the chloroplast types during isolation, different degrees of purity of the chloroplast types (cross contamination during isolation), or difference in assay conditions. Bundle sheath chloroplasts of NADP-ME species have a smaller photosynthetic unit (low level of light-harvesting Chl per reaction centre) than mesophyll chloroplast (Table 11.6). Therefore *in vivo* relatively low light may saturate noncyclic electron flow in mesophyll chloroplasts while high light may be

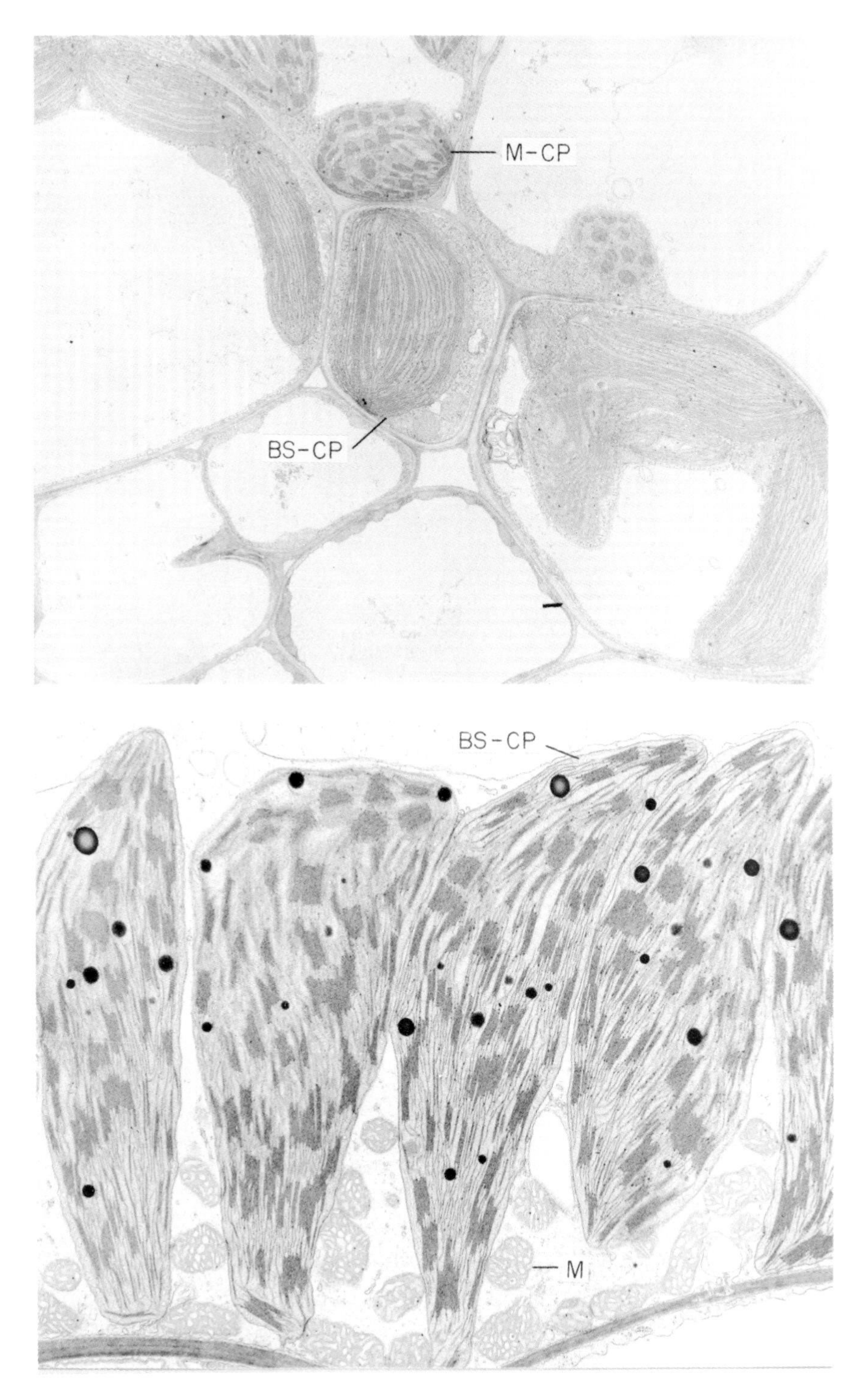


**Fig. 11.7.** a. Electron micrograph illustrating agranal bundle sheath chloroplasts (BS-CP) and granal mesophyll chloroplasts (M-CP) of *Sorghum sudanense*, NADP-ME species, × 6000 Micrograph by S. E. Frederick, courtesy of E. H. Newcomb, University of Wisconsin. b. Electron micrograph showing granal bundle sheath chloroplasts (BS-CP) in young leaf of *Portulaca oleracea*, NAD-ME species × 16 300. Note the prominent mitochondria (M) which are the site of malate decarboxylation in this $C_4$ subgroup (see Section 11.2). (Micrograph by Mary E. Doohan, courtesy of E. H. Newcomb, University of Wisconsin.)

**Table 11.5.** Noncyclic electron flow with isolated chloroplasts from NADP-malic enzyme species. Various oxidants have been used to measure oxygen evolution in noncyclic flow with the chloroplast types. These include methyl viologen, ferricyanide, p-benzoquinone and NADP.

| Species | Common name | % distribution of noncyclic electron flow: Mesophyll chloroplasts | Bundle sheath chloroplasts | Grana development |
|---|---|---|---|---|
| *Saccharum officinarum* | sugarcane | 100–97 | 0–3 | agranal |
| *Sorghum bicolor* | sorghum | 100–95 | 0–5 | agranal |
| *Zea mays* | corn | 85–70 | 15–30 | rudimentary |
| *Digitaria sanguinalis* | crabgrass | 70 | 30 | rudimentary |

required for saturation with bundle sheath chloroplasts. Recognizing some variation in activities reported, the percentages in Table 11.5 are an indication of the distribution of non-cyclic electron flow between the cell types in some NADP-ME species.

Another correlation exists between p-benzoquinone dependent oxygen evolution and capacity for $CO_2$ fixation by isolated bundle sheath cells of several NADP-ME species (Table 11.7). These rates of Hill reaction and $CO_2$ fixation with sugarcane and sorghum are only about 1 % of the rates of leaf photosynthesis in $C_4$ plants. There are two possible explanations. Either the results show that sorghum and sugarcane bundle sheath chloroplasts (which tend to be

**Table 11.6.** Ratio of Chl/*P700*, Chl *a*/*b* ratio, and liquid nitrogen fluorescence emission (F685/730) of chloroplast types of NADP-malic enzyme species in comparison to $C_3$ chloroplasts. (After Edwards *et al*, 1976; Mayne *et al*, 1974.)

| Species | F685/730 Mesophyll | F685/730 Bundle sheath | Chl *a*/*b* Mesophyll | Chl *a*/*b* Bundle sheath | Chl/*P700* Mesophyll | Chl/*P700* Bundle sheath |
|---|---|---|---|---|---|---|
| NADP-malic enzyme | | | | | | |
| *Saccharum officinarum* | 1.13 | 0.25 | 3.70 | 5.56 | 412 | 193 |
| *Sorghum bicolor* | 1.14 | 0.23 | 3.73 | 6.60 | 418 | 191 |
| *Zea mays* | 1.14 | 0.18 | 3.30 | 6.51 | 413 | 218 |
| *Digitaria sanguinalis* | 0.76 | 0.20 | 3.58 | 5.80 | 322 | 188 |
| *Setaria lutescens* | 0.85 | 0.28 | 3.54 | 5.50 | 428 | 264 |
| $C_3$ species | | | | | | |
| *Avena sativa* | 0.87 | — | 3.45 | — | 444 | — |
| *Hordeum vulgare* | 0.76 | — | 3.66 | — | 478 | — |
| *Triticum aestivum* | 1.57 | — | 3.41 | — | 439 | — |

**Table 11.7.** Capacity of Hill reaction using p-benzoquinone versus light-dependent $CO_2$ fixation with isolated bundle sheath cells of various NADP-malic enzyme species. Cells were isolated enzymatically. (After Edwards *et al*, 1976.)

| Species | Hill reaction (1 mM p-benzoquinone) | $CO_2$ fixation |
|---|---|---|
| | μmol $O_2$ evolved $mg^{-1}$ Chl $h^{-1}$ | μmol $mg^{-1}$ Chl $h^{-1}$ |
| *Saccharum officinarum* | 3 | 2 |
| *Sorghum bicolor* | 3 | 1 |
| *Zea mays* | 21 | 7 |
| *Digitaria sanguinalis* | 84 | 22 |
| *Setaria lutescens* | 62 | 33 |

agranal) do not have functional non-cyclic electron flow *in vivo*, or the isolated bundle sheath cells are susceptible to loss of photosystem II activity, particularly those with agranal chloroplasts. Downton and co-workers (1970) used an in-vivo method for assaying non-cyclic electron flow. This involved blue tetrazolium dye as a Hill reagent, which becomes opaque upon reduction. They concluded that bundle sheath chloroplasts of sorghum lack non-cyclic electron flow, that those of maize have low activity, and that those of *Panicum miliaceum* (granal bundle sheath chloroplasts) have relatively high activity. As can be seen, the subsequent studies of non-cyclic electron flow with isolated chloroplasts and bundle sheath cells support these results. Although there is some correlation between structure and function in these chloroplasts of $C_4$ species, a tendency towards agranal chloroplasts in photosynthetic tissue cannot always be taken as evidence for a deficiency in PSII (e.g. red algal chloroplasts have single unappressed lamellae).

On average, about 90% of the delayed light emission comes from mesophyll chloroplasts, and 10% from bundle sheath chloroplasts of NADP-ME species. Again, this strongly suggests PSII deficiency in these bundle sheath chloroplasts. In NADP-ME species, a slightly higher level of the Chl has generally been found in the mesophyll cells (roughly 60%), which corresponds to the primary site of $NADP^+$ reduction. (See Section 4.13)

Measurements on Chl *a*/*b* ratio, liquid nitrogen fluorescence emission spectra and *P700* content all indicate that bundle sheath chloroplasts of NADP-ME species have a high ratio of photosystem I/II. Mesophyll chloroplasts of these species have a Chl *a*/*b* ratio similar to that of $C_3$ plants. The bundle sheath chloroplasts have a much higher Chl *a*/*b* ratio suggesting a high PSI capacity (Table 11.6). In $C_3$ chloroplasts the light-harvesting Chl of PSI contains Chl *a* and the light harvesting

Chl of PSII is thought to contain equal amounts of Chl *a* and *b* (Section 4.13). The majority of the Chl of the leaf is the light-harvesting Chl and a much lower level is in the reaction centres. Thus from the Chl *a*/*b* ratio an estimate can be made of the distribution of the light-harvesting Chl between the two photosystems. For example, chloroplasts with a Chl *a*/*b* ratio of 3 would have roughly 2 Chl *a* in PSI, and 1 Chl *a* and 1 Chl *b* in PSII. This gives an approximately equal distribution of Chl between the photosystems. A Chl *a*/*b* ratio of 6, as in bundle sheath chloroplasts of some NADP-ME species, would suggest roughly 5 Chl *a* in PSI and 1 Chl *a* and 1 Chl *b* in PSII (or about 70% of the Chl in PSI and 30% of the Chl associated with PSII). If acceptors to PSII are limiting, the Chl *a*/*b* protein may also donate energy to PSI (Section 4.11). This would allow bundle sheath Chl to trap energy primarily for PSI.

The bundle sheath chloroplasts of NADP-ME species also have a relatively low ratio of F685/730 (Table 11.6, Fig. 11.8) which indicates a high ratio of photosystem

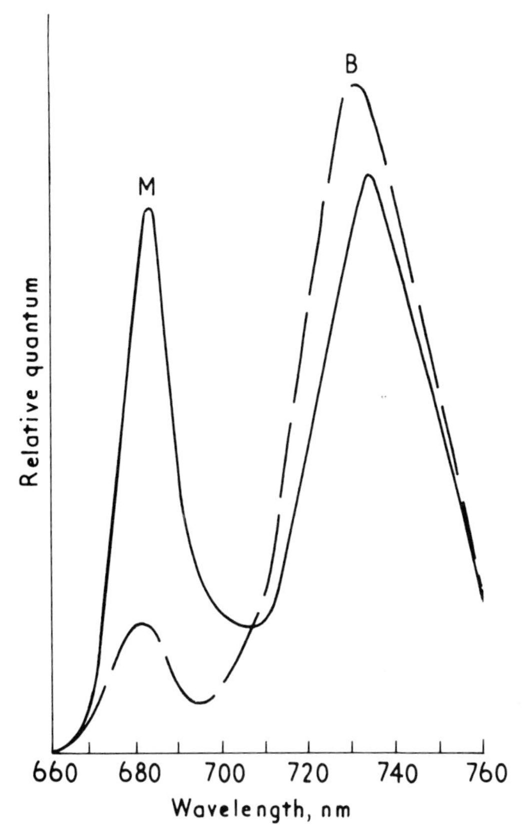


**Fig. 11.8.** Fluorescence emission spectra at 77° *K* of chloroplast fragments of mesophyll (*M*) and bundle sheath cells (B) of *Sorghum bicolor* (NADP-ME species). (After Mayne *et al*, 1974)

I/II. The ratio of Chl/*P700* in bundle sheath chloroplasts in these species is about half that of mesophyll chloroplasts and is similar to that of PSI particles of $C_3$ chloroplasts. The Chl/*P700* of the $C_4$ mesophyll chloroplasts of NADP-ME species is similar to that of $C_3$ chloroplasts (Table 11.6). Taken together, the photochemical evidence suggests that bundle sheath chloroplasts of these species have a high level of PSI, which presumably supports cyclic photosphorylation, while the mesophyll chloroplasts are the primary site of $NADP^+$ reduction.

In most NADP-ME species it is likely that bundle sheath chloroplasts contribute some reductive power. Sorghum and sugar-cane represent extreme cases where photochemical data suggest that mesophyll chloroplasts provide all of the reductive power (photochemical data suggests 95–100%). Consider that two PGA molecules are formed for each $CO_2$ fixed by RBP carboxylase in the bundle sheath chloroplasts. With one turn of the $C_4$ cycle, one NADPH would be transferred from mesophyll to bundle sheath chloroplasts with malate transport and decarboxylation, which could be used to reduce one molecule of PGA. The other PGA would be reduced by the mesophyll chloroplasts through a shuttle of PGA and triose phosphate between the cell types (Section 11.6). In this case the dark reactions of the reductive phase would be functioning at equal rates in the two cell types with all of the reductive power being generated in the mesophyll chloroplasts.

### (c) NAD-ME and PEP-CK species

In NAD-ME and PEP-CK species both mesophyll and bundle sheath chloroplasts contain grana. The granal bundle sheath chloroplasts of an NAD-ME species (*Portulaca oleracea*) are shown in Fig. 11.7b. In these two groups from 2/3 to 3/4 of the leaf Chl is in bundle sheath chloroplasts. This contrasts with NADP-ME species which have a high percentage of the leaf Chl in mesophyll cells. During anatomical investigations, Walter Brown noted that species of the Eragrostoideae, a subfamily of the Gramineae (now known to have NAD-ME and PEP-CK species), had the major portion of the leaf Chl in bundle sheath cells.

Based on Hill reaction activity and delayed light emission both mesophyll and bundle sheath chloroplasts of NAD-ME and PEP-CK species have PSII activity. However, a higher proportion of the reductive power may be in bundle sheath cells due to the higher percentage of leaf Chl in these cells. Thus, in these species more than half of the PGA may be reduced in the bundle sheath chloroplasts, rather than an equal reduction by the two cell types as suggested in the example in Table 11.4.

The previous calculations of energy requirements for NAD-ME species (Table 11.4) suggest a slightly higher ATP/NADPH ratio in mesophyll chloroplasts (3/1) than in bundle sheath chloroplasts (2/1). In support of this, mesophyll chloroplasts

**Table 11.8.** Chl/*P700* ratios and delayed light emission of mesophyll and bundle sheath chloroplasts of NAD-ME and PEP-CK species. (After Edwards *et al*, 1976.)

| | Chl/*P700* | | % distribution of delayed light emission | |
|---|---|---|---|---|
| Group | Mesophyll | Bundle sheath | Mesophyll | Bundle sheath |
| NAD-ME | | | | |
| *Panicum capillare* | 271 | 376 | 13 | 87 |
| *Panicum miliaceum* | 207 | 292 | 17 | 83 |
| PEP-CK | | | | |
| *Panicum maximum* | 403 | 242 | 42 | 58 |
| *Panicum texanum* | 428 | 342 | 35 | 65 |

of NAD-ME species have a lower Chl/*P700* ratio than do bundle sheath chloroplasts. On the other hand, mesophyll chloroplasts of PEP-CK species have a relatively high Chl/*P700* ratio (Table 11.8) which is consistent with these chloroplasts having a low ATP/NADPH ratio (Table 11.4).

Currently, the photochemical characteristics of the chloroplast types can only be compared in a general way with the major flow of carbon expected in a given $C_4$ subgroup. The energy requirements of a given $C_4$ cycle, the transport of aspartate versus malate, and the extent of PGA reduction by mesophyll chloroplasts will all influence the energy utilization between the cell types. The photochemical provision of energy is thus coordinated with the reactions of carbon assimilation.
[Refs **5, 9, 10, 15, 17, 21, 22, 28**]

## 11.8 Taxonomy

$C_4$ plants have only been identified among terrestrial angiosperms (Section 10.7). Sixteen of approximately 300 families of flowering plants are known to possess $C_4$ species. These are Gramineae (grasses), Cyperaceae (sedges) and 14 dicotyledonous families; Acanthaceae, Aizoaceae, Amaranthaceae, Asteraceae, Boraginaceae, Capparidaceae, Caryophyllaceae, Chenopodiaceae, Euphorbiaceae, Nyctaginaceae, Polygonaceae, Portulacaceae, Scrophulariaceae, and Zygophyllaceae. $C_4$ photosynthesis must have evolved on many separate occasions. No family is comprised entirely of $C_4$ species. However, a particular genus, tribe or subfamily within a family may have only $C_4$ species. The $C_4$ syndrome is particularly prominent among species in the families Gramineae, Cyperaceae and Chenopodiaceae. In one survey of more than 500 grasses, using carbon isotope fractionation for identification (Section 15.12), Bruce Smith and Walter Brown found about half of the species examined were $C_4$.

Many of the worst weeds of the world are $C_4$ species (Table 11.9) whilst most crops are $C_3$. The most economically important $C_4$ crops are sorghum, maize and sugar-cane (belonging to the NADP-ME subgroup), millets (*Pennisetum typhoides*, *Panicum miliaceum*, *Setaria* millets, and *Eleusine coracana*), and a number of pasture grasses. Also, many important prairie grasses are $C_4$ (Table 11.10).

There has been a relatively extensive study of the family Gramineae with respect to identifying species as $C_3$ or $C_4$ and classifying species into $C_4$ subgroups. There are three major subfamilies in the Gramineae as illustrated by a survey of native American species (Table 11.11).

Most species of the subfamily Panicoideae are $C_4$, all species of subfamily Eragrostoideae are $C_4$, whilst all species of the subfamily Festucoideae are $C_3$.

**Table 11.9.** Occurrence of $C_4$ plants among weeds.

A. World's ten worst weeds. (After Holm, 1969, also see Black *et al*, 1969)

| Species | Common name | Subsequent classification |
|---|---|---|
| *Cynodon dactylon* | bermuda grass | $C_4$ (NAD-ME) |
| *Cyperus rotundus* | purple nutsedge | $C_4$ (NADP-ME) |
| *Echinochloa colonum* | jungle rice | $C_4$ |
| *Echinochloa crusgalli* | barnyard grass | $C_4$ (NADP-ME) |
| *Eleusine indica* | goose grass | $C_4$ (NAD-ME) |
| *Imperata cylindrica* | congo grass | $C_4$ |
| *Panicum maximum* | Guinea grass | $C_4$ (PEP-CK) |
| *Sorghum halepense* | Johnson grass | $C_4$ (NADP-ME) |
| *Eichhornia crassipes* | water hyacinth | $C_3$ |
| *Lantana camara* | lantana | $C_3$ |

B. The ten worst weeds of field crops in U.S. (After Doresch, 1970)

| Species | Common name | Subsequent classification |
|---|---|---|
| *Amaranthus* spp. | pigweed | $C_4$ (NAD-ME) |
| *Cyperus* spp. | nutsedge | $C_4$ (NADP-ME) |
| *Echinochloa crusgalli* | barnyard grass | $C_4$ (NADP-ME) |
| *Panicum* spp. | | $C_4$ (primarily NAD-ME) |
| *Setaria* spp. | foxtail | $C_4$ (NADP-ME) |
| *Sorghum halepense* | Johnson grass | $C_4$ (NADP-ME) |
| *Agropyron repens* | quack grass | $C_3$ |
| *Chenopodium album* | lambs quarter | $C_3$ |
| *Ipomoea* spp. | morning glory | $C_3$ |
| *Xanthium* spp. | cocklebur | $C_3$ |

**Table 11.10.** Important $C_4$ prairie grasses.

| Species | Common name | Subgroup |
|---|---|---|
| *Andropogon scoparius* | little bluestem | (NADP-ME) |
| *Andropogon geradii* | big bluestem | (NADP-ME) |
| *Bouteloua gracilis* | blue grama | (NAD-ME) |
| *Bouteloua curtipendula* | sideoats grama | (PEP-CK) |
| *Buchloe dactyloides* | buffalo grass | (NAD-ME) |
| *Panicum virgatum* | switch grass | (NAD-ME) |
| *Sorghastrum nutans* | indian grass | (NADP-ME) |

**Table 11.11.** Major subfamilies of the family Gramineae. (After Gould, 1968)

| Subfamily | % of U.S. native species |
|---|---|
| Panicoideae | 33 |
| Eragrostoideae | 28.5 |
| Festucoideae | 36 |
| Total | 97.5 |

Species of subfamilies Panicoideae and Eragrostoideae are generally best adapted to warm climates and, in particular, some species among the Eragrostoideae are noted for drought tolerance.

Fig. 11.9 represents a classification of genera of the Gramineae according to a survey of representative species for their type of decarboxylating mechanism. NADP-ME species are only found among the subfamilies Panicoideae (including sorghum and sugarcane in tribe Andropogoneae; maize in tribe Maydeae) and Aristoideae. Species of each $C_4$ subgroup among the Gramineae have distinguishing anatomical features (Fig. 11.10). The bundle sheath chloroplasts of NADP-ME species are located in a centrifugal position (towards the outside of the cell) and the bundle sheath chloroplasts tend to be agranal (also see Fig. 10.7a & Fig. 11.7a). Except for some *Panicum* species, NAD-ME species are found only in genera of the subfamily Eragrostoideae. In fact the only true *Panicum* species may be NAD-ME types, e.g. *Panicum miliaceum*, while other so called *Panicums* may be more appropriately classified in other genera. The bundle sheath chloroplasts of NAD-ME species contain grana and are located in a centripetal position (towards the vascular tissue).

PEP-CK species are found only among some genera of the subfamily Eragrostoideae except for a few genera in subfamily Panicoideae (*Urochloa*,

*Eriochloa*, and *Brachiaria*). A few *Panicum* species, e.g. *Panicum maximum*, which are PEP-CK types, might now be more correctly placed in the genus *Brachiaria*. PEP-CK species of the Gramineae have granal bundle sheath chloroplasts, located primarily in a centrifugal position.

It has already been noted that the agranal chloroplasts of NADP-ME species have a deficiency of non-cyclic electron transport (Section 11.7). Otherwise no functional-structural relationships have been established based on differences in the $C_4$ subgroups of the Gramineae shown in Fig. 11.10. Furthermore, in the $C_4$ dicots, no relationship has been found between the decarboxylation mechanism and position of bundle sheath chloroplasts. At least a few of the $C_4$ dicots are noted to have bundle sheath chloroplasts in the centripetal position whether belonging to the NAD-ME or NADP-ME subgroup (Fig. 10.7c). Considerable diversity exists in some of the general features of Kranz anatomy between various families and possible correlations between biochemical and anatomical features remain to be studied.
[Refs. **3, 6, 13, 14, 16, 27**]

## 11.9 $C_3/C_4$ intermediates

There are a few species which cannot be classified by standard criteria as either $C_3$ or $C_4$ because they have some characteristics of both groups. This was first recognized with hybrids obtained between $C_3$ and $C_4$ *Atriplex spp*. The $F_1$ and $F_2$ generations had a range of activities of PEP carboxylase and varying degrees of development of Kranz anatomy. The $CO_2$ compensation points of all progeny were $C_3$-like. Interestingly, there were some hybrids which had carboxylase levels and anatomy very similar to $C_4$ plants and yet did not function biochemically as $C_4$. The progeny generally had lower photosynthetic rates than either that of the $C_3$ or $C_4$ parent. These studies showed that $C_4$ photosynthesis only functions when all parameters of the $C_4$ syndrome are present and properly coordinated.

The genus *Mollugo* (Aizoaceae) contains $C_3$, $C_4$ and species with intermediate characteristics. *Mollugo verticillata* is an intermediate type. It has both $C_3$ and $C_4$ acids as primary initial products of $CO_2$ fixation, intermediate levels of photorespiration, and bundle sheath cells containing chloroplasts. The mesophyll tissue consists of palisade and spongy parenchyma. Pulse-chase studies ($^{14}CO_2$ followed by $^{12}CO_2$, Section 10.2) show metabolism of the $C_4$ acids during the chase. Ecotypes of this species show varying degrees of intermediacy between $C_3$ and $C_4$. Some *Panicum* species (Laxa section) have also been identified as $C_3/C_4$ intermediates. *Panicum milioides* of this type (independently identified by Brown and Brown, Kanai and Kashiwagi, 1975) has a $CO_2$ compensation point of 20–25 ppm (in comparison

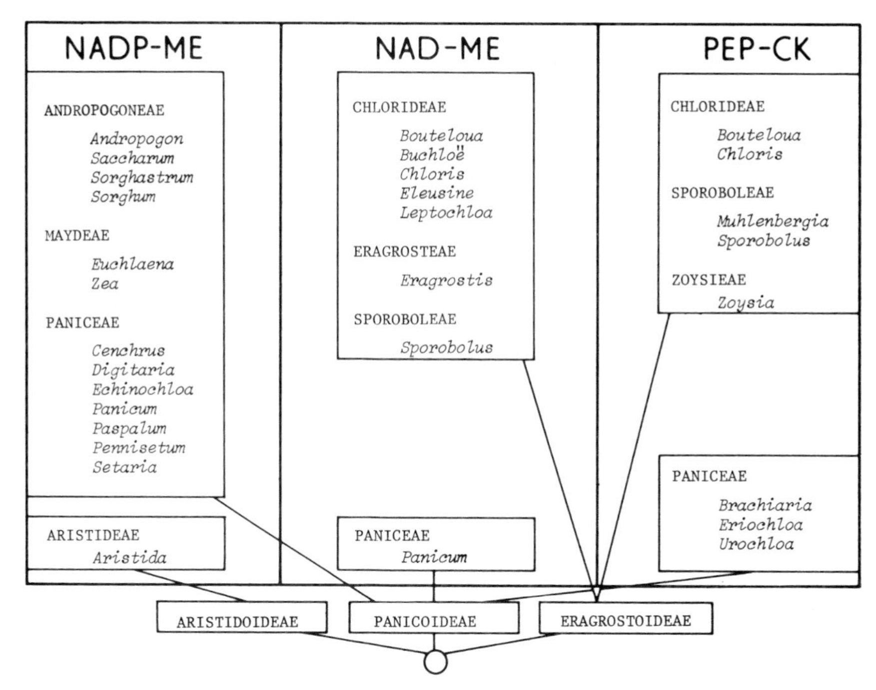


**Fig. 11.9** Phylogenetic scheme of $C_4$ groups in the Gramineae based on biochemical and cytological examinations of species of the subfamilies Aristidoideae, Eragrostoideae and Panicoideae. (After Gutierrez *et al*, 1974; 1976)

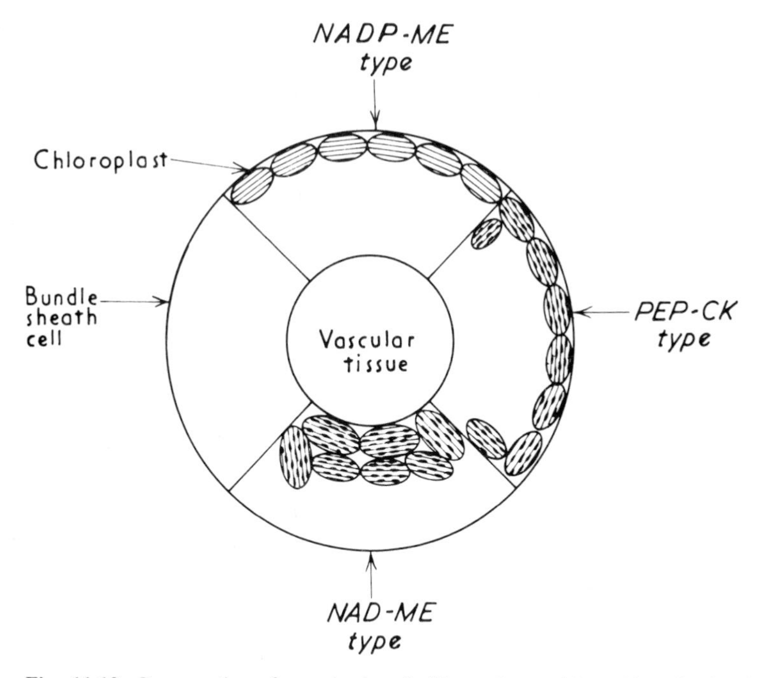


**Fig. 11.10.** Cross section of vascular bundle illustrating position of bundle sheath chloroplasts in different $C_4$ subgroups of the Gramineae. Note NADP-ME and PEP-CK types are distinguished by agranal versus grana-containing chloroplasts.

to 40–60 ppm for $C_3$ species), and low levels of $O_2$ inhibition of photosynthesis. The anatomy is Kranz (very distinct layer of chloroplast-containing bundle sheath cells surrounding vascular tissue, Section 10.4) although the number of mesophyll cells between the vascular bundles is much higher than in $C_4$ species. PEP carboxylase activity is substantially higher than in $C_3$ plants and, like *Mollugo verticillata*, there is some evidence for $^{14}CO_2$ fixation into the $C_4$ acids malate and aspartate. *Moricandia arvensis* is another species having intermediate $C_3/C_4$ characteristics based on $CO_2$ compensation point and $O_2$ inhibition of photosynthesis.

In *Mollugo verticillata*, *P. milioides* and *Moricandia arvensis*, no strict differential compartmentation of PEP carboxylase and RBP carboxylase has been found between different photosynthetic cell types as occurs in $C_4$ plants. Pyruvate,Pi dikinase is present in leaves of *P. milioides* suggesting that PEP may be generated in part from pyruvate rather than from 3-phosphoglycerate. However, whether $CO_2$ is fixed into $C_4$ acids through a secondary carboxylation (substrate for PEP carboxylase originating from PGA formed through RBP carboxylase, see Section 10.1) or a primary carboxylation in these species has not been ascertained. It should be noted that in all species reported to have some characteristics intermediate to those of $C_3$ and $C_4$, the carbon isotope composition (expressed as $\delta^{13}C$, see Section 15.12) is similar to that of $C_3$ plants. This indicates that under atmospheric conditions direct fixation of external $CO_2$ through PEP carboxylase is relatively small compared to fixation through RBP carboxylase. Thus a few plants have been identified having photosynthetic characteristics intermediate to those of $C_3$ and $C_4$ with some evidence of Kranz-like bundle sheath cells. However, no species have been identified which exhibit $C_4$ photosynthesis but lack intercellular compartmentation of metabolism (Section 10.7).

[Refs. **1, 2, 7, 11, 19, 20, 24, 26**]

### General Reading—see chapter 10

### Specific citations


1 Apel P. & Ohle H. (1979) $CO_2$-Kompensationspunkt und Blattanatomie bei Arten der Gattung *Moricandia DC.* (Cruciferae). *Biochem. Physiol. Pflanzen*, **174**, 68–75.

2 Björkman O., Nobs M. A. & Pearcy R. W. (1971) Hybrids between *Artiplex* species with and without B-carboxylation photosynthesis. *Carnegie Inst. Wash. Year Book*, **69**, 624–48.

3 Black C. C., Chen T. M. & Brown R. H. (1969) Biochemical basis for plant competition. *Weed Sci.* **17**, 338–44.

4 Black C. C. & Mayne B. C. (1970) *P700* activity and chlorophyll content of plants with different photosynthetic carbon dioxide fixation cycles. *Plant Physiol.* **45**, 738–41.

5 Brown W. V. (1958) Leaf anatomy in grass systematics. *Bot. Gaz.* **119**, 170–8.

6 Brown W. V. (1977) The Kranz syndrome and its subtypes in grass systematics. *Memoirs Torrey Bot. Club.* **23**, 1–97.

7 Brown R. H. & Brown W. V. (1975) Photosynthetic characteristics of *Panicum milioides*, a species with reduced photorespiration. *Crop Sci.* **15**, 681–5.
8 Doresch R. (1970) The 10 worst weeds of field crops. *Crops and Soils*, **1**, 14.
9 Downton W. J. S., Berry J. A. & Tregunna E. B. (1970) $C_4$-photosynthesis: non cyclic electron flow and grana development in bundle sheath chloroplasts. *Z. Pflanzenphysiol.* **63**, 194–9.
10 Edwards G. E., Huber S. C., Ku S. B., Gutierrez M., Rathnam C. K. M. & Mayne B. C. (1976) Variation in photochemical activities in $C_4$ plants in relation to $CO_2$ fixation. In *Metabolism and Productivity of Plants.* (eds. R. H. Burris and C. C. Black) pp. 83–112. Baltimore, Maryland: University Park Press.
11 Golstein L. D., Ray T. B., Kestler D. P., Mayne B. C., Brown R. H. & Black C. C. (1976) Biochemical characterisation of *Panicum* species which are intermediate between $C_3$ and $C_4$ photosynthesis plants. *Plant Sci. Lett.* **6**, 85–90.
12 Gould F. W. (1968) *Grass Systematics.* p. 97. New York: McGraw-Hill.
13 Gutierrez M., Edwards G. E. & Brown W. V. (1976) PEP carboxykinase containing species in the *Brachiaria* group of the subfamily Panicoideae. *Biochem. System. Ecol.* **4**, 47–9.
14 Gutierrez M., Gracen V. E. & Edwards G. E. (1974) Biochemical and cytological relationships in $C_4$ plants. *Planta*, **119**, 279–300.
15 Hardt H. & Kok B. (1978) Comparison of photosynthetic activities.of spinach chloroplasts with those of corn mesophyll and corn bundle sheath tissue. *Plant Physiol.* **62**, 59–63.
16 Hatch M. D., Kagawa T. & Craig S. (1975) Subdivision of $C_4$-pathway species based on differing $C_4$ acid decarboxylating systems and ultrastructural features. *Aust. J. Plant Physiol.* **2**, 111–28.
17 Hatch M. D. & Osmond C. B. (1976) Compartmentation and transport in $C_4$ photosynthesis. In *Transport in plants. Encyclopedia of Plant Physiology.* Vol. III (eds. C. R. Stocking and U. Heber.) pp. 144–84. Berlin: Springer-Verlag.
18 Holm L. (1969) Weed problems in developing countries. *Weed Sci.* **17**, 113–8.
19 Kanai R. & Kashiwagi M. (1975) *Panicum milioides*, a Gramineae plant having a Kranz anatomy without $C_4$ photosynthesis. *Plant Cell Physiol.* **16**, 669–79.
20 Ku S. B., Edwards G. E. & Kanai R. (1976) Distribution of enzymes related to $C_3$ and $C_4$ pathway of photosynthesis between mesophyll and bundle sheath cells of *Panicum hians* and *Panicum milioides*. *Plant Cell Physiol.* **17**, 615–20.
21 Ku S. B., Gutierrez M., Kanai R. & Edwards G. E. (1974) Photosynthesis in mesophyll protoplasts and bundle sheath cells of $C_4$ plants. II. Chlorophyll and Hill reaction studies. *Z. Pflanzenphysiol.* **72**, 320–37.
22 Mayne B. C., Dee A. M. & Edwards G. E. (1974) Photosynthesis in mesophyll protoplasts and bundle sheath cells of various types of $C_4$ plants. Fluorescence emission spectra, delayed light emission and *P700* content. *Z. Pflanzenphysiol.* **74**, 275–91.
23 Randall D. D. & Tolbert N. E. (1971) Two phosphatases associated with photosynthesis and the glycolate pathway. In *Photosynthesis and photorespiration.* (eds. M. D. Hatch, C. B. Osmond and R. O. Slayter) pp. 259–66. New York: Wiley Interscience.
24 Rathnam C. K. M. & Chollet R. (1978) $CO_2$ donation by malate and aspartate reduces photorespiration in *Panicum milioides*, a $C_3/C_4$ intermediate species. *Biochem. Biophys. Res. Commun.* **85**, 801–8.
25 Rathnam C. K. M. & Edwards G. E. (1977) $C_4$ acid decarboxylation and $CO_2$ donation to photosynthesis in bundle sheath strands and chloroplasts from species representing three groups of $C_4$ plants. *Arch. Biochem. Biophys.* **182**, 1–13.
26 Sayre R. T. & Kennedy R. A. (1977) Ecotypic differences in the $C_3$ and $C_4$ photosynthetic activity in *Mollugo verticillata*, a $C_3/C_4$ intermediate. *Planta*, **134**, 257–62.
27 Smith B. N. & Brown W. V. (1973) The Kranz syndrome in the Gramineae as indicated by carbon isotopic ratios. *Amer. J. Bot.* **60**, 505–13.
28 Walker G. H. & Izawa S. (1979) Photosynthetic electron transport in isolated maize bundle sheath cells. *Plant Physiol.* **63**, 133–8.

# Chapter 12
# Integration of Functions in $C_4$ Photosynthesis



## SUMMARY


The biochemistry of $C_4$ photosynthesis and the intracellular localization of enzymes is sufficiently understood to sketch the sequence of carbon flow during $CO_2$ fixation within the leaf of the three $C_4$ subgroups. In-vitro studies of carbon metabolism by mesophyll and bundle sheath preparations provide clear evidence for division of labour between the photosynthetic cell types. $C_4$ photosynthesis requires coordination between the $C_4$ and the RPP pathways and extensive metabolite transport. These are evaluated and obvious gaps in current understanding are outlined. Other processes closely linked to photosynthesis, namely starch and sucrose synthesis and nitrate metabolism, are evaluated with respect to their compartmentation in $C_4$


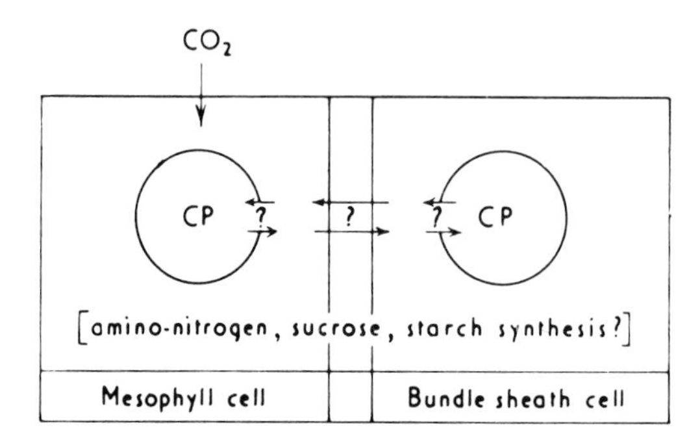


**Fig. 12.1.** The $C_4$ cycle functions between two photosynthetic cell types in the leaf. Mechanisms must exist for metabolite transport between cells and across chloroplast envelopes (CP). By isolating chloroplasts from the cells, their role in $C_4$ photosynthesis can be studied. Coordination of metabolism between the cells extends beyond the $C_4$ cycle and RPP pathway to synthesis of sucrose, starch, and amino-nitrogen.

plants. In the first part of this chapter evidence for the transport of $C_4$ cycle metabolites within (intracellular) mesophyll and bundle sheath cells and $C_4$ metabolism *in vivo* will be considered (Sections 12.1–12.5). Approaches to studying intracellular metabolite transport include: the intracellular localization of enzymes; metabolism of substrates by isolated organelles; shrinkage-swelling measurements; and monitoring uptake into or exchange of metabolites out of organelles, for example, using silicone oil centrifugation techniques (Section 8.5).

## 12.1 Intracellular localization of enzymes of the $C_4$ cycle in mesophyll cells

Mesophyll chloroplasts, prepared both mechanically and via protoplasts have been used to determine the intracellular localization of enzymes of the $C_4$ pathway. In early studies, Slack and Hatch using nonaqueous methods with maize (Section 10.2), suggested that pyruvate,Pi dikinase, pyrophosphatase, adenylate kinase, and NADP-malate dehydrogenase were associated with the mesophyll chloroplasts. The location of PEP carboxylase was uncertain since it was sometimes associated with the chloroplasts and at other times with the cytosol. [The cytosol is the soluble fraction of the cell in which the organelles are imbedded.]

There are two advantages of using protoplasts for enzyme localization studies: pure preparations of the mesophyll tissue are obtained, and a high percentage (80–99 %) of intact organelles can be isolated. The chloroplasts, and mitochondria + microbodies can be isolated. The chloroplasts, and mitochondria + microbodies can be separated from the cytosol by differential centrifugation at 400 and 10 000 g, respectively. Alternatively, the total protoplast extract can be layered on a sucrose density gradient and the organelles separated by isopynic centrifugation techniques whereby organelles equilibrate according to their buoyant densities (Section 13.7).

When aqueous methods are used to isolate $C_4$ mesophyll chloroplasts, PEP carboxylase is mainly in the cytosolic fraction, although a small percentage is often associated with the chloroplastic fraction. For example, with chloroplasts isolated from mesophyll protoplasts of maize, NADP-triose phosphate dehydrogenase is associated with the chloroplastic fraction, while 97 % of the PEP carboxylase is in the cytosol (Table 12.1).

When chloroplasts isolated from $C_4$ mesophyll protoplasts are separated on sucrose density gradients, more than 80 % of the NADP-triose dehydrogenase is associated with the band of intact chloroplasts, while all of the PEP carboxylase is at the top of the gradient in the cytosolic fraction. Therefore, PEP carboxylase is restricted to the cytosol of the cell. *In vivo* it might be loosely associated with the outer part of the chloroplast envelope, but in considering metabolite transport, this would still place the enzyme outside the chloroplast.

**Table 12.1.** Localization of PEP carboxylase in cytosol of mesophyll protoplast extracts of maize.

Total activity of NADP-triose-P dehydrogenase and PEP carboxylase was 705 and 864 $\mu$mol mg$^{-1}$ Chl h$^{-1}$ respectively. (After Gutierrez *et al.* 1975).

| NADP-triose-P dehydrogenase | | PEP carboxylase | |
|---|---|---|---|
| 400 g pellet | supernatant | 400 g pellet | supernatant |
| % distribution | | | |
| 82 | 18 | 3 | 97 |

In all three $C_4$ subgroups, there appears to be a similar localization of enzymes in the mesophyll cells for metabolism of pyruvate and $CO_2$ to malate. All of the enzymes for this part of the $C_4$ pathway are in the mesophyll chloroplast except for PEP carboxylase. The localization of PEP carboxylase is critical in determining the metabolite transport required by the chloroplast. The presence of this enzyme in the cytosol suggests that PEP, Pi and oxaloacetate, as well as pyruvate and malate, are metabolites which must be transported across the chloroplast envelope (Fig. 12.2).

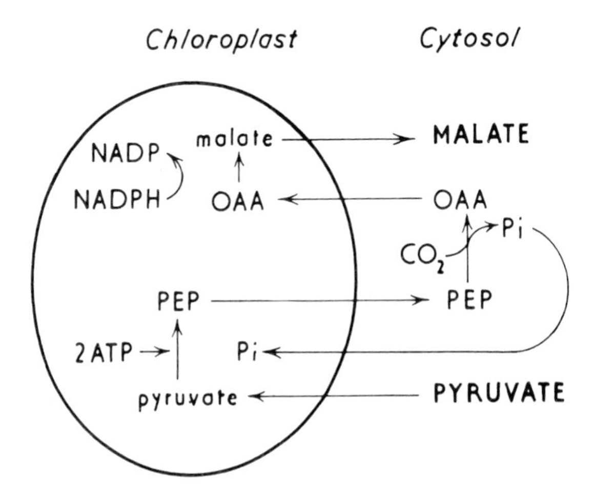


**Fig. 12.2.** Malate synthesis in $C_4$ plants.

### (a) NADP-ME SPECIES

In NADP-ME species, malate and pyruvate are considered the primary transport metabolites between mesophyll and bundle sheath cells in the $C_4$ cycle (Sections 11.2

& 11.3). Therefore, the conversion of $CO_2$ + pyruvate to malate is the major function of the $C_4$ pathway in these mesophyll cells.

A secondary shuttle may occur through aspartate-alanine. High levels of aspartate aminotransferase are found in mesophyll chloroplasts of NADP-ME species. In crabgrass, high levels of alanine aminotransferase appear both in the chloroplast and cytosol. Thus during aspartate synthesis in the mesophyll chloroplast of these species, it is uncertain which amino acid (but most likely glutamate or alanine) is taken up by the chloroplasts as the amino donor. If alanine is taken up by the chloroplasts, then the transport metabolites would be alanine, aspartate, Pi, PEP, and oxaloacetate (Fig. 12.3).

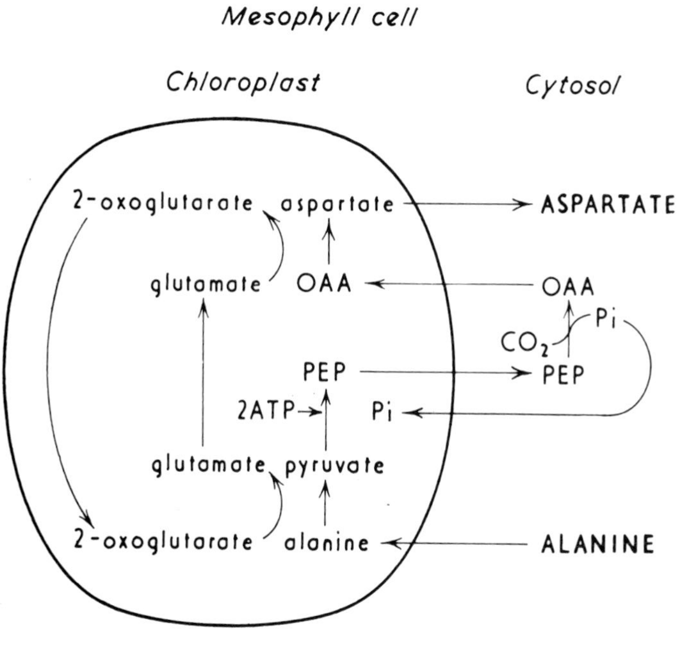

**Fig. 12.3.** NADP-ME species. Aspartate synthesis.

### (b) NAD-ME SPECIES

In these species, aspartate + alanine appear to be the main intercellular transport metabolites in the $C_4$ cycle (Sections 11.2 & 11.3). There is high activity of aspartate- and alanine aminotransferase in mesophyll cells of these species and the enzymes are localized in the cytosol. Therefore, in conversion of alanine and $CO_2$ to aspartate,

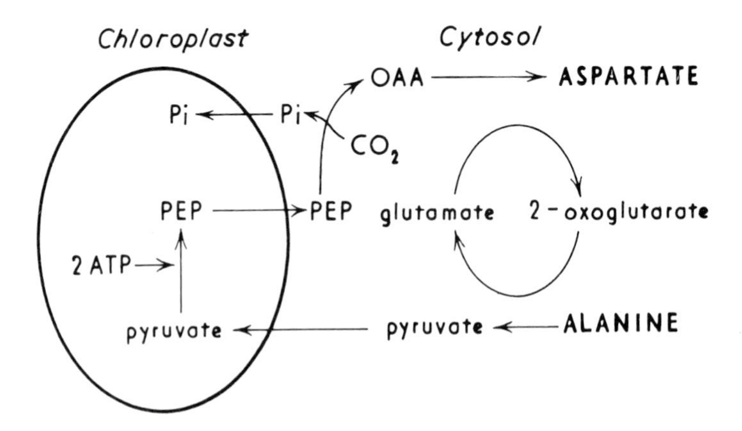


Fig. 12.4. NAD-ME species. Aspartate synthesis.

only PEP, Pi, and pyruvate would be transported across the chloroplast envelope. The shuttle of malate-pyruvate is low and considered secondary in these species (see Fig. 12.4).

### (c) PEP-CK SPECIES

In PEP-CK species, high activity of alanine- and aspartate aminotransferase are found in the cytosol of the mesophyll cell. Three $C_4$ shuttles may operate in these species, the primary shuttle being aspartate-PEP.

1. Aspartate-PEP through PEP-carboxykinase. There would be no metabolite transport across the mesophyll chloroplast in this cycle. For discussion of intercellular transport between cells through plasmadesmata see Section 12.6 (Fig. 12.5).
2. Malate-PEP through PEP-carboxykinase (Fig. 12.6).
3. A minor shuttle of aspartate-alanine through NAD-malic enzyme as in NAD-ME species.

NADP-malate dehydrogenase, alanine- and aspartate aminotransferase are found in mesophyll cells of all three groups of $C_4$ plants. In NADP-ME species, NADP-malate dehydrogenase activity exceeds maximum activity of leaf photosynthesis while in NAD-ME and PEP-CK species, the level of the enzyme is lower and below the measured rates of leaf photosynthesis. The highest levels of aspartate- and alanine aminotransferase are found in the mesophyll tissue of NAD-ME and PEP-

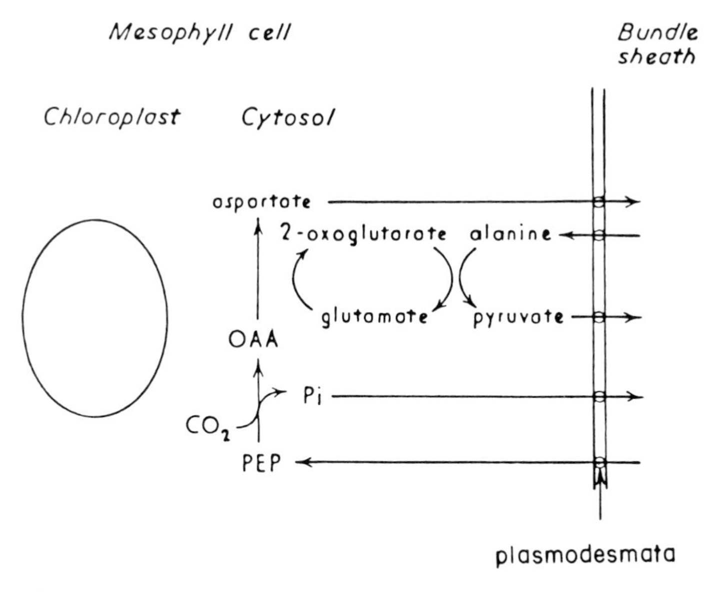


**Fig. 12.5.** PEP-CK species. Aspartate synthesis.

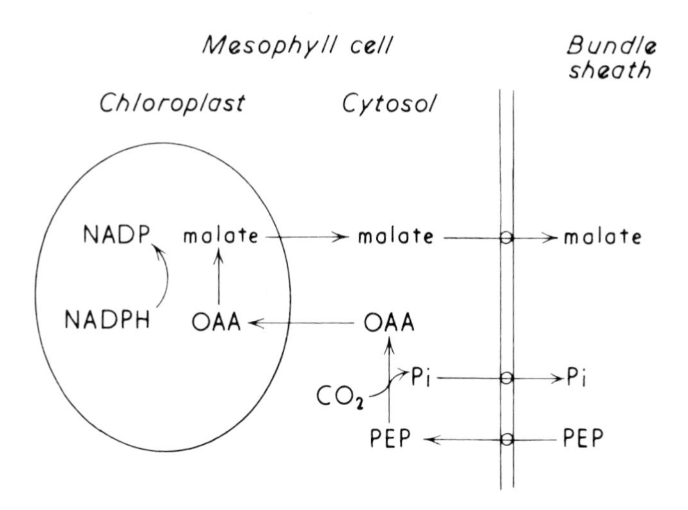


**Fig. 12.6.** PEP-CK species. Malate synthesis.

CK species, exceeding that of leaf photosynthesis, and the activities are many times higher than that found in $C_3$ plants. These results are consistent with the conclusion that malate is the primary product and transport metabolite in NADP-ME species, whilst aspartate is in NAD-ME and PEP-CK species.

## 12.2 Enzyme localization and intracellular metabolite transport in the $C_4$ cycle in bundle sheath cells.

Both aspartate and malate are intercellular transport metabolites in the $C_4$ cycle (Sections 11.2, 11.3 & 11.6). In all three $C_4$ groups malate is considered a transport metabolite to bundle sheath chloroplasts and aspartate a transport metabolite to bundle sheath mitochondria. The basis for this hypothesis is that aspartate aminotransferase is located in the bundle sheath mitochondria in all three groups, with no detectable activity in bundle sheath chloroplasts. The bundle sheath chloroplasts have the enzymes to metabolize malate either through NADP-malic enzyme, as in NADP-ME species, or through malate dehydrogenase in the other species. Initial metabolism of malate by the chloroplasts would allow direct utilization of the reductive power by the RPP pathway. As will be discussed, the initial metabolism of aspartate by the mitochondria, or malate by the chloroplasts, does not necessarily mean that decarboxylation occurs in the same organelle. At present, the schemes of intracellular transport in the bundle sheath cells in relation to the $C_4$ cycle are based on a few studies with several species. To date, there have been difficulties in isolating intact organelles from bundle sheath cells for such studies.

### (a) NADP-ME species

The intracellular transport of metabolites of the $C_4$ cycle in bundle sheath cells appears to be simplest in NADP-ME species. NADP-malic enzyme is found in bundle sheath chloroplasts (studies with maize). Malate and pyruvate are the main intercellular transport metabolites of the $C_4$ cycle in these species (Sections 11.2 & 11.3). This requires transport of malate and pyruvate across the bundle sheath chloroplast envelope (and hydroxyl transfer from bundle sheath to mesophyll cells, see Section 11.4). The mechanisms for this transport are unknown. (Fig. 12.7)

Transport of aspartate-alanine might occur as a minor shuttle. This has been suggested to occur through NADP-malic enzyme in *Gomphrena globosa* (See Fig. 12.8). It was initially thought that NADP-ME species may have substantial levels of NAD-malic enzyme. However this appears to have been due to NADP-malic enzyme recognizing NAD as a co-factor to a limited extent with a pH optimum around 7.5.

### (b) NAD-ME species

In these plants, the intracellular transport in bundle sheath cells through the $C_4$ cycle involves carbon metabolism in the cytosol and the mitochondria. The primary

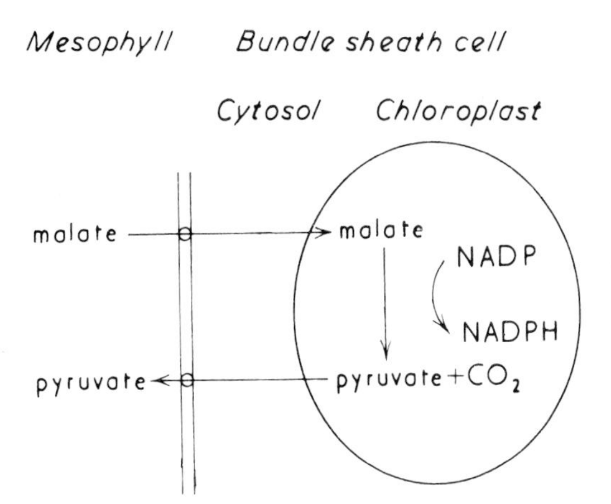


**Fig. 12.7.** NADP-ME species. Malate decarboxylation.

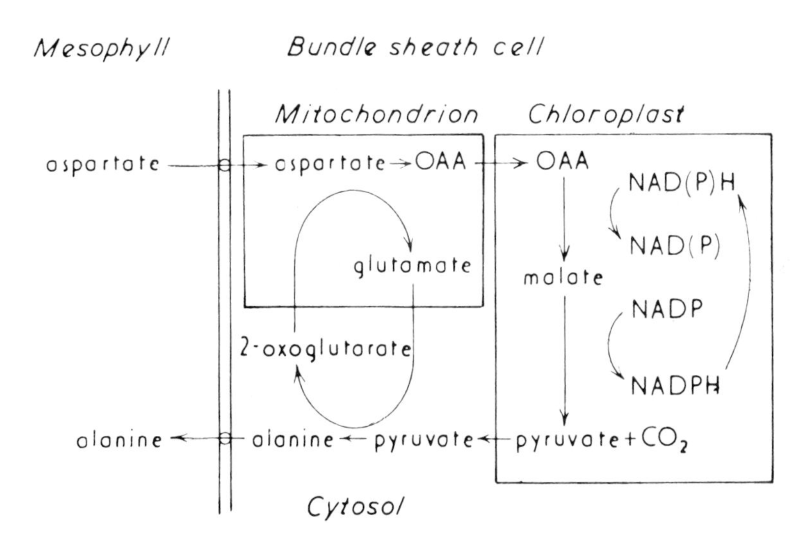


**Fig. 12.8.** NADP-ME species. Possible mechanism of aspartate decarboxylation

shuttle is aspartate-alanine. Transport of aspartate, pyruvate, 2-oxoglutarate, and glutamate by the bundle sheath mitochondria is required (Fig. 12.9). These mitochondria are very prominent (e.g., Fig. 11.7b). A minor cycle in NAD-ME species may be through malate-pyruvate. This could require malate-oxaloacetate transport by chloroplasts as illustrated in Fig. 12.10.

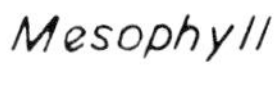


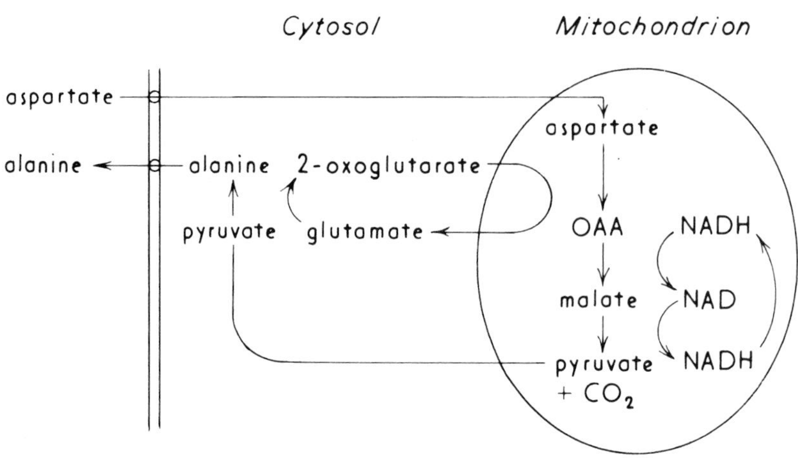


**Fig. 12.9.** NAD-ME species. Aspartate decarboxylation.

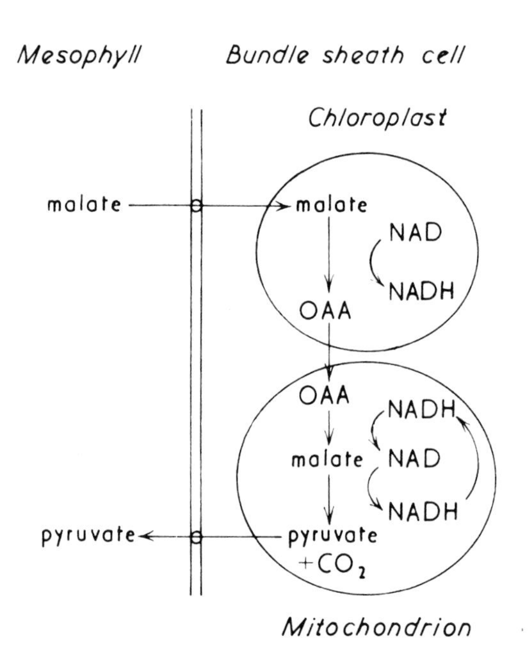


**Fig. 12.10.** NAD-ME species. Malate decarboxylation.

### (c) PEP-CK SPECIES

Decarboxylation appears to be confined to the cytoplasm since PEP carboxykinase is reported to be in the cytoplasm of bundle sheath cells of several species of this subgroup. The major shuttle in the $C_4$ cycle may be aspartate-PEP. If aspartate is converted to oxaloacetate in the bundle sheath mitochondria (as in NAD-ME types), transport of aspartate, oxaloacetate, 2-oxoglutarate and glutamate by mitochondria would be required (Fig. 12.11).

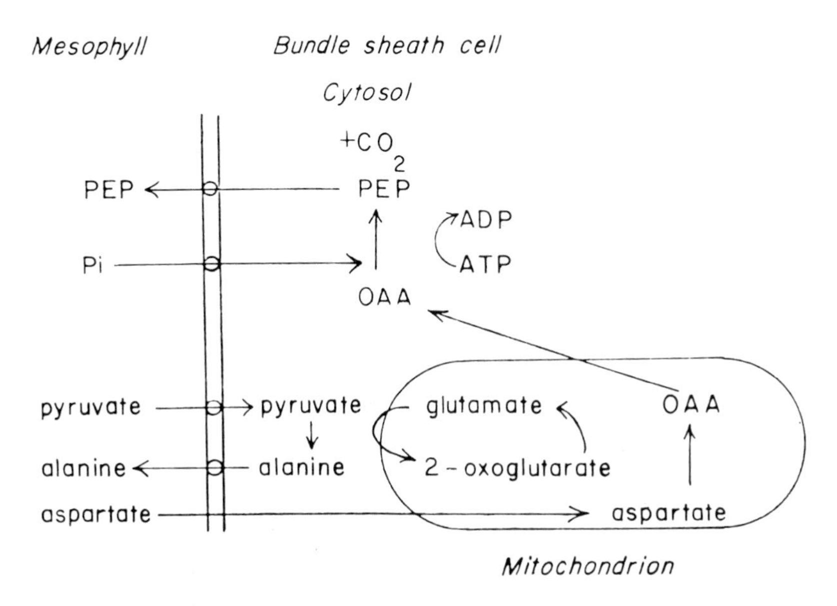


**Fig. 12.11.** PEP-CK species. Aspartate decarboxylation.

In a shuttle of malate-PEP between the cell types decarboxylation through PEP carboxykinase could be entirely cytoplasmic (Fig. 12.12) although the initial step, malate conversion to oxaloacetate, may be chloroplastic.

A shuttle of aspartate-alanine may also occur to a limited extent in PEP-CK species through NAD-malic enzyme.
[Ref. **38**]

## 12.3 Evidence for photosynthetic functions of mesophyll cells

In order to understand the mechanism of photosynthesis in $C_4$ plants, component reactions must be studied in isolated cells and subcellular fractions. To make

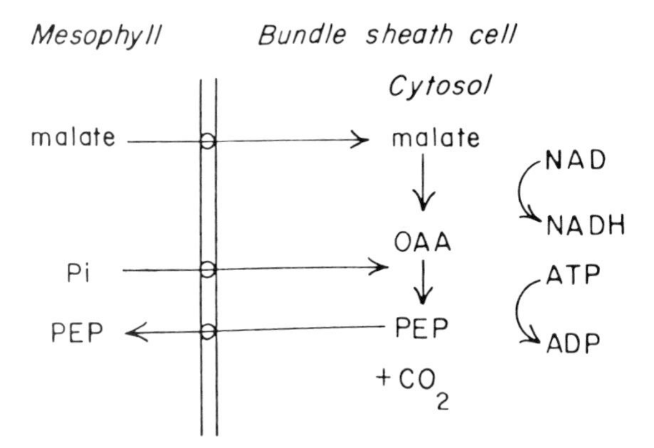


**Fig. 12.12.** PEP-CK species. Possible mechanism of malate decarboxylation.

comparisons of these preparations with the parent tissue, it is useful to remember that $C_4$ plants photosynthesize at rates of about 200 $\mu$mol $CO_2$ assimilated $mg^{-1}$ Chl $h^{-1}$. With approximately equal distribution of leaf chlorophyll between the two cell types, this would give rates of 400 $\mu$mol $mg^{-1}$ Chl $h^{-1}$ in mesophyll cells in the $C_4$ pathway and 400 $\mu$mol $mg^{-1}$ Chl $h^{-1}$ in bundle sheath cells in the RPP pathway since all net fixation occurs in the latter pathway!

In initial studies, isolated $C_4$ mesophyll cells of crabgrass (*Digitaria sanguinalis*, NADP-ME species) fixed $CO_2$ in the light if provided with pyruvate (about 15 $\mu$mol $CO_2$ $mg^{-1}$ Chl $h^{-1}$). This suggests that $CO_2$ fixation by these cells is dependent on the precursor to the $C_4$ pathway. High rates of $CO_2$ fixation were obtained with addition of PEP which is consistent with PEP carboxylase localization in these cells. In principal PEP-induced $CO_2$ fixation is little more than assay of PEP carboxylase in the cells since $CO_2$ is fixed both in the light and dark and the primary product is oxaloacetate. Addition of PEP does give a measure of the carboxylation potential of the $C_4$ pathway to compare with pyruvate- or alanine- dependent $CO_2$ fixation. The latter depends on photochemically generated energy and a totally functional $C_4$ carboxylation phase in the mesophyll cells.

Isolated $C_4$ mesophyll chloroplasts have little or no capacity for light-dependent $CO_2$ fixation. They lack RBP carboxylase and other enzymes of the RPP pathway, and PEP carboxylase is in the cytosol. It is not surprising that these chloroplasts fail to fix $CO_2$ since they lack these carboxylating enzymes. Very low rates of light-enhanced $CO_2$ fixation by the $C_4$ mesophyll chloroplasts have been observed when PEP is present. This appears to be due to a small percentage of the PEP carboxylase of the cell which remains with the chloroplasts (Section 12.1). This $CO_2$ fixation may be stimulated by light through the reducing power generated, resulting in the conversion of oxaloacetate (which is a strong inhibitor of PEP carboxylase) to malate.

There are several requirements which must be met before $C_4$ mesophyll chloroplasts are able to bring about high rates of light-dependent $CO_2$ fixation. Orthophosphate must be provided as well as a three carbon precursor, pyruvate or alanine, and the chloroplasts must be combined with the cytosolic fraction. The latter can be effectively accomplished by preparing extracts from protoplasts. If protoplasts are gently broken, the organelles remain intact and the total extract is used in the assay. Addition of pyruvate and $CO_2$ to protoplast extracts in the light results in the synthesis of oxaloacetate and malate. With protoplast extracts of NAD-ME and PEP-CK species, the addition of alanine and 2-oxoglutarate in the light results in the synthesis of oxaloacetate as well as aspartate. Rates of $CO_2$ fixation into $C_4$ acids of 250–350 $\mu$mol mg$^{-1}$ Chl h$^{-1}$ at 40°C have been achieved in such preparations. The $CO_2$ fixation is dependent on the chloroplasts converting pyruvate + orthophosphate to PEP, using energy generated photochemically, and on PEP carboxylase being in the cytosol. The dependence of $CO_2$ fixation on pyruvate is shown with mesophyll protoplast extracts of species representing the three $C_4$ groups. As expected, photosynthesis by the chloroplasts from wheat, a $C_3$ species, is not dependent on pyruvate (Table 12.2).

**Table 12.2.** Light-dependent $CO_2$ fixation by mesophyll protoplast extracts of various species, (After Edwards *et al*, 1976b).

| Group | | | + 5 mM pyruvate | |
|---|---|---|---|---|
| | Light | Dark | Light | Dark |
| | | $\mu$mol mg$^{-1}$ Chl h$^{-1}$ | | |
| NADP-ME | | | | |
| *Digitaria sanguinalis* | 4 | 1 | 267 | 1 |
| NAD-ME | | | | |
| *Eleusine indica* | 6 | 1 | 146 | 1 |
| PEP-CK | | | | |
| *Urochloa panicoides* | 4 | 1 | 210 | 1 |
| $C_3$ | | | | |
| *Triticum aestivum* | 168 | 1 | 148 | 1 |

If the $C_4$ mesophyll chloroplasts are separated from the protoplast extracts, pyruvate-dependent rates of $CO_2$ fixation are quite low. The residual activity is thought to be due to a small amount of PEP carboxylase remaining with the chloroplasts following differential centrifugation of protoplast extracts (see Table 12.1). As expected, the chloroplasts (free of the cytosolic fraction) and protoplast

**Table 12.3.** Light-dependent $CO_2$ fixation by protoplast extracts and chloroplasts. (After Edwards *et al*, 1976a).

Chloroplasts were separated from mesophyll protoplast extracts by centrifugation at 600 g for 3 min, resuspended and washed once. Induction of $CO_2$ fixation with *D. sanguinalis* included 5 mM pyruvate and 0.5 mM oxaloacetate.

| Species | Protoplast extract | Chloroplasts |
|---|---|---|
| | $\mu$mol $CO_2$ fixed $mg^{-1}$ Chl $h^{-1}$ | |
| NADP-ME | | |
| *Digitaria sanguinalis* | 325 | 10 |
| $C_3$ | | |
| *Hordeum vulgare* | 126 | 114 |

extracts of barley, a $C_3$ plant, have similar rates of photosynthesis since the carboxylase of the RPP pathway resides in the chloroplast (Table 12.3).

There is a several-fold increase in pyruvate-dependent $CO_2$ fixation with $C_4$ mesophyll protoplast extracts when low concentrations (0.5 mM) of oxaloacetate are added. The addition of oxaloacetate induces high rates of noncyclic electron flow as reductive power is utilized to convert oxaloacetate to malate. This provides a large source of ATP from noncyclic photophosphorylation for conversion of pyruvate to PEP. Without the addition of oxaloacetate, fixation of $CO_2$ by pyruvate is largely dependent on cyclic and pseudocyclic photophosphorylation until sufficient oxaloacetate is synthesized.

PGA induces high rates of $CO_2$ fixation in the light or dark ($> 200$ $\mu$mol $mg^{-1}$ Chl $h^{-1}$) with $C_4$ mesophyll protoplast extracts. In the light, with PGA as a substrate, oxaloacetate, malate and aspartate are products of $CO_2$ fixation. Apparently, PGA is a precursor to PEP formation through phosphoglyceromutase and enolase. The extent to which PGA may serve as a natural precursor to the pathway *in vivo* is unknown. For the $C_4$ cycle to function efficiently *in vivo*, it is essential that its metabolites be maintained at a certain level. The $C_4$ cycle is thought to catalyse a stoichiometric transfer of carbon from the atmosphere to the RPP pathway. This would not change the total carbon pool of the $C_4$ pathway. If there were no net loss of carbon from the cycle, carbon input would not be required except to allow for growth. However, some net loss of carbon from the $C_4$ cycle *in vivo* is expected (for example, translocation of amino acids, precursors to synthesis of macromolecules, respiration and changing pool size in response to environmental fluctuations). Since net synthesis of $CO_2$ occurs through RBP carboxylase, metabolism of PGA to PEP would be a means of adding carbon to the $C_4$ cycle. The rate of carbon input from PGA might be low in comparison to the rate of flow of $CO_2$ through the cycle, but without it, the cycle could be depleted of its intermediates

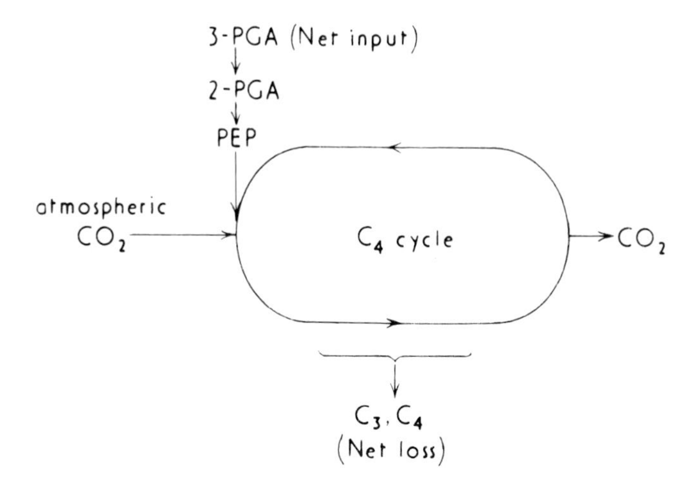


**Fig. 12.13.** Net input of carbon to the $C_4$ cycle from PGA.

and would be incapable of increasing its yield in relation to changing metabolic demand (see Fig. 12.13).

In labelling studies with whole leaves of sugarcane, Hatch and Slack followed the distribution of label in malate versus PGA with time (Table 12.4).

If the $C_4$ cycle was operating as a completely closed shuttle, only labelling in carbon 4 of malate would be expected. However, if PGA served as a partial precursor to the cycle, malate would become labelled in a position other than the C–4 with labelling in the C–1 of malate expected before labelling in C–2 + C–3. Initially, PGA is labelled in the C–1 position with carboxylation of $^{14}CO_2$ through RBP carboxylase. As part of the $^{14}C$–PGA is metabolized in the RPP pathway, RBP becomes labelled. This results in PGA becoming labelled in the C–2 and C–3

**Table 12.4.** Distribution of label in malate and PGA after $^{14}CO_2$ feeding with sugar-cane leaves (After Hatch & Slack, 1970).

| Seconds after exposure to $^{14}CO_2$ | % Distribution of label | | | | |
|---|---|---|---|---|---|
| | Malate | | | PGA | |
| | C–1 | C–2 + C–3 | C–4 | C–1 | C–2 + C–3 |
| 3 | 10 | 0 | 90 | 98 | 2 |
| 45 | 29 | 4 | 67 | 61 | 39 |
| 150 | 33 | 25 | 42 | 45 | 55 |

positions. With PGA as a partial precursor to the $C_4$ cycle, labelling in C–1 of PGA would appear in C–1 of malate, while labelling in C–2 + C–3 of PGA would appear in C–2 + C–3 of malate, consistent with the labelling pattern observed (Table 12.4). Alternatively, labelling of C–1 in malate could occur by partial randomization of label by equilibration with fumarate through fumarase activity (see Section 15.8). Labelling in C—1 and C–2 + C–3 of malate could occur by equilibration of PGA with PEP through phosphoglyceromutase and enolase without net carbon input from PGA (Fig. 12.14).

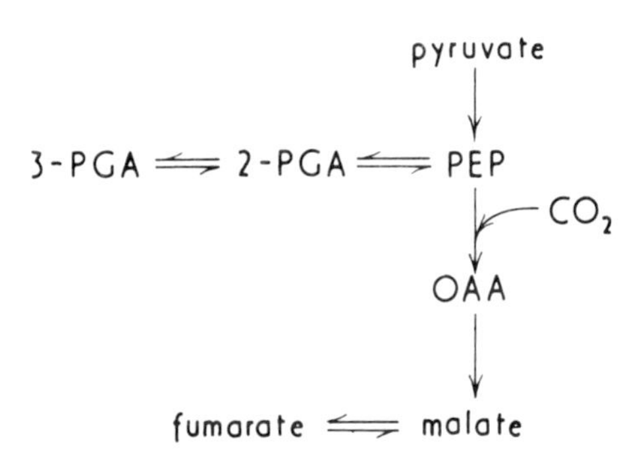


**Fig. 12.14**

Although $C_4$ mesophyll chloroplasts, when free of the cytosolic fraction, fix little $CO_2$, they do show light-dependent $O_2$ evolution with addition of certain substrates such as PGA or pyruvate + oxaloacetate. The dependence of $O_2$ evolution on PGA is consistent with the presence of the reductive phase of the RPP pathway and it indicates that PGA is readily taken up by these chloroplasts. As in $C_3$ chloroplasts, both ATP and NADPH would be utilized in reduction of PGA to triose phosphates through phosphoglycerate kinase and NADP-triose phosphate dehydrogenase.

$C_4$ mesophyll chloroplasts from crabgrass show high rates of both $O_2$ evolution and non-cyclic electron flow following addition of pyruvate and low levels of oxaloacetate but not if oxaloacetate alone is added. This indicates that non-cyclic electron flow is coupled to ATP synthesis and that utilization of both ATP and NADPH are required for maximum rates. (See Section 11.1 for utilization of ATP for pyruvate conversion to PEP). The pyruvate + oxaloacetate-dependent $O_2$ evolution, unlike $CO_2$ fixation with $C_4$ mesophyll chloroplasts, occurs without the cytosolic fraction since PEP carboxylase is not required (see Fig. 12.15).
[Refs. **2**, **11–14**, **19**, **24**, **28**]

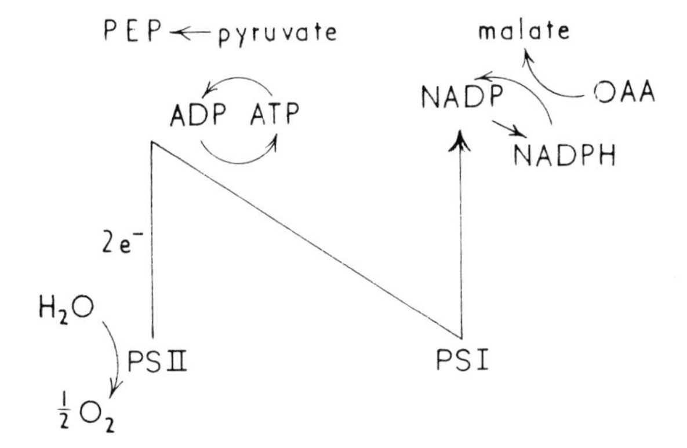


**Fig. 12.15.** Use of assimilatory power from non-cyclic electron flow in mesophyll chloroplasts in $C_4$ photosynthesis.

## 12.4 Evidence for photosynthetic functions of bundle sheath cells

In early experiments, mechanically prepared bundle sheath cells of crabgrass (NADP-ME species) showed substantial rates of $CO_2$ fixation if ribose-5-phosphate were included in the medium. The large stimulation of $CO_2$ fixation by ribose-5-phosphate may be an outcome of some damage caused during isolation, such that it rapidly enters the cells. A relatively small uptake of ribose-5-phosphate may be sufficient to replenish the metabolite pools of the chloroplast, which would shorten the induction period and allow maximum rates (Section 7.7). A limitation of light-dependent $CO_2$ fixation by isolated bundle sheath cells of NADP-ME species, due to a deficient photosystem II, has now become apparent (Section 11.6). Subsequently, rates in excess of 100 $\mu$mol $CO_2$ fixed $mg^{-1}$ Chl $h^{-1}$ were obtained with bundle sheath cells isolated from some NAD-ME and PEP-CK species without addition of organic substrates.

DCMU inhibits $CO_2$ fixation by isolated bundle sheath cells by preventing reduction of $NADP^+$ through non-cyclic electron flow. Malate can partially overcome the DCMU inhibition with cells isolated from all three $C_4$ groups. This indicates that malate can donate reductive power to the bundle sheath cells, presumably through NADP-malic enzyme in NADP-ME species and NAD-malate dehydrogenase in NAD-ME and PEP-CK species. Glyceraldehyde, an inhibitor of the RPP pathway (see Chapter 8) inhibits $CO_2$ fixation in the bundle sheath cells. It has less effect on pyruvate-dependent $CO_2$ fixation in mesophyll preparations.

Two methods have been used for studying the donation of carbon from $C_4$ acids into the RPP pathway with isolated bundle sheath cells or chloroplasts:

(a) By following $C_4$ acid-dependent $O_2$ evolution with malate, aspartate, or oxaloacetate in the light. Oxaloacetate is labile and slowly decarboxylates to pyruvate and $CO_2$ nonenzymatically and undergoes rapid decarboxylation in the presence of divalent cations and certain proteins. Thus, partly artificial donation of $CO_2$ to the RPP pathway must be considered when using this substrate *in vitro.*

(b) By following light-dependent incorporation of label from $^{14}C$–4 malate or aspartate into metabolites of the RPP pathway (i.e., PGA, triose phosphate, and hexose phosphate) through chromatographic analysis. This provides direct evidence for donation of carbon into the RPP pathway. In practice, this is more laborious than the other method. Using these methods, rates of carbon donation from $C_4$ acids to the RPP pathway of 100–300 $\mu$mol $CO_2$ donated or $O_2$ evolved $mg^{-1}$ Chl $h^{-1}$ occur.

Maximum rates of aspartate-dependent $O_2$ evolution by bundle sheath cells of NAD-ME or PEP-CK species require the addition of pyruvate (1 mM) or 2-oxoglutarate (10 mM). In both of these $C_4$ subgroups the initial step of the decarboxylation phase of the $C_4$ pathway requires 2-oxoglutarate. Pyruvate may stimulate aspartate decarboxylation through a coupling of aspartate- and alanine-aminotransferase (Fig. 12.16).

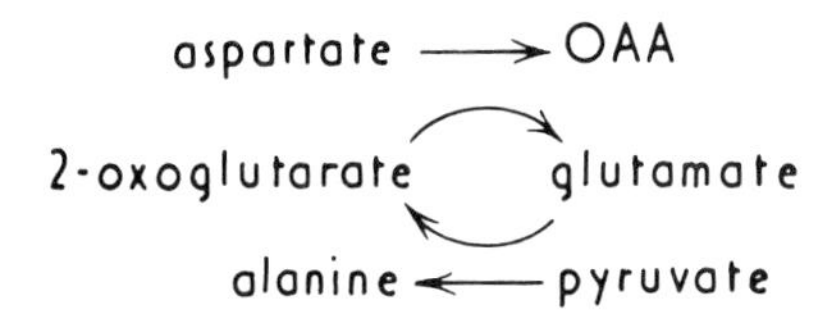


**Fig. 12.16**

A further method has been used to follow decarboxylation of $C_4$ acids by isolated bundle sheath cells. $^{14}C$–4 dicarboxylic acids are added and decarboxylation is followed in the light by measuring $^{14}CO_2$ released in the presence of D, L-glyceraldehyde. Glyceraldehyde prevents the $^{14}CO_2$ from being fixed in the RPP pathway. Under these conditions, with isolated bundle sheath cells of species from all three $C_4$ groups, PGA causes a large enhancement of malate decarboxylation. This indicates that malate oxidation is coupled to the reductive phase of the RPP pathway. With bundle sheath cells of PEP-CK species, 0.1 mM 3-mercaptopicolinic acid (an inhibitor of PEP-CK), inhibits malate decarboxylation. This inhibitor has no effect on malate decarboxylation in NAD-ME and NADP-ME species. Oxalic acid (0.1 mM) specifically inhibits malate decarboxylation by bundle sheath cells in NADP-ME species (crabgrass). This suggests that malate decarboxylation is occurring via different mechanisms in different $C_4$ subgroups. Similar concentrations of 3-mercatopicolinic and oxalic acid have no effect on $^{14}CO_2$ fixation by the isolated bundle sheath cells, indicating their specificity for inhibition of $C_4$ acid decarboxylation.

Decarboxylation of $^{14}C$–4 aspartate by bundle sheath cells in the presence of D L-glyceraldehyde requires the addition of 2-oxoglutarate or pyruvate for maximum rates. 3-mercaptopicolinic acid strongly inhibits asparate decarboxylation in PEP-CK species (but not in NAD- and NADP-ME species). Thus, the main pathway of aspartate decarboxylation in these species appears to be through PEP carboxykinase. Decarboxylation of $^{14}C$–4 aspartate by bundle sheath cells of PEP-CK species is stimulated by light. The reason for this is uncertain; perhaps the ATP required for decarboxylation through PEP-carboxykinase is generated photochemically. In NAD-ME species the rates of aspartate decarboxylation are similar in the light or dark. This is expected since energy is not required for aspartate decarboxylation through NAD-malic enzyme.

High rates of aspartate decarboxylation occur in bundle sheath mitochondria of NAD-ME species when malate, Pi, and 2-oxoglutarate are added. Metabolism of aspartate to pyruvate and $CO_2$ in the mitochondria requires uptake of aspartate and 2-oxoglutarate in exchange for glutamate and pyruvate. Malate and Pi serve as catalysts for metabolism of aspartate to pyruvate $+CO_2$.

In mammalian mitochondria and mitochondria from non-photosynthetic plant tissues, malate uptake can be linked to a phosphate translocators. In the bundle sheath mitochondria, a high $NADH/NAD^+$ ratio would favour aspartate conversion to malate since NADH is required. Malate uptake, coupled to phosphate and dicarboxylate translocators, could change the redox state in the mitochondria allowing maximum rates of aspartate conversion to pyruvate and $CO_2$ (Fig. 12.17).

In some mitochondria, a carrier is proposed for the exchange of malate with 2-oxoglutarate. Thus, malate and Pi may also facilitate uptake of 2-oxoglutarate. The mechanism of transport in these bundle sheath mitochondria remains to be studied. [Refs. **2, 11, 14, 24, 45, 46**].

## 12.5 Mechanism of intracellular metabolite transport

Photosynthesis in $C_4$ plants requires rapid transport of certain metabolites of the $C_4$ cycle across the envelopes of chloroplasts and mitochondria. Little is known about the mechanism of metabolite transport, particularly in bundle sheath chloroplasts and mitochondria. Metabolite transport in relation to the $C_4$ cycle has been studied with $C_4$ mesophyll chloroplasts. This is done by measuring substrate-induced

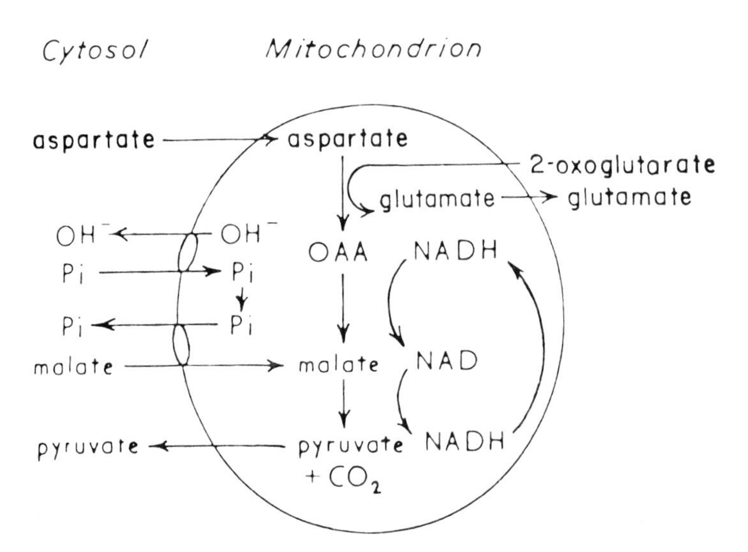


**Fig. 12.17.** Proposed metabolite transport in mitochondria during aspartate decarboxylation via NAD-malic enzyme.

osmotic changes spectrophotometrically at 535 nm and uptake or exchange of metabolites out of the chloroplasts using silicone oil centrifugation techniques.

Osmotic swelling has been used to study metabolite transport with mammalian mitochondria, plant mitochondria from nonphotosynthetic tissue, $C_3$ (spinach) chloroplasts, and $C_4$ (crabgrass) chloroplasts. By measuring osmotic changes, net uptake of solutes, but not metabolite exchange (except for hydroxyl-anion exchange using ammonium salts), across membranes can be detected (see Chapter 8 and Fig. 12.19).

With silicone oil centrifugation techniques, exchange mechanisms for transport can be identified and the kinetics of uptake can be determined. Furthermore, uptake by diffusion can be distinguished from carrier-mediated uptake using physiological concentrations of metabolites. In contrast to uptake by diffusion, carrier-mediated transport shows saturation kinetics with substrate (usually within 10 mM), is extremely temperature sensitive and shows that the metabolites which share a given carrier compete with each other for uptake. Carrier-mediated uptake may be inhibited by certain sulphydryl reagents. [Sulphydryl reagents react with sulphydryl groups on proteins which may then prevent the protein from carrying out its catalytic function.] Inhibition of uptake (e.g., by specific substrate analogs) also provides evidence that uptake is carrier-mediated.

*Pyruvate carrier in $C_4$ mesophyll chloroplasts*

$C_4$ mesophyll chloroplasts (crabgrass) have an electrogenic carrier (metabolite transport on carrier is not electroneutral) for pyruvate uptake. Intact chloroplasts with sorbitol as osmoticum (0.25 M) immediately shrink upon addition of 0.1 M potassium pyruvate. The chloroplasts will swell to their original volume if there is net uptake of the solute (i.e., if both cation and anion permeate). The addition of valinomycin, which facilitates $K^+$ transport, causes rapid swelling of the chloroplasts. This indicates that pyruvate is a permeant anion. However, in chloroplasts of spinach, valinomycin in the presence of potassium pyruvate does not induce swelling, showing that pyruvate is not a permeant anion in $C_3$ chloroplasts. In $C_4$ mesophyll chloroplasts, there is some swelling with the addition of 0.1 M ammonium pyruvate but this is not as rapid as with potassium pyruvate + valinomycin. This suggests that a pyruvate–hydroxyl exchange is limiting. [With ammonium salts, uptake of the anion in exchange for hydroxyl ions can be followed by observing chloroplast swelling. $NH_3$, which readily permeates membranes, is taken up and equilibrates to $NH_4^+$. Where an appropriate carrier exists, the hydroxyl ions generated in the chloroplasts exchange for the external anion, resulting in the uptake of the ammonium salt.] Silicone oil centrifugation techniques have shown that the pH gradient across the chloroplast envelope has little influence on the uptake of $^{14}C$-pyruvate. This also suggests that uptake of pyruvate by hydroxyl exchange is limited. (see Figs 12.18 and 12.19).

The velocity of $^{14}C$-pyruvate uptake as determined by these techniques follows Michaelis-Menten type kinetics with an apparent $Km$ for pyruvate of 0.7 to 1 mM. This shows that the uptake of pyruvate is carrier-mediated rather than by diffusion.

Uptake of $^{14}C$-pyruvate by the $C_4$ mesophyll chloroplasts is inhibited by mersalyl, a sulfhydryl reagent. Mersalyl also inhibits chloroplast swelling in the presence of potassium pyruvate + valinomycin. This inhibition suggests that pyruvate transport is carrier-mediated.

Derivatives of $\alpha$-cinnamic acids are potent inhibitors of carrier-mediated pyruvate uptake in animal mitochondria. Uptake of pyruvate by $C_4$ mesophyll chloroplasts is specifically inhibited by a number of pyruvate analogs and $\alpha$-cinnamic acid derivatives.

*PEP-Pi exchange on phosphate carrier in $C_4$ mesophyll chloroplasts*

$C_4$ mesophyll chloroplasts (crabgrass) have a phosphate translocator (membrane localized carrier which exchanges Pi and/or anions of organic phosphates analogous to that in $C_3$ chloroplasts, Chapter 8) which catalyses exchange of PEP and

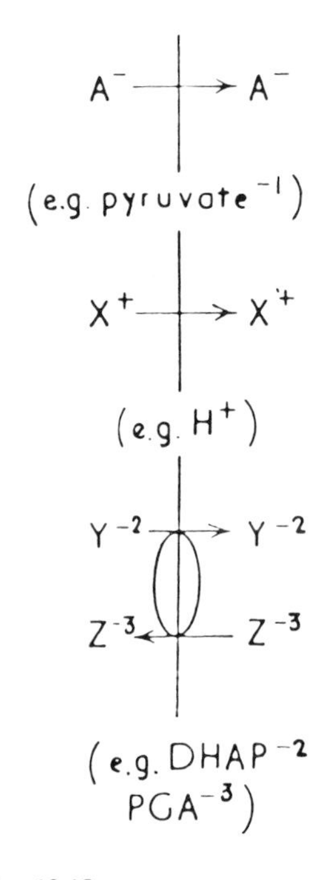


Fig. 12.18

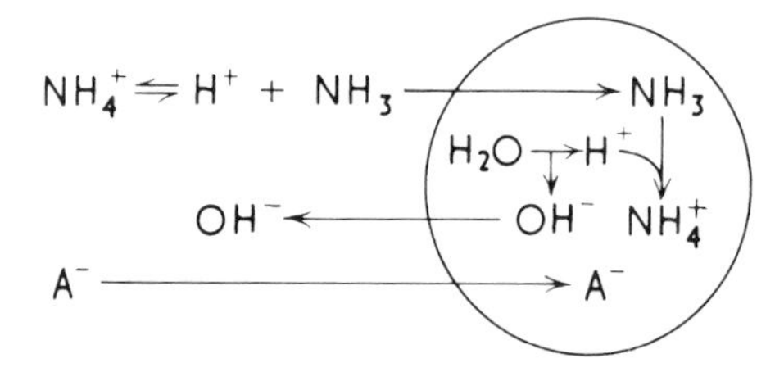


**Fig. 12.19.** Illustration of uptake of an ammonium salt through a hydroxyl-anion ($A^-$) exchange.

orthophosphate. The isolated chloroplasts have a pool of PEP which can be exported by exchange with Pi or PGA. PEP is not a permeant anion since the chloroplasts do not swell in potassium phosphoenolpyruvate with or without valinomycin.

When using silicone oil centrifugation with the $C_4$ mesophyll chloroplasts, PEP is a competitive inhibitor ($Ki = 0.45$ mM) of $^{32}$Pi uptake ($Km = 0.2$ mM). This suggests that PEP has a relatively high affinity for the phosphate transporter in these chloroplasts. In comparison, the $Ki$ for PEP inhibition of $^{32}$Pi uptake with spinach chloroplasts is about ten-fold higher than in the $C_4$ chloroplasts, indicating that PEP is a poor transport metabolite in $C_3$ chloroplasts.

Synthesis of PEP by $C_4$ chloroplasts in the light with pyruvate, but in the absence of external Pi, is slow and the PEP concentration in the chloroplast reaches a high level (internal/external concentration = 30). The addition of low levels of Pi (0.1 mM) rapidly reduces the level of PEP in the chloroplasts (internal/external concentration = 2). Therefore, Pi is required both for the synthesis of PEP and for its exchange out of the chloroplasts. Transport on the pyruvate carrier and phosphate carrier may each be electrogenic. As a consequence, the transport of the three metabolites PEP, Pi and pyruvate would be electroneutral (Fig. 12.20).

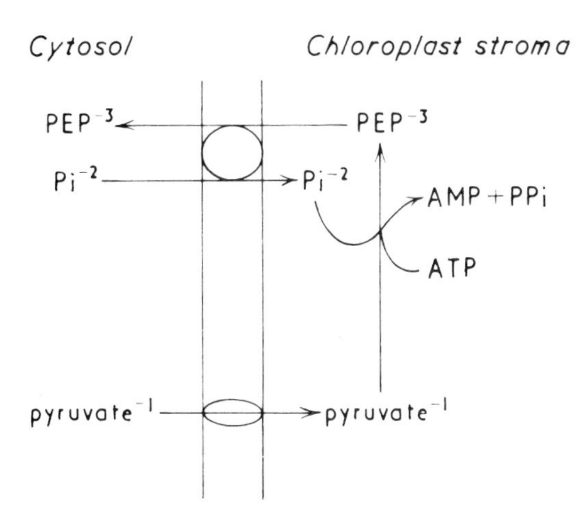


**Fig. 12.20.** Transport of pyruvate, PEP and Pi in $C_4$ mesophyll chloroplasts.

There is also a dicarboxylate translocator in $C_4$ mesophyll chloroplasts (crabgrass) which has a $Km$ for oxaloacetate of 0.13 mM. Uptake of oxaloacetate is competitively inhibited by malate ($Ki = 0.26$ mM) and aspartate ($Ki = 0.2$ mM), suggesting that transport of the dicarboxylic acids involves the same carrier.

The carriers are thought to be localized on the inner membrane of the chloroplasts. Sucrose permeates the outer envelope of the $C_4$ mesophyll chloroplasts but not the inner membrane as indicated by studies on uptake of $^3H_2O$

and $^{14}C$-sucrose. When the osmolarity of the medium is increased by varying the sorbitol concentration, the total chloroplast space remains constant (measured by uptake of $^3H_2O$), while the intermembrane space (measured by uptake of $^{14}C$-sucrose) increases. Thus, increasing sorbitol concentrations decreases the stromal space and increases the intermembrane space, showing that the inner membrane represents a permeability barrier to certain solutes (Fig. 12.21).

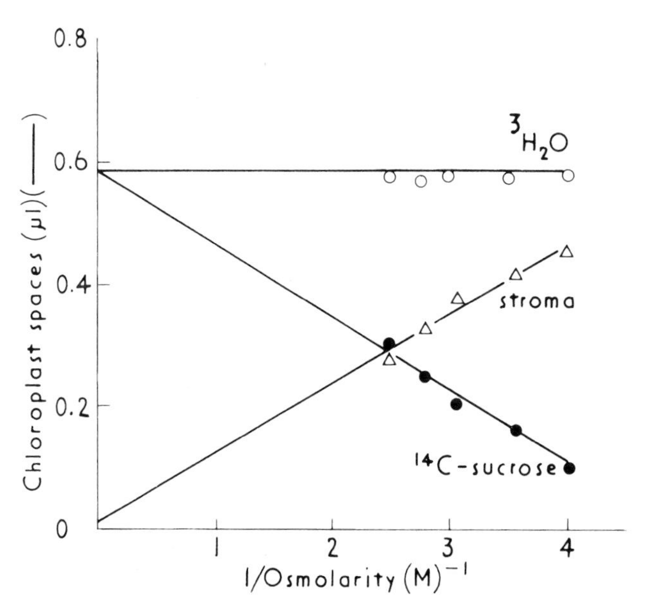


**Fig. 12.21.** Influence of varying sorbitol concentration. on the intermembrane and stromal space of $C_4$ mesophyll chloroplasts of crabgrass. Stromal space is determined by the difference between the total chloroplast space ($^3H_2O$) and the intermembrane space ($^{14}C$-sucrose). Each reaction contained 6μg Chl (After Huber & Edwards, 1977b).

Thus, there is evidence for carriers on the inner membrane of $C_4$ mesophyll chloroplasts which catalyse the metabolite exchange proposed in Section 12.1. Studies on mechanisms of metabolite transport in bundle sheath chloroplasts and mitochondria have not yet been made.
[Refs. **34, 35**]

## 12.6 Mechanism of intercellular metabolite transport

Photosynthesis in $C_4$ plants requires a rapid bidirectional movement of metabolites between mesophyll and bundle sheath cells. It is generally thought that this occurs by a symplastic pathway through plasmodesmata. Symplastic transport between cells occurs through the cytoplasm via pores (plasmodesmata), which allow continuity of cytoplasm without the barrier of a membrane or cell wall. Apoplastic transport is via the cell wall and entails movement across the plasma membrane.

There are a few reports that plasmodesmata interconnect mesophyll and bundle sheath cells. For example, there is substantial evidence for plasmodesmata in primary pitfields between these cell types in maize. A high frequency of primary pit-fields between the mesophyll and bundle sheath cells of a $C_4$ species is seen from the eloquent scanning electron microscopy of Craig and Goodchild (Fig. 12.22). The simplest hypothesis is that movement of metabolites occurs between the cells along diffusion gradients through the plasmodesmata.

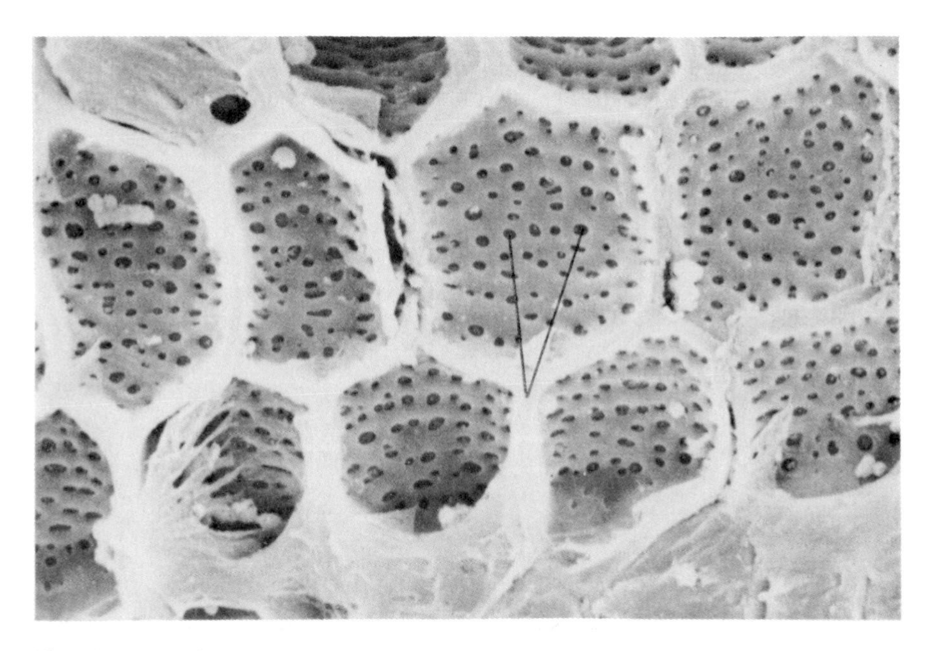

**Fig. 12.22.** Scanning electron micrograph of pronase-digested leaf tissue of *Triodia irritans* illustrating primary pit fields of bundle sheath walls viewed from the bundle sheath side, × 1500. Arrows indicate primary pitfields in a cell wall. Each pit-field contains numerous plasmodesmata (After Craig & Goodchild, 1977).

Part of the problem in determining the transport between the cells is a lack of understanding of the structure of the plasmodesmata. It is uncertain whether plasmodesmata represent open channels between cells or if there is restricted or selective movement of ions and molecules. Robards conceives the structure of

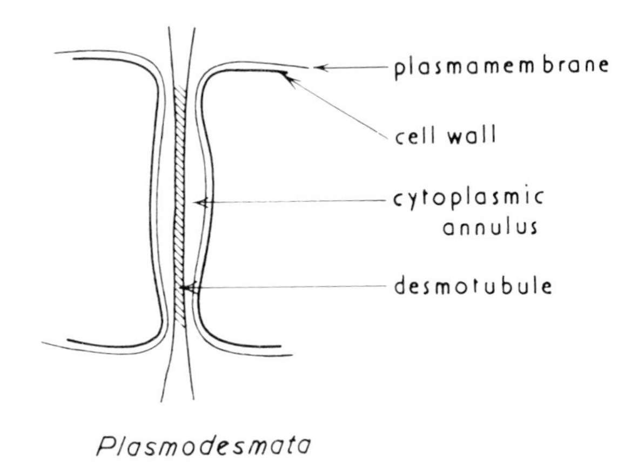


**Fig. 12.23.** Illustration of the structure of plasmodesmata (Adapted from Robards, 1975).

plasmodesmata as having a thin strand of desmotubule (modified endoplasmic reticulum) running through the pore. In addition, estimates of the actual size of the open area of plasmodesmata and their frequency is problematical (Fig. 12.23).

This model raises the question whether transport occurs through the endoplasmic reticulum (ER), the cytoplasmic annulus, or both. ER is a complex network of lipoprotein membranes forming tubules or sacs and extending from the nucleus and possibly other organelles throughout the cytoplasm of cells. It may have multiple functions including transport and protein synthesis, and acting as a precursor in the synthesis of other membranes. From an electron microscopy study of the plasmodesmata between mesophyll and bundle sheath cells of maize, Evert *et al.* suggested that flow via the cytoplasmic annulus may be restricted and that metabolite transport may be through desmotubules, and hence through ER. Assuming that the ER is a network of tubules, intercellular transport would require exchange of metabolites across the ER membranes in each cell. In this case, free non-selective diffusion of metabolites through the plasmodesmata is unlikely. Evert *et al.* also saw occasional evidence for ER attachment to the outer envelope of chloroplasts and suggested the connection of mesophyll and bundle sheath chloroplasts by ER. At least for $C_4$ mesophyll chloroplasts, the outer membrane is freely permeable to solutes like sorbitol and sucrose, whilst the inner membrane is the limiting one for transport (Section 12.5). ER connected to the outer chloroplast membrane might thus have direct access to the solutes of the cytosol, rather than those of the chloroplast.

In the $C_4$ cycle, intercellular transport of metabolites would occur from site of synthesis in one cell to site of utilization in the adjacent cell. If transport of $C_4$ acids depends on diffusion alone, the concentration gradient between mesophyll and bundle sheath cells necessary for flux commensurate with measured rates of photosynthesis can be calculated. Such calculations require estimates of plasmodesmata frequency, pore size of the plasmodesmata, and the assumption that free movement of metabolites occurs through the pores. These calculations suggested that concentration gradients between mesophyll and bundle sheath cells of 10–30 mM of the $C_4$ acids would be required. Also, from calculations on the active pools of $C_4$ acids in the leaf, concentrations of 40–60 mM were proposed if all of the $C_4$ acid was confined to the cytoplasm of the mesophyll cells. On the other hand, malate and aspartate at 1–10 mM levels inhibit PEP carboxylase and pyruvate-dependent $CO_2$ fixation with $C_4$ mesophyll preparations (Section 12.8). Whether PEP carboxylase is equally sensitive *in vivo*, or whether calculations on the pool size of $C_4$ acids in the mesophyll cytosol are an overestimate, is uncertain. The foregoing considerations emphasize that current views on the mechanisms of intercellular metabolite transport in $C_4$ photosynthesis are largely derived from speculation. [Refs. **16, 24, 26, 43, 50**].

## 12.7 $C_4$ metabolism linked to cyclic, pseudocyclic and noncyclic photophosphorylation

Mesophyll chloroplasts of $C_4$ plants have pyruvate,Pi dikinase which, coupled with adenylate kinase, utilizes two ATP for conversion of pyruvate to PEP (Section 11.1). With rates of leaf photosynthesis of 200 $\mu$mol mg$^{-1}$ Chl h$^{-1}$, the rate of metabolism of pyruvate to PEP per unit Chl in the mesophyll tissue is about 400 $\mu$mol mg$^{-1}$ Chl h$^{-1}$. This corresponds to a rate of ATP utilization of 800 $\mu$mol mg$^{-1}$ Chl h$^{-1}$ in mesophyll cells. Obviously pyruvate metabolism by mesophyll chloroplasts places a high requirement on ATP generation through photophosphorylation. This presents an ideal system for determining the potential for cyclic, pseudocyclic, and non-cyclic photophosphorylation with intact chloroplasts since the ATP requirement is high and non-cyclic flow can be controlled by addition of oxaloacetate or PGA.

With $C_4$ mesophyll protoplast extracts, the photophosphorylation capacity can be measured indirectly by following pyruvate dependent fixation of $^{14}CO_2$. With $CO_2$ fixation coupled through pyruvate,Pi dikinase, pyrophosphatase, adenylate kinase, and PEP carboxylase, two ATP are required per $CO_2$ fixed (Section 11.1). This stoichiometry has been experimentally verified by measuring pyruvate-dependent $CO_2$ fixation in the dark with ATP (rates of 40–70 $\mu$mol mg$^{-1}$ Chl h$^{-1}$). ATP is taken up by the chloroplasts on an adenylate translocator and 2 ATP are utilized per $CO_2$ fixed.

The potential for cyclic photophosphorylation can be determined with protoplast extracts in the light by measuring pyruvate-dependent $CO_2$ fixation at 2% $O_2$ (to prevent pseudocyclic photophosphorylation through the Mehler reaction) and by maintaining a low Chl concentration in order to minimize the accumulation of oxaloacetate. Otherwise, oxaloacetate would utilize NADPH through NADP-malate dehydrogenase and induce non-cyclic electron flow. Rates of cyclic photophosphorylation at 2% $O_2$ are about 60–80 $\mu$mol $mg^{-1}$ Chl $h^{-1}$. This photophosphorylation is sensitive to antimycin A (10 $\mu$M) and DBMIB (6 $\mu$M), but not to DCMU (for proposed sites of inhibition, see Fig. 5.2). In fact, DCMU at low concentrations (0.2 $\mu$M) causes a large stimulation of cyclic photophosphorylation, apparently by preventing 'over reduction' (Section 5.4).

Pseudocyclic electron flow provides a major source of ATP when the reactions are run under high $O_2$ levels, with antimycin A to block cyclic and with low Chl concentrations to minimize oxaloacetate accumulation. Rates of pseudocyclic photophosphorylation under 21% $O_2$ are similar to those of cyclic photophosphorylation (60–80 $\mu$mol ATP $mg^{-1}$ Chl $h^{-1}$). Pseudocyclic photophosphorylation reaches half of its maximum rate at an $[O_2]$ of about 30% (Figs. 5.2 and 5.3).

Non-cyclic photophosphorylation is a major source of ATP for pyruvate-dependent $CO_2$ fixation in the presence of 0.5 mM oxaloacetate. Rates of non-cyclic photophosphorylation in the presence of antimycin A (which blocks cyclic) and at 2% $O_2$ (which blocks pseudocyclic photophosphorylation) of 500–700 $\mu$mol ATP $mg^{-1}$ Chl $h^{-1}$ are obtained. Both pseudocyclic and non-cyclic photophosphorylation are inhibited by DCMU and DBMIB (Figs. 5.2 and 5.3).

Since $C_4$ plants require 5 ATP and 2 NADPH per $CO_2$ fixed (Section 11.6), it is to be expected that some ATP is provided through cyclic and pseudocyclic photophosphorylation. A $P/2e^-$ ratio of 1–2 only provides 2–4 ATP per NADPH (Section 5.14). The potential for cyclic photophosphorylation in mesophyll and bundle sheath chloroplasts of different $C_4$ subgroups has been inferred from the photochemical characteristics (Section 11.7), but the relative contribution from cyclic and pseudocyclic photophosphorylation is unknown. Particularly when aspartate is a major product of $C_4$ photosynthesis, cyclic and pseudocyclic photophosphorylation may be important sources of ATP in mesophyll cells since alanine + $CO_2$ → aspartate requires ATP (Section 11.1).

[Refs. **15, 30–33**]

## 12.8 Regulation of enzymes of the $C_4$ cycle

### (a) ILLUMINATION OF ETIOLATED TISSUE

Enzymes which are involved in photosynthesis often show low activity in etiolated tissue and show a many-fold increase upon greening. This is so for several enzymes of the $C_4$ cycle including pyruvate,Pi dikinase, adenylate kinase, PEP carboxylase, alanine- and aspartate aminotransferases, and NAD-malic enzyme. In maize and sorghum, both chloramphenicol and cycloheximide (inhibitors of protein synthesis) inhibit the light-dependent increase in activity of PEP carboxylase, pyruvate,Pi dikinase and adenylate kinase. This suggests the increase in activity during greening is a result of protein synthesis rather than light-activation of pre-existing protein. Isocitrate dehydrogenase and glucose-6-phosphate dehydrogenase, both non-photosynthetic enzymes, do not increase in activity when the etiolated tissue is illuminated. From such studies it is inferred that certain enzymes function in the $C_4$ cycle. For example, there are low activities of PEP carboxylase in etiolated tissue. A distinctly different isozyme appears in green leaf tissue. Also, low levels of isozymes of the $C_4$ cycle are expected to occur in the leaf with a function unrelated to $C_4$ photosynthesis.

In photosynthesizing leaves, enzymes of the $C_4$ cycle show several types of regulation. These include: light modulation of enzyme activity, regulation of allosteric enzymes by effectors (Section 12.8c), and regulation by substrate levels, $NADPH/NADP^+$ ratio, adenine nucleotides, pH and divalent cations.

### (b) LIGHT ACTIVATION OF PHOTOSYNTHETIC ENZYMES

Two enzymes of the $C_4$ cycle, pyruvate,Pi dikinase and NADP-malate dehydrogenase, are activated when the leaves are in the light and inactivated when leaves are in the dark.

$$\begin{array}{ccccccc} \text{pyruvate} & \rightarrow & \text{PEP} & \rightarrow & \text{oxaloacetate} & \rightarrow & \text{malate} \\ & \uparrow & & \uparrow & & & \\ & 2\,\text{ATP} & & CO_2 & \text{NADPH} & \curvearrowright & \text{NADP}^+ \end{array}$$

The degree of enzyme activation depends on the light intensity with high light required for full activation. Hatch and Slack placed $C_4$ leaves in full sunlight or its equivalent to obtain maximum activation of pyruvate,Pi dikinase just before extracting the enzyme.

The light-activation of NADP-malate dehydrogenase is thought to be mediated by reduction of disulphide groups on the protein. The exact nature of the disulphide reduction is uncertain but a simple possibility is illustrated. The reductive power for

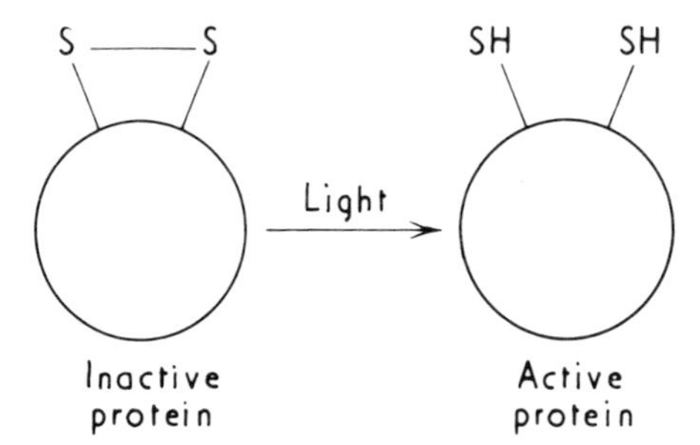


**Fig. 12.24.** Illustration of how reduction of a disulfide group may occur on a protein.

this conversion probably comes from non-cyclic electron flow. A reductant (modulator) formed in the terminal part of non-cyclic electron flow then donates electrons to the enzyme (see Fig. 12.24).

In like manner, NADP-malate dehydrogenase and enzymes of the RPP pathway, NADP-triose phosphate dehydrogenase, fructose 1,6-bisphosphatase, sedoheptulose 1,7-bisphosphatase and phosphoribulokinase of $C_3$ plants are activated in the light by reduction of disulphide groups on the protein (Chapter 9). In $C_3$ plants, this reduction appears to be mediated by a soluble, small molecular weight protein, called thioredoxin, which in turn is reduced by ferredoxin-thioredoxin reductase. In-vitro activation can be mediated by non-enzymatic reduction of thioredoxin by dithiothreitol.

The activation of pyruvate,Pi dikinase requires Pi, a large molecular weight heat labile protein and is inhibited by AMP and ADP. Inactivation requires ADP. Activation can occur *in vitro* in the absence of reducing agents suggesting that reduction of disulphide groups on the protein is not required (T. Sugiyama and M. D. Hatch, personal communication). In-vivo activation may in part be mediated by conversion of AMP and ADP to ATP in the chloroplasts in the light.

### (c) Regulation of PEP carboxylase

Other types of regulation exist among particular enzymes of the $C_4$ cycle. Substantial research has been done on the kinetic properties of PEP carboxylase from $C_4$ plants and its regulation. Like the bacterial PEP carboxylase, PEP carboxylase in $C_4$ plants exhibits sigmoidal kinetics (from analysis of plots of [PEP] versus activity) and allosteric properties, but in comparison to bacteria, there are some differences in effectors. With allosteric enzymes, metabolites bind at non-substrate sites and cause conformational changes in the protein. These changes either increase (activator) or decrease (inhibitor) the affinity of the enzyme for one or more of its substrates. For example, malate and aspartate are negative effectors with PEP carboxylase from $C_4$ plants, causing an increase in the *K*m for PEP. Glucose-6-phosphate (G6P) is an allosteric activator increasing the affinity of the enzyme for PEP (decreasing *K*m). In addition, these effectors have little effect on the *V*max (Fig. 12.25). Oxaloacetate is also a strong inhibitor of PEP carboxylase from $C_4$ plants.

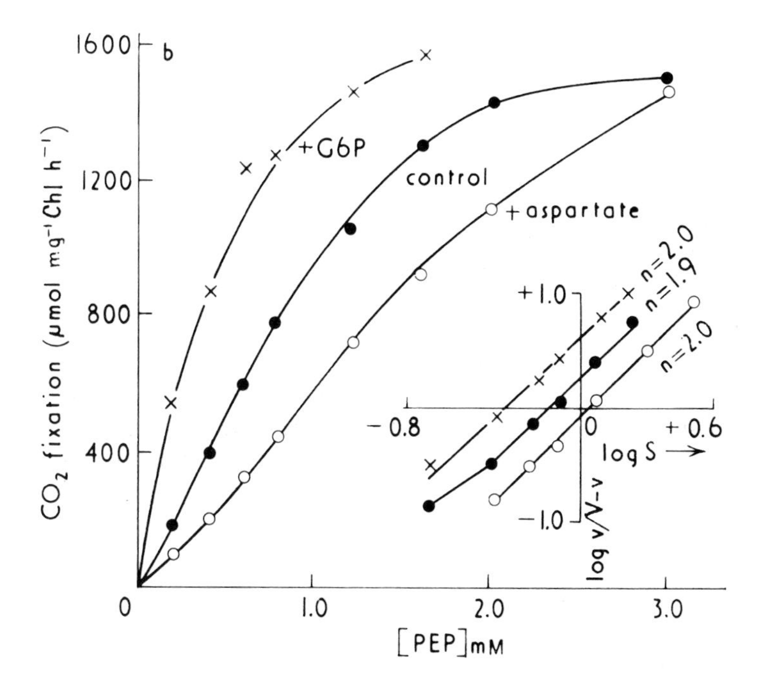


**Fig. 12.25.** Kinetics of PEP carboxylase of a $C_4$ plant without effector (control), with positive effector (2 mM G6P), and with negative effector (1 mM aspartate). Inset is Hill plot. Note the most control over enzyme activity by effectors is at limiting concentration of substrate (After Huber & Edwards, 1975b).

Sigmoidal kinetics (by the enzyme) indicate that an enzyme has more than one substrate binding site. As the substrate concentration is increased and the enzyme reacts with more than one substrate, it is easier for subsequent substrate to interact. Hill plots [(log $v/V - v$ versus log S) where $v$ = velocity, $V$ = *V*max, S = substrate] are used to determine the degree of cooperativity in substrate binding. The slope of the Hill plot (n) indicates the minimum number of substrate sites on the protein (i.e., n = 2 indicates at least two substrate binding sites). With normal Michaelis-Menten kinetics, n = 1.

An allosteric enzyme with cooperative kinetics may lose its regulatory properties (for example, during purification, change in temperature, pH, etc.), but maintain its catalytic properties, thus responding with typical Michaelis-Menten kinetics. Failure to see allosteric control or sigmoidal kinetics is not strict evidence that the enzyme lacks this type of regulation.

A combination of allosteric control and sigmoidal kinetics gives a high degree of regulation by effectors at limiting levels of substrate. The *V*max of PEP carboxylase from $C_4$ plants is several-fold higher than the rates of leaf photosynthesis. The levels of PEP are likely to be rate limiting *in vivo*, which would result in conditions for maximum regulation by allosteric effectors.

With PEP carboxylase from $C_4$ plants, the maximum regulation by aspartate, malate, and G6P occurs at relatively low pH and low $Mg^{2+}$ (as well as low PEP). *In vivo* the pH, level of $Mg^{2+}$, PEP, $C_4$ acids and G6P control the activity of PEP carboxylase. The rate of utilization of $C_4$ acids for carbon donation to the RPP pathway may in turn regulate carboxylation in the $C_4$ pathway by feedback inhibition of PEP carboxylase. G6P is a key metabolite in gluconeogenesis leading to sucrose and starch synthesis and other macromolecules. In $C_4$ mesophyll cells, PEP could be used in G6P synthesis as well as in $C_4$ acid formation. Activation of PEP carboxylase by G6P may prevent loss of carbon from the $C_4$ cycle due to PEP metabolism to G6P.

PEP carboxylase generally occurs in all plants although the activity of the enzyme is many-fold higher from leaves of $C_4$ plants and CAM plants than in $C_3$ plants. It has different kinetic constants for PEP and $Mg^{2+}$ depending on its source ($C_3$ plants, $C_4$ plants, CAM plants, nonphotosynthetic tissue).

An important consideration with carboxylases is the form of inorganic carbon fixed ($HCO_3^-$ or free $CO_2$) and the *K*m for this substrate. This is particularly relevant for making comparisons between $C_3$ and $C_4$ photosynthesis and considering the significance of PEP carboxylase versus RBP carboxylase in the initial fixation of atmospheric $CO_2$. A discussion of this is given in Chapter 14.

### (d) Regulation of the decarboxylating enzymes

All three $C_4$ acid decarboxylating enzymes require divalent cations for activation: NADP-malic enzyme ($Mg^{2+}$ or $Mn^{2+}$), NAD-malic enzyme ($Mn^{2+}$) and PEP carboxykinase ($Mn^{2+}$, highest activities with $Mn^{2+} + Mg^{2+}$). Bicarbonate inhibits the decarboxylation by NADP- and NAD-malic enzyme ($CO_2$ may be the active species). This could regulate activity of the $C_4$ cycle *in vivo*.

High $[CO_2]$ in bundle sheath cells $\rightarrow$ Inhibits decarboxylation and formation of $C_3$ precursor $\rightarrow$

Increases $C_4$ acids $\rightarrow$ Inhibits PEP carboxylase

The cycle must function at rates which maintain an elevated $[CO_2]$ in the bundle sheath cells. However, excessive rates may result in considerable loss of $CO_2$ and unnecessary expenditure of energy. Additional control may exist through NADP-malic enzyme as the enzyme is inhibited by pyruvate and NADPH, the products of decarboxylation. The relative concentration of $NADP^+$ and NADPH will regulate the activity of the enzyme. Decarboxylation through NADP-malic enzyme is dependent on the reductive phase of photosynthesis for regenerating the $NADP^+$.

$$\text{malate} + NADP^+ \rightarrow \text{pyruvate} + NADPH + H^+ + CO_2 \qquad \text{Eqn. 12.1}$$

$$RBP + CO_2 \rightarrow 2\,PGA \qquad \text{Eqn. 12.2}$$

$$2\,PGA \xrightarrow[2\,ATP \;\rightarrow\; 2\,ADP]{} 2(1{,}3\,DPGA) \xrightarrow[2\,NADPH + 2H^+ \;\rightarrow\; 2NADP^+]{} 2 \text{ triose phosphate} \qquad \text{Eqn. 12.3}$$

In the RPP pathway, twice as much NADPH is required to reduce the PGA as is generated by malate decarboxylation. This assures the availability of $NADP^+$ for the decarboxylating enzyme. Bundle sheath chloroplasts of NADP-ME species have low levels of non-cyclic electron flow (Section 11.7). An excessive photochemical potential for generation of NADPH in bundle sheath chloroplasts might be a disadvantage due to competition with malic enzyme for $NADP^+$ and subsequent limitation of decarboxylation.

NAD-malic enzyme (from *Atriplex spongiosa* and *Amaranthus edulis*) has allosteric properties and shows cooperative kinetics with acetyl Co A, Co A and fructose 1,6 bisphosphate serving as positive effectors and $HCO_3^-$ as a negative effector. Normally, pyruvate is metabolized to acetyl Co A + $CO_2$ in the TCA cycle. Acetyl Co A + OAA then combine to form isocitrate. This could lead to respiratory depletion of metabolites of the $C_4$ pathway. It may be significant that NAD-malic enzyme is activated by acetyl Co A, stimulating reversed flow in the cycle (OAA $\rightarrow$ malate), and thereby assuring recycling of pyruvate (product of $C_4$ decarboxylation) to the mesophyll cells.

Decarboxylation through PEP carboxykinase has a high specificity for ATP in comparison to other nucleoside triphosphates. Fructose bisphosphate, PGA, and DHAP (1–5 mM) partially inhibit the enzyme.

### (e) Conclusions

Several enzymes of the $C_4$ cycle are inactivated in the dark. Therefore, a useless shuttle of $CO_2$ at the expense of respiratory energy is prevented. In the light, enzymes of the $C_4$ cycle may be activated by metabolites in other pathways which represent 'sinks' for metabolites of the $C_4$ cycle (i.e. G6P for PEP carboxylase, acetyl Co A for NAD-malic enzyme). In this respect, PEP and pyruvate are metabolites which may otherwise be metabolized in gluconeogenesis and respiratory metabolism.

Certain metabolites of the RPP pathway may influence the activity of the $C_4$ cycle, and light-dark changes in [$Mg^{2+}$] and pH in certain cellular compartments may control certain enzymes. Understanding the regulation of enzymes in a cycle which spans two cells and a number of cellular compartments is a considerable challenge. Above all, caution is necessary in considering the possible physiological significance of any factor which changes enzyme activity *in vitro*. Certainly, some of these factors may have no physiological meaning whatsoever.
[Refs. **2, 5, 17, 18, 21–23, 29, 36, 48, 51, 52, 55, 56**]

## 12.9 Nitrogen assimilation—$C_4$ versus $C_3$ plants

The reduction and incorporation of nitrogen from nitrate into amino acids in leaves of plants depends on energy provided by the chloroplast. Nitrate would first be converted to ammonia by nitrate reductase and nitrite reductase (Eqns. 12.4 and 12.5). For years ammonia conversion to amino nitrogen was considered to occur through glutamate dehydrogenase (Eqn. 12.6). There is an NAD-glutamate dehydrogenase in mitochondria and NAD(P)-glutamate dehydrogenase in chloroplasts. More recently, glutamine synthetase—glutamate synthase (Eqns. 12.7, 12.8) has been discovered as an alternate mechanism, and there is considerable evidence that the major pathway for nitrogen assimilation involves these enzymes. Glutamine is often the major initial product of assimilation of $^{15}NO_3$ or $^{15}NH_4$ in photosynthetic tissue followed by appearance of label in glutamate which is consistent with the sequence of ammonia incorporation through Equations 12.7 and 12.8. Methionine sulphoximine (MSO) is an inhibitor of glutamine synthetase and azaserine is an inhibitor of glutamine amide transfer reactions, including glutamate synthase. Their inhibitory effects on the conversion of nitrogen (in blue-green algae), nitrate, or ammonia to glutamate in a number of studies has been taken as evidence for assimilation through the glutamine synthetase–glutamate synthase pathway. For example, metabolism of ammonia or nitrate + 2-oxoglutarate to glutamate by isolated chloroplasts of peas ($C_3$) is inhibited by MSO and azaserine. In the presence of azaserine, ammonia or nitrate is incorporated into glutamine but not into glutamate, consistent with its site of inhibition through glutamate synthase. The $K$m for ammonia of the higher plant glutamate dehydrogenase is about two orders of magnitude higher than that of glutamine synthetase, making the latter the more favourable reaction under relatively low concentrations of ammonia.

$NO_3^- + NADH + H^+ \rightarrow NO_2^- + NAD^+$ (nitrate reductase) Eqn. 12.4

$NO_2^- + 6$ Fd (reduced) $\rightarrow NH_4^+ + 6$ Fd (oxidized) (nitrite reductase) Eqn. 12.5

$NH_3 + NAD(P)H + H^+ +$ 2-oxoglutarate $\rightarrow NAD(P)^+ +$ glutamate $+ H_2O$
(glutamate dehydrogenase) Eqn. 12.6

$NH_3 + ATP +$ glutamate $\rightarrow$ glutamine $+ ADP + Pi$
(glutamine synthetase) Eqn. 12.7

glutamine + 2-oxoglutarate + 2 Fd (reduced) $\rightarrow$
2 glutamate + 2 Fd (oxidized) (glutamate synthase) Eqn. 12.8

There is no known direct link between nitrogen assimilation and $C_4$ photosynthesis. However, conversion of nitrate to amino nitrogen requires energy, and the location of this metabolism in $C_4$ leaves is of interest in view of the two photosynthetic cell types. Studies with species of all three $C_4$ subgroups have shown that 80% or more of the enzyme activity for reduction of nitrate to ammonia (nitrate reductase, nitrite reductase) is localized in the mesophyll cells. Some studies show glutamine synthetase and glutamate synthase are mainly localized in mesophyll cells; others indicate substantial activity of these enzymes in both cell types. Even so, the mesophyll cells can be regarded as the primary site of nitrogen assimilation in $C_4$ plants.

In mesophyll cells of both $C_3$ and $C_4$ plants, there appears to be a similar pathway of nitrate conversion to amino nitrogen. Nitrate reductase is localized in the cytosol or on the outer chloroplast envelope. In either case, reductive power may be shuttled out of the chloroplast. There are several alternatives (Fig. 12.26).

In the first alternative (A), reductive power is shuttled to the cytosol through malate dehydrogenase with transport of the $C_4$ acids across the chloroplast envelope facilitated by the dicarboxylate translocator (Sections 8.24 & 12.5). In the second and third options (B, C), DHAP export from the chloroplast is coupled with PGA uptake on the phosphate translocator (Section 8.23). In both cases the DHAP is oxidized to PGA in the cytoplasm, generating reductive power. With the irreversible NADP-triose phosphate dehydrogenase (option B), PGA is the product and only reductive power is formed (Eqn. 12.9).

$G3P + NADP^+ \rightarrow 3\text{-}PGA + NADPH + H^+$ Eqn. 12.9

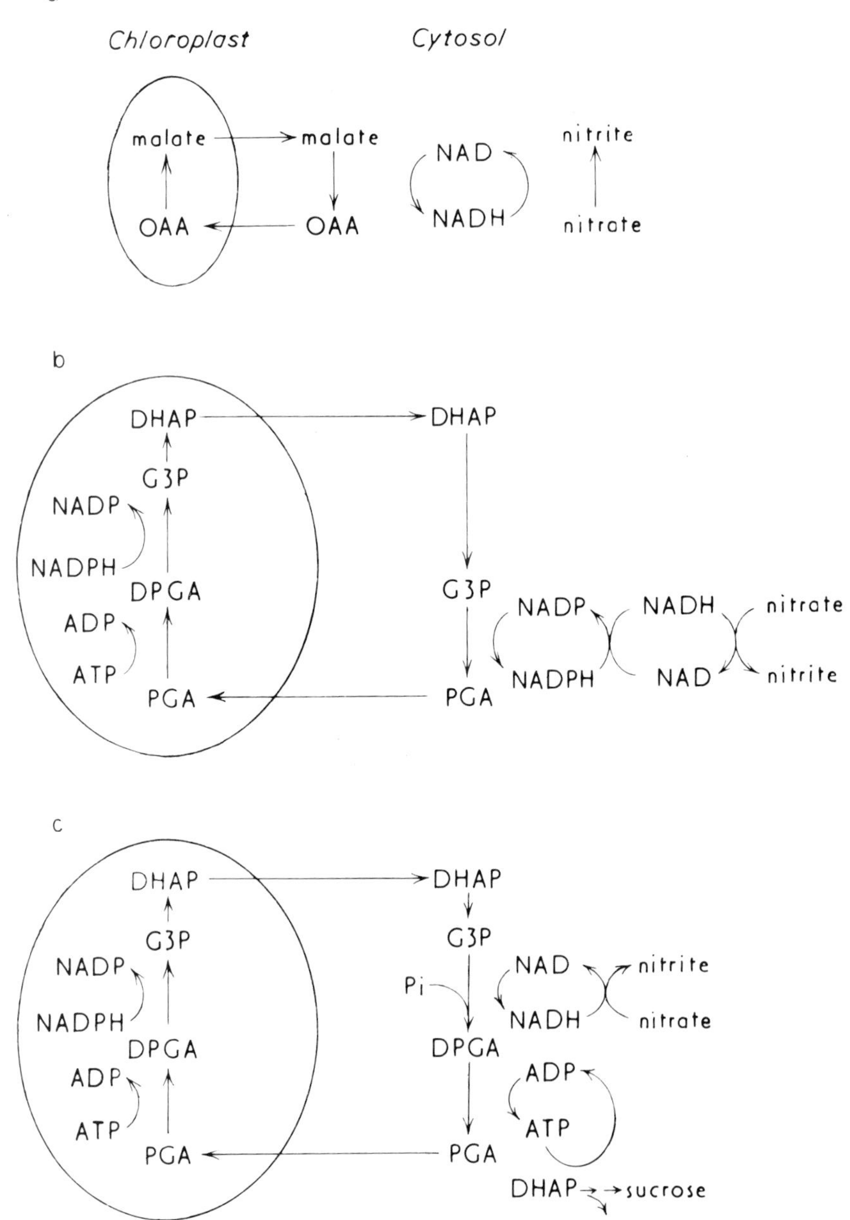


**Fig. 12.26.** Means for shuttling reductive power from the chloroplasts to cytosol for nitrate reduction.

a. Oxaloacetate-malate shuttle through malate dehydrogenase.
b. Shuttle through irreversible NADP-triose phosphate dehydrogenase.
c. Shuttle through reversible NAD-triose phosphate dehydrogenase.

With the reversible NAD-triose phosphate dehydrogenase, 1,3 DPGA is the product of the oxidation of glyceraldehyde-3-phosphate. Subsequent synthesis of PGA through phosphoglycerate kinase leads to ATP synthesis. The ATP may be utilized in other metabolism in the cytosol, e.g., for sucrose synthesis (Chapter 9).

Other enzymes for metabolism of nitrite to glutamate are localized in the chloroplast (Fig. 12.27). [Some glutamine synthetase may be localized in the cytosol and there is an NAD-glutamate dehydrogenase in the mitochondria. Their function in nitrogen metabolism is uncertain.] Nitrite reductase requires reduced ferredoxin and 6 $e^-$ are involved in the reduction of $NO_2^-$ to $NH_4^+$ (equivalent to 3 NADPH). For each nitrate converted to amino nitrogen through glutamine synthetase-glutamate synthase, 1 ATP and 5 NADPH (equivalents of reducing power) are required. Nitrate metabolism to amino nitrogen through glutamate dehydrogenase would

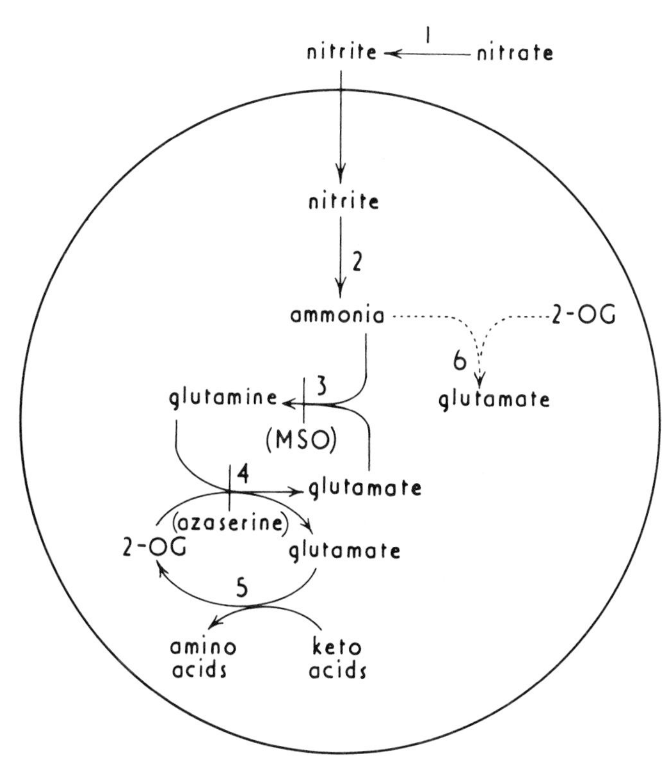


**Fig. 12.27.** Suggested scheme of nitrite metabolism in $C_3$ and $C_4$ mesophyll chloroplasts (1) nitrate reductase; (2) nitrite reductase; (3) glutamine synthetase, (MSO) methionine sulphoximine inhibits; (4) glutamate synthase, azaserine inhibits; (5) transaminases; (6) NADP-glutamate dehydrogenase (broken line suggests minor pathway). 2 – OG = 2-oxoglutarate.
(See Rathnam & Edwards, 1976; Lea & Miflin, 1978; Moore & Black, 1979)

require an equivalent amount of reductive power but no ATP. The energy requirements for nitrogen assimilation in $C_4$ plants can be considered in relation to the energetics of $C_4$ photosynthesis (for comparison with $C_3$ plants, see Section 13.6). In $C_4$ plants, $CO_2$ assimilation requires 5 ATP/2 NADPH per $CO_2$ fixed (Section 11.6). The nitrogen content of leaves is about 6% that of carbon (atom : atom) so that the rate of nitrogen assimilation can be assumed to be 6% of the rate of photosynthesis. Accordingly, the energy requirement per $CO_2$ fixed through the $C_4$ cycle–RPP pathway plus the energy requirement for nitrate to amino nitrogen conversion in the leaf would be:

$$(5\,ATP + 2\,NADPH/CO_2) + 0.06\ (1\ ATP + 5\ NADPH) = 5.1\ ATP/2.3\ NADPH$$

This suggests that about 13% of the reductive power would be used for nitrogen assimilation during photosynthetic assmilation of carbon. Both the synthesis of proteins from amino acids and the synthesis of carbohydrates such as starch and sucrose will require some additional energy.

The assimilation of nitrate may be more expensive to the carbon economy of the plant. Nitrate uptake and reduction generates hydroxyl ions which may be balanced by synthesis of organic acids (commonly malate). This requires at least one carboxyl equivalent per nitrate reduced to ammonia. With a stoichiometry of malic acid $+ 2KNO_3 \rightarrow K_2$malate $+ 2(-NH_2)$, two organic carbons would be synthesized per nitrate reduced (Section 16.5). A large part of the malate stored in vacuoles in plants may be to compensate for alkalization during nitrate conversion to amino nitrogen. The implications are that the more protein a plant synthesizes utilizing nitrate as the nitrogen source, the more carbon it must partition into organic anions.

Some studies indicate that the enzymes for nitrogen assimilation are primarily localized in mesophyll cells in $C_4$ plants. Therefore, the reductive power generated in mesophyll cells will be used for malate synthesis, reduction of PGA transported from bundle sheath cells and reduction of nitrate to amino nitrogen. The reductive power generated in bundle sheath cells (through malate decarboxylation or photochemically) is therefore used primarily in the RPP pathway.

About 80% of the NAD-glutamate dehyrogenase of $C_4$ leaves has been found in mitochondria of bundle sheath cells. It is uncertain whether this enzyme functions in amination or deamination; although glutamate dehydrogenases are noted for having an extremely high $K$m for ammonia. In $C_3$ plants, glycine decarboxylation (2 glycine $\rightarrow$ serine $+ CO_2 + NH_3$) in the glycolate pathway occurs in the mitochondria. At least some metabolism in the glycolate pathway is thought to occur in bundle sheath cells of $C_4$ plants (see Section 13.9). The ammonia released during glycine decarboxylation may be reassimilated either by the glutamate

dehydrogenase of the mitochondria or by the glutamine synthetase of the bundle sheath choloroplast.
[Refs. **1, 20, 37, 39–42, 44**]

### 12.10 Starch and sucrose synthesis in $C_4$ plants

Both mesophyll and bundle sheath chloroplasts of $C_4$ plants can store starch, but the bundle sheath chloroplasts are the primary site of storage. It has been suggested that bundle sheath chloroplasts function as amyloplasts, although it is now obvious that they have an essential carboxylating function in photosynthesis. In some species such as maize, starch is also found in the mesophyll chloroplasts if the plants are kept in continuous light. In maize and several other $C_4$ species examined, the enzymes of starch metabolism (ADPG pyrophosphorylase, ADPG starch synthetase and phosphorylase) are primarily found in bundle sheath cells. In particular, the capacity of mesophyll chloroplasts for starch synthesis may vary considerably between species. Whether phosphorylase is involved in starch synthesis or breakdown is uncertain.

De Fekete and Vieweg found that isolated bundle sheath cells of maize synthesized starch in the light when provided with glucose-1-phosphate (G1P) but not when provided with $CO_2$. This led them to suggest that starch synthesis *in vivo* is dependent on mesophyll cells to provide G1P. However, the results do not indicate that $CO_2$ is converted to hexose phosphates in mesophyll cells. Isolated bundle sheath cells of maize incorporate $^{14}CO_2$ primarily into PGA although some label appears in starch. Rates of $CO_2$ fixation are relatively low due to limiting reductive power caused by a deficiency in photosystem II. In-vivo the reductive power for synthesis of triose phosphates in these cells is provided by malate through NADP-malic enzyme. Thus, it is logical that the isolated bundle sheath cells would be capable of metabolizing DHAP or G1P to starch, but not PGA or $CO_2$ (see Section 11.7).

The enzymes of sucrose synthesis, UDPG pyrophosphorylase, sucrose-synthetase and sucrose-P phosphatase have been found in both mesophyll and bundle sheath cells of several $C_4$ species. In some species such as maize and crabgrass, however, a higher proportion of these enzymes is in mesophyll cells. Downton and Hawker suggested the mesophyll cell as the primary site of sucrose synthesis in maize.

The level of fructose bisphosphate aldolase (Chapter 6) in bundle sheath cells of certain $C_4$ species is about an order of magnitude higher than in mesophyll cells. If the aldolase levels in the mesophyll cells is limiting for metabolizing triose phosphates to sucrose, then hexose phosphate may be provided by the bundle sheath. However, for every hexose-P synthesized through the aldolase and metabolized to sucrose, at least five hexose-P must be synthesized in the chloroplast in the regenerative phase of the RPP pathway, which is localized in bundle sheath cells (Section 11.5). Thus, it is not surprising that aldolase is primarily found in bundle sheath cells. The hexose-P synthesized in bundle sheath cells through fructose bisphosphate aldolase may be used for the regenerative phase of the RPP pathway, starch synthesis, and some sucrose synthesis. Hexose-P in mesophyll cells may be used primarily for sucrose synthesis. In addition, transport of reduced carbon from mesophyll to bundle sheath tissue could occur in the form of hexose-P or triose-P.

In $C_3$ plants, the Pi concentration in the cytosol seems to have a regulatory role in controlling carbon flow to sucrose (Chapter 9). In $C_4$ plants enzymes for the terminal steps of sucrose synthesis such as UDPG pyrophosphorylase and sucrose-P synthetase are localized in the cytoplasm. Thus sucrose synthesis apparently occurs in the cytoplasm as suggested for $C_3$ plants. In the bundle sheath cells the cytosolic level of Pi may influence the rate of exchange of PGA or triose phosphate out of the chloroplast. With the subsequent metabolism of the PGA or triose phosphate to sucrose in the cytosol of either mesophyll or bundle sheath cells, Pi would be released and returned to bundle sheath chloroplasts (see Section 11.6). In the $C_4$ mesophyll chloroplast, Pi is taken up in exchange for PEP (Section 12.5). During photosynthesis in $C_3$ plants, the primary exchange on the phosphate transporter is Pi-DHAP with relatively low export of PGA. Thus, the cytosolic level of PGA during photosynthesis may be relatively low. In $C_4$ plants, export of PGA from bundle sheath chloroplasts is suggested with its subsequent reduction in mesophyll chloroplasts. Therefore, during $C_4$ photosynthesis, the cytosolic level of PGA may be relatively high. In the mesophyll cells, the level of PGA in the cytosol may partially control the rate of sucrose synthesis through a PGA-DHAP exchange by the chloroplasts. In contrast to conditions in a $C_3$ cell, high levels of cytosolic Pi in the $C_4$ mesophyll tissue would favour PEP exchange and inhibit sucrose synthesis by competitively inhibiting PGA uptake by the chloroplasts.

In $C_3$ plants, the synthesis of sucrose is thought to be, in part, controlled by export of triose phosphates from the chloroplasts in exchange for Pi on the phosphate transporter. Thus, limited cytosolic Pi may reduce export of carbon from the chloroplasts, reduce sucrose synthesis and stimulate starch synthesis. In fact, mannose sequesters Pi as mannose phosphate in some tissues and has the effect of stimulating photosynthetic starch formation in certain $C_3$ species (Section 9.7). In contrast, mannose feeding to leaf tissue of $C_4$ species in the light inhibits starch synthesis (see Fig. 9.12). The bundle sheath chloroplasts are generally the main site of starch synthesis in $C_4$ plants. Reduced levels of Pi in the cytosol of bundle sheath

cells (e.g., sequestering with fed mannose) might be expected to reduce export of PGA or dihydroxyacetone phosphate from the chloroplasts and thus stimulate starch synthesis as in $C_3$ species. However, photosynthesis in $C_4$ species is linked to the $C_4$ cycle. This cycle requires Pi uptake by mesophyll chloroplasts in exchange for PEP (Section 12.5). Thus the sequesteration of cytoplasmic Pi with mannose in leaf tissue of $C_4$ species causes a marked inhibition of both photosynthesis and starch synthesis.
[Refs. **3, 4, 7–10, 27, 53, 54**]

### General reading—See Chapter 10.

### Specific citations


1 BEEVERS L. & R. H. HAGEMAN. (1972). The role of light in nitrate metabolism in higher plants. In *Photophysiology* (ed. A. C. Giese) **7**, 85–113. New York: Academic Press.

2 BLACK C. C. (1973) Photosynthetic carbon fixation in relation to net $CO_2$ uptake. *Ann. Rev. Plant Physiol.*, **24**, 253–86.

3 BUCKE C. & OLIVER I. R. (1975) Location of enzymes metabolizing sucrose and starch in the grasses *Pennisetum purpureum* and *Muhlenbergia montana. Planta*, **122**, 45–52.

4 CHEN T. M., DITTRICH P., CAMPBELL W. H. & BLACK C. C. (1974) Metabolism of epidermal tissues, mesophyll cells, and bundle sheath strands resolved from mature nutsedge leaves. *Arch. Biochem. Biophys.* **163**, 246–62.

5 COOMBS J. (1976) Interactions between chloroplasts and cytoplasm in $C_4$ plants. In *The Intact Chloroplast*, (ed. J. Barber.) pp. 279–313. The Netherlands: Elsevier.

6 CRAIG S. & GOODCHILD D. J. (1977) Leaf ultrastructure of *Triodia irritans*: A $C_4$ grass possessing an unusual arrangement of photosynthetic tissue. *Aust. J. Bot.*, **25**, 277–90.

7 DAVIES D. R. (1974) Some aspects of sucrose metabolism. In *Plant Carbohydrate Biochemistry*. (ed. J. B. Pridham) pp. 61–81. New York: Academic Press.

8 DE FEKETE M. A. R. & VIEWEG G. H. (1974) Starch metabolism: Synthesis versus degradation pathways. In *Plant Carbohydrate Biochemistry*. (ed. J. B. Pridham) pp. 127–144. New York: Academic Press.

9 DOWNTON W. J. S. & HAWKER J. S. (1973) Enzymes of starch and sucrose metabolism in *Zea mays* leaves. *Phytochemistry*, **12**, 1551–6.

10 EDWARDS G. E. & BLACK C. C. (1971) Photosynthesis in mesophyll cells and bundle sheath cells isolated from *Digitaria sanguinalis* leaves. In *Photosynthesis and Photorespiration*. (eds. M. D. Hatch, C. B. Osmond, and R. O. Slatyer) pp. 153–68. New York: Wiley-Interscience.

11 EDWARDS G. E., LEE S. S., CHEN T. M. & BLACK C. C. (1970) Carboxylation reactions and photosynthesis of carbon compounds in isolated bundle sheath cells of *Digitaria sanguinalis* (L.) Scop. *Biochem. Biophys. Res. Commun.*, **39**, 389–95.

12 EDWARDS G. E., HUBER S. C. & GUTIERREZ M. (1976a) Photosynthetic properties of plant protoplasts. In *Microbial and Plant Protoplasts*. (eds. J. F. Peberdy, A. H. Rose, H. J. Rogers and E. C. Cocking) pp. 299–322. New York: Academic Press.

13 EDWARDS G. E., HUBER S. C., KU S. B., GUTIERREZ M., RATHNAM C. K. M. & MAYNE B. C. (1976b) Variation in photochemical activities in $C_4$ plants in relation to $CO_2$ fixation. In *$CO_2$ Metabolism and Productivity of Plants*. (eds. R. H. Burris and C. C. Black) pp. 83–112. Baltimore Maryland: University Park Press.

14 EDWARDS G. E. & HUBER S. C. (1979) $C_4$ metabolism in isolated cells and protoplasts. In *Encyclopedia of Plant Physiology*, New series. Vol. 6. Photosynthesis II. Photosynthetic carbon

metabolism and related processes. (eds. M. Gibbs and E. Latzko) pp. 102–12 New York: Springer-Verlag.
15 EDWARDS G. E. & HUBER S. C. (1978) Usefulness of isolated cells and protoplasts for photosynthetic studies. In *Fourth International Congress on Photosynthesis.* (eds. D. O. Hall, J. Coombs and T. W. Goodwin) pp. 95–106. London: The Biochemical Society.
16 EVERT R. F., ESCHRICH W. & HEYSER W. (1977) Distribution and structure of the plasmodesmata in mesophyll and bundle sheath cells of *Zea mays L. Planta,* **136**, 77–89.
17 GOATLEY M. B. & SMITH H. (1974) Differential properties of phosphoenolpyruvate carboxylase from etiolated and green sugarcane. *Planta,* **117**, 67–73.
18 GRAHAM D., HATCH M. D., SLACK C. R. & SMILLIE R. M. (1970) Light-induced formation of enzymes of the $C_4$ dicarboxylic acid pathway of photosynthesis in detached leaves. *Phytochemistry,* **9**, 521–32.
19 GUTIERREZ M., HUBER S. C., KU S. B., KANAI R. & EDWARDS G. E. (1975) Intracellular carbon metabolism in mesophyll cells of $C_4$ plants. In *Third International Congress on Photosynthesis Research.* (ed. M. Avron) pp. 1219–30. The Netherlands: Elsevier.
20 HAREL E., LEA P. J. & MIFLIN B. J. (1977) The localization of enzymes of nitrogen assimilation in maize leaves and their activities during greening. *Planta,* **134**, 195–200.
21 HATCH M. D. (1978) Regulation of enzymes in $C_4$ photosynthesis. In *Current Topics in Cellular Regulation.* (eds. B. L. Horecker and E. K. Stadtman) Ch. 14, pp. 1–28. New York: Academic Press.
22 HATCH M. D. & MAU S. (1977) Properties of phosphoenolpyruvate carboxylase in $C_4$ pathway photosynthesis. *Aust. J. Plant Physiol.,* **4**, 207–16.
23 HATCH M. D., MAU S. & KAGAWA T. (1974) Properties of leaf NAD malic enzyme from plants with $C_4$ pathway photosynthesis. *Arch. Biochem. Biophys.,* **165**, 188–200.
24. HATCH M. D. & OSMOND C. B. (1976) Compartmentation and transport in $C_4$ photosynthesis. In *Encyclopedia of Plant Physiology,* New series. Vol. III. Transport in plants. (eds. C. R. Stocking and U. Heber) pp. 144–84. New York: Springer-Verlag.
25 HATCH M. D. & SLACK C. R. (1970) The $C_4$ carboxylic acid pathway of photosynthesis. In *Progress in phytochemistry.* (eds. L. Reinhold and Liwschitz) pp. 35–106. London: Wiley Interscience.
26 HATTERSLEY P. W., WATSON L. & OSMOND C. B. (1976) Metabolite transport in leaves of $C_4$ plants: specification and speculation. In *Transport and Transfer Processes in Plants.* (eds. I. F. Wardlaw and J. B. Passioura) pp. 191–201. New York: Academic Press.
27 HEROLD A., LEWIS D. H. & WALKER D. A. (1976) Sequestration of cytoplasmic orthophosphate by mannose and its differential effect on photosynthetic starch synthesis in $C_3$ and $C_4$ species. *New Phytol.,* **76**, 397–407.
28 HUBER S. C. & EDWARDS G. E. (1975a) Regulation of oxaloacetate, aspartate and malate formation in mesophyll protoplast extracts of several $C_4$ plants *Plant Physiol.,* **56**, 324–31.
29 HUBER S. C. & EDWARDS G. E. (1975b) Inhibition of phosphoenolpyruvate carboxylase from $C_4$ plants by malate and aspartate. *Canad. J. Bot.,* **53**, 1925–33.
30 HUBER S. C. & EDWARDS G. E. (1975c) Effect of DBMIB, DCMU and antimycin A on cyclic and noncyclic electron flow in $C_4$ mesophyll chloroplasts. *FEBS Letters,* **58**, 211–4.
31 HUBER S. C. & EDWARDS G. E. (1975d) The effect of oxygen on $CO_2$ fixation by mesophyll protoplast extracts of $C_3$ and $C_4$ plants. *Biochem. Biophys. Res. Commun.,* **67**, 28–34.
32 HUBER S. C. & EDWARDS G. E. (1976) Studies on cyclic photophosphorylation in $C_4$ mesophyll chloroplasts. *Biochim. Biophys. Acta,* **449**, 420–33.
33 HUBER S. C. & EDWARDS G. E. (1977a) The effect of reducing conditions on inhibition of cyclic photophosphorylation by DBMIB, EDAC and antimycin A by $C_4$ mesophyll chloroplasts. *FEBS Letters,* **77**, 207–11.
34 HUBER S. C. & EDWARDS G. E. (1977b) Transport in $C_4$ mesophyll chloroplasts. Characterization of the pyruvate carrier. *Biochim. Biophys. Acta,* **462**, 583–602.
35 HUBER S. C. & EDWARDS G. E. (1977c) Transport in $C_4$ mesophyll chloroplasts. Evidence for an exchange of inorganic phosphate and phosphoenolpyruvate *Biochim. Biophys. Acta,* **462**, 603–12.

36 KAGAWA T. & HATCH M. D. (1977) Regulation of $C_4$ photosynthesis: characterization of a protein factor mediating the activation and inactivation of NADP-malate dehydrogenase. *Arch. Biochem. Biophys.*, **184**, 290–7.
37 KELLY G. J. & GIBBS M. (1973) A mechanism for the indirect transfer of photosynthetically reduced nicotinamide adenine dinucleotide phosphate from chloroplasts to the cytoplasm. *Plant Physiol.*, **52**, 674–6.
38 KU M. S. B., SPALDING M. H. & EDWARDS G. E. (1980) Intracellular localization of phosphoenolpyruvate carboxykinase in leaves of $C_4$ and CAM plants. *Plant Sci. Letters* **19**, 1–8.
39 LEA P. J. & MIFLIN B. J. (1979) Photosynthetic ammonia assimilation. In *Encyclopedia of Plant Physiology*. New series. Vol. 6. Photosynthesis II. Photosynthetic carbon metabolism and related processes. (eds. M. Gibbs and E. Latzko) pp. 445–55. New York: Springer-Verlag.
40 LEECH R. M. & MURPHY D. J. (1976) The cooperative function of chloroplasts in the biosynthesis of small molecules. In *The Intact Chloroplasts*. (ed. J. Barber) pp. 365–401. Amsterdam: Elsevier/North-Holland Biomedical Press.
41 MIFLIN B. J. & LEA P. J. (1977) Amino acid metabolism. *Ann. Rev. Plant Physiol.*, **28**, 299–332.
42 MOORE R. & BLACK C. C. (1979) Nitrogen assimilation pathways in leaf mesophyll and bundle sheath cells of $C_4$ photosynthesis plants formulated from comparative studies with *Digitaria sanguinalis (L)* Scop. *Plant Physiol.*, **64**, 309–13.
43 OSMOND C. B. & SMITH F. A. (1976) Symplastic transport of metabolites during $C_4$ photosynthesis. In *Intercellular Communication in Plants: Studies on Plasmodesmata*. (eds. B. E. S. Gunning and A. W. Robards) pp. 229–40. New York: Springer-Verlag.
44 RATHNAM C. K. M. & EDWARDS G. E. (1976) Distribution of nitrate assimilating enzymes between mesophyll protoplasts and bundle sheath cells in leaves of three groups of $C_4$ plants. *Plant Physiol.*, **57**, 730–3.
45 RATHNAM C. K. M. & EDWARDS G. E. (1977) $C_4$ acid decarboxylation and $CO_2$ donation to photosynthesis in bundle sheath strands and chloroplasts from species representing three groups of $C_4$ plants. *Arch. Biochem. Biophys.*, **183**, 1–13.
46 RATHNAM C. K. M. & EDWARDS G. E. (1977) $C_4$-dicarboxylic-acid metabolism in bundle sheath chloroplast, mitochondria and strands of *Eriochloa borumensis*, a PEP-carboxykinase type $C_4$ species. *Planta*, **133**, 135–44.
47 RAY T. B. & BLACK C. C. (1976a) Inhibition of oxaloacetate decarboxylation during $C_4$ photosynthesis by 3-mercaptopicolinic acid. *J. Biol. Chem.*, **251**, 5824–6.
48 RAY T. B. & BLACK C. C. (1976b) Characterization of phosphoenolpyruvate carboxykinase from *Panicum maximum*. *Plant Physiol.*, **58**, 603–7.
49 REPO E. & HATCH M. D. (1976) Photosynthesis in *Gomphrena celosioides* and its classification amongst $C_4$-pathway plants. *Aust. J. Plant Physiol.*, **3**, 863–76.
50 ROBARDS A. W. (1975) Plasmodesmata. *Ann. Rev. Plant Physiol.*, **26**, 13–29.
51 SUGIYAMA T. (1974) Proteinaceous factor reactivating an inactive form of pyruvate,Pi dikinase isolated from dark treated maize leaves. *Plant Cell Physiol.*, **15**, 723–6.
52 UEDAN K. & SUGIYAMA T. (1976) Purification and characterization of phosphoenolpyruvate carboxylase from maize leaves. *Plant Physiol.*, **57**, 906–10.
53 USUDA H. & EDWARDS G. E. (1980) Localization of glycerate kinase and some enzymes for sucrose synthesis in $C_3$ and $C_4$ plants. *Plant Physiol.*, **65**, 1017–22.
54 USUDA H., KANAI R. & MIYACHI S. (1975) Carbon dioxide assimilation and photosystem II deficiency in bundle sheath strands isolated from $C_4$ plants. *Plant Cell Physiol.*, **16**, 485–94.
55 UTTER M. E. & KOLENBRANDER H. M. (1972) Formation of oxaloacetate by $CO_2$ fixation on phosphoenolpyruvate. In *The Enzymes*. Vol. VI (ed. P. D. Boyer) pp. 117–68. New York: Academic Press.
56 WALKER D. A. (1960) Physiological studies on acid metabolism. 7. Malic enzyme from *Kalanchoe crenata*; effects of carbon dioxide concentration. *Biochem. J.* **74**, 216–23.

# Chapter 13
# Photorespiration



## SUMMARY


Photorespiration, as the name implies, is respiration in the light. It can be detected in $C_3$ plants by a number of methods, and these indicate that it is apparently absent in $C_4$ plants. It is difficult to measure absolute levels of photorespiration, since under steady-state conditions in the light, photosynthesis and photorespiration are


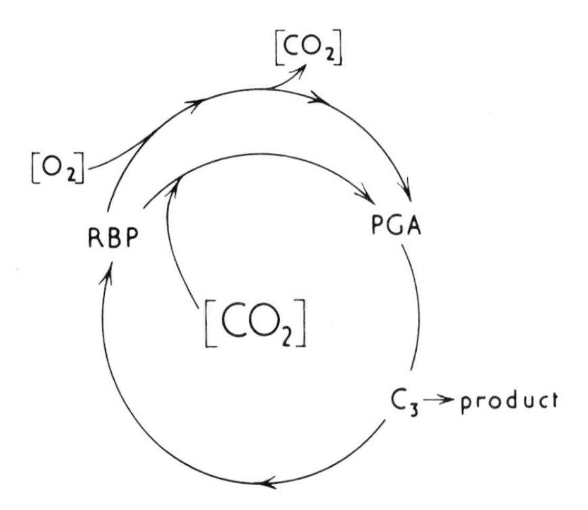


**Fig. 13.1** Photorespiration and $O_2$ inhibition of photosynthesis. $O_2$ and $CO_2$ are in competition for reacting with RBP (ribulose bisphosphate) of the RPP pathway. This provides two routes for carbon flow. Reaction with $CO_2$ leads to carbon assimilation while reaction with $O_2$ bypasses carboxylation and leads to loss of $CO_2$ in the light (photorespiration). The magnitude of photosynthesis versus $O_2$ inhibition of photosynthesis depends on the $CO_2/O_2$ ratio in the leaf and the relative affinity of the reactions for the two gases. $O_2$ inhibits photosynthesis by competing with $CO_2$ for RBP and by loss of $CO_2$ from photorespiration. In $C_4$ plants, the increased $[CO_2]$ in bundle sheath cells increases photosynthesis and reduces photorespiration.

occurring simultaneously. Glycolate is considered the primary substrate of this 'light-respiration.' The enzyme RBP carboxylase-oxygenase is a branch point between photorespiratory and photosynthetic metabolism. Oxygen, reacting with RBP, leads to glycolate synthesis and photorespiration whilst $CO_2$, reacting with RBP, leads to photosynthesis. If the $O_2/CO_2$ ratio becomes sufficiently high, plants will photo-respire as rapidly as they photosynthesize (this occurs, for example, at the $CO_2$ compensation point). Conversely, when the $CO_2/O_2$ ratio is high, as is proposed in bundle sheath cells during $C_4$ photosynthesis, the photorespiratory activity will be relatively low. Oxygen and $CO_2$ show competitive interactions as substrates for RBP carboxylase-oxygenase (i.e. $O_2$ inhibits the carboxylase, $CO_2$ inhibits the oxygenase). Thus, the $O_2$ inhibition of photosynthesis in $C_3$ plants is due to $O_2$ inhibition of carboxylation and to glycolate synthesis which leads to photo-respiration.

The energy requirement per $CO_2$ fixed increases in $C_3$ plants with increasing $O_2$ (i.e. $O_2$ decreases the quantum yield). Whether $O_2$ or $CO_2$ serves as a substrate for RBP carboxylase-oxygenase, PGA is an ultimate product which is reduced and metabolized through the RPP pathway with regeneration of RBP. Oxygen causes a shunt around the carboxylation step and as more $O_2$ reacts more energy is required per $CO_2$ fixed. The primary reason for the decreased quantum yield with $O_2$ is the additional flow of carbon through the RPP pathway per $CO_2$ fixed, rather than energy utilized in the glycolate pathway. $C_3$ and $C_4$ plants have a similar quantum yield for $CO_2$ fixation under atmospheric levels of $O_2$ at 30° C. $C_3$ plants use additional energy per $CO_2$ fixed due to the effects of $O_2$ inhibition, while $C_4$ plants require additional energy in the $C_4$ cycle.

During photorespiratory metabolism, it is proposed that a massive, but poorly understood, intracellular transport of metabolites takes place between chloroplasts, peroxisomes and mitochondria. The role of photorespiration is uncertain. $C_3$ plants clearly benefit when photorespiration is experimentally reduced by $CO_2$ enrichment.

## 13.1 Introduction

In the dark, the majority of green plants take up $O_2$ and evolve $CO_2$. In CAM this net exchange is masked by the extent of dark $CO_2$ fixation. Uptake of $O_2$ and evolution of $CO_2$ results from respiratory consumption of metabolites. In the typical green leaf, the principal respiratory substrate is likely to be hexose phosphate derived from starch, but in seeds and other long-term storage organs, lipids and proteins may also be consumed. Sugar phosphates and the like are degraded by glycolysis, the oxidative pentose phosphate pathway and the tricarboxylic acid cycle. Reduced coenzymes are reoxidized and as electrons pass to $O_2$ through a series of cytochromes and other carriers, energy is conserved as ATP and made available for other metabolic processes. Similarly, metabolites are drained from the various pathways to be used in the synthesis of new cellular components. In the light, some of these processes undoubtedly continue although it is unlikely that they do so entirely unchanged and uninfluenced by the photosynthetic processes which are then occuring in adjacent compartments. In addition, there are new reactions in which $O_2$ is taken up and $CO_2$ is evolved. These occur only in the light and are collectively referred to as 'photorespiration' to distinguish them from 'dark respiration' although the former term is sometimes used to embrace all of the respiratory events which take place in the light. Just as the gaseous exchange resulting from dark respiration can be modified or masked by other metabolic processes, photo-respiratory gaseous exchange diminishes the $O_2$ evolution and net $CO_2$ fixation associated with photosynthesis. Particularly in whole plants, cells, or protoplasts, the term 'apparent photosynthesis' may be applied to gaseous exchange which has not been corrected for respiratory contributions. 'True photosynthesis' is then applied to values for which such corrections have been made.

## 13.2 Recognition

### (a) $CO_2$-BURSTS

Photorespiration is not as readily recognized as 'dark' respiration because of the masking effect of photosynthesis. An obvious procedure is to terminate photo-synthesis at its maximum and to see whether 'dark' respiration resumes unchanged or whether there are transient readjustments. When this was done by Decker, and subsequently by others, they observed a $CO_2$ burst. Thus, if photosynthesis is allowed to proceed in a closed vessel until the $CO_2$ concentration has reached a steady-state and the illuminated plant is then plunged into darkness, $CO_2$ is initially released at a high rate and only falls to the level associated with normal 'dark' respiration after an interval. This clearly suggests the operation of some process in the light which does not stop immediately upon darkening. In itself, however, this could be as readily interpreted as an effect of light on 'dark' respiration as the temporary persistence of a 'light' process. The view that it is, in fact, the latter gains credence from the fact that it is greatly diminished if the surrounding $O_2$ concentration is lowered to about 2%. Dark respiration is largely unaffected by decreases in $O_2$ of this magnitude since cytochrome oxidase, which plays a major role in 'dark' respiration has a very high affinity for $O_2$. It may be noted, however, that in this sort of experiment, the preferred conditions (e.g., low $CO_2$) are often those which are believed to favour photorespiration. Under conditions of high $CO_2$

when photorespiration is thought to be repressed (and $CO_2$ out-bursts reduced), transient 'gulps' in $O_2$ can still be observed in passing from light to darkness. Their cause remains to be established but, *a priori*, there seems to be no reason to doubt that they reflect either a transient readjustment in the pool size of metabolites or an effect of light on 'dark' respiration or both. This emphasizes the dangers of interpreting all such transients in terms of photorespiration.

### (b) Release of $CO_2$

If a plant is allowed to fix $CO_2$ in the light and the atmosphere is then replaced by $CO_2$-free air, some $CO_2$ is released into the atmosphere. Although the release of $CO_2$ may decline after some minutes, it is usually faster than in the dark. Such experiments may also be carried out with $^{14}CO_2$. The specific activity of the $^{14}CO_2$ released is then greater than it is in the dark implying that the substrate is different and more newly formed (on the assumption that the tracer will be progressively more diluted as it passes from the photosynthetic cycle into existing storage pools). This, like most methods, is only a qualitative measure of photorespiration since transferring plants to low $CO_2$ in the light favours photorespiration over photosynthesis.

### (c) $CO_2$ Compensation point ($\Gamma$)

When $C_3$ plants are illuminated in small enclosed spaces, they will diminish the $CO_2$ concentration to somewhere in the region of 50 ppm (compared to the 326 ppm in normal air). At this concentration of $CO_2$, the compensation point, the photosynthetic uptake is believed to exactly balance, or 'compensate' for, the respiratory loss. In low $O_2$ (2% or less), the $CO_2$ is decreased to near zero. The difference in compensation point between the two conditions may therefore be regarded as an indication of photorespiration. Photorespiratory release of $CO_2$ in high $O_2$ is assumed to be sufficiently vigorous to displace the compensation point in an upward direction whereas dark respiration, which is less $O_2$-sensitive, can be almost entirely masked even when photosynthesis is doing little more than refixing $CO_2$ which would otherwise be released to the exterior. $C_4$ plants, which do not seem to photorespire, have very low $CO_2$ compensation points (usually 0–5 ppm), and these are largely independent of the external $O_2$ concentration.

### (d) $O_2$ uptake

In early studies on $^{18}O_2$ uptake with algae and higher plants, relatively high $CO_2$ (well above atmospheric levels) and below atmospheric levels of $O_2$ were often used.

These conditions are now known to be unfavourable for photorespiratory metabolism. Recently $^{18}O_2$ has been used to study the mechanism and the pathway of photorespiration. $^{18}O_2$ is incorporated into the carboxyl group of glycolate and $^{18}O$ is subsequently found in the carboxyl group of glycine and serine. The rate of $^{18}O_2$ uptake in the light is much higher than the rate in the dark. As the $[CO_2]$ is increased, the amount of $^{18}O_2$ incorporated during photosynthesis of $C_3$ plants is decreased, which is consistent with suggestions that $CO_2$ represses the rate of photorespiration (Section 13.10). Uptake of $^{18}O_2$ during photorespiration should occur through RBP oxygenase, glycolate oxidase and glycine oxidase when the latter is coupled to mitochondrial phosphorylation. Considerable uptake of $^{18}O_2$ also occurs in the light in $C_4$ plants such as maize although this may not be linked to photorespiratory metabolism. In algae during the induction period of photosynthesis Radmer and Kok found that the uptake of $^{18}O_2$ and evolution of $^{16}O_2$ occurred at linear and equivalent rates (resulting in no net $O_2$ exchange) in the absence of $CO_2$ fixation. As $CO_2$ fixation reached a steady-state rate, the incorporation of $^{18}O_2$ was repressed. Such results were taken as evidence for pseudocyclic electron transport, with $O_2$ and $CO_2$ competing for reductive power. It is, therefore, difficult to equate the rate of release of photorespiratory $CO_2$ to the uptake of $O_2$ in the light, since $^{18}O_2$ uptake may also be linked to the electron transport reactions (the Mehler reaction, Section 5.8) and to mitochondrial respiration (not linked to the glycolate pathway).

### (e) Apparent photosynthesis as a function of $[CO_2]$

Above the $CO_2$ compensation point and under otherwise favourable conditions, the rate of photosynthesis increases with increasing $[CO_2]$. At first, there is a near linear relationship and if this line is extrapolated backwards, the intercept on the vertical axis is negative. Its value indicates the occurrence of photorespiration. At concentrations of $CO_2$ above those in air, the rate of photosynthesis in $C_3$ plants continues to increase until $CO_2$ saturation is reached. Conversely, photosynthesis in $C_4$ plants is much less influenced by $[CO_2]$ in excess of 300 ppm; under $N_2$, the difference between $C_3$ and $C_4$ is much less marked (Sections 14.3 & 14.4). Similarly, photosynthesis in many $C_3$ plants is photosaturated at about 30% of full sunlight whereas the rate of photosynthesis displayed by $C_4$ plants continues to increase more or less indefinitely with increasing light. Again, the difference between the two groups tends to become much less marked under $N_2$ and again, all of these effects may be interpreted as a consequence of photorespiration in $C_3$ species.

Thus, there are a number of methods which, taken together, provide evidence for photorespiration. A quantitative measure of the process under steady-state

photosynthesis is extremely difficult (Section 13.10). Transients in exchange of gases under non-steady-state conditions, or under non-physiological levels of $CO_2$ and $O_2$, may exaggerate or underestimate the magnitude of photorespiration in comparison to that during normal photosynthesis.
[Refs. 1, 2, 4, **13, 14, 21, 23, 56**]

### 13.3 Origin of glycolate

Glycolate was recognized as a product of photosynthesis in the studies of Calvin and colleagues around 1950. It is synthesized in the light and is a product of carbon assimilation in the RPP pathway. There have been multiple suggestions both for the synthesis and the subsequent metabolism of glycolate during photorespiration (see references on general reading at end of Chapter). However, only one mechanism of synthesis and one pathway of glycolate metabolism (Section 13.4) appear to account for most of the known aspects of photorespiration in higher plants. Thus, Bowes, Ogren and Hageman found that RBP carboxylase could also serve as an 'oxygenase', giving rise to phosphoglycolate as well as phosphoglycerate. Phosphoglycolate would then be converted to glycolate by phosphoglycolate phosphatase, a chloroplastic enzyme.

$O_2 + \text{RBP} \rightarrow \text{P-glycolate} + \text{PGA}$ (RBP oxygenase) Eqn. 13.1

$\text{P-glycolate} \rightarrow \text{Pi} + \text{glycolate}$ (phosphoglycolate phosphatase) Eqn. 13.2

$CO_2 + H_2O + \text{RBP} \rightarrow 2\ \text{PGA}$ (RBP carboxylase) Eqn. 13.3

It is important to note that reactions 13.1 and 13.3 are catalyzed by the same enzyme. $O_2$ is both a competitive inhibitor of RBP carboxylase and a substrate for the enzyme through RBP oxygenase activity. In the same manner, $CO_2$ is a competitive inhibitor of the oxygenase reaction and a substrate for the enzyme through RBP carboxylase. The lowest reported values for $K\text{m}_{(O_2)}$ for the oxygenase and $K\text{i}_{(O_2)}$ for the carboxylase are around 200–400 $\mu$M $O_2$. The lowest reported values of $K\text{m}_{(CO_2)}$ for carboxylase and $K\text{i}_{(CO_2)}$ for the oxygenase reaction are 6–15 $\mu$M. [*K*m is the Michaelis constant. It is the concentration of substrate at which the velocity of a reaction is $\frac{1}{2}V\text{max}$. *K*i is the dissociation constant for *EI*, the enzyme-inhibitor complex.

$$EI \rightleftarrows E + I$$

$$K\text{i} = \frac{[E]\,[I]}{[EI]}$$

Thus the greater the affinity of the inhibitor for the enzyme, the lower is the value of

*K*i]. These concentrations of $CO_2$ and $O_2$ for the respective *K*m values are within the range of solubility of atmospheric $CO_2$ and $O_2$ in water (Section 14.2). The relative *V*max of the oxygenase appears to be about a quarter of that of the carboxylase. Thus with $O_2$ and $CO_2$ concentrations around the *K*m values for the enzyme, the carboxylation reaction would be about four-fold higher than the oxygenase reaction. Glycolate synthesis is known to be favoured by high $O_2$ and relatively low concentrations of $CO_2$. These same conditions are most favourable for photo-respiration and $O_2$ inhibition of photosynthesis.

Through RBP carboxylase-oxygenase activity, glycolate synthesis is dependent on both $CO_2$ and $O_2$. Some fixation of $CO_2$ is required in order to regenerate RBP, the substrate for the oxygenase reaction. Without $CO_2$ fixation, a coupling of the glycolate pathway and RPP pathway would lead only to photorespiration and rapid metabolite depletion in the pathways (Fig. 13.2). (See Section 13.6 for details of metabolism in the glycolate pathway).

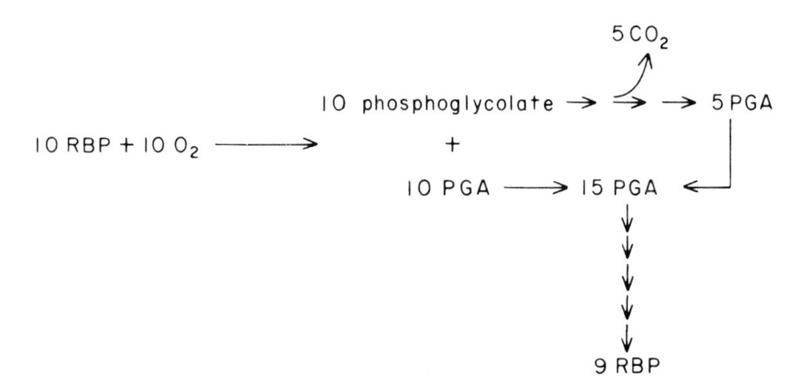


**Fig. 13.2.** Loss of carbon from RPP pathway by photorespiration.

Under-water stress where the intercellular $[CO_2]$ may be low, some stored carbon in the plant may be metabolized into the RPP and glycolate pathways to prevent substrate depletion. This will result in net loss of carbon from the plant through photorespiration of its endogenous carbon. If a $C_3$ plant and $C_4$ plant are both put in a closed chamber in the light, the $C_3$ plant will die from loss of carbon via photorespiration while the $C_4$ plant will scavenge the $CO_2$ and remain healthy at least for a limited period.

With chloroplasts isolated from $C_3$ plants, glycolate is a major product of photosynthesis under 21% $O_2$ and low $CO_2$ (e.g. 0.4 mM $HCO_3^-$ corresponding to about 8 μM $CO_2$ at pH 8.0). Both RBP oxygenase and phosphoglycolate phos-

phatase are chloroplastic enzymes. The enzymes for further metabolism of glyolate are found outside the chloroplast in the peroxisomes and mitochondria (Section 13.7).
[Refs. 1–8, **11, 12, 18–20, 22, 32, 40, 48**].

## 13.4 The glycolate pathway in $C_3$ plants

The glycolate pathway is the proposed route for metabolism of glycolate to triose phosphate. Initial methods which led to the present concept of the pathway involved the use of labelled substrates with algae or whole leaf tissue. Such studies were made primarily in the 1950s and 60s by Tolbert, Zelitch, Waygood, Whittingham and their colleagues. $^{14}C$-labelled glycolate, glycine, or serine was fed to leaf tissue and the products of their metabolism were analyzed. The products of metabolism of $^{14}C$-glycolate in the light included glycine, serine, glycerate, and sucrose. Products of $^{14}C$-glycine metabolism in the light included serine, glycerate, and sucrose. All three of the labelled metabolites were rapidly metabolized to sucrose and starch. This led to the conclusion that the glycolate pathway has a gluconeogenic function. [Gluconeogenesis is the synthesis of sugars, e.g., glucose, via reversed glycolysis.]

In other studies, products of $^{14}CO_2$ fixation have been determined in the presence or absence of potential inhibitors of glycolate metabolism. For example, Zelitch found a light-dependent accumulation of glycolic acid in tobacco leaf tissue when $\alpha$-hydroxysulfonate, an inhibitor of glycolate oxidase, was added. In the presence of isonicotinyl hydrazine (INH, an inhibitor of transaminases), Whittingham, *et al.* found up to an eight-fold increase in incorporation of $^{14}CO_2$ into glycolate and glycine by *Chlorella* without an effect on the rate of $CO_2$ fixation. INH is now known to strongly inhibit conversion of glycine to serine (Eqns. 13.8 & 13.9). Aminoacetonitrile is a competitive inhibitor of glycine oxidation with isolated mitochondria. Under low concentrations of $H^{14}CO_3^-$ it inhibits photosynthesis of wheat protoplasts and causes a large increase of labelling of glycine, while at high $HCO_3^-$ concentrations it has no effect on photosynthesis. Butyl 2-hydroxy-3-butynoate is an irreversible inhibitor of glycolate oxidase which causes glycolate accumulation and inhibits photosynthesis under 21% $O_2$ but not under nitrogen. Sometimes an inhibitor, at first thought to be relatively specific, is later found to have multiple sites of action and considerable effort is often expended in determining its side effects.

More recent studies show that $^{18}O_2$ is incorporated into the carboxyl position of glycolate and not into PGA through RBP oxygenase. $^{18}O_2$ fed to whole leaves in the light is rapidly incorporated into the carboxyl position of glycine and serine. These

results are consistent with the proposed synthesis and subsequent metabolism of glycolate (Sections 13.5 & 13.6).
[Refs. 1–8, **45, 52, 58, 63, 66, 69**]

## 13.5 Simplified scheme of carbon flow in the glycolate pathway

From studies with intact organisms using labelled metabolites and from studies on enzymes, a condensed scheme of carbon flow in the glycolate pathway in higher plants can be given (Fig. 13.3) (see Section 13.6 for details on enzymology). The main characteristics of this pathway are:

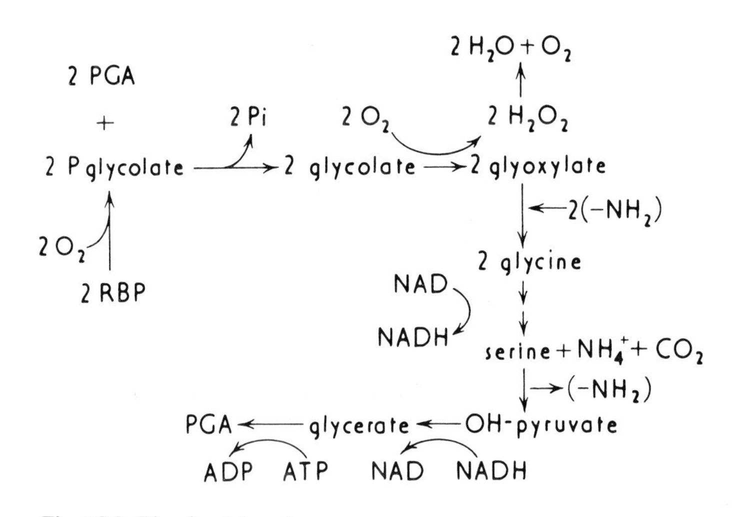


**Fig. 13.3.** The glycolate pathway.

1. $CO_2$ is generated during conversion of two molecules of glycine to serine.
2. Oxygen is consumed, both in glycolate synthesis and in its subsequent metabolism (i.e., through RBP oxygenase and glycolate oxidase). Also, if the NADH formed from glycine decarboxylation in the mitochondria is oxidized, this will result in further $O_2$ consumption.
3. The loss of $CO_2$ and consumption of $O_2$ occurring through glycolate metabolism in the light are, by definition, photorespiration.
4. Free ammonia is generated during metabolism in the pathway.
5. The glycolate pathway is gluconeogenic. PGA is a product which leads to sucrose and starch synthesis.

## 13.6 Reactions of the glycolate pathway and energy requirements

In the glycolate pathway, reactions resulting in the metabolism of 2 molecules of glycolate to PGA, a precursor for sucrose synthesis in higher plants, are as follows:

2 glycolate + 2 $O_2$ → 2 glyoxylate + 2 $H_2O_2$ (glycolate oxidase) Eqn. 13.4

2 $H_2O_2$ → 2 $H_2O$ + $O_2$ (catalase) Eqn. 13.5

glyoxylate + glutamate → glycine + 2-oxoglutarate Eqn. 13.6
(glycine aminotransferase)

glyoxylate + serine → glycine + hydroxypyruvate Eqn. 13.7
(serine-glyoxylate aminotransferase)

glycine + THF* + $NAD^+$ → NADH + $H^+$ + methylene-THF + $CO_2$ + $NH_3$
(glycine synthase) Eqn. 13.8

methylene-THF + glycine + $H_2O$ → serine + THF Eqn. 13.9
(serine hydroxymethyltransferase)

hydroxypyruvate + NADH + $H^+$ → $NAD^+$ + glycerate Eqn. 13.10
(glyoxylate reductase)**

glycerate + ATP → PGA + ADP * (glycerate kinase) Eqn. 13.11

---

**Sum:** 2 glycolate + $O_2$ + glutamate + ATP →
PGA + ADP + $CO_2$ + $NH_3$ + 2-oxoglutarate + $H_2O$ Eqn. 13.12

*THF = tetrahydrofolate

** This enzyme which is NAD-specific can catalyze reduction of either glyoxylate or hydroxypyruvate. According to the Enzyme Commission it is named glyoxylate reductase.

It is of interest that glycolate oxidase from *Pisum sativum* has a $Km_{(O_2)}$ of 170 $\mu$M. Likewise, RBP oxygenase has a high *Km* for $O_2$ (Section 13.3). Thus, relatively high $[O_2]$, e.g., 21% (see Section 14.2 for solubility of $O_2$), is required for glycolate synthesis and metabolism in comparison to dark respiration which tends to be saturated at low $O_2$ i.e. about 2%. For the pathway to continue, $NH_4^+$ must be reincorporated into glutamate. This might occur through glutamate dehydrogenase or a coupling of the glutamine synthetase-glutamate synthase reactions. In either case, the equivalent of one NADPH is required. The glutamine synthetase-glutamate synthase is the more likely pathway, as glutamate dehydrogenase has a very high *Km* for ammonia. (see also Section 12.9)

glutamate + ATP + $NH_3$ → glutamine + ADP + Pi Eqn. 13.13
(glutamine synthetase)

glutamine + 2-oxoglutarate + 2 Fd (reduced) →
2 glutamate + 2 Fd (oxidized) (glutamate synthase) Eqn. 13.14

[Ferredoxin (Fd) accepts one electron upon reduction. Thus, two electrons or 2 Fd (reduced) are required in converting ammonia to amino nitrogen.]

---

**Sum:** 2-oxoglutarate + $NH_3$ + 2 Fd (reduced) + ATP →
glutamate + 2Fd (oxidized) + ADP + Pi  Eqn. 13.15

Taking 2 Fd (reduced) as equivalent in energy to one NADPH, then the sum of Equations 13.12 and 13.15 give:

2 glycolate + $O_2$ + 2 ATP + NADPH + $H^+$ →
PGA + $H_2O$ + 2ADP + Pi + $CO_2$ + $NADP^+$  Eqn. 13.16

Thus, the overall energy requirements for metabolizing 2 glycolate to PGA by these reactions would be 1 NADPH and 2 ATP. However, the actual expense of energy in the glycolate pathway must be considered from its consequences on the energetics of $CO_2$ fixation in the RPP pathway. By way of an example, under atmospheric conditions, consider a ratio of carboxylase/oxygenase activity of 3.5, e.g., 3.5 $CO_2$/1 $O_2$ reacting with RBP, or 7 $CO_2$/2 $O_2$. (Note at the $CO_2$ compensation point the ratio would be 1 $CO_2$/2$O_2$', See Fig. 6.11).

7 $CO_2$ + 7 RBP + 7 $H_2O$ → 14 PGA  Eqn. 13.3
2 $O_2$ + 2 RBP → 2 PGA + 2 Pglycolate  Eqn. 13.1
2 Pglycolate + 2 $H_2O$ → 2 Pi + 2 glycolate  Eqn. 13.2
2 glycolate + $O_2$ + 2 ATP + NADPH + $H^+$ →  Eqn. 13.16
PGA + $H_2O$ + 2 ADP + Pi + $CO_2$ + $NADP^+$

---

**Sum:** 6 $CO_2$ + 9 RBP + 8 $H_2O$ + 3 $O_2$ + 2 ATP + NADPH + $H^+$ →
17 PGA + 2 ADP + $NADP^+$ + 3 Pi  Eqn. 13.17

17 PGA + 17 ATP + 17 NADPH + 17 $H^+$ →
17 TP* + 17 ADP + 17 $NADP^+$ + 17 Pi  Eqn. 13.18

15 TP + 9 ATP + 6 $H_2O$ → 9 RBP + 9 ADP + 6 Pi  Eqn. 13.19

Then: (Sum of Eqns. 13.17 to 13.19) =
6 $CO_2$ + 3 $O_2$ + 28 ATP + 18 NADPH + 18 $H^+$ + 14 $H_2O$ →
2 TP + 28 ADP + 26 Pi + 18 $NADP^+$  Eqn. 13.20

18 $NADP^+$ + 18 $H_2O$ → 18 NADPH + 18 $H^+$ + 9 $O_2$  Eqn. 13.21
28 ADP + 28 Pi → 28 ATP + 28 $H_2O$  Eqn. 13.22

---

**Sum:** (Eqns. 13.20 to 13.22) =
6 $CO_2$ + 2 Pi + 4 $H_2O$ → 2 TP + 6 $O_2$  Eqn. 13.23

*TP = triose phosphate

In this example, as summarized in Fig. 13.4, the energy requirements are:

28 ATP, 18 NADPH per 6 $CO_2$ or 4.67 ATP, 3 NADPH per $CO_2$ assimilated.

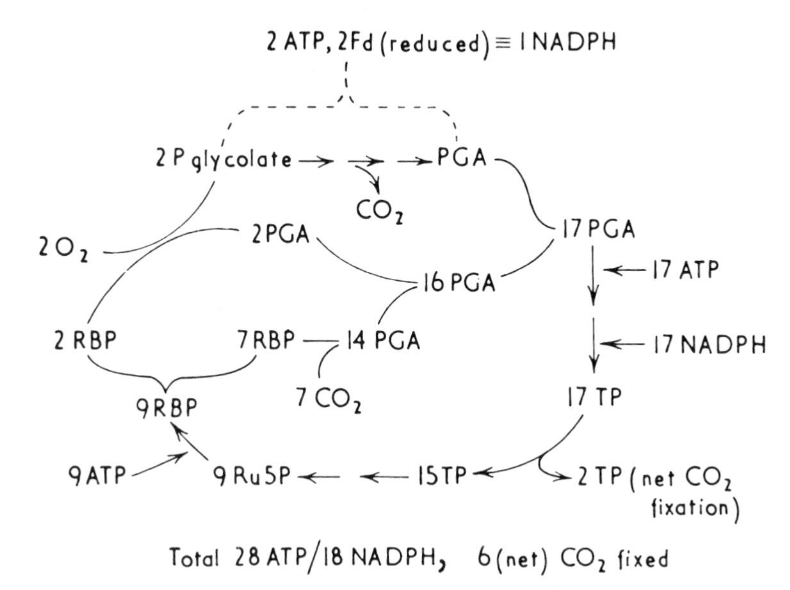


**Fig. 13.4.** Theoretical energy requirements during photosynthesis in a $C_3$ plant with a stoichiometry of 7 $CO_2/2O_2$ reacting with RBP carboxylase-oxygenase under atmospheric levels of $CO_2$ and $O_2$. See text for details.

It is of interest to consider the rates of ammonia assimilation and the energy requirements in relation to that for carbon assimilation. In the above example, for each 6 $CO_2$ fixed (net fixation), one ammonia would be released through photorespiration. The energy required for reassimilating this ammonia to glutamate is 1 ATP and equivalent of 1 NADPH, or 0.17 ATP, 0.17 NADPH per $CO_2$ fixed (through glutamine synthetase-glutamate synthase, Eqns. 13.13 & 13.14). This is a small part of the energy required for $CO_2$ fixation (Table 13.1). The competitive inhibition of RBP carboxylase by $O_2$ and $CO_2$ loss by photorespiration are the *major* causes for increased energy requirements in $C_3$ photosynthesis, rather than the energy required for reassimilating the ammonia.

Recycling ammonia in photorespiration results in no net gain of amino nitrogen to the plant. Plants utilize additional energy in converting nitrate to amino nitrogen for protein synthesis. The nitrogen content in organic compounds in leaves is about 6 % of the total carbon content (atom to atom basis). Conversion of nitrate to amino

Table 13.1. Summary of theoretical energy requirements in the RPP pathway and for nitrate assimilation in $C_3$ plants. Figures in parentheses represent the percentage of the total energy required for reassimilating ammonia and for nitrate metabolism.

| Metabolism | Requirements per $CO_2$ fixed ATP | NADPH (or equivalent) |
|---|---|---|
| RPP pathway (no $O_2$) | 3 | 2 |
| RPP pathway (21% $O_2$) | 4.67 | 3 |
| RPP pathway + nitrate metabolism to glutamate (21% $O_2$) | 4.73 | 3.3 |
| Proportion due to ammonia reassimilation from photorespiration (21% $O_2$) | 0.17 (3.6%) | 0.17 (5.2%) |
| Proportion due to nitrate metabolism to glutamate (21% $O_2$) | 0.06 (1.3%) | 0.30 (9.1%) |

acids through nitrate reductase, nitrite reductase, and glutamine synthetase-glutamate synthase requires 1 ATP and the equivalent of 5 NADPH (Section 12.9). With a nitrogen/carbon ratio in the leaf of 0.06/1, the energy requirements for $C_3$ photosynthesis plus the energy requirements for the nitrate metabolism would be: (4.67 ATP + 3 NADPH/$CO_2$) + 0.06(1 ATP + 5 NADPH/$NO_3^-$) = 4.73 ATP/3.3 NADPH per $CO_2$ fixed. The energy requirements for nitrate metabolism to amino nitrogen are thus relatively small in comparison to that required for $CO_2$ fixation (Table 13.1). Rates of ammonia conversion to amino nitrogen in $C_3$ leaves may be higher than that in $C_4$ leaves due to recycling of the ammonia released in photorespiration. With the example given, the total rate of ammonia assimilation in $C_3$ would be about 23% of the rate of $CO_2$ fixation (where the ratio of ammonia released/$CO_2$ fixed equals 1/6 or 17% and the rate of nitrate reduction is 6% of the rate of $CO_2$ fixation). With rates of photosynthesis in $C_3$ leaves of 100 to 200 $\mu$mol. $CO_2$ fixed $mg^{-1}$ Chl $h^{-1}$, corresponding rates of ammonia assimilation would be 23–46 $\mu$mol $mg^{-1}$ Chl $h^{-1}$. With a carboxylase/oxygenase ratio of 7/2, 6 $CO_2$ are fixed in the presence of $O_2$ (summarized in Fig. 13.4) whereas approximately 9 $CO_2$ could be fixed in the absence of $O_2$. This is a reasonable estimate of the increase in carboxylation when $O_2$ is removed under atmospheric conditions. The number of additional molecules of $CO_2$ reacting in the absence of $O_2$ is not necessarily equal to the decrease in oxygenase activity. The relative magnitude of the velocity of the carboxylase with and without $O_2$ and the velocity of the oxygenase depends on the concentration of the gases and kinetic constants (*V*max and *K*m) as indicated in Section 13.10 (Eqns. 13.33, 13.36 & 13.39). In this illustration the percentage inhibition of photosynthesis by $O_2$ would be:

$$\frac{9-6}{9}(100) = 33\%$$

This is similar to the percentage inhibition of leaf photosynthesis and the percentage decrease in quantum yield by 21 % $O_2$ compared to 2 % $O_2$ (Sections 13.11 & 13.12). In this case under atmospheric conditions photorespiration would account for 1/3 and competitive inhibition for 2/3 of the total inhibition of photosynthesis by $O_2$. Although it is difficult to measure the relative magnitude of the two components *in vivo*, this is a good approximation (see Section 13.10).

With a given absorbance of energy by the leaf, $O_2$ decreases $CO_2$ fixation. This results in an increased energy requirement per $CO_2$ fixed. If 3 ATP and 2 NADPH are taken as the energy requirements per $CO_2$ fixed in the RPP pathway under 2 % $O_2$, then a 33 % decrease in quantum yield (or inhibition of photosynthesis by $O_2$, Section 13.12) by 21 % $O_2$ would increase the energy requirements to:

3 ATP/2 NADPH per 0.67 $CO_2$ fixed or 4.48 ATP/3 NADPH per $CO_2$ fixed.

In this case, 1/3 of the energy is apparently diverted to metabolism associated with $O_2$ inhibition of photosynthesis.

Thus, there is close agreement between these two means of estimating the energy requirements, even though the balance between energy utilization and energy generation in the glycolate pathway is not fully understood (Section 13.7). While it is certain that $O_2$ increases the energy requirement per $CO_2$ fixed in $C_3$ plants (Section 13.12), the exact nature of the loss of energy is uncertain. The loss of absorbed energy in the presence of $O_2$ could occur by increased fluorescence (perhaps due to competitive inhibition of photosynthesis by $O_2$) or through utilization of photochemically generated energy in photorespiration. Whole leaf measurements of $O_2$ inhibition of quantum yield do not distinguish between these two means of energy loss.
[Refs. 5, **26, 33, 40, 41**]

## 13.7 Intracellular localization of enzymes of the glycolate pathway

The glycolate pathway in plants was proposed in the early 1960s. In 1968, Tolbert and associates isolated organelles from leaves which they called peroxisomes. Peroxisomes were found to contain some key enzymes of the glycolate pathway, including glycolate oxidase, catalase, and glyoxylate reductase (the latter functioning to reduce hydroxypyruvate to glycerate in the peroxisome, E.C. 1.1.1.26). These organelles are classified as a type of microbody. Peroxisomes are 0.5–1.5 microns in diameter and in comparison to mitochondria and chloroplasts, are surrounded by a single rather than a double membrane (Fig. 13.5). Centrifugation techniques are used to separate these three types of organelles.

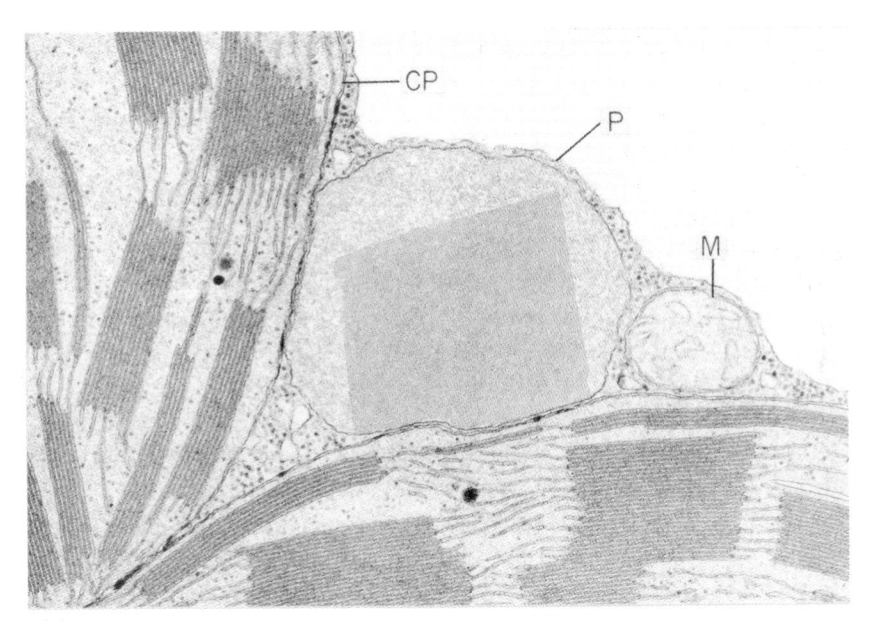


**Fig. 13.5.** Electron micrograph showing leaf peroxisome (P), mitochondrion (M) and parts of two chloroplasts (CP) in the mesophyll cell of a young leaf of tobacco (*Nicotiana tabacum* L.) × 48 000. The large crystalline inclusion in the peroxisome may be catalase. Micrograph by Sue E. Frederick, courtesy of E. H. Newcomb, University of Wisconsin.

Chloroplasts, due to their greater size, have a higher sedimentation velocity than mitochondria or peroxisomes. Thus, by differential centrifugation (sequential centrifugation of leaf or protoplast extracts at increasing gravitational forces), a fraction enriched in chloroplasts can be obtained (500–1000 g for several minutes). Similarly, if leaf or protoplast extracts are layered on a sucrose density gradient and centrifuged for a relatively short period (e.g., 15 min), intact chloroplasts can be separated from mitochondria + peroxisomes. Chloroplasts, having a higher sedimentation velocity, will move further into the gradient than these other Organelles (e.g., see gradients developed by Leech and by Miflin and Beevers).

However, this method is not satisfactory for separating peroxisomes from mitochondria since their sedimentation velocities are similar.

A further method for separating organelles uses equilibrium-density gradient (isopynic) centrifugation. The leaf extracts are layered onto a sucrose density gradient and centrifuged until the organelles have equilibrated in the gradient according to their buoyant density. Thus, the separation of the organelles is based on differences in density rather than differences in sedimentation velocities. Peroxisomes have a higher buoyant density than intact chloroplasts while intact chloroplasts have a higher buoyant density than intact mitochondria. Typically, leaf extracts are layered onto a sucrose density gradient (about 20–60% w/w) which is ultracentrifuged with a swinging bucket rotor for 3–4 hours at approximately 100 000 g. Peroxisomes equilibrate at densities around 1.25, intact chloroplasts at 1.20–1.22, and mitochondria at densities of 1.18–1.20. The higher density of peroxisomes may result from a partial permeability to sucrose. The gradient is then fractionated (about 1ml fractions of a total of about 40 ml) and enzymes of interest are assayed. Marker enzymes can be used for each organelle, for example, cytochrome *c* oxidase for mitochondria, glycolate oxidase for peroxisomes, and NADP-triose phosphate dehydrogenase for intact chloroplasts. Equilibrium-density centrifugation was used by Breidenbach and Beevers in the discovery of glyoxysomes (a type of microbody found in seeds which store fats) and by Tolbert *et al.*, to isolate and purify peroxisomes.

Sucrose density gradients are useful for separating organelles and studying their enzyme compositions, but techniques for the purification of *functional* (i.e., metabolically active) peroxisomes, mitochondria, and chloroplasts from leaf tissue are limited.

The mitochondria have a key role in glycolate metabolism since they contain the enzymes which bring about decarboxylation of glycine to serine + $CO_2$. Therefore, the loss of $CO_2$ associated with photorespiration occurs in this organelle. The synthesis of glycolate and its subsequent metabolism depends on partial reactions of the pathway in chloroplasts, peroxisomes, and mitochondria and therefore on substantial metabolite transport between them.

The peroxisomes have the enzymes needed to convert glycolate to glycine (Eqns. 13.4–13.6) and serine to hydroxypyruvate to glycerate (Eqns. 13.7–13.10). Glycerate kinase is localized in the chloroplasts and is apparently the site of glycerate conversion to PGA.

In the pathway, glycolate is synthesized in the chloroplasts, transported to peroxisomes, and there converted to glycine. Glycine is then transported to mitochondria and decarboxylated to serine + $CO_2$ + $NH_3$ through glycine synthase—serine hydroxymethyltransferase (Eqns. 13.8 & 13.9). The serine is further metabolized to glycerate in the peroxisomes and the glycerate is metabolized to triose phosphate in the chloroplasts. Metabolite exchange of glycolate, glycine, serine and glycerate is proposed to occur in the peroxisomes and metabolite exchange of glycine, serine and $CO_2$ is proposed to occur in the mitochondria (Fig. 13.6). Whether such metabolite transport between peroxisomes and mitochondria is accomplished by diffusion or a carrier has not been determined. The glycolate pathway results in a net transfer of nitrogen from the peroxisomes to the

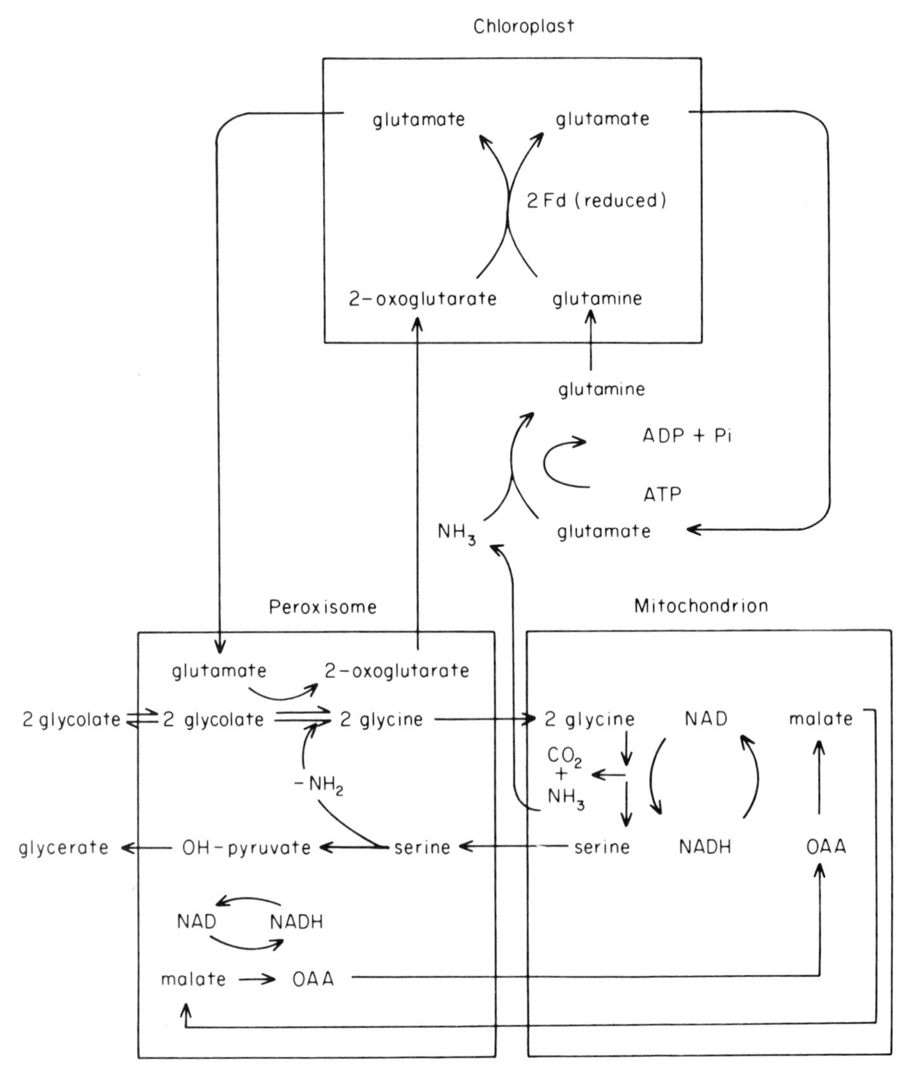


**Fig. 13.6.** Possible shuttle of metabolites between peroxisomes, mitochondria, and chloroplasts during metabolism in the glycolate pathway (see Section 13.6 for complete list of enzymes involved).

mitochondria. Thus, nitrogen must be recycled to the peroxisome in the form of an amino acid. The equivalent of one NAD(P)H is needed for amination of $NH_3$ to glutamate. One NADH is generated in the mitochondria through glycine decarboxylation and one NADH is utilized in the peroxisome through glyoxylate reductase. Thus, metabolism of two glycolate to glycerate requires a net input of reductive power (equivalent to one NADPH) which presumably comes from the chloroplasts. If additional reductive power for the glycolate pathway were generated from mitochondrial respiration, further loss of $CO_2$ would result. Therefore, the chloroplasts may provide reducing power either to convert hydroxypyruvate to glycerate or to convert ammonia to amino nitrogen.

Plant mitochondria have a NAD-glutamate dehydrogenase which may convert ammonia back to amino nitrogen. The reductive power generated from glycine decarboxylation could then be used to reassimilate ammonia.

$$2\text{ glycine} + NAD^+ + H_2O \rightarrow NADH + H^+ + \text{serine} + CO_2 + NH_3 \quad \text{Eqns. 13.8, 13.9}$$

$$NADH + H^+ + \text{2-oxoglutarate} + NH_3 \rightarrow NAD^+ + \text{glutamate} + H_2O \quad \text{Eqn. 13.24}$$

A more likely route for ammonia assimilation is via glutamine synthetase—glutamate synthase (also see Section 12.9), since the *K*m for ammonia of NAD-glutamate dehydrogenase is very high. There is evidence that glutamine synthetase is localized both in the chloroplast and cytoplasm whereas glutamate synthase is localized only in the chloroplast. Thus, as shown in Fig. 13.6, ammonia may be reassimilated in the cytoplasm and glutamate synthesized in the chloroplast. A transfer of assimilated nitrogen could then occur through a shuttle of glutamate–2-oxoglutarate between chloroplasts and peroxisomes. Reducing power generated in the mitochondria by glycine decarboxylation may be transferred to the peroxisomes for hydroxypyruvate reduction via an oxaloacetate–malate shuttle (Fig. 13.6).

There is evidence with mitochondrial preparations that the reductive power formed by glycine decarboxylation can be linked to oxidative phosphorylation. It is uncertain whether the reductive power formed from glycine decarboxylation *in vivo* is used directly to drive other reactions of the pathway, or whether it provides a source of ATP for alternative metabolic processes. In the latter case, the chloroplast might provide reductive power both for assimilation of ammonia and for conversion of hydroxypyruvate to glycerate.

It has been suggested that the glycolate pathway may constitute a major route for synthesis of sucrose. Sucrose synthesis is believed to occur in the cytosol of $C_3$ plants. Metabolism of triose phosphate to sucrose in the cytosol releases Pi, which can be taken up by the chloroplasts on the phosphate translocator in exchange for triose phosphate (Section 8.23). Glycerate, a product of the glycolate pathway, is thought

to be metabolized to triose phosphate in the chloroplasts. Triose phosphate synthesized in this manner would form a common pool with the triose phosphate synthesized directly in the chloroplasts through RBP carboxylase. Thus the triose phosphate synthesized during glycolate metabolism would be eventually used in the regenerative phase of the RPP pathway, for starch or for sucrose synthesis. This would not provide a direct or alternative route for sucrose synthesis from the PGA derived through glycolate metabolism.
[Refs. 4, 5, **28, 34, 41, 43, 47, 60, 61, 65, 68**]

## 13.8 Metabolism of isolated organelles in relation to the glycolate pathway

Isolated chloroplasts, mitochondria and peroxisomes can be used to test their proposed role in glycolate metabolism, including the mechanism of metabolite exchange and regulation of partial reactions of the pathway. Most studies on metabolism of the organelles have been performed with only partially purified preparations (see Section 13.7 for technical limitations).

### Chloroplasts

Chloroplasts synthesize glycolate and metabolize glycerate to triose phosphates. In low $HCO_3^-$ (0.4–1.0 mM) and high $O_2$ (21–100%), glycolate is a major product of chloroplast photosynthesis. There is no evidence that glycolate is photorespired or further metabolized by chloroplasts. When $^{14}C$-glycolate is supplied to isolated chloroplasts in the light, only occasional traces of label appear in glycine and serine. This may be due to cytoplasmic contamination of the chloroplasts. In addition, $O_2$ inhibits photosynthesis by isolated chloroplasts at low $HCO_3^-$ concentration, whilst high levels $HCO_3^-$ prevent $O_2$ inhibition. Thus, isolated chloroplasts, like photosynthetic cells or leaf tissue, show $O_2$ inhibition of photosynthesis although the chloroplasts lack photorespiration (see also Section 13.10).

Isolated chloroplasts of spinach ($C_3$ plant) show some capacity to metabolize glycerate. The chloroplasts show glycerate-dependent $O_2$ evolution in the light when assays are performed in the absence of $HCO_3^-$ to prevent $CO_2$ dependent $O_2$ evolution. The conversion of glycerate to triose phosphate requires 2 ATP and 1 NADPH.

$$\text{glycerate} + \text{ATP} \rightarrow \text{PGA} + \text{ADP} \qquad \text{Eqn. 13.11}$$
$$\text{PGA} + \text{ATP} \rightarrow \text{DPGA} + \text{ADP} \qquad \text{Eqn. 13.25}$$
$$\text{DPGA} + \text{NADPH} + \text{H}^+ \rightarrow \text{NADP}^+ + \text{TP} + \text{Pi} \qquad \text{Eqn. 13.26}$$

$$glycerate + 2\,ATP + NADPH + H^+ \rightarrow TP + Pi + 2\,ADP + NADP^+ \qquad \text{Eqn. 13.27}$$

There is no evidence for transport of glycolate or glycerate across the chloroplast envelope, other than by simple diffusion of the protonated acid (Chapter 8).

### PEROXISOMES

The fragility of isolated peroxisomes and the use of isopynic centrifugation in high osmolarity sucrose for their purification has made it difficult to isolate them intact and suitable for metabolic studies. There is some evidence that isolated peroxisomes metabolize $^{14}C$-glycolate to glyoxylate and glycine. Glycine becomes the major product of glycolate metabolism when amino donors glutamate, alanine, or serine are added.

### MITOCHONDRIA

Glycine decarboxylation with a stoichiometry of 2 glycine to 1 $CO_2$ + 1 serine + 1 $NH_3$ occurs in mitochondria isolated from spinach leaves (stoichiometry of Eqns. 13.8 + 13.9). It can be linked to electron transport with a stoichiometry of $CO_2$ evolution to $O_2$ uptake of 2/1, and oxidative photophosphorylation with a $P/2e^-$ ratio approaching 3 which is consistent with glycine decarboxylation being linked to $NAD^+$ reduction. The addition of oxaloacetate stimulates glycine decarboxylation by mitochondria, and this decarboxylation is not coupled to electron transport. This activity is apparently linked to reoxidation of NADH through malate dehydrogenase. In-vivo a dicarboxylate shuttle may transfer the reductive power from the mitochondria to the peroxisomes (Fig. 13.6)

$$2\ glycine + NAD^+ \rightarrow NADH + H^+ + serine + CO_2 + NH_3 \qquad \text{Eqns. 13.8 \& 13.9}$$

$$oxaloacetate + NADH + H^+ \rightarrow NAD^+ + malate \qquad \text{Eqn. 13.28}$$

There is currently no evidence that the reductive power formed during glycine respiration can be used to reassimilate the ammonia in the mitochondria through glutamate dehydrogenase.
[Refs. 4, **10, 24, 64, 68**]

## 13.9 The glycolate pathway in $C_4$ plants

All the methods for detecting photorespiration (Section 13.2) from whole leaf measurements indicate that photorespiration is absent or very low in $C_4$ plants. However, this is not due to a lack of enzymes of the glycolate pathway. Rather, it appears that metabolism through the glycolate pathway is repressed in comparison to $C_3$ plants. It is proposed that, as a result of the $C_4$ cycle, a relatively high level of $CO_2$ exists in bundle sheath cells during photosynthesis (Section 12.4). This would give a higher $CO_2/O_2$ ratio than that in $C_3$ plants, favouring photosynthesis and tending to repress synthesis of glycolate. However, some glycolate synthesis is likely through RBP oxygenase, since the bundle sheath compartment is expected to contain some $O_2$. This is especially true in $C_4$ species which have high photosystem II and $O_2$ evolution capacity in bundle sheath chloroplasts (Section 11.7). Oxygen inhibits light-dependent $CO_2$ fixation by isolated bundle sheath cells of $C_4$ plants. Thus, in the absence of the $C_4$ cycle, photosynthesis of bundle sheath chloroplasts, like that of $C_3$ plant chloroplasts, is sensitive to $O_2$.

In various $C_4$ species examined, about 90% or more of the glycolate oxidase and glyoxylate reductase are localized in bundle sheath cells. Phosphoglycolate phosphatase and catalase are also largely confined to bundle sheath cells. The activity of these enzymes in $C_3$ plants is, on a chlorophyll basis, several-fold higher than the activity in bundle sheath cells of $C_4$ species. Any $CO_2$ released from the glycolate pathway in $C_4$ plants would appear to occur primarily in bundle sheath cells. The $CO_2$ may then be refixed by RBP carboxylase in bundle sheath cells or by PEP carboxylase in the mesophyll cells. However, glycerate kinase is localized in mesophyll cells of $C_4$ plants. It appears that any glycerate synthesized via the glycolate pathway in bundle sheath cells must be transported to mesophyll cells for conversion to triose phosphates. In $C_4$ plants the RPP pathway and glycolate pathway are not directly linked as they are in $C_3$ plants.

Leaf extracts of $C_3$ plants have a high ratio of phosphoglycolate phosphatase/PGA phosphatase. Conversely, leaf extracts of $C_4$ plants have a high ratio of PGA phosphatase/phosphoglycolate phosphatase. In $C_4$ plants, the PGA phosphatase is primarily localized in mesophyll cells while the phosphoglycolate phosphatase is largely found in bundle sheath cells. The role of PGA phosphatase in $C_4$ plants is uncertain. One suggestion is that it facilitates shuttle of PGA from bundle sheath to mesophyll cells (Section 11.6). An alternative suggestion is that, together with glyoxylate reductase, it leads to serine synthesis in $C_4$ plants. A major part of the label which appears in serine following $^{14}CO_2$ fixation in $C_4$ plants is in the carboxyl position. Initially, during fixation of $^{14}CO_2$, PGA would be primarily carboxyl labelled. Subsequent metabolism of PGA to serine through glycerate and hydroxypyruvate would result in label in the carboxyl group of serine. In $C_3$ plants, the label from $^{14}CO_2$ fixation tends to give uniformly labelled serine and glycine. This is consistent with their synthesis through the glycolate pathway. During fixation of $^{14}CO_2$ in the RPP pathway, RBP becomes uniformly labelled. Reaction of $O_2$ with RBP then results in label in both carbons of glycolate. It is important to note that substantial labelling of the carboxyl carbon of serine, as in $C_4$ plants, does not establish serine synthesis from PGA as a major pathway. As $^{14}CO_2$ is fixed, the specific activity of PGA will be much higher than that of glycolate, such that labelling of serine from these two precursors would not be representative of the relative flux of carbon to serine through the two pathways.

Since the quantum yield is the same at 21% $O_2$ versus 2% $O_2$ in $C_4$ plants (Section 13.12), it can be argued that, under atmospheric levels of $CO_2$, there is little metabolism through the glycolate pathway. When $C_4$ plants are exposed to water stress which results in increased stomatal resistance, or at high temperatures, the intercellular $[CO_2]$ may decrease. At high temperatures the $[CO_2]$ in the aqueous phase will decrease due to lower solubility. In either case photosynthesis will be reduced due to a limited shuttle of $CO_2$ from the atmosphere to the bundle sheath cells. This is likely to increase the $O_2/CO_2$ ratio in these cells, increase the proportion of carbon flow to glycolate, and lead to more recycling of $CO_2$ (refixation through PEP or RBP carboxylase).

Perhaps under these conditions, with limiting $CO_2$, the quantum yield in $C_4$ would decrease. Even though there is no apparent loss of $CO_2$ from the $C_4$ leaf, a large amount of recycling requires additional energy per net $CO_2$ fixed. This should result in a measurable decrease in quantum yield.
[Refs. 1, 2, **31**, **36**, **50**, **57**, **64**]

### 13.10 $O_2$ inhibition of photosynthesis and its two components

In 1920, Warburg found that $O_2$ inhibits photosynthesis in *Chlorella*. Studies by Björkman, Downes and Hesketh and Krotkov *et al.* in the 1960's showed that the atmospheric level of $O_2$ was inhibitory to photosynthesis in certain higher plants (later to become known as $C_3$ plants) but not in others (later to become known as $C_4$ plants). In 1966, Forrester *et al.* and Tregunna *et al.* suggested that $O_2$ inhibition of photosynthesis has two components. One component was defined as $O_2$ inhibition through photorespiration. This was suggested from the fact that increasing $[O_2]$ causes an increase in the post-illumination burst and a linear increase in the $CO_2$ compensation point ($\Gamma$) (Fig. 13.7). The second component was defined as an $O_2$ inhibition of the efficiency of $CO_2$ fixation (carboxylation efficiency, *CE*). *CE* is determined from the initial slope of a plot of apparent photosynthesis (APS) versus increasing $[CO_2]$ (Fig. 13.8).

$$CE = \frac{\text{APS}}{[CO_2] - \Gamma} \qquad \text{Eqn. 13.29}$$

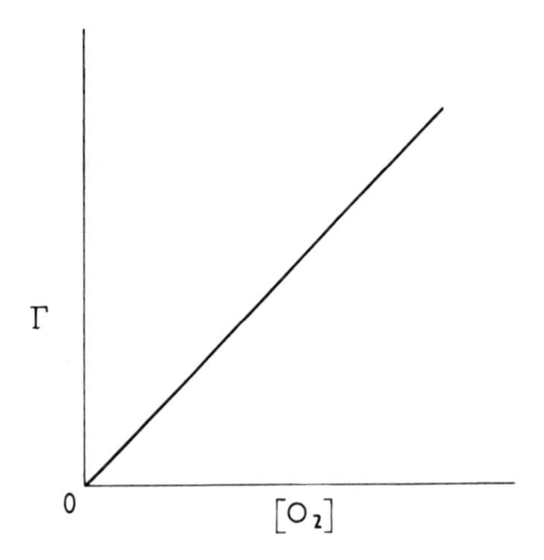


**Fig. 13.7.** Dependence of the $CO_2$ compensation point (Γ) on $O_2$ in a $C_3$ plant.

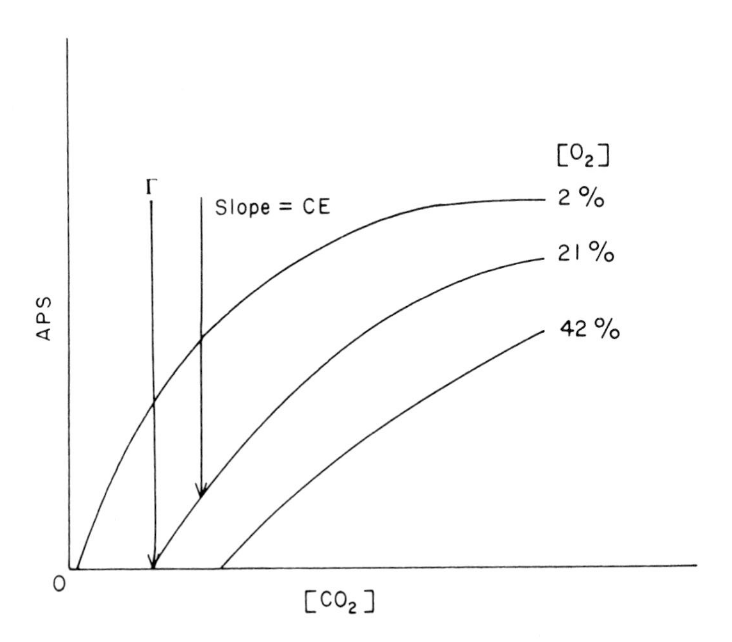


**Fig. 13.8.** Carboxylation efficiency (CE) in a $C_3$ plant.

Thus, the *CE* decreases with increasing $O_2$ concentration in a negative curvelinear manner (Fig. 13.9). If APS is plotted against intercellular $[CO_2]$ and *CE* is determined, the reciprocal of $CE(1/CE)$ is equal to $r_m$ (mesophyll resistance).

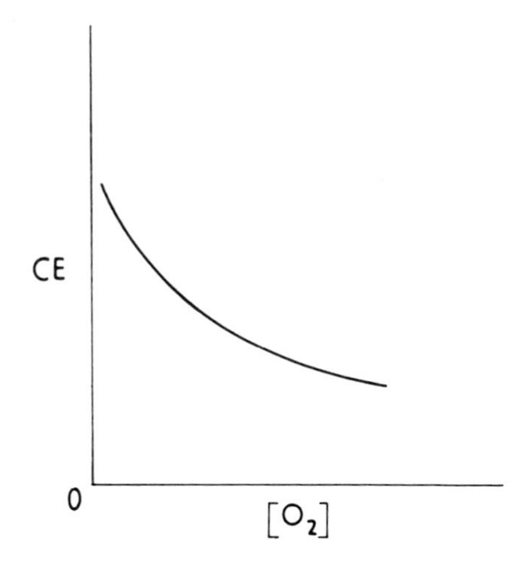


**Fig. 13.9.** Dependence of carboxylation efficiency (CE) on $[O_2]$ in a $C_3$ plant.

$$APS = \frac{[CO_2]_{int} - [CO_2]_{CP}}{r_m} \qquad \text{Eqn. 13.30}$$

$[CO_2]_{int}$ = intercellular $CO_2$ concentration in air space inside the leaf.
$[CO_2]_{CP}$ = concentration of $CO_2$ in chloroplast.

$r_m$ = mesophyll resistance (composed of transfer resistance through aqueous phase + carboxylation resistance).

If $[CO_2]_{CP}$ is taken as approximately $\Gamma$, then

$$APS = \frac{[CO_2]_{int} - \Gamma}{r_m} \qquad \text{Eqn. 13.31}$$

$$r_m = \frac{[CO_2]_{int} - \Gamma}{APS} = 1/CE \qquad \text{Eqn. 13.32}$$

Therefore, increasing $O_2$ causes a decrease in the *CE* or an increase in the mesophyll resistance. This reciprocal relationship between *CE* and $r_m$ only holds when *CE* is calculated on the basis of the intercellular $[CO_2]$ and when measurements of photosynthesis are made at relatively low $[CO_2]$ where the initial slope of photosynthesis versus $[CO_2]$ is linear.

Support for the idea of two components of $O_2$ inhibition was provided several years before biochemical evidence was obtained. One component is the direct inhibition of the carboxylase by $O_2$ while the second component is due to $CO_2$ loss in the glycolate pathway (Section 13.3).

Chemical inhibition of the glycolate pathway has been considered as a means to repress photorespiration and thus to stimulate photosynthesis. However, there is no evidence that inhibition of the glycolate pathway will reduce the competitive inhibition of photosynthesis by $O_2$. Although several inhibitors have been found to block certain steps of the pathway (Section 13.4), there is no conclusive evidence for increased photosynthesis when glycolate metabolism is inhibited. For example, butyl 2-hydroxy-3-butynoate, an effective inhibitor of glycolate oxidase, causes glycolate accumulation and inhibition of photosynthesis (Section 13.4). It may be essential that once glycolate is formed, it is largely metabolized to triose phosphates, thus returning carbon to the RPP pathway. Some scientists believe that a chemically modified RBP carboxylase-oxygenase with kinetic constants ($Km$ and $V$max), even less favourable for $O_2$ binding reactions, would result in higher rates of photosynthesis in $C_3$ plants. Others continue to think that photorespiration may offer some benefits under present environmental conditions. These aspects will be discussed in Section 13.13.

Now that the reason for $O_2$ inhibition of photosynthesis is better understood, leaf photosynthesis at varying $[CO_2]$ and $[O_2]$ can be analyzed in relation to biochemical mechanisms. Expressions for leaf photosynthesis can be written which include kinetic constants for the carboxylase and the two components of $O_2$ inhibition of photosynthesis, and these can be considered in relation to experimental measurements of leaf photosynthesis (see Section 14.3). For example, in the absence of $O_2$, if leaf photosynthesis at varying $[CO_2]$ follows Michaelis–Menten kinetics, then:

$$\mathrm{TPS}_{0\%\mathrm{O}_2} = \frac{V\mathrm{max}[\mathrm{CO}_2]}{[\mathrm{CO}_2] + K\mathrm{m}_{(\mathrm{CO}_2)}} \qquad \text{Eqn. 13.33}$$

where TPS = the true rate of photosynthesis, $V$max = maximum velocity of the carboxylase, $[CO_2]$ = concentration of $CO_2$ and $Km_{(CO_2)}$ = Michaelis Constant of the carboxylase for $CO_2$. In reciprocal form, Eqn. 13.33 can be written:

$$1/\mathrm{TPS}_{0\%\mathrm{O}_2} = 1/V\mathrm{max} + \frac{1}{[\mathrm{CO}_2]} \cdot \frac{K\mathrm{m}_{(\mathrm{CO}_2)}}{V\mathrm{max}} \qquad \text{Eqn. 13.34}$$

In double reciprocal plots of $TPS_{0\%O_2}$ versus $[CO_2]$ (Lineweaver-Burk plots), the intercept on the abscissa, when $1/TPS_{0\%O_2}$ is zero, is equal to $-1/Km$. An analysis of leaf photosynthesis of a $C_3$ plant at low $O_2$ and of a $C_4$ plant indicate that this approach can be used to estimate the $Km_{(CO_2)}$ (Chapter 14, Fig. 14.5).

In the presence of $O_2$ which inhibits $C_3$ photosynthesis, the following equation can be derived:

$$APS = V\text{max} \cdot \frac{[CO_2] - \Gamma}{[CO_2] + Km_{(CO_2)}(1 + [O_2]/Ki_{(O_2)})} \qquad \text{Eqn. 13.35}$$

This equation follows from relationships between APS, the velocity of the carboxylase ($Vc$), velocity of the oxygenase ($Vo$) and $\Gamma$ (see Farquahar *et al.*) where

$$Vc = \frac{V\text{max}[CO_2]}{[CO_2] + Km_{(CO_2)}(1 + [O_2]/Ki_{(O_2)})} \qquad \text{Eqn. 13.36}$$

$$Vc = \frac{V\text{max}[CO_2]}{[CO_2] + Km_{app(CO_2)}} \qquad \text{Eqn. 13.37}$$

$Km_{app(CO_2)}$ is the apparent $Km$ for $CO_2$ at varying $O_2$ concentrations where

$$Km_{app(CO_2)} = Km_{(CO_2)}(1 + [O_2]/Ki_{(O_2)}) \qquad \text{Eqn. 13.38}$$

In a similar manner

$$Vo = \frac{Vo\text{max}[O_2]}{[O_2] + Km_{(O_2)}(1 + [CO_2]/Ki_{(CO_2)})} \qquad \text{Eqn. 13.39}$$

where $Vo$ = velocity of oxygenase, $Vo\text{max}$ = velocity of oxygenase at saturating levels of substrate, $Km_{(O_2)}$ = Michaelis constant for $O_2$, and $Ki_{(CO_2)}$ = inhibitor constant for $CO_2$ (see Section 13.3). It follows that

$$APS = Vc - 0.5\ Vo \qquad \text{Eqn. 13.40}$$

At $\Gamma$, $Vc = 0.5\ Vo$, since one $CO_2$ is fixed in the carboxylase reaction per two glycolate formed in the oxygenase reaction. Metabolism of two molecules of glycolate in the glycolate pathway results in one $CO_2$ released.

Also, it follows that $\Gamma = \frac{0.5\ Vo}{Vc}[CO_2]$, and $0.5\ Vo = \frac{\Gamma Vc}{[CO_2]}$

Then $APS = Vc - \frac{\Gamma Vc}{[CO_2]}$

By substitution from equation 13.37 for $Vc$

$$APS = \frac{V\text{max}[CO_2]}{[CO_2] + Km_{app(CO_2)}} - \frac{\Gamma \frac{V\text{max}[CO_2]}{[CO_2] + Km_{(app(CO_2)}}}{[CO_2]}$$

Rearranging: $$APS = V\text{max} \times \frac{[CO_2] - \Gamma}{[CO_2] + Km_{app(CO_2)}} \qquad \text{Eqn. 13.41}$$

This equation accounts for the two forms of $O_2$ inhibition, one the photorespiratory component, by substracting $\Gamma$ in the numerator, and the other, the direct, competitive inhibition of the carboxylase by including $Km_{app(CO_2)}$ in the denominator. It tends to be valid where the concentration of RBP is saturating or remains constant.

In reciprocal form Eqn. 13.41 can be written as:

$$1/\text{APS} = \frac{1}{V\text{max}} + \frac{1}{[CO_2] - \Gamma} \times \frac{\Gamma + Km_{app(CO_2)}}{V\text{max}} \qquad \text{Eqn. 13.42}$$

Then in double reciprocal plots of 1/APS versus $1/([CO_2] - \Gamma)$, the intercept on the abscissa, when 1/APS is zero, is equal to $-1/(Km_{app(CO_2)} + \Gamma)$ (Fig. 13.10).

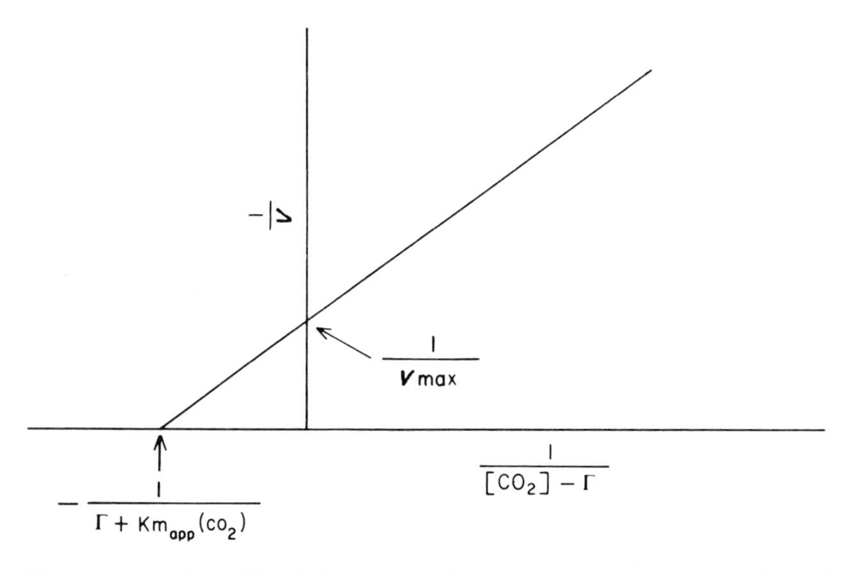


**Fig. 13.10.** Model describing leaf photosynthesis in a $C_3$ plant (from Eqn. 13.42 where 1/APS = $1/V$).

The total inhibition of photosynthesis by $O_2$, competitive inhibition by $O_2$ and inhibition due to photorespiration can be expressed as follows:

$TPS_{0\% O_2} - APS_{21\% O_2}$ = total inhibition by 21% $O_2$

$TPS_{0\% O_2} - TPS_{21\% O_2}$ = competitive inhibition by 21% $O_2$

$TPS_{21\% O_2} - APS_{21\% O_2}$ = inhibition due to photorespiration at 21% $O_2$

where: $TPS_{0\%O_2}$ = true rate of photosynthesis at 0% $O_2$
$TPS_{21\%O_2}$ = true rate of photosynthesis at 21% $O_2$
$APS_{21\%O_2}$ = apparent rate of photosynthesis at 21% $O_2$

In the form of equations:

$$TPS_{0\%O_2} = \frac{V\max[CO_2]}{[CO_2]+Km_{(CO_2)}} \qquad \text{Eqn. 13.33}$$

$$TPS_{21\%O_2} = \frac{V\max[CO_2]}{[CO_2]+Km_{app(CO_2)}} \qquad \text{Eqn. 13.43}$$

$$APS_{21\%O_2} = V\max\frac{[CO_2]-\Gamma}{[CO_2]+Km_{app(CO_2)}} \qquad \text{Eqn. 13.41}$$

From these equations response curves for APS and TPS versus $[CO_2]$ can be generated. Values of $V$max and $Km_{(CO_2)}$ can be determined from direct measurements on the carboxylase or calculated from plots of 1/APS versus $1/([CO_2]-\Gamma)$, while $\Gamma$ can be experimentally measured. Plots of 1/APS versus $1/([CO_2]-\Gamma)$ (analogous to 1/V versus 1/S in enzyme kinetics) at varying $[O_2]$ show that $O_2$ is a competitive inhibitor (see Fig. 14.5b) and that the $Km_{app(CO_2)}$ increases with increasing $[O_2]$ as expected for a competitive inhibitor (see Figs. 14.5b and 14.6). Deviations in whole leaf photosynthesis may occur at high $[CO_2]$ such that photosynthesis never reaches the $V$max of the carboxylase due to limitations on energy from photochemical reactions (see Fig. 14.6).

Experimentally it is easy to measure $APS_{21\%O_2}$ and roughly measure $TPS_{0\%O_2}$. The difference between the two measurements represents the total inhibition of photosynthesis by $O_2$. For $TPS_{0\%O_2}$ measurements are usually made at 1 or 2% $O_2$ since the stomata of some species tend to close without some $O_2$ in the atmosphere.* On the contrary, $TPS_{21\%O_2}$ is difficult to measure directly. If photorespiration could be measured, then $APS_{21\%O_2}$ + the rate of photorespiration would equal $TPS_{21\%O_2}$.

Most methods of measuring photorespiration are only of qualitative value and do not indicate actual rates of photorespiration under steady-state conditions at atmospheric levels of $CO_2$. One of the problems is that part of the photorespiratory $CO_2$ is being refixed by photosynthesis. Ludwig and Canvin developed the use of a double isotope method using $^{12}CO_2$ and $^{14}CO_2$ to measure the *apparent* rate of photorespiration under steady-state photosynthesis. $^{12}CO_2$ is fed under steady-state

* It should be noted that the $[O_2]$ in the chloroplast will be very close to that in the atmosphere even though $O_2$ is being evolved during photosynthesis. The efflux of $O_2$ out of the leaf will equal the rate of influx of $CO_2$. The maximum differential in concentration of $CO_2$ between the atmosphere and the chloroplast is about 0.03%. Therefore, considering that the path of resistance for diffusion of $CO_2$ and $O_2$ are the same, at most the $[O_2]$ in the leaf would be 0.03% higher than that in the atmosphere.

conditions and uptake measured with an infra-red gas analyzer to determine $APS_{21\% O_2}$. $^{14}CO_2$ is introduced and its rate of uptake measured simultaneously over 30 s in order to determine $TPS_{21\% O_2}$. This short period of $^{14}CO_2$ fixation is considered insufficient time for photorespiration of the $^{14}CO_2$ products. The difference between rates of $^{14}CO_2$ and $^{12}CO_2$ fixation is taken as the rate of photorespiration. This method indicates there are two components of $O_2$ inhibition of photosynthesis, one due to direct inhibition and the other due to photorespiration which may contribute about equally to the total inhibition of photosynthesis by $O_2$ under atmospheric conditions. However, the degree of internal recycling of photorespired $^{12}CO_2$ and photorespiration of assimilated $^{14}CO_2$ during the assay under varying conditions is uncertain. Thus, the procedure is probably of limited value for determining the relative magnitude of photorespiration, for example, at varying $[CO_2]$, varying $[O_2]$ and temperatures. In some reports with this method, the rate of photorespiration under 21 % $O_2$ was suggested to remain unchanged from low up to atmospheric $[CO_2]$. However, measurements of the post-illumination burst and calculations of rates to account for a non-steady-state system indicate that the magnitude of photorespiration progressively decreases with increasing $[CO_2]$ above $\Gamma$. $TPS_{21\% O_2}$ can then be estimated by adding the rate of the post-illumination burst at varying $[CO_2]$ to the rate of $APS_{21\% O_2}$. At atmospheric levels of $CO_2$, photorespiration was then estimated to account for about 1/3 of the total inhibition of photosynthesis by $O_2$. Experiments on the rate of $^{18}O_2$ uptake in the light also indicate that photorespiration decreases with increasing $[CO_2]$. Since $CO_2$ competes with $O_2$ for reaction through RBP carboxylase-oxygenase, it is logical that high $CO_2$ levels should inhibit photorespiration. During photo-respiration in the presence of $^{18}O_2$, an atom of $O_2$ is incorporated into the carboxyl group of phosphoglycolate (Eqn. 13.1) and subsequently into carboxyl groups of other metabolites of the glycolate pathway. It should be noted that label from $^{18}O_2$ can also be incorporated during photosynthesis by the Mehler reaction (see Section 5.8). Thus, uptake of $^{18}O_2$ in the light cannot be assumed to occur exclusively through photorespiration.

It is necessary to reconsider the meaning of carboxylation efficiency (*CE*) relative to Equations 13.33 and 13.41. As originally defined, *CE* is determined from the initial slope of the plot of APS versus $[CO_2]$ and as already noted, *CE* decreases with increasing $[O_2]$. In the absence of $O_2$ it follows from Equation 13.33 that at low $[CO_2]$, the concentration of substrate becomes negligible relative to $Km_{(CO_2)}$ such that $TPS_{0\% O_2} = V\max[CO_2]/Km_{(CO_2)}$. For this reason, the initial slope of photosynthesis at low $[CO_2]$ appears linear and $CE = TPS_{0\% O_2}/[CO_2] = V\max/Km_{(CO_2)}$. The decrease in *CE* with increasing $[O_2]$ is considered to be due to a direct competitive inhibition of photosynthesis by $O_2$. By definition, a competitive inhibitor decreases the initial slope of a velocity versus substrate curve by increasing the $Km$, such that $CE = V\max/Km_{app}$, where $Km_{app}$ increases with increasing concentration of inhibitor. In the presence of $O_2$, from Equation 13.41 the slope of the response curve for $APS_{21\% O_2}$ versus $[CO_2]-\Gamma = V\max/([CO_2] + Km_{app(CO_2)})$. As $[CO_2]$ approaches $\Gamma$ the slope is equal to $V\max/(\Gamma + Km_{app(CO_2)})$. Since $\Gamma$ is small relative to $Km_{app(CO_2)}$, the initial slope is indicative of the $CE$ [i.e. $V\max/Km_{app(CO_2)}$] (see Fig. 13.11).

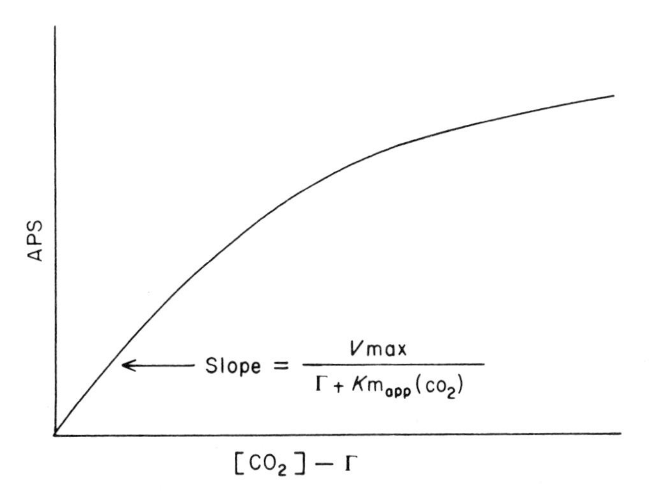


**Fig. 13.11.** Carboxylation efficiency, $APS/([CO_2]-\Gamma)$, as the $[CO_2]$ decreases and approaches $\Gamma$. (developed from Eqn. 13.41, see text).

Double reciprocal plots of photosynthesis versus $[CO_2]-\Gamma$ indicate that the $CO_2$ response curve follows Michaelis–Menten kinetics until high $[CO_2]$ are reached (Fig. 14.5). Also, analyses of whole leaf photosynthesis have been made which indicate that $O_2$ decreases $CE$ (or increases $r_m$) in a manner expected of a competitive inhibitor. The CE (or $V\max/Km_{app}$) at various $[O_2]$ can be determined from plots of 1/v versus $1/([CO_2]-\Gamma)$ as in Fig. 13.10. Plots of 1/CE versus $[O_2]$ are linear and the $Ki_{(O_2)}$ can be determined (Fig. 13.12) where

$$1/CE = \frac{Km_{app(CO_2)}}{V\max} = \frac{Km_{(CO_2)}}{V\max} + [O_2] \times \frac{Km_{(CO_2)}}{V\max\, Ki_{(O_2)}}$$

For further evaluation of whole leaf photosynthesis relative to carboxylation and photochemical potential of $C_3$ and $C_4$ plants see Chapter 14.
[Refs. 4, **13, 14, 15, 23, 26, 27, 29, 37, 46, 62**]

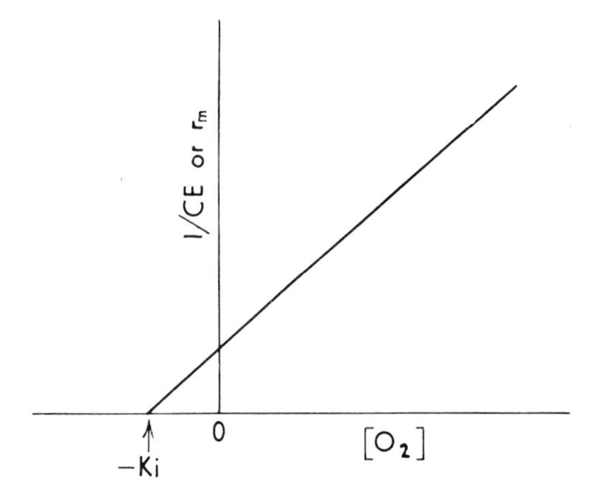


**Fig. 13.12.** Determining $Ki_{(O_2)}$ for photosynthesis *in vivo* in a $C_3$ plant.

## 13.11 Percentage inhibition of photosynthesis by $O_2$

$$\frac{APS_{2\%\,O_2} - APS_{21\%\,O_2}}{APS_{2\%\,O_2}}(100) = \%\text{ inhibition of photosynthesis by } O_2.$$

This is a good measure of the magnitude of $O_2$ inhibition of photosynthesis under varying conditions. With increasing $[CO_2]$,the percentage inhibition of photosynthesis by $O_2$ decreases from 100% at Γ, to nearly zero at 2- to 3-fold the atmospheric levels of $CO_2$ (approximately 0.1% $CO_2$). With increasing temperature, this inhibition by $O_2$ increases steadily (from 20–30% at 20°C to 40–60% at 35°C). This shows that photosynthesis of $C_3$ plants is more sensitive to $O_2$ at higher temperatures. The reason for this is uncertain, but it may be due to an increased $O_2/CO_2$ solubility ratio or a change in the kinetic constants for RBP carboxylase-oxygenase which favours the oxygenase reaction (see Chapter 14). $C_4$ photosynthesis, with its $CO_2$ concentrating mechanism, is adapted to overcome this high temperature sensitivity of $C_3$ plant photosynthesis to $O_2$ (Section 14.5). With increased light intensity, the absolute level of $O_2$ inhibition increases as does photosynthesis. However, the percentage inhibition of photosynthesis by $O_2$ shows little change, or a slight decrease, with increasing light intensity.
[Refs. **23, 27, 37, 39**]

## 13.12 $O_2$ effect on quantum yield

Typically, $O_2$ inhibition of photosynthesis in $C_3$ plants is measured at high light intensities. The influence of $O_2$ on quantum yield (QY) of photosynthesis (mol $CO_2$

fixed/mol quanta absorbed, Section 3.9) has also been studied. In the latter case, measurements are made at low light intensities of less than 5 % of full sunlight. At a given [$CO_2$] and temperature, the percentage inhibition of photosynthesis by $O_2$ measured at high light is equal to the percentage inhibition of QY. This shows that the percentage inhibition by $O_2$ is not more severe at high light intensities, as might be expected if its function was to dissipate excess energy. Even at very low light, $O_2$ decreases the efficiency of utilization of the absorbed energy. Increasing [$CO_2$] above atmospheric levels overcomes the $O_2$ inhibition of QY and the $O_2$ inhibition of photosynthesis (under high light). This indicates that the influence of $O_2$ is very dependent on the relative levels of $O_2$ and $CO_2$ (see Figs. 13.13 and 13.14). The decrease in QY by $O_2$ may be due to a loss of energy in $O_2$-dependent fluorescence and $O_2$-dependent photorespiration. An increased fluorescence emission in the presence of $O_2$ could result from $O_2$ inhibition of enzymes of the RPP pathway, which then reduces the utilization of photochemical energy.

With increasing temperature, the percentage inhibition of QY by $O_2$ increases in $C_3$ species. This is similar to the increased percentage inhibition of photosynthesis by $O_2$ seen at high light with increasing temperature. Similarly, this greater inhibition of QY at higher temperatures may be due to the increased $O_2/CO_2$ solubility ratio with increasing temperature or a change in the kinetic constants for $O_2$ and $CO_2$ favouring reactions with $O_2$ (Fig. 13.15) (Chapter 14). In $C_4$ species, the

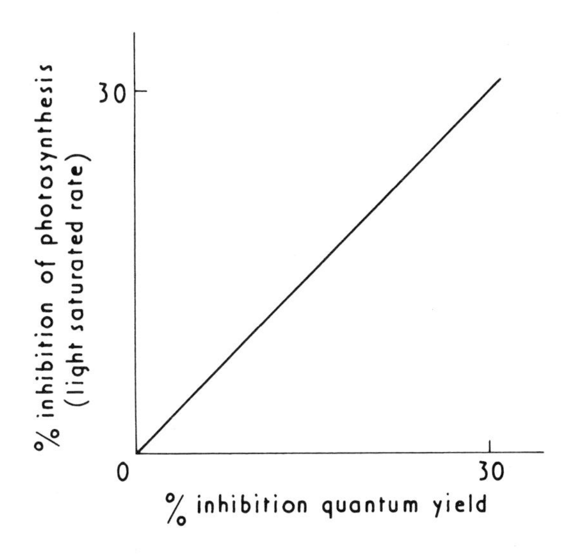


**Fig. 13.13.** Relationship between the percentage inhibition of quantum yield and percentage inhibition of photosynthesis under high light with increasing [$O_2$]. (After Björkman, 1966).

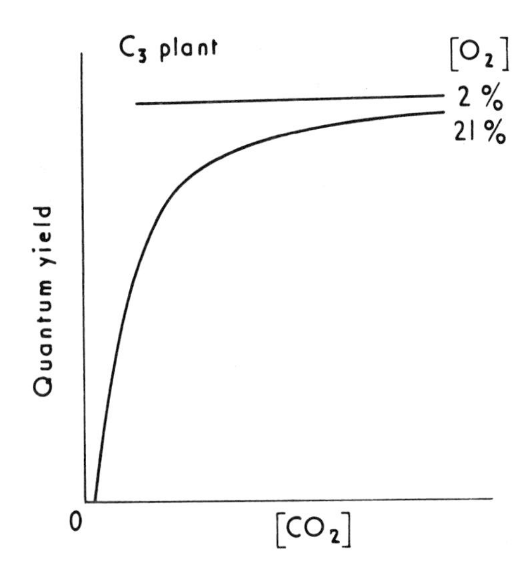


Fig. 13.14. High $[CO_2]$ overcomes $O_2$ inhibition of quantum yield. (after Ehleringer & Björkman, 1977).

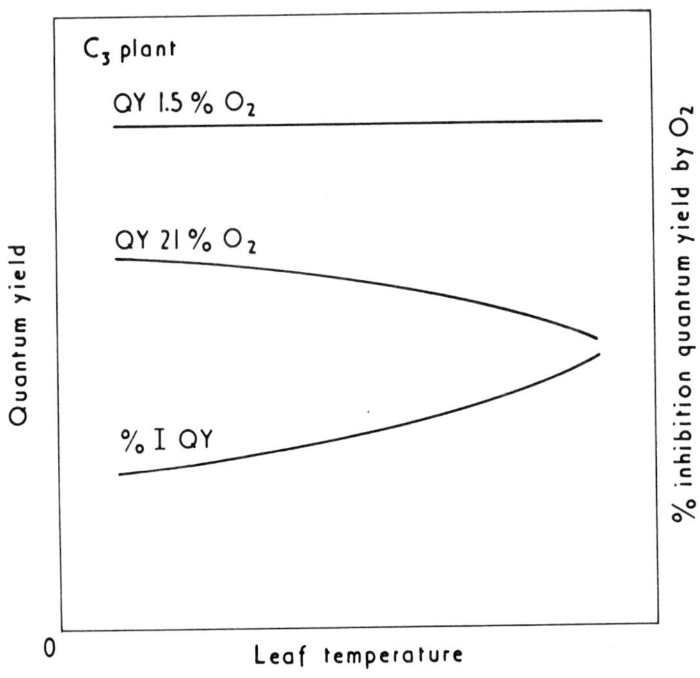


Fig. 13.15. Dependence of the quantum yield on temperature in $C_3$ plants. (After Ku & Edwards, 1978).

QY is the same at low (2%) or high (21%) $O_2$, and temperatures of 15–35°C have no effect on the QY. This is consistent with a lack of $O_2$ inhibition of $C_4$ photosynthesis.

At 30°C, 21% $O_2$, and with atmospheric [$CO_2$], the values of the QY of $C_3$ and $C_4$ species are very similar. At 2% $O_2$, the QY of $C_3$ plants increases while the QY of $C_4$ plants are the same at 2% versus 21% $O_2$ (Table 13.2). Thus, at 30°C, low light and 2% $O_2$, $C_3$ plants are more efficient (36% from Table 13.2) in using absorbed energy than $C_4$ plants. In general, these differences can be explained as follows. At 2% $O_2$, photosynthesis in neither $C_3$ nor $C_4$ plants is inhibited by $O_2$. The theoretical energy requirements would then be 3 ATP/2 NADPH per $CO_2$ fixed in the RPP pathway in $C_3$ plants and 5 ATP/2 NADPH per $CO_2$ in $C_4$ plants. Thus, the $C_4$ plants are expected to have a lower quantum efficiency. However, at 21% $O_2$, the quantum efficiency decreases in $C_3$ plants due to $O_2$ inhibition of photosynthesis. Take, for example, a 30% inhibition of photosynthesis by $O_2$.

**Table 13.2.** Average quantum yield of several $C_3$ and $C_4$ species at 30° C, atmospheric $CO_2$ and 2% versus 21% $O_2$. (After Ehleringer & Björkman, 1977)

| | Quantum yield | | Quanta required/$CO_2$ fixed | |
|---|---|---|---|---|
| | 21% $O_2$ | 2% $O_2$ | 21% $O_2$ | 2% $O_2$ |
| $C_3$ species | 0.0524 | 0.0733 | 19.1 | 13.6 |
| $C_4$ species | 0.0534 | 0.0538 | 18.7 | 18.6 |

$$\text{Relative rate of photosynthesis}\quad 2\%\ O_2 = 100$$
$$21\%\ O_2 = \ 70$$

$$\text{Then } \%\text{ inhibition of photosynthesis by } O_2 = \frac{100-70}{100}(100) = 30\%$$

With a constant absorption of energy at 2% and 21% $O_2$, the energy requirements to fix $CO_2$ to the level of triose phosphates would increase from 3 ATP/2 NADPH per $CO_2$ fixed at 2% $O_2$ to 4.3 ATP/2.9 NADPH (3 ATP/2 NADPH per 0.7 $CO_2$ fixed) per $CO_2$ fixed at 21% $O_2$. Under atmospheric $O_2$, then $C_3$ plants would require 43 ATP and 29 NADPH to fix 10 $CO_2$ in comparison to 50 ATP and 20 NADPH to fix 10 $CO_2$ in $C_4$ plants (Section 11.6). Therefore, at 21% $O_2$ the total energy required/$CO_2$ fixed becomes similar with $C_3$ and $C_4$ plants.

The measured quantum requirement is higher for both $C_3$ and $C_4$ species than might be calculated theoretically. This could be due to energy requirements for other processes outside of $CO_2$ fixation to the level of triose phosphates; such as nitrogen assimilation, synthesis of macromolecules, respiration, as well as some nonspecific absorption of energy (e.g. cell walls) by the leaves. For example, take 4 quanta/NADPH and a $P/2e^- = 1.33$ (where the $H^+/ATP = 3$, see Section 5.14). Then the theoretical quantum requirement for $C_3$ plants at 2% $O_2$ would be 9 and that for $C_4$ would be 15.

For $C_3$ at 2% $O_2$ (3 ATP/2 NADPH)

2 NADPH, 2.66 ATP = 8 quanta
3 ATP − 2.66 ATP = (0.34 ATP)(3 $H^+$/ATP) = 1 $H^+$ = 1 quanta

---

Total = 9 quanta (compared to 13.6 measured, Table 13.2)

Similar calculations for a $C_3$ plant under 21% $O_2$, considering 4.3 ATP/2.9 NADPH per $CO_2$ fixed, give a quantum requirement of approximately 13 (compared to 19.1 measured, Table 13.2)

For $C_4$ at 2% or 21% $O_2$ (5 ATP/2 NADPH)

2 NADPH, 2.66 ATP = 8 quanta
5 ATP − 2.66 ATP = (2.34 ATP)(3 $H^+$/ATP) = 7 $H^+$ = 7 quanta

---

Total = 15 quanta (compared to 18.6 measured Table 13.2)

Under atmospheric conditions and as temperature decreases below 30°C, there is a decrease in $O_2$ inhibition of the QY in $C_3$ plants. This results in the QY of $C_3$ species rising above that of $C_4$ species. Therefore, under limiting light and low temperature, $C_3$ plants are more efficient in the use of energy than $C_4$ plants. Conversely at high temperatures and low light, $C_4$ plants are more efficient than $C_3$ species (due to the decrease in QY with increasing temperatures in $C_3$).

While the degree of inhibition of photosynthesis by $O_2$ is similar at varying light intensities, $CO_2$ is most limiting for $C_3$ photosynthesis at high light. Thus at high light, atmospheric $CO_2$ may be less than saturating for $C_3$ photosynthesis even without $O_2$ inhibition of photosynthesis (Section 14.5). For this reason, under high light $C_4$ plants by the $CO_2$ concentrating mechanism will be more efficient than $C_3$ plants in utilizing the available energy. Temperature, light, $CO_2$, and $O_2$ are all important in determining the relative efficiency of these two photosynthetic groups. [Refs. **25, 38**].

## 13.13 Suggested roles for photorespiration.

Why some plants photorespire and others do not has long been open to debate. One idea is that photorespiration dissipates excess energy produced photochemically preventing photoinhibition of $CO_2$ fixation resulting from photooxidative destruction of the photochemical apparatus. Illuminating photosynthetic cells under nitrogen in the absence of $CO_2$ causes irreversible damage to their capacity for $CO_2$ fixation. Oxygen tends to protect against such photochemical damage. Of course little photoinhibition would occur under low light, and under this condition where energy is rate-limiting rather than $CO_2$, little $O_2$ inhibition of photosynthesis might be expected. However, the percentage inhibition of photosynthesis by $O_2$ is just as severe under low light as it is under high light. Related to the above suggestion is the idea that photorespiration protects the plant from photodestruction when $CO_2$ is most limiting—during periods of water stress and/or high temperatures. For example at very low $[CO_2]$, i.e., the $CO_2$ compensation point, energy is dissipated without a net fixation of $CO_2$ (see Fig. 6.11).

Perhaps the glycolate pathway has a role in leading to sucrose synthesis and the synthesis of amino acids, glycine, and serine. However, the rate of carbon flow through the glycolate pathway (on average estimated at 30–40 $\mu$mol $CO_2$ evolved $mg^{-1}$ Chl $h^{-1}$) is much higher than requirements for amino acid synthesis. The significance of sucrose synthesis through the glycolate pathway is uncertain since the pathway of sucrose synthesis from triose phosphates is more direct. Also, in $C_4$ plants, the glycolate pathway functions at a much lower rate which indicates that high rates of sucrose and amino acid synthesis by this means are unnecessary.

Some hypotheses are based on the idea that photorespiration may be of no present benefit. One suggestion is that RBP carboxylase-oxygenase evolved millions of years ago when the $CO_2/O_2$ ratio was higher and therefore much more favourable for photosynthesis relative to photorespiration. The RPP pathway is the only known carboxylating mechanism which can function autocatalytically (as a breeder reaction, Section 6.3). The susceptibility of RBP to oxidation may be an unavoidable consequence. With the changed environmental conditions (lower $CO_2/O_2$ ratio), plant metabolism can function better without this oxidation as demonstrated by $C_4$ plants. A more effective, $O_2$-insensitive carboxylase (PEP carboxylase) evolved in $C_4$ and CAM plants but PEP carboxylase cannot function independently of RBP carboxylase because it lacks an independent autocatalytic cycle.

Goldsworthy considered that the glycolate pathway may have been an early adaptation in algae which had a symbiotic relationship with bacteria. In a localized environment of low $[CO_2]$ and high $[O_2]$, synthesis of glycolate by the alga could have been of mutual advantage. It may have attracted and provided a source of carbon for the bacteria and at the same time, the bacteria may have supplied the primitive alga with some essential metabolite, for example an amino acid. When photosynthetic organisms became self-sufficient, they no longer needed to

synthesize glycolate but they retained this capacity through RBP carboxylase, an essential enzyme of the RPP pathway.
[Refs. 1, 2, 4–8, **30, 35, 44, 49, 59**].

## 13.14 Improved growth of $C_3$ plants under low $O_2$ or enriched $CO_2$ environments

Either decreasing [$O_2$] to 2% or increasing [$CO_2$] three-fold above atmospheric levels eliminates $O_2$ inhibition of photosynthesis. Under these conditions glycine synthesis is repressed and sucrose synthesis increases (Table 13.3) and these biochemical changes in whole leaf photosynthesis are consistent with a diminishing photorespiratory metabolism. If photorespiration involves a loss of newly fixed $CO_2$ and affords little advantage in other regards (it may be supposed, for example, that normal photosynthesis and 'dark' respiration may meet the plant's requirements for ATP and metabolites), then improved growth might be predicted in atmospheres having an increased $CO_2/O_2$ ratio. For many years, growth of plants in greenhouses

**Table 13.3.** Effect of $O_2$ and $CO_2$ concentration on some labelled products (glycine and sucrose) of $^{14}CO_2$ assimilation.

| A. Varying $CO_2$ with 21% $O_2$[a] (After Lee & Whittingham, 1974) | Percentage label among total products | |
|---|---|---|
| $CO_2$ (ppm) | Photosynthesis relative rates | Glycine | Sucrose |
| 100 | 0.08 | 36.5 | 3.6 |
| 300 | 0.20 | 15.6 | 12.5 |
| 950 | 0.63 | 4.9 | 32.4 |

| B. Varying $CO_2$ and $O_2$[b] (After Osmond & Björkman, 1972) | | | Percentage total products | |
|---|---|---|---|---|
| $CO_2$ (ppm) | $O_2$ (%) | Photosynthesis ($\mu$mol $^{14}CO_2$ fixed dm$^{-2}$ leaf area 5 min$^{-1}$) | Glycine | Sucrose |
| 302 | 21 | 28.2 | 8.0 | 21.6 |
| 302 | 2 | 42.4 | 2.1 | 30.9 |
| 1068 | 20 | 86.0 | 2.1 | 36.9 |
| 1038 | 2 | 82.0 | 0.3 | 42.3 |

[a] Products were determined after 3 min of fixation of $^{14}CO_2$ in the light with the $C_3$ species *Lycopersicon esculentum*.
[b] Products were determined after 5 min of fixation of $^{14}CO_2$ in the light with the $C_3$ species *Atriplex patula*.

**Table 13.4.** Effect of $CO_2$ and $O_2$ on vegetative and reproductive development in *Sorghum* and *Glycine max* and symbiotic $N_2$ fixation in *Glycine max*. (After Quebedeaux & Hardy, 1973; 1975; Quebedeaux *et al*, 1975)

| $O_2$ (%) | $CO_2$ (ppm) | Vegetative growth (leaves, stems, roots) (g dry weight/plant) | Reproductive growth (seed) (g dry weight/plant) |
|---|---|---|---|
| A. *Sorghum* ($C_4$ plant), Vegetative and reproductive growth[a] | | | |
| 5 | 300 | 95.7 | 0.0 |
| 10 | 300 | 88.7 | 1.6 |
| 21 | 300 | 109.9 | 7.7 |
| B. *Glycine max* ($C_3$ plant), Vegetative and reproductive growth[b] | | | |
| 5 | 300 | 55.3 | 0.0 |
| 21 | 300 | 10.0 | 12.9 |
| 5 | 2000 | 117.2 | 0.0 |
| 21 | 2000 | 102.0 | 20.1 |

C. *Glycine max* ($C_3$ plant), $N_2$ assimilation[c]

| $O_2$ (%) | $CO_2$ (ppm) | mg $N_2$ fixed/plant after 83 days |
|---|---|---|
| 21 | 300 | 155 |
| 5 | 300 | 350 |
| 21 | 1200 | 700 |

[a] Measured at 168 days of age. Roots were exposed to normal air while the aerial portion of plants were exposed to $CO_2$ and $O_2$ as indicated. Day and night temperatures were 30° and 21° C, respectively.
[b] Measured at 83 days with aerial portion of plants exposed to indicated gases from 14–83 days while roots were exposed to air. Day and night temperatures were 30° and 21° C, respectively.
[c] The authors integrated data from mg $N_2$ fixed plant$^{-1}$ day$^{-1}$ to give total mg $N_2$ fixed per plant. Plants were grown at day and night temperatures of 29° and 18° C, respectively.

under enriched atmospheres of $CO_2$ has been practised with up to several-fold increases in yield. These are all horticultural or floricultural plants now known to be $C_3$ species. Except for maize, sorghum, sugar-cane, millets and some pasture grasses which are $C_4$, all major crops are $C_3$ species, e.g., soybeans, potatoes and small grains such as rice, barley and wheat, (Section 11.8). With the discovery of $C_3$ and $C_4$ photosynthetic pathways and the obvious differences in response to atmospheric $O_2$ and $CO_2$, comparative studies on growth of $C_3$ and $C_4$ species in different atmospheres were of interest. For example, Björkman and his colleagues at the Carnegie Institution have shown 1.5–2-fold increases in dry weight when a *Phaseolus* species and *Mimulus* species ($C_3$) were grown under low $O_2$, whereas *Zea mays* ($C_4$) was unaffected. Similarly, in extensive comparisons of $C_3$ and $C_4$ species, Akita and Tanaka in Japan found a 2–5-fold increase in dry weight of $C_3$ species and little or no effect on $C_4$ species with $CO_2$ enrichment up to 1000 or 2500 ppm $CO_2$.

Growth of soybeans ($C_3$) in enriched atmospheres of $CO_2$ causes dramatic increases in vegetative and reproductive growth and in $N_2$ assimilation (Table 13.4). The enhanced $N_2$ assimilation with either $CO_2$ enrichment or reduced $O_2$ is apparently due to increased photosynthesis and more photosynthate being available for the $N_2$-fixing symbiotic bacteria in the roots.

Decreased atmospheric $[O_2]$ also increased vegetative growth and $N_2$ fixation in the soybeans. An unexpected effect of reduced $O_2$ was the inhibition of reproductive growth in both soybeans and sorghum ($C_4$). The reason for this high $O_2$ requirement for reproductive growth is not understood but it seems unrelated to photorespiratory metabolism.

$CO_2$ concentrations about 2-fold higher than present atmospheric levels largely overcome any repression of growth due to $O_2$ inhibition of photosynthesis. The degree of increase in growth by reducing photorespiration naturally depends on other environmental conditions such as light, temperature, and nutrition. From 1860 to 1960, the average levels of $CO_2$ in the atmosphere increased continually from about 285–325 ppm and continues to increase at about 0.7 ppm/year, primarily due to the burning of fossil fuels. At least some improvement in crop yields might be due to the higher $[CO_2]$. The buffering capacity of the ocean in absorbing $CO_2$ and the enhanced photosynthesis with higher $[CO_2]$ are factors which will influence the potential rise in atmospheric $CO_2$ levels in the future.
[Refs. **9, 16, 17, 51, 53–55, 67**]

## General Reading


1 Chollet R. & Ogren W. L. (1975) Regulation of photorespiration in $C_3$ and $C_4$ species. *Bot. Rev.*, **41**, 137–79.
2 Jackson W. A. & Volk R. J. (1970) Photorespiration. *Ann. Rev. Plant Physiol.*, **21**, 385–432.
3 Lorimer G. H., Woo K. C., Berry J. A. & Osmond C. B. (1978) The $C_2$ photorespiratory carbon oxidation cycle in leaves of higher plants: pathway and consequences. In *Fourth International Congress on Photosynthesis* (eds. D. O. Hall, J. Coombs and T. W. Goodwin) pp. 311–22. London: The Biochemical Society.
4 Schnarrenberger C. & Fock H. (1976) Interactions among organelles involved in photorespiration. *Encyclopedia of Plant Physiology*. Vol. 111. Transport in Plants. (eds. C. R. Stocking and U. Heber) pp. 185–234. New York: Springer-Verlag.
5 Tolbert N. E. (1971) Microbodies—peroxisomes and glyoxysomes. *Ann. Rev. Plant Physiol.* **22**, 45–74.
6 Whittingham C. P. (1971) *Photosynthesis*. London: Oxford University Press.

7 Wolf F. T. (1970) Photorespiration, the $C_4$ pathway of photosynthesis and related phenomena. *Adv. Frontiers Plant Sci.* **26**, 161–231.
8 Zelitch I. (1971) *Photosynthesis, Photorespiration and Plant Productivity*. New York: Academic Press.

## Specific citations

9 Akita S. & Tanaka I. (1973) Studies on the mechanisms of differences in photosynthesis among species. IV. The differential response in dry matter production between $C_3$ and $C_4$ species to atmospheric carbon dioxide enrichment. *Proc. Crop Sci. Soc. Japan.* **42**, 288–95.
10 Arron G. P., Spalding M. H. & Edwards G. E. (1979) Stoichiometry of carbon dioxide release and oxygen uptake during glycine oxidation in mitochondria isolated from spinach (*Spinacia oleracea*) leaves. *Biochem J.*, **184**, 457–60.
11 Badger M. R. & Andrews T. J. (1974) Effects of $CO_2$, $O_2$ and temperature on a high-affinity form of ribulose diphosphate carboxylase-oxygenase from spinach. *Biochem. Biophys. Res. Commun.*, **60**, 204–10.
12 Badger M. R. & Lorimer G. H. (1976) Activation of ribulose-1,5-bisphosphate oxygenase. The role of $Mg^{2+}$, $CO_2$ and pH. *Arch. Biochem. Biophys.*, **175**, 723–9.
13 Berry J. A. & Badger M. R. (1979) Direct measurement of photorespiration as a function of $CO_2$ concentration. Carnegie Inst. Wash. Year Book 1978, pp. 175–8.
14 Berry J. A., Osmond C. B. & Lorimer G. H. (1978) Fixation of $^{18}O_2$ during photorespiration. Kinetic and steady-state studies of the photorespiratory carbon oxidation cycle with intact leaves and isolated chloroplasts of $C_3$ plants. *Plant Physiol.*, **62**, 954–67.
15 Björkman O. (1966) The effect of oxygen concentration on photosynthesis in higher plants. *Physiol. Plant.*, **19**, 618–33.
16 Björkman O., Gauhl E., Hiesey W. M., Nicholson F. & Nobs M. A. (1969) Growth of *Mimulus*, *Marchantia* and *Zea* under different oxygen and carbon dioxide levels. Carnegie Inst. Wash. Year book **67**, 477–9.
17 Bolin B. & Keeling C. D. (1963) Large scale atmospheric mining as deduced from the seasonal and meridianal variations of carbon dioxide. *J. Geophys. Res.*, **68**, 3899–920.
18 Bowes G., Ogren W. L. & Hageman R. H. (1971) Phosphoglycolate production catalyzed by ribulose diphosphate carboxylase. *Biochem. Biophys. Res. Commun.*, **45**, 716–22.
19 Bowes G., Ogren W. L. & Hageman R. H. (1975) pH dependence of the $Km(CO_2)$ of ribulose 1,5-diphosphate carboxylase. *Plant Physiol.*, **56**, 630–3.
20 Canvin D. T., Fock H. & Lloyd N. D. H. (1978) Environmental treatments and carbon flow in the glycolate pathway. *Fourth International Congress on Photosynthesis.* (eds. D. O. Hall, J. Coombs and T. W. Goodwin.) pp. 323–334 *The Biochemical Society*. London.
21 Decker J. P. (1955) A rapid, post-illumination deceleration of respiration in green leaves. *Plant Physiol.*, **30**, 82–4.
22 Delaney M. E. & Walker D. A. (1978) A comparison of the kinetic properties of ribulose bisphosphate carboxylase in chloroplast extracts of spinach, sunflower and four other reductive pentose phosphate pathway species. *Biochem. J.*, **171**, 477–82.
23 Doehlert D. C., Ku M. S. B. & Edwards G. E. (1979) Dependence of the post-illumination burst of $CO_2$ on temperature, light, $CO_2$, and $O_2$ concentration in wheat (*Triticum aestivum*). *Physiol. Plant.*, **46**, 299–306.
24 Douce R., Moore A. L. & Neuburger M. (1977) Isolation and oxidative properties of intact mitochondria isolated from spinach. *Plant Physiol.*, **60**, 625–8.
25 Ehleringer J. & Björkman D. (1977) Quantum yields for $CO_2$ uptake in $C_3$ and $C_4$ plants. Dependence on temperature, $CO_2$, and $O_2$ concentration. *Plant Physiol.*, **56**, 86–90.

26 Farquhar G. D., von Caemmerer S. & Berry J. A. (1980) A biochemical model of photosynthetic $CO_2$ assimilation in leaves of $C_3$ species. *Planta*, **149**, 78–90.

27 Forrester M. L., Krotkov G. & Nelson C. D. (1966) Effect of oxygen on photosynthesis, photorespiration and respiration in detached leaves. I. Soybean. *Plant Physiol.*, **41**, 422–7.

28 Frederick S. E. & Newcomb E. H. (1969) Microbody-like organelles in leaf cells. *Science*, **163**, 1353–5.

29 Gerbaud A. & Andre M. (1979) Photosynthesis and photorespiration in whole plants of wheat. *Plant Physiol.*, **64**, 735–8.

30 Goldsworthy A. (1969) Riddle of photorespiration. *Nature*, **224**, 501–2.

31 Huang A. H. C. & Beevers H. (1972) Microbody enzymes and carboxylases in sequential extracts from $C_4$ and $C_3$ leaves. *Plant Physiol.*, **50**, 242–248.

32 Jensen R. G. & Bahr J. T. (1977) Ribulose 1,5-bisphosphate carboxylase-oxygenase. *Ann. Rev. Plant Physiol.*, **28**, 379–400.

33 Kerr M. W. & Groves D. (1975) Purification and properties of glycolate oxidase from *Pisum sativum* leaves. *Phytochemistry*, **14**, 359–62.

34 Keys A. J., Bird I. F., Cornelius M. J., Lea P. J., Wallsgrove R. M. & Miflin B. J. (1978) Photorespiratory nitrogen cycle. *Nature*, **275**, 741–3.

35 Krause G. H., Lorimer G. H., Heber U. & Kirk M. R. (1978) Photorespiratory energy dissipation in leaves and chloroplasts. In *Fourth International Congress on Photosynthesis*. (eds. D. O. Hall, J. Coombs and T. W. Goodwin) London: Biochemical Society. pp. 299–310.

36 Ku S. B. & Edwards G. E. (1975) Photosynthesis in mesophyll protoplasts and bundle sheath cells of various types of $C_4$ plants. IV. Enzymes of respiratory metabolism and energy utilizing enzymes of carbon assimilation. *Z. Pflanzenphysiol.*, **77**, 16–32.

37 Ku S. B. & Edwards G. E. (1977) Oxygen inhibition of photosynthesis. II. Kinetic characteristics as affected by temperature. *Plant Physiol.*, **59**, 991–9.

38 Ku S. B. & Edwards G. E. (1978) Oxygen inhibition of photosynthesis. III. Temperature dependence of quantum yields and its relation to $O_2/CO_2$ solubility ratio. *Planta*, **140**, 1–6.

39 Ku S. B., Edwards G. E. & Tanner C. B. (1977) Effects of light, carbon dioxide, and temperature on photosynthesis, oxygen inhibition of photosynthesis, and transpiration in *Solanum tuberosum*. *Plant Physiol.*, **59**, 868–72.

40 Laing W. A., Ogren W. L. & Hageman R. H. (1974) Regulation of soybean net photosynthetic $CO_2$ fixation by the interaction of $CO_2$, $O_2$ and ribulose-1,5-diphosphate carboxylase. *Plant Physiol.* **54**, 678–85.

41 Lea P. J. & Miflin B. J. (1979) Photosynthetic ammonia assimilation. In *Encyclopedia of Plant Physiology*. New series. Vol. 6. Photosynthesis II. Photosynthetic carbon metabolism and related processes. (eds. M. Gibbs and E. Latzko) pp. 445–455 New York: Springer-Verlag.

42 Lee R. B. & Whittingham C. P. (1974) The influence of partial pressure of carbon dioxide upon carbon metabolism in the tomato leaf. *J. Exp. Bot.*, **24**, 277–87.

43 Leech R. M. (1964) The isolation of structurally intact chloroplasts. *Biochim. Biophys. Acta*, **79**, 637–9.

44 Lorimer G. H. & Andrews T. J. (1973) Plant photorespiration—an inevitable consequence of the existence of an oxygen atmosphere. *Nature*, **243**, 359–60.

45 Lorimer G. H., Andrews T. J. & Tolbert N. E. (1973) Ribulose diphosphate oxygenase. II. Further proof of reaction products and mechanism of action. *Biochemistry*, **12**, 18–23.

46 Ludwig L. J. & Canvin D. T. (1971) An open gas-exchange system for the simultaneous measurement of the $CO_2$ and $^{14}CO_2$ fluxes from leaves. *Can. J. Bot.*, **49**, 1299–313.

47 Miflin B. J. & Beevers H. (1974) Isolation of intact plastids from a range of plant tissues. *Plant Physiol.*, **53**, 870–4.

48 Ogren W. L. & Bowes G. (1971) Ribulose diphosphate carboxylase regulates soybean photorespiration. *Nature, New Biol.*, **230**, 159–60.

49 Osmond C. B. & Björkman O. (1972) Simultaneous measurements of oxygen effects on net photosynthesis and glycolate metabolism in $C_3$ and $C_4$ species of *Atriplex*. *Carnegie Inst. Wash. Year book 1971*. 141–8.
50 Osmond C. B. & Harris B. (1971) Photorespiration during $C_4$ photosynthesis. *Biochim. Biophys. Acta*, **234**, 270–2.
51 Plass G. N. (1959) Carbon dioxide and climate. *Sci. Amer.*, **201**, 41–5.
52 Pritchard G. G., Whittingham C. P. & Griffin W. J. (1961) Effect of isonicotinyl hydrazide on the path of carbon in photosynthesis. *Nature*, **190**, 553–4.
53 Quebedeaux B. & Hardy R. W. F. (1973) Oxygen as a new factor controlling reproductive growth. *Nature*, **243**, 477–9.
54 Quebedeaux B. & Hardy R. W. F. (1975) Reproductive growth and dry matter production of *Glycine max* (L.) Merr. in response to oxygen concentrations. *Plant Physiol.*, **44**, 102–7.
55 Quebedeaux B., Havelka U. D., Livak K. L. & Hardy R. W. F. (1975) Effect of altered $pO_2$ in the aerial part of soybean on symbiotic $N_2$ fixation. *Plant Physiol.*, **56**, 761–4.
56 Radmer, R. J. & Kok B. (1976) Photoreduction of $O_2$ primes and replaces $CO_2$ assimilation. *Plant Physiol.*, **58**, 336–40.
57 Randall D. D. & Tolbert N. E. (1971) Two phosphatases associated with photosynthesis and the glycolate pathway. In *Photosynthesis and Photorespiration*. (eds., M. D. Hatch, C. B. Osmond and R. O. Slatyer.) pp. 259–66 New York: Wiley-Interscience.
58 Servaites J. C. & Ogren W. L. (1977) Chemical inhibition of the glycolate pathway in soybean leaf cells. *Plant Physiol.*, **60**, 461–6.
59 Smith B. N. (1976) Evolution of $C_4$ photosynthesis in response to changes in carbon and oxygen concentrations in the atmosphere through time. *Biosystems*, **8**, 24–32.
60 Ting I. P., Rocha V., Mukerji S. K. & Curry R. (1971) On the localization of plant cell organelles. In *Photosynthesis and Photorespiration*. (eds., M. D. Hatch, C. B. Osmond and R. O. Slatyer) pp. 534–40 New York: Wiley-Interscience.
61 Tolbert N. E. (1971) Isolation of leaf peroxisomes. *Methods in Enzymol.*, **23**, 665–82.
62 Tregunna E. B., Krotkov G. & Nelson C. D. (1966) Effect of oxygen on the rate of photorespiration in detached tobacco leaves. *Plant Physiol.*, **19**, 723–33.
63 Usuda H., Arron G. P. & Edwards G. E. (1981) Inhibition of glycine decarboxylation by aminoacetonitrile and its effect on photosynthesis in wheat. *J. Exp. Bot.*, **31**, 1477–83.
64 Usuda H. & Edwards G. E. (1980) Localization of glycerate kinase and some enzymes for sucrose synthesis in $C_3$ and $C_4$ plants. *Plant Physiol.*, **65**, 1017–22.
65 Wallsgrove R. M., Lea P. J. & Miflin B. J. (1979) Distribution of the enzyme of nitrogen assimilation within the pea leaf cell. *Plant Physiol.*, **63**, 232–6.
66 Whittingham C. P., Hiller R. G. & Bermingham M. (1963) The production of glycolate during photosynthesis. In *Photosynthetic Mechanisms of Green Plants*. pp. 675–683. Publ. 1145, Natl. Acad. *Sci*—Natl. Res. Council, Washington, D.C.
67 Wittwer S. H. & Robb W. (1964) Carbon dioxide enrichment of greenhouse atmospheres for food crop production. *Econ. Bot.*, **18**, 34–56.
68 Woo K. C. & Osmond C. B. (1976) Glycine decarboxylation in mitochondria isolated from spinach leaves. *Aust. J. Plant Physiol.*, **3**, 771–85.
69 Zelitch I. (1959) The relationship of glycolic acid to respiration and photosynthesis in tobacco leaves. *J. Biol. Chem.*, **234**, 3077–81.

# Chapter 14
# Primary Carboxylases and Environmental Regulation of Photosynthesis and Transpiration



## SUMMARY


The biochemistry of $C_3$ and $C_4$ photosynthesis as well as $O_2$ effects on photosynthesis in $C_3$ and $C_4$ plants must be considered in relation to the atmospheric levels of $CO_2$ and $O_2$. Based on characteristics of PEP and RBP carboxylase, models for photosynthesis in these species can be derived as a function of $CO_2$ and $O_2$ levels. RBP carboxylase utilizes $CO_2$ ($K$m = 5–15$\mu$M), while PEP carboxylase fixes


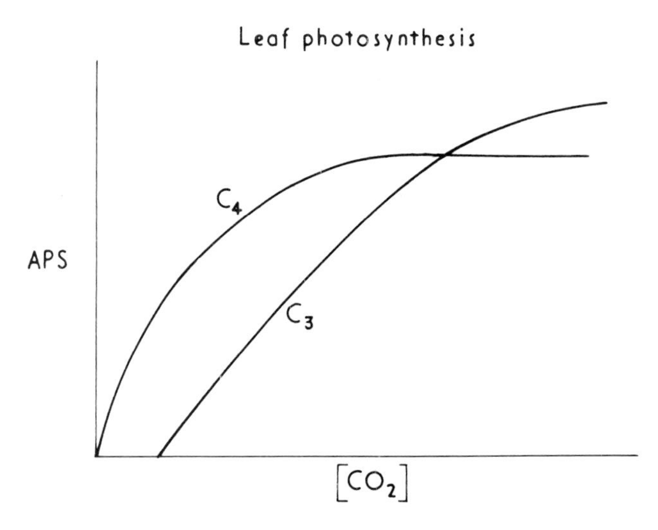


**Fig. 14.1.** General response of apparent photosynthesis (APS) to varying $[CO_2]$ in $C_3$ and $C_4$ plants.
$C_3$ and $C_4$ plants have very different response curves for photosynthesis plotted against $[CO_2]$. The $CO_2$ concentrating mechanism of $C_4$ plants allows maximum rates of photosynthesis at relatively low external $[CO_2]$. $C_3$ plants require high $[CO_2]$ for saturation of photosynthesis, and at low $[CO_2]$, photosynthesis is severely limited due to photorespiration (Chapter 13). Water stress, which leads to stomatal closure, and temperature will control the $[CO_2]$ available to the photosynthetic tissue of $C_3$ and $C_4$ plants.

bicarbonate ($Km = 0.3$–$0.4$mM or lower). $CO_2$ is hydrated to bicarbonate with bicarbonate by far the predominant species at high pH. Under low $O_2$ (e.g. 2 %) at pH 8.0, the relative affinity of the two carboxylases for atmospheric $CO_2$ may be similar. Under atmospheric $O_2$ (21 %) RBP carboxylase is at a disadvantage due to an $O_2$ dependent increase in the $Km_{(CO_2)}$.

The amount of $CO_2$ available to the primary carboxylase in the $C_3$ or $C_4$ leaf is a function of the atmospheric concentration, the stomatal resistance, the leaf temperature, and the transfer resistance through the aqueous phase. PEP carboxylase is localized in the cytosol of $C_4$ plants as is carbonic anhydrase, whereas in $C_3$ plants, carbonic anhydrase is localized in the chloroplasts and perhaps also in the cytoplasm. In both $C_3$ and $C_4$ species, carbonic anhydrase may minimize the transfer resistance by facilitating rapid equilibration of $CO_2$ to $HCO_3^-$ and providing a high concentration of the latter for diffusion.

When leaf temperature and the intercellular [$CO_2$] in the leaf are known, the concentration of the gas in solution at the surface of the mesophyll cells can be calculated. Models for leaf photosynthesis in $C_3$ and $C_4$ can then be considered in relation to the [$CO_2$] available to the cells and the levels of the carboxylases.

At atmospheric levels of $CO_2$, photosynthesis of $C_3$ plants is less than the carboxylation potential due to $O_2$ inhibition of RBP carboxylase and $O_2$ dependent respiration of glycolate (photorespiration). At saturating levels of $CO_2$, photosynthesis increases as the $O_2$ inhibition is relieved. However, the maximum carboxylation potential at saturating $CO_2$ is not achieved due to limiting assimilatory power for regenerating RBP.

In $C_4$ plants under atmospheric $CO_2$ and $O_2$, RBP carboxylase may operate near maximum capacity due to the $CO_2$ concentrating mechanism of the $C_4$ cycle. The lack of $O_2$ inhibition of PEP carboxylase and the $CO_2$ concentrating mechanism allows relatively high rates of photosynthesis even under low [$CO_2$] as occurs with increased temperature or increased stomatal resistance with water stress. The relatively low rates of photosynthesis in some $C_4$ species at low temperatures could be due to limiting levels of RBP carboxylase, the cold lability of pyruvate,Pi dikinase or restrictions on metabolite transport.

Increasing temperature may have a negative effect on $C_3$ photosynthesis by a decrease in the solubility of $CO_2$, an increase in the $O_2/CO_2$ solubility ratio, and possibly by a change of the kinetic constants of RBP carboxylase-oxygenase. Likewise, under water stress intercellular [$CO_2$] may become limiting and this will put photosynthesis in $C_3$ plants particularly, at a distinct disadvantage.

Mesophyll resistance, which includes the carboxylation component of $CO_2$ fixation, is a partial resistance to $CO_2$ fixation but not to transpiration. Thus, other things being equal, a low mesophyll resistance will give a low transpiration/photosynthesis ratio, as is found in $C_4$ plants.

## 14.1 Introduction

What advantages, if any, does the $C_4$ mechanism of photosynthesis provide? Briefly, external $[CO_2]$ is rate limiting for photosynthesis. It is particularly limiting at high temperature which reduces the solubility of $CO_2$ and under stress conditions (water or salt) which may increase stomatal resistance and make atmospheric $CO_2$ less available. Photosynthetically, $C_4$ plants are more efficient at relatively high temperatures, while $C_3$ plants are more efficient in photosynthesis at relatively low temperatures. $O_2$ inhibits photosynthesis in $C_3$ plants but $[CO_2]$ above atmospheric levels will overcome the $O_2$ inhibition (Chapter 13). The $C_4$ pathway offsets the effects of low $[CO_2]$ by providing a $CO_2$-enriched environment within the bundle sheath cells.

## 14.2 Composition of atmosphere and solubility of gases

The differences between $C_3$ and $C_4$ photosynthesis can best be understood in relation to physical aspects in the environment, particularly $CO_2$, $O_2$ concentration, and temperature. Firstly, the $[CO_2]$ in the atmosphere is very low, on average about 0.0325 % (325$\mu$l/litre, 325 $\mu$l/$10^6\mu$l, 325 ppm), with some small fluctuations geographically.

COMPOSITION OF ATMOSPHERE BY VOLUME (%)

| | |
|---|---|
| $N_2$ | 78.084 |
| $O_2$ | 20.946 |
| $CO_2$ | 0.033 |
| Ar | 0.934 |
| | 99.997 |

The solubility of a gas in water depends on its absorption coefficient ($\alpha$ or Bunsen coefficient, which is the volume of gas, absorbed by one volume of water when the gas is at one atmosphere of pressure). As noted by Strang (1981), in determining the affinity of an enzyme for a substrate which is a gas it is only correct to do so in terms of the true concentration of the gas in contact with the enzyme in molar units rather than that in the gas phase. As the temperature increases, the solubility of the gas

decreases (Bunsen coefficient decreases). For example, at 15°C and 325 ppm $CO_2$, the concentration of free $CO_2$ in water is 14.8 μM while at 35°C, the $[CO_2]$ in water would be 8.59 μM.
To calculate μmol $CO_2$/litre water at 15°C at a standard pressure of 1 atm:

$$CO_2\ (\mu mol/litre) = \frac{P_{CO_2}(\alpha)(10^6)}{22.4}$$

where 22.4 is in units of litres/mol at standard temperature and pressure.

$$P_{CO_2} = \text{partial pressure of } CO_2 = 0.0325\,\% = \frac{0.0325}{100} = 0.000325$$

When temperature of water is 15°C, $\alpha = 1.019$ ml gas/ml water.

$$\text{Then } CO_2\ (\mu mol/litre) = \frac{(0.000325)(1.019)(10^6)}{22.4} = 14.78\ \mu M$$

At a given temperature, the amount of a gas dissolved in water will be proportional to its partial pressure above the solution (Henry's law). Deviations can occur at pressures above atmospheric levels or at high solute concentrations. The solubility of $O_2$ or $CO_2$ at the surface of mesophyll cells at a given partial pressure in the intercellular space should be close to that in water, since salts found in plant sap (about 0.1 M) only depress the solubility of the gases by 5–10% (Rabinowitch, 1945). Since the absorption coefficient decreases with increasing temperature, there will be a proportional decrease in the chemical activity (or active mass) of the gas and its concentration in solution. So for practical purposes, Henry's law can be applied and the concentration of $CO_2$ at the surface of mesophyll cells calculated, provided $[CO_2]$ in the air space of the leaf and leaf temperature are determined.

The relative concentration of $O_2/CO_2$ in the atmosphere at 21% $O_2$ and 0.0325% $CO_2$ is 646/1 while in water at 25°C, the ratio is 24/1 (264 μM $O_2$/11μm $CO_2$). This is because $CO_2$ is more soluble in water than $O_2$ (i.e., $CO_2$ has a higher absorption coefficient). Note that with increasing temperature, the solubility of $CO_2$ decreases relatively faster than the solubility of $O_2$ (Fig. 14.2). This results in an increased solubility ratio of $O_2/CO_2$ at higher temperatures. The amount of $HCO_3^-$ in equilibrium with the soluble $CO_2$ is dependent on pH, with a p$K$ for the first dissociation constant around 6.33 and p$K_2$ of 10.38.

$$H_2O + CO_2 \underset{pK_1}{\rightleftarrows} H^+ + HCO_3^- \underset{pK_2}{\rightleftarrows} H^+ + CO_3^{-2}$$

[Just as 'pH' means negative logarithm of the hydrogen ion concentration (Section 1.7), 'p$K$' means the negative logarithm of $K$, where $K$ is the apparent equilibrium constant (Section 2.5) for the dissociation of an acid (HA) according to $HA \rightleftarrows H^+$

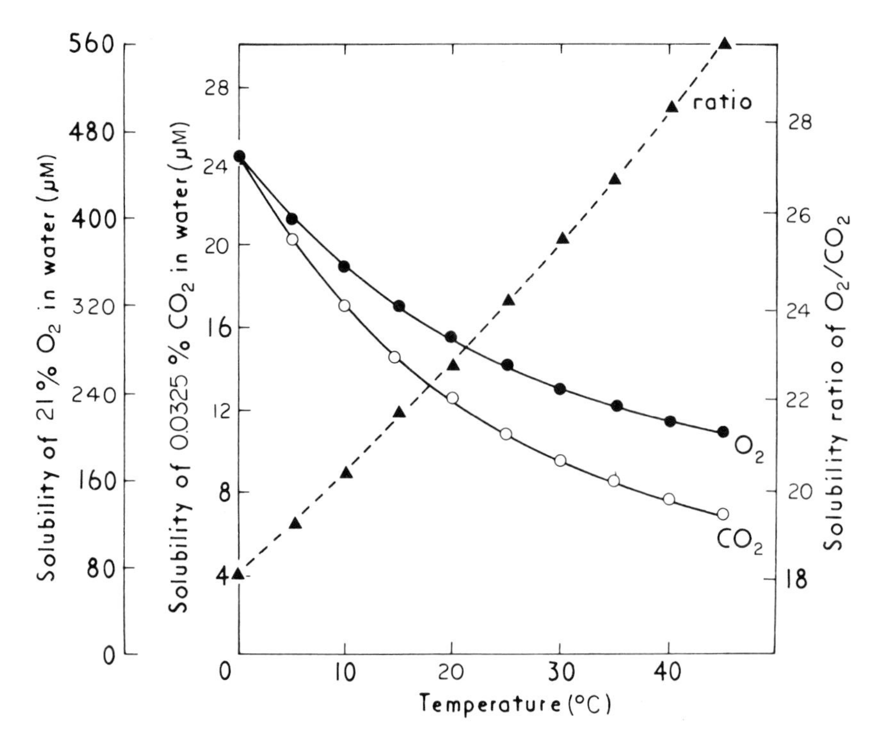


**Fig. 14.2.** Effect of temperature on solubility of atmospheric $CO_2$ (0.0325%) and $O_2$ (21%) and ratio of $O_2/CO_2$ in water. Data is calculated from Bunsen coefficients of $O_2$ and $CO_2$ at varying temperatures (After Ku & Edwards, 1977a).

$+A^-$. Thus,

$$K = \frac{[H^+][A^-]}{[HA]} \text{ and } pK = -\log K \text{ or } \log 1/K.$$

The terms p$K$ and pH are related by the Henderson–Hasselbach equation,

$$pH = pK + \log \frac{[A^-]}{[HA]}$$

It will be seen from this equation that the p$K$ of an acid is the pH at which it is half dissociated, i.e., when $[A^-] = [HA]$, the Henderson–Hasselbach equation simplifies to pH = p$K$.]

Some variation in p$K_1$ cited for bicarbonate can, in part, be accounted for by changes in p$K_1$ with temperatures. Umbreit *et al.* considered a change in the p$K_1$ of about 0.0005 units/°C between 20° and 40°C. These values will be used in examples to follow.

| Temp °C | α $CO_2$ | $pK_1$ |
|---|---|---|
| 20 | 0.878 | 6.392 |
| 25 | 0.759 | 6.365 |
| 30 | 0.665 | 6.348 |
| 35 | 0.592 | 6.328 |
| 40 | 0.530 | 6.312 |

(After Umbreit *et al.*, 1972).

These differences in p$K$ would make a small difference in the calculation of the $Km_{(CO_2)}$ for RBP carboxylase *in vitro* since the enzyme uses free $CO_2$ as the substrate. At pH 8.0, 35°C, with 0.4 mM total inorganic carbon ($CO_2 + HCO_3^-$) in the assay medium and with $pK_1 = 6.328$, the calculated concentration of soluble $CO_2$ is 8.33 μM.

[*Illustration*

$$pH = pK + \log\frac{[HCO_3^-]}{[CO_2]}$$

$$8.0 = 6.328 + \log\left(\frac{0.4\ \text{mM} - X}{X}\right)$$

Where X = concentration of $CO_2$

$$1.672 = \log\left(\frac{0.4\ \text{mM}}{X} - 1\right)$$

$$1.672 = \log 47$$

$$\frac{0.4\ \text{mM}}{X} - 1 = 47$$

X = 0.00833 mM = 8.33 μM]

In calculations at pH 8.0 and below, $pK_2$ can be ignored as there is very little $CO_3^{2-}$. In-vivo, the $[CO_2]$ available to the mesophyll cells is a function of the partial pressure of $CO_2$ in the intercellular air space of the leaf and is independent of pH. However, the equilibration of the $CO_2$ to bicarbonate in the leaf will be very dependent on pH. At 0.0325% atmospheric $CO_2$ and 30°C, the amount of dissolved $CO_2$ is 9.65 μM (see Fig. 14.3).

Then with a $pK_1 = 6.348$, $pH = 6.348 + \log\frac{[HCO_3^-]}{[CO_2]}$

At pH 6.348, the amount of $HCO_3^-$ = 9.65 μM
At pH 7.348, the amount of $HCO_3^-$ = 96.5 μM
At pH 8.0, the amount of $HCO_3^-$ = 432 μM

The significance of these physical characteristics of the gases will become apparent as the enzymology of the carboxylases in $C_3$, $C_4$ photosynthesis is considered. [Refs. **29, 47, 50**]

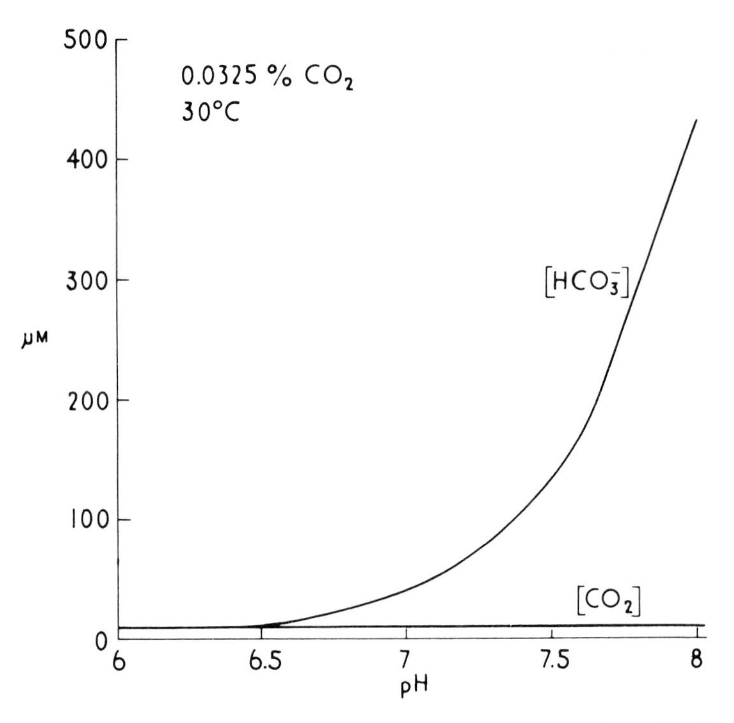


**Fig. 14.3.** Concentration of $CO_2$ and bicarbonate in water at various pH values at 30°C.

## 14.3 $C_3$ plants—RBP carboxylase

The form of $CO_2$ fixed by RBP carboxylase is free $CO_2$, as opposed to $HCO_3^-$ (Section 8.12). It is uncertain whether carbonic anhydrase is involved in catalyzing the transfer of $CO_2$ from the surface of the mesophyll cell to its site of fixation in the $C_3$ chloroplasts or is serving in other capacities. This enzyme, which catalyzes the hydration of $CO_2$, may be localized in both the chloroplasts and the cytoplasm of the $C_3$ mesophyll cell.

$$CO_2 + H_2O \rightleftarrows H^+ + HCO_3^- \qquad \text{Eqn. 14.1}$$

$CO_2$ will equilibrate to $H^+$ and $HCO_3^-$ with the equilibrium concentration of $HCO_3^-$ depending on the pH of the solution. Carbonic anhydrase, as a catalyst, allows a much faster equilibration.

As an example of how carbonic anhydrase may function in $C_3$ mesophyll cells in relation to RBP carboxylase, consider a temperature of 30°C, intercellular $[CO_2]$ of 250 ppm, and pH in the cytosol and chloroplast stroma of 8.0. The $[CO_2]$ as it dissolves at the surface of the mesophyll cells would be 7.4 $\mu$M. Carbonic anhydrase would give rapid equilibration of $CO_2$ with $HCO_3^-$ giving a final concentration of 331 $\mu$M $HCO_3^-$. $CO_2$ or $HCO_3^-$ diffusion to and within the chloroplasts must be along a gradient from high to low concentration (there is no evidence for active transport). Diffusion is dependent on the concentration gradient and the diffusivity of the molecules. Since the $[HCO_3^-]$ is about 45 times higher than the $[CO_2]$, there is a large potential for diffusion of $HCO_3^-$. However, the $HCO_3^-$ being a charged molecule has a much higher resistance for diffusion across the chloroplast envelope than free $CO_2$. Within the cytosol and within the chloroplast, the rapid equilibration of $CO_2$ to $HCO_3^-$ by carbonic anhydrase may facilitate their diffusion within the respective compartments. This may prevent localized depletion of $CO_2$ within the chloroplast during photosynthesis (Section 8.12, Fig. 14.4). The $Km_{(CO_2)}$ for RBP

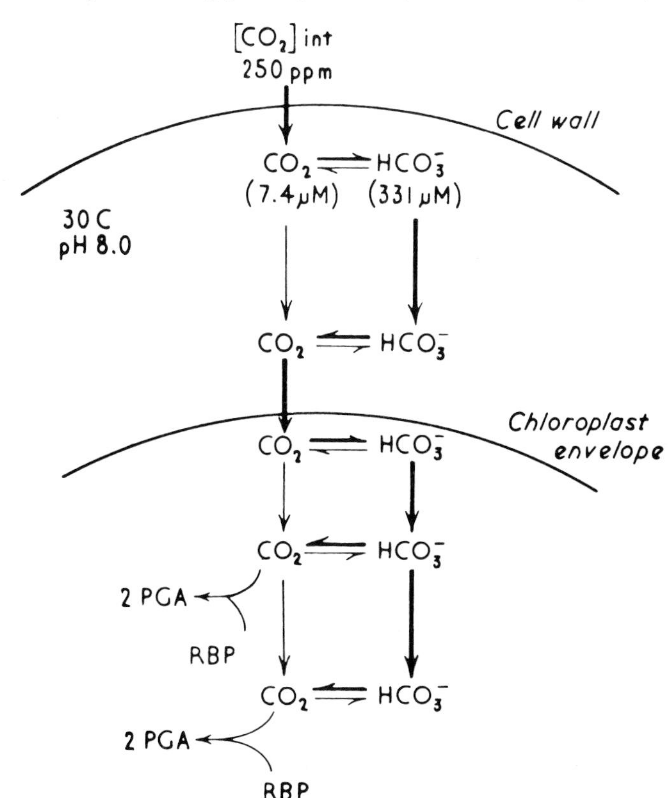


**Fig. 14.4.** Proposed primary path of transfer of $CO_2$ (heavy lines) in a $C_3$ leaf as facilitated by carbonic anhydrase.

$CO_2 \rightleftharpoons HCO_3^-$ represents $CO_2 + H_2O \rightleftharpoons HCO_3^- + H^+$ as catalyzed by carbonic anhydrase.

carboxylase has become progressively lower with improved techniques of assay (Section 6.5). Current values, around 10 $\mu$M (under $N_2$), are about equal to the solubility of 0.0325% $CO_2$ at 30°C. The problem with RBP carboxylase no longer seems to lie in its affinity for $CO_2$ or its maximum activity ($V$max), but in the fact that $O_2$ has been found to be a competitive inhibitor (Section 13.3).

The exact values of $Km_{(CO_2)}$ and $Ki_{(O_2)}$ (the concentration of $CO_2$ required for half maximum activity and the kinetic constant for $O_2$ as an inhibitor respectively) for RBP carboxylase are uncertain. The lowest $Km_{(CO_2)}$ values range from 6–16 $\mu$M and the lowest $Ki_{(O_2)}$ values range from about 200–400 $\mu$M. Both of these constants are similar to the solubility of atmospheric levels of the gases in water. As discussed in Chapter 13, $O_2$ is both a competitive inhibitor of RBP carboxylase and a substrate for the enzyme in the sense that the enzyme also functions as RBP oxygenase. In order to relate the kinetic constants of the carboxylase to leaf photosynthesis, the kinetics of leaf photosynthesis as a function of $[CO_2]$ must be analyzed. For this comparison, several factors must be considered with whole leaf measurements: boundary layer resistance; stomatal resistance (for example stomatal resistance often increases with increasing concentrations of atmospheric $CO_2$); photorespiration, which causes the $CO_2$ response curve to intercept at the $CO_2$ compensation point ($\Gamma$) rather than at the origin; transfer resistance of $CO_2$ through the aqueous phase in the photosynthetic tissue; and possible photochemical limitations on the response at high $[CO_2]$. None of these factors are involved when the carboxylase is assayed *in vitro*. In order to estimate the $Km_{(CO_2)}$ for RBP carboxylase *in vivo*, the following procedure can be used.

(a) Measure the $CO_2$ response curve under 2% $O_2$ to repress photorespiration and lower $\Gamma$ to near zero.

(b) Plot rate of photosynthesis versus the calculated soluble $[CO_2]$ at the surface of mesophyll cells, by measuring the intercellular $[CO_2]$ and leaf temperature. For each external $[CO_2]$, the $[CO_2]$ in the intercellular space in the leaf around mesophyll cells can be calculated according to Gaastra (1959):

$[CO_2]_{int} = [CO_2]_{ext} - APS\ (r_a + r_s)$ where  Eqn. 14.2

$[CO_2]_{ext} = [CO_2]$ in ambient air,

APS = rate of apparent photosynthesis,

$r_a$ = boundary layer resistance to diffusion of $CO_2$, and

$r_s$ = stomatal resistance to diffusion of $CO_2$.

These four parameters can be measured at each $[CO_2]_{ext}$ (Section 14.6) so that $[CO_2]_{int}$ can be calculated. The equation is derived from:

$$APS = \frac{[CO_2]_{ext} - [CO_2]_{CP}}{r_a + r_s + r_m} \qquad \text{Eqn. 14.3}$$

$$APS = \frac{[CO_2]_{ext} - [CO_2]_{int}}{r_a + r_s} = \frac{[CO_2]_{int} - [CO_2]_{CP}}{r_m} \qquad \text{Eqn. 14.4}$$

(where $[CO_2]_{CP} = [CO_2]$ in chloroplast and $r_m$ = mesophyll resistance).
(c) Make a reciprocal plot of the data as 1/APS versus $1/(CO_2 - \Gamma)$ for determining $Km_{(CO_2)}$ (also see Section 13.10).

Such an experiment with wheat at 30°C is illustrated in Fig. 14.5a, b, with a resulting $Km_{(CO_2)}$ of approximately 5 $\mu$M which corresponds to about 190 ppm $CO_2$ in the intercellular space. There is deviation from linearity in the reciprocal plot at high $[CO_2]$ which results in the experimental *V*max (Fig. 14.5a) being lower than the calculated *V*max (Fig. 14.5b). This may be due to photochemical limitations which have been previously suggested in a similar analysis with chloroplasts. Any transfer resistance (resistance to diffusion of $CO_2$ through the aqueous phase) would appear to be a minor component or else the $Km_{(CO_2)}$ of RBP carboxylase is even lower than is presently believed. If similar experiments are conducted at varying concentrations of $O_2$ and a plot is made of 1/APS against $1/(CO_2 - \Gamma)$ [where $(CO_2 - \Gamma)$ substracts out the photorespiratory component, Section 13.10], the $Ki_{(O_2)}$ and the increase in the apparent $Km_{(CO_2)}$ by $O_2$ can be determined. Such measurements suggest that the $Ki_{(O_2)}$ *in vivo* is about 200 $\mu$M (see Section 13.10). In the presence of 21% $O_2$, the apparent $Km_{(CO_2)}$ increased to about 15 $\mu$M (Fig. 14.5b).

Based on whole leaf analysis of $CO_2$ fixation and kinetic properties of RBP carboxylase-oxygenase, a general model of leaf photosynthesis in a $C_3$ species can be given (Fig. 14.6). The exact values of $Km_{(CO_2)}$, $Ki_{(O_2)}$, and *V*max RBP carboxylase in various species *in vitro* and *in vivo* is open to further analysis. In Fig. 14.6, curve A represents the theoretical capacity (based on the carboxylation potential) of photosynthesis in the absence of $O_2$. Curve B represents the corresponding capacity at 21% $O_2$. Curve C represents the rate of apparent photosynthesis at 21% $O_2$ in the absence of photochemical limitations. Curve D reflects the rate of photosynthesis *in vivo* at varying $[CO_2]$. In this example, the solubility of atmospheric levels of $CO_2$ at the surface of mesophyll cells is 7.4 $\mu$M which corresponds to a rate of leaf photosynthesis of about 250 $\mu$mol $CO_2$ fixed $mg^{-1}$ Chl $h^{-1}$. $O_2$ inhibition of photosynthesis at atmospheric $CO_2$ (i.e. at 7.4 $\mu$M in this example) is due both to $O_2$ inhibition of RBP carboxylase (Curve A – Curve B) and photorespiratory loss of $CO_2$ (Curve B – Curve C). The two components are linked since 1 $CO_2$ will be released in photorespiration for each 2 $O_2$ reacting with RBP (Section 13.10). However, the relative magnitude of the two components in relation to photosynthesis does not remain constant at varying $[CO_2]$. This is the case since the number of additional $CO_2$ reacting with RBP in the absence of $O_2$ is not equal to the

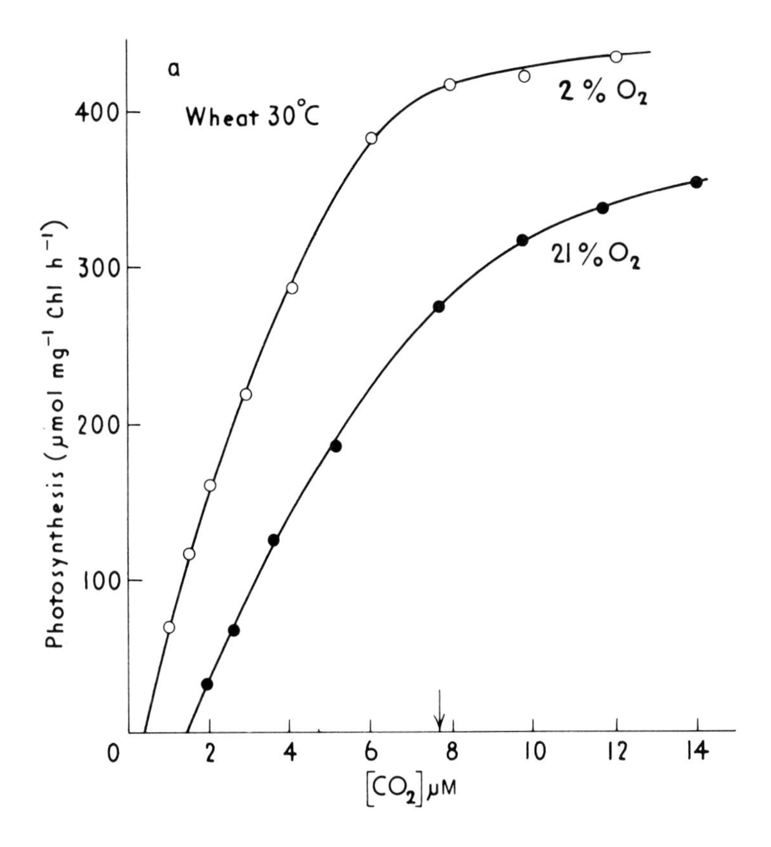


Fig. 14.5(a)

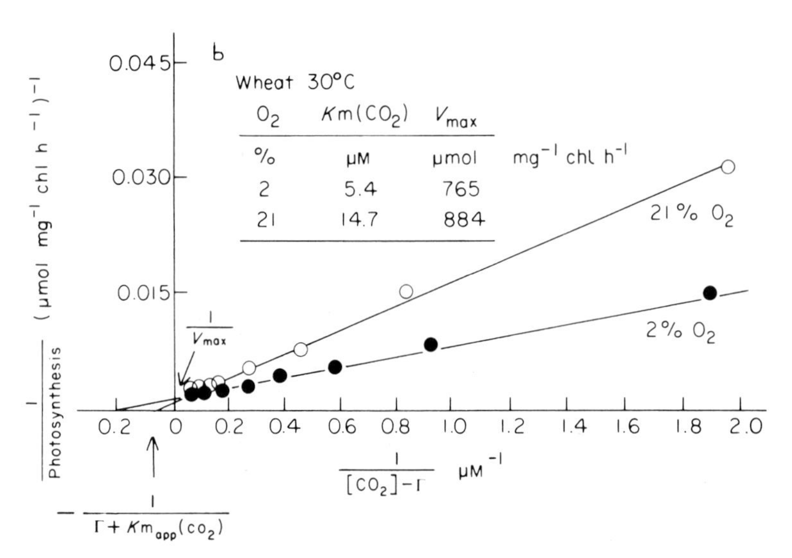


Fig. 14.5(b)

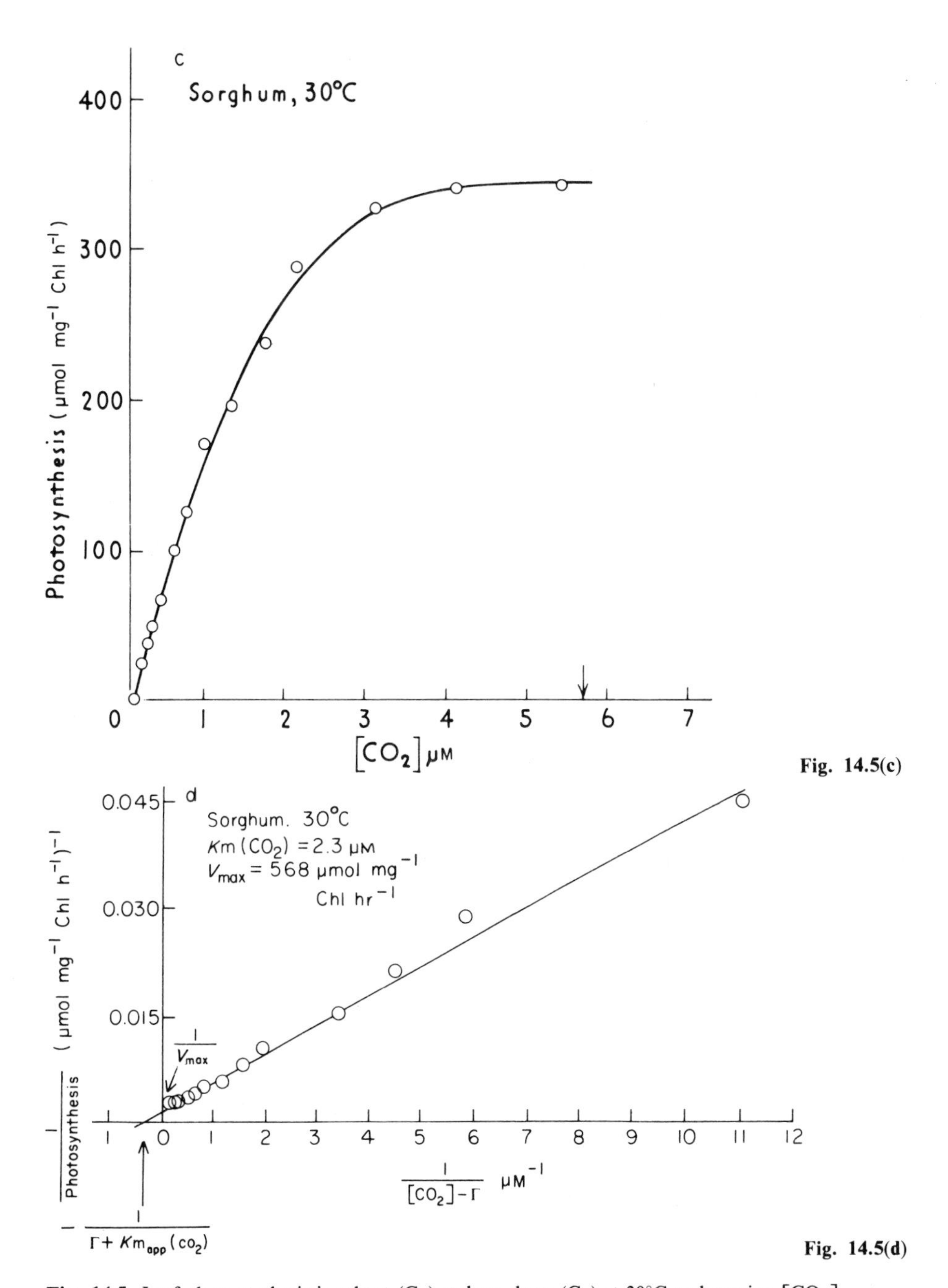


Fig. 14.5(c)

Fig. 14.5(d)

**Fig. 14.5.** Leaf photosynthesis in wheat ($C_3$) and sorghum ($C_4$) at 30°C and varying [$CO_2$]. Gaseous $CO_2$ was varied and the [$CO_2$] at the surface of leaf mesophyll cells was calculated from intercellular values of [$CO_2$] and leaf temperature. (a) Leaf photosynthesis of wheat at 2 and 21% $O_2$. (b) Double reciprocal plot of leaf photosynthesis of wheat versus [$CO_2$] (also see Section 13.10). (c) Leaf photosynthesis of sorghum at 21% $O_2$. (d) Double reciprocal plot of leaf photosynthesis of sorghum versus intercellular [$CO_2$]. Quantum flux density = 145 nmol quanta $cm^{-2}$ $s^{-1}$ (400–700 nm). Chl per unit leaf area = 0.3 g $cm^{-2}$. Arrow on the abscissa in (a) and (b) is the internal [$CO_2$] obtained at normal atmospheric level of $CO_2$. (Data kindly provided by S. B. Ku; also see Ku & Edwards, 1977b; 1978).

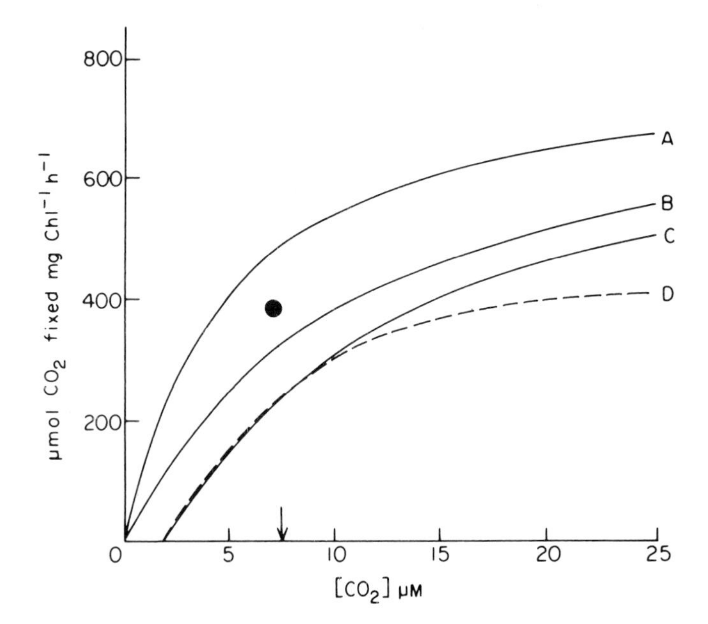


**Fig. 14.6** Model for leaf photosynthesis in a $C_3$ plant in relation to RBP carboxylase potential. (A) RBP carboxylase activity in absence of $O_2$, where

$$v = \frac{V\max[CO_2]}{[CO_2] + Km_{(CO_2)}}$$

(B) RBP carboxylase activity at 21 $\%O_2$, where

$$v = \frac{V\max[CO_2]}{[CO_2] + Km_{(CO_2)}(1 + [O_2]/Ki_{(O_2)})}$$

(C) Whole leaf photosynthesis at 21 % $O_2$ at a given $[CO_2]$ at the surface of mesophyll cells, without any limitation on assimilatory power, where

$$APS_{21\%O_2} = V\max \times \frac{[CO_2] - \Gamma}{[CO_2] + Km_{(CO_2)}(1 + [O_2]/Ki_{(O_2)})} \qquad \text{(see Eqn. 13.35)}$$

(D) Whole leaf photosynthesis at 21 % $O_2$ with limitations on generation of assimilatory power at high $[CO_2]$. Such a limitation would in effect limit the rate of regeneration of RBP as a substrate for the carboxylase with deviations from curve C at high $[CO_2]$.
(●) Rate of whole leaf photosynthesis at 2 % $O_2$ and atmospheric $[CO_2]$.
The arrow on the abscissa is equivalent to atmospheric $[CO_2]$ of 0.0325 %. See Section 13.10 for details of equations for carboxylase and leaf photosynthesis relative to $O_2$ inhibition of photosynthesis.
*Input:* atmospheric $CO_2$ = 0.0325 %; intercellular $CO_2$ = 0.025 %; leaf temperature = 30° C; solubility of $CO_2$ at the surface of mesophyll cells = 7.4μM; solubility of 21 % $O_2$ in the leaf = 245μM; Γ, $CO_2$ compensation point = 60 ppm = 2μM $CO_2$; RBP carboxylase, $Km_{(CO_2)}$ = 5μM

oxygenase activity in the presence of $O_2$. This would only occur if the kinetic constants (*K*m, *K*i, and *V*max; Equations 13.36, 13.39) were the same for $O_2$ and $CO_2$. The number of additional $CO_2$ reacting with the carboxylase in the absence of $O_2$ depends on the [$CO_2$] and the kinetic constants for the carboxylase. At high [$CO_2$], the photorespiratory component diminishes faster than the competitive component indicating that at high [$CO_2$], the oxygenase activity is relatively small compared to the increased capacity for carboxylation in the absence of $O_2$. As found experimentally with $C_3$ plants, maximum rates of photosynthesis can be achieved *in vivo* by either increasing [$CO_2$] above atmospheric levels or by decreasing [$O_2$] (Fig. 14.5). It is also apparent that the $CO_2/O_2$ ratio in the leaf will be significant in regulating rates of photosynthesis.

The actual photosynthetic response of the leaf (D) deviates from RBP carboxylase potential at 21% $O_2$ (B) for two reasons. At low [$CO_2$] photorespiration decreases the rate of photosynthesis and this is most severe as [$CO_2$] approaches the $CO_2$ compensation point. At [$CO_2$] above atmospheric levels, photosynthesis becomes saturated prior to the carboxylase due to insufficient assimilatory power. Under atmospheric [$CO_2$] the proportion of photochemical apparatus to carbon assimilating potential may be such that there is sufficient assimilatory power when light intensity is relatively high. However, at saturating [$CO_2$], the assimilatory power may be limiting even at full sunlight. Providing supplemental light above that of full sunlight may overcome this limitation of energy, unless some carrier of the electron transport is limiting or the maximum absorption of light is obtained at full sunlight. Although experimentally the effect of varying $CO_2$ has been studied over a wide range of concentrations, light intensity has usually been considered at the equivalent of full sunlight or less. Insufficient assimilatory power at high [$CO_2$] would, in effect, reduce the [RBP] such that it would become limiting for carboxylation.

In a few studies in which the level of RBP has been measured in cells or leaves, it has been suggested that the [RBP] may fall below the concentration of binding sites on the enzyme in the chloroplast. This in itself does not show that RBP is rate limiting since at least part of the substrate binding sites are presumably occupied by a 6 carbon intermediate of the reaction. The concentration of substrate binding sites on RBP carboxylase in the chloroplasts seems unusually high, e.g. about 3 mM (consider about 5 mg fraction I protein per mg Chl, a mol. wt. of 550 000; a stromal

---

(value from kinetic analysis with leaves, typical value with isolated enzyme is about 10$\mu$M, Section 13.3); $Ki_{(O_2)}$ = 200$\mu$M; *V*max RBP carboxylase = 800$\mu$mol $CO_2$ fixed $mg^{-1}$ Chl $h^{-1}$. This is taken as the *V*max of RBP carboxylase *in vivo* at saturating $CO_2$. The actual *V*max of the extracted enzyme may be higher than that *in vivo* if conditions *in vivo* are not optimum (e.g., RBP and $Mg^{2+}$ concentrations, pH).

volume of about 25 $\mu$l/mg Chl; and 8 binding sites per molecule of enzyme, i.e. one binding site per large subunit). The $Km_{(CO_2)}$ of about 10 $\mu$M is below the concentration of binding sites for the substrates on **RBP** carboxylase in the chloroplasts. If one added 10 $\mu$M $CO_2$ to a concentration of **RBP** carboxylase equivalent to that in the stroma of the chloroplasts, obviously most of the binding sites would not be occupied. However, since the enzyme has a relatively high affinity for $CO_2$, during steady-state photosynthesis, a large number of the binding sites can be occupied where the rate of influx of $CO_2$ from the atmosphere greatly exceeds the turnover rate of the enzyme. Then, $\mu$M levels of free $CO_2$ would occur in the chloroplasts while mM levels could be bound to the enzyme.

The degree of activation of **RBP** carboxylase *in vivo* under varying conditions is unknown. Naturally if the degree of activation of this or other enzymes changes with varying [$CO_2$] or [$O_2$], it could influence the kinetics of photosynthesis.

There are both empirical and mechanistic models for whole leaf photosynthesis in response to varying environmental factors. It is possible to derive mathematically one or more equations which are consistent with experimental data but they are not necessarily meaningful in relation to biochemistry. For example, the model of Tenhunen *et al.* predicts the response of photosynthesis of a $C_3$ plant (wheat) to varying [$CO_2$] and [$O_2$]. However, the authors made assumptions which resulted in a calculated value of the $Km_{(CO_2)}$ for the carboxylase as low as 0.1$\mu$M which is about two orders of magnitude lower than that of the lowest measured value. It was assumed that the experimental $V$max for photosynthesis is equivalent to the $V$max of the carboxylase (as did Jones and Slatyer). However, the carboxylase potential is now known to be much higher than the experimental $V$max, probably due to a photochemical limitation at high [$CO_2$]. Also, the $V$max of the carboxylase (rate at saturating $CO_2$) was assumed to be equal to $V$max of the oxygenase (rate at saturating $O_2$); however, the $V$max of the carboxylase is several fold higher than the $V$max of the oxygenase. The transfer resistance to $CO_2$ through the aqueous phase was considered to be significant such that the [$CO_2$] in the chloroplast would be much lower than that at the cell surface. Although some gradient is required for diffusion, there is no evidence that the transfer resistance is significant. If carbonic anhydrase functions to facilitate diffusion of $CO_2$, the [$CO_2$] at the carboxylation site could be close to the concentration at the surface of the cell (Fig. 14.4).

As models are based on certain assumptions and are sometimes empirically derived, there are inherent limitations in their interpretation relative to the biochemistry of photosynthesis. Further knowledge of the mechanism of photosynthesis in different species and of internal responses to changing environmental factors can provide useful information for modelling photosynthesis.
[Refs. 3, 6, **20, 22, 26, 29, 30, 31, 36, 48, 51**]

## 14.4 $C_4$ plants—PEP and RBP carboxylase

The properties of RBP carboxylase in $C_4$ plants are apparently similar to those in $C_3$ plants in that the enzyme is inhibited by $O_2$ and has a relatively low $Km_{(CO_2)}$. However, the activity of the enzyme and amount of the protein (fraction I protein) in $C_4$ plants may be considerably lower than in $C_3$ plants (per unit Chl, leaf area or as percentage of total soluble protein). For example, maize and *Panicum miliaceum* ($C_4$) have about 1mg fraction I protein/mg leaf chlorophyll while wheat, tobacco and potato ($C_3$) have 3–6 mg fraction I protein/unit leaf chlorophyll. Whether these differences reflect differences in the carboxylation potential *in vivo* is uncertain. In the models for leaf photosynthesis, a 2-fold higher carboxylation potential/unit chlorophyll for the $C_3$ relative to the $C_4$ plant has been used (Fig. 14.6 versus 14.8).

At one time it was thought that the ability of $C_4$ plants to fix $CO_2$ at very high rates (particularly in high light and at high temperatures) might be attributable to the high affinity for $CO_2$ displayed by PEP carboxylase. At present, however, the reported $Km_{(CO_2/HCO_3^-)}$ values for the two carboxylases are such (cf Sections 6.5 and 13.3) that this argument is no longer tenable (see below).

The characteristics of PEP carboxylase from $C_4$ plants are: (a) it uses $HCO_3^-$ as the substrate and not free $CO_2$, (b) it appears to have a $Km_{(HCO_3^-)}$ around 0.2–0.4 mM, (c) its *V*max with saturating PEP and $HCO_3^-$ is 800– 1200 $\mu$mol mg$^{-1}$ Chl h$^{-1}$ at 30°C, and (d) unlike RBP carboxylase, it is not inhibited by $O_2$. Thus, the advantage of PEP carboxylase over RBP carboxylase in efficiency of carboxylation may lie in its lack of oxygenase activity and possibly higher *V*max, although the actual *V*max of RBP carboxylase is still subject to some question. Certainly, the *V*max of PEP carboxylase at saturating PEP and $HCO_3^-$ is well above the rates of leaf photosynthesis.

Does a $Km_{(HCO_3^-)}$ of 0.2–0.4 mM for PEP carboxylase imply that the enzyme has a low or high affinity for atmospheric $CO_2$? Several factors influence the concentration of $HCO_3^-$ available to PEP carboxylase *in vivo*. These include the concentration of $CO_2$ in the intercellular air space of the leaf, the leaf temperature, the pH of the cytosol, and the possible role of carbonic anhydrase.

In $C_4$ plants, more than 90% of the carbonic anhydrase is in mesophyll cells where it is localized in the cytoplasm. Thus, both PEP carboxylase and carbonic anhydrase are localized in the same compartment. Once $CO_2$ enters the cytosol of the cell, it rapidly hydrates to $HCO_3^- + H^+$. Both $CO_2$ and $HCO_3^-$ diffuse throughout the cytosol along concentration gradients. As the potential gradient for $HCO_3^-$ is much larger than that for $CO_2$, the former is readily available as a substrate for PEP carboxylase (Fig. 14.7). If the catalytic capacity of carbonic anhydrase greatly exceeds that of PEP carboxylase, the final concentration of

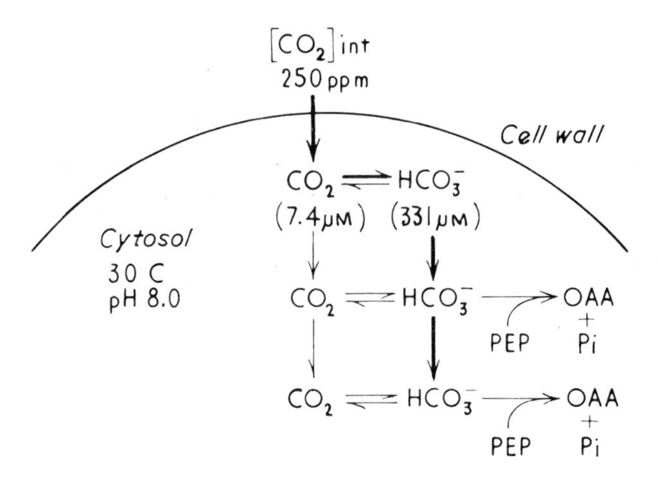


**Fig. 14.7.** Proposed primary path of transfer of $CO_2$ (heavy lines) in $C_4$ leaf as facilitated by carbonic anhydrase.
$CO_2 \rightleftarrows HCO_3^-$ represents $CO_2 + H_2O \rightleftarrows HCO_3^- + H^+$ as catalyzed by carbonic anhydrase.

$HCO_3^-$ in the cytosol is then dependent on the intercellular $[CO_2]$ in the air space around the mesophyll cells, the temperature and the pH of the cytosol. For example, a concentration of 250 ppm $CO_2$ in the leaf air space at a leaf temperature of 30°C would give a solubility of free $CO_2$ of 7.4 $\mu$M. At a pH of 8.0 in the cytosol, this $CO_2$ would be in equilibrium with 331 $\mu$M $HCO_3^-$, a value close to the $Km_{(HCO_3^-)}$ for PEP carboxylase. This could allow rates of about 500 $\mu$mol $CO_2$ fixed $mg^{-1}$ Chl $h^{-1}$ at saturating PEP. Thus PEP carboxylase can have a rather high affinity for atmospheric $CO_2$. There will naturally be variation in the $[HCO_3^-]$ in the mesophyll cell depending on stomatal resistance, leaf temperature, and pH of the cytosol. A major unknown is the cytosolic pH, although the maximum activity of PEP carboxylase from $C_4$ plants *in vitro* is around pH 8.0 with very low activity at pH 7.0.

In contrast to $C_3$ plants, photosynthesis in $C_4$ plants is close to saturation at atmospheric levels of $CO_2$. When rates of photosynthesis are considered in relation to intercellular $[CO_2]$ and the kinetics analyzed by double reciprocal plots of APS versus $(CO_2 - \Gamma)$, sorghum, a $C_4$ species, has an apparent $Km_{(CO_2)}$ of about 2 $\mu$M at 30°C, which corresponds to 68 ppm $CO_2$ in the intercellular space (Fig. 14.5c, d). Therefore, the $C_4$ leaf may have a slightly higher affinity for $CO_2$ than the $C_3$ leaf, even when comparisons are made with the $C_3$ leaf under low $O_2$. At 30°C, 2 $\mu$M $CO_2$ would be in equilibrium with 0.09 mM $HCO_3^-$ (assuming that the pH of the cytosol of mesophyll cells is 8.0). This is less than the $Km_{(HCO_3^-)}$ generally reported for the extracted PEP carboxylase.

At atmospheric levels of $CO_2$ and at 30°C, the $C_4$ plant has a higher rate of photosynthesis than the $C_3$ plant (Fig. 14.5). This can be accounted for by oxygen inhibition of $C_3$ photosynthesis. However, the $C_3$ leaf (wheat) has a higher $CO_2$ saturated rate of photosynthesis than the $C_4$ leaf, which may be due to higher levels of RBP carboxylase in $C_3$ leaves. By assigning kinetic constants to PEP and RBP carboxylase from $C_4$ plants, a general model of leaf photosynthesis can be given (Fig. 14.8). As shown in the illustration, PEP carboxylase has a large carboxylation potential at low [$CO_2$]. It fixes atmospheric $CO_2$ at higher rates than RBP carboxylase in $C_3$ plants, whose carboxylation capacity is inhibited by $O_2$. The $C_4$ cycle is believed to provide a high [$CO_2$] in the bundle sheath cells. Studies with $^{14}CO_2$ fixation suggest that the total pool of $CO_2$ in leaves of $C_4$ plants (on a fresh weight basis) is about 10-fold higher than that in $C_3$ plants, with estimates of up to 60 $\mu$M $CO_2$ in bundle sheath cells during steady-state photosynthesis. In Fig. 14.8 at atmospheric levels of $CO_2$ (indicated by the arrow on the horizontal axis), leaf photosynthesis is saturated at a rate of 310 $\mu mol\ mg^{-1}\ Chl\ h^{-1}$ which corresponds to a [$CO_2$] concentration in bundle sheath cells of about 40 $\mu$M.

The actual rate of $C_4$ photosynthesis is below that of PEP carboxylase activity, particularly at higher [$CO_2$]. PEP is probably not saturated for the enzyme *in vivo* because its regeneration is limited by the rate of the RPP pathway. At these non-saturating levels of the substrate, the enzyme is under allosteric control (Section 12.8).

At 30°C and similar stomatal resistances, $C_4$ plants have higher rates of photosynthesis than $C_3$ plants (e.g., compare rates at 6 $\mu$M $CO_2$, Figs. 14.5, 14.6, 14.8). It is also clear that as [$CO_2$] in the leaf decreases due to increased stomatal resistance or increased temperature, the rates of photosynthesis in $C_4$ plants can readily become 2- to 3-fold higher than in $C_3$ plants.

Whether there is a strict one-to-one relationship between carbon flow through the $C_4$ cycle and the RPP pathway is uncertain. If all the atmospheric $CO_2$ fixed in the RPP pathway in the $C_4$ leaf is linked to the $C_4$ cycle, then the $C_4$ cycle must function at rates at least equivalent to that of the RBP pathway. Even the slightest diffusion of $CO_2$ out of the bundle sheath cells would require the $C_4$ cycle to function at a higher rate than the RPP pathway. Since there is probably a high resistance for diffusion of $CO_2$ out of the bundle sheath cells, the loss of $CO_2$ and additional cycling required through the $C_4$ pathway may be insignificant. Following the illumination of $C_4$ leaves, the $C_4$ cycle must function at a faster rate than the RPP pathway in order that the [$CO_2$] in the bundle sheath can rise above that in the mesophyll cells. Reference has occasionally been made to the $C_4$ cycle as a '$CO_2$-pump'. Like more conventional pumps, energy is used in the process.
[Refs. 4–6, **28–31, 34**]

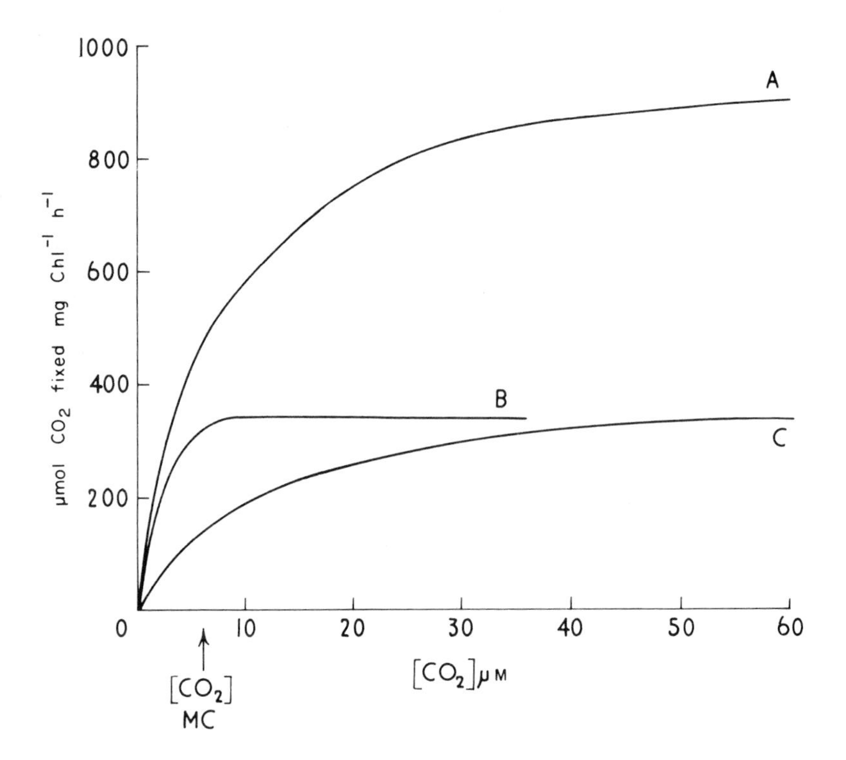


**Fig. 14.8.** Model for leaf photosynthesis in a $C_4$ plant in relation to PEP and RBP carboxylase potential (see Hatch & Osmond, 1976). (a) Potential rate of PEP carboxylase according to Michaelis–Menten kinetics at saturating PEP, cytoplasmic pH of 8.0 where $Km_{(HCO_3^-)}$ for PEP carboxylase equals 0.3 mM equivalent to 6.6 μM $CO_2$ (value could be as low as 2 μM according to kinetic analysis on whole leaf photosynthesis, Fig. 14.5d). (b) Actual rate of leaf photosynthesis in response to [$CO_2$] in mesophyll cells (plot based on experimental data, e.g. Fig. 14.5c). (c) Rate of $CO_2$ fixation through RBP carboxylase at 21% $O_2$ in response to varying [$CO_2$] in bundle sheath cells where

$$V = \frac{V\max[CO_2]}{[CO_2] + Km_{(CO_2)}(1 + [O_2]/Ki_{(O_2)})}$$

(see Chapter 13, Equation 13.36). The arrow on the abscissa represents the [$CO_2$] available to mesophyll cells under atmospheric levels of $CO_2$. At any given rate of photosynthesis, the corresponding [$CO_2$] from curves B and C would represent the [$CO_2$] in mesophyll and bundle sheath cells respectively.

Input: PEP carboxylase, $Km_{(HCO_3^-)} = 0.3$ mM, $V\max = 1000$ μmol mg$^{-1}$ Chl h$^{-1}$; RBP carboxylase, $Km_{(CO_2)} = 5$ μM, $Ki_{(O_2)} = 200$ μM, $V\max = 400$ μmol mg$^{-1}$ Chl h$^{-1}$; $[CO_2]_{int} = 0.025\%$; leaf temperature = 30°C; solubility of $CO_2$ in mesophyll = 7.4 μM $CO_2$, 331 μM $HCO_3^-$; solubility $O_2$ = 245 μM.

## 14.5 Influence of temperature, light, and water stress on carboxylation

$C_4$ species grow particularly well at temperatures of 25–35°C, while $C_3$ species are more temperate and grow best at 15–25°C. This is illustrated by the data of Kawanabe who compared the relative growth of temperate and tropical-origin grasses under three temperature regimes (Table 14.1). One of the earliest and most comprehensive studies on whole leaf photosynthesis in relation to temperature, light and [$CO_2$] using the infra-red gas analyzer was by Gaastra in the 1950s. His and subsequent studies on species such as tomatoes, spinach, and cucumber (now known to be $C_3$ species) show that photosynthesis is limited by [$CO_2$] particularly at higher temperatures (see Figs. 14.9 and 14.10).

In $C_3$ plants at relatively high temperature, the *V*max of RBP carboxylase would be high (typically the *V*max of enzymes increase about 2-fold with each 10°C increase in temperature, i.e. $Q_{10} = 2$). However, the limiting solubility of $CO_2$ and the increased $O_2/CO_2$ ratio at high temperature limits photosynthesis. The greater sensitivity of

**Table 14.1.** Relative growth rates of seedling plants of several warm- and cool-season grass species grown in three day/night temperature regimes. (Adapted from Kawanabe, 1968.)

| Subfamily and species | Temperature regime C* 15/10 | 27/22 | 36/31 |
|---|---|---|---|
| *$C_3$ grasses* | | | |
| Festucoideae | | | |
| *Agropyron trichophorum* (Link) Richt. | 71 | 100 | 62 |
| *Bromus inermis* Leyss. | 77 | 100 | 66 |
| *Festuca arundinacea* Schreb. | 77 | 100 | 56 |
| *Poa pratensis* L. | 69 | 100 | 63 |
| *Trisetum spicatum* (L.) Richt. | 58 | 100 | 64 |
| *Phalaris arundinacea* L. | 71 | 100 | 66 |
| *Stipa hyalina* Nees. | 54 | 100 | 71 |
| *$C_4$ grasses* | | | |
| Eragrostoideae | | | |
| *Chloris gayana* Kunth. | 22 | 91 | 100 |
| Panicoideae | | | |
| *Setaria sphacelata* Stapf. and Hubbard | 5 | 88 | 100 |
| *Cenchrus ciliaris* L. | 11 | 86 | 100 |
| *Paspalum dilatatum* Poir. | 23 | 90 | 100 |
| *Panicum coloratum* Walt. | 35 | 82 | 100 |
| *Panicum maximum* Jacq. | 23 | 89 | 100 |
| *Digitaria argyrogranta* (Nees) Stapf. | 32 | 86 | 100 |
| *Sorghum almum* Parodi | 26 | 77 | 100 |

* Day-night temperatures; data for each $C_3$ species normalized to 100 at 27/22° C, data for each $C_4$ species normalized to 100 at 36/31° C.

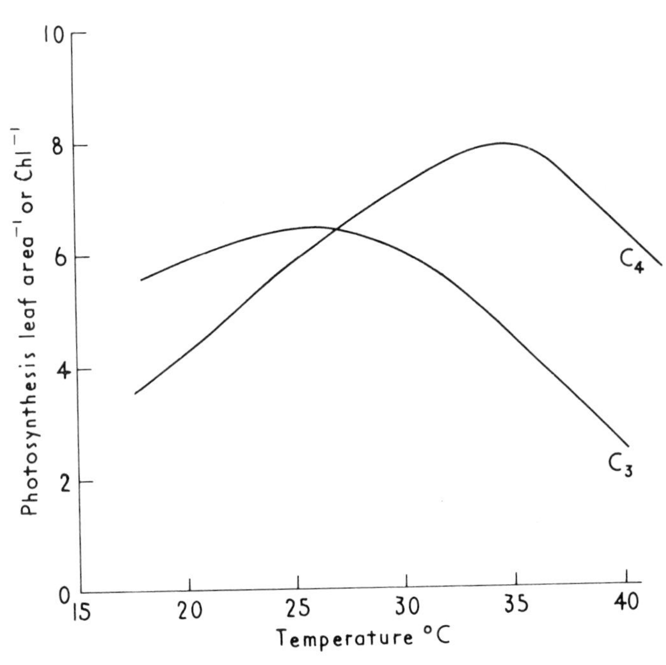


**Fig. 14.9.** Typical temperature response curve for photosynthesis in $C_3$ and $C_4$ plants.

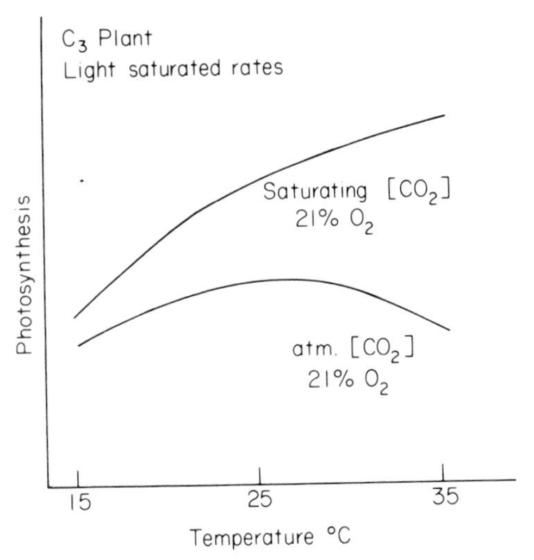


**Fig. 14.10.** Typical temperature response of photosynthesis in a $C_3$ plant under atmospheric levels and saturating levels of $CO_2$.

photosynthesis of $C_3$ plants to $O_2$ at higher temperature may be in part due to a physical effect due to an increased $O_2/CO_2$ solubility ratio with increasing temperature (Fig. 14.2). It has also been suggested that a biochemical change could occur in the kinetic constants of RBP carboxylase-oxygenase which would favour the oxygenase reaction at higher temperatures. However, there are conflicting reports, possibly due to the technical difficulty in simultaneously assaying the oxygenase-carboxylase activities under *known* concentrations of $O_2$ and $CO_2$ at different temperatures. Differential temperature effects on either the activation energies which influence the relative change in the $V$max of the oxygenase to that of the carboxylase or on the kinetic constants $Km_{(CO_2)}$ for the carboxylase, $Km_{(O_2)}$ for the oxygenase have been reported. In another study, temperature had a similar effect on the $Km_{(O_2)}$, $Km_{(CO_2)}$ when calculated on the basis of concentration in the aqueous phase (indicating there is no differential effect of temperature at the biochemical level). However, when calculated from the respective partial pressures of $O_2$ and $CO_2$ in the gas phase, the $Km_{(CO_2\text{, gas phase})}$ increased relatively more than $Km_{(O_2\text{, gas phase})}$ with increasing temperature (indicating a temperature effect due to solubility properties of the gases). (see Refs. **10, 11, 29, 35**).

Dark respiration increases with increasing temperature and may well reduce the net rate of photosynthesis in the light. However, since dark respiration is saturated at low [$O_2$] (Section 13.6), the increased sensitivity of photosynthesis to 21% $O_2$ compared to 2% $O_2$ is probably not due to increased levels of dark respiration.

Obviously, photosynthesis of the plant as a whole responds to the concentration of gases ($CO_2$ and $O_2$) in the atmosphere. However, the biochemical response of photosynthesis in relation to these gases can only be properly analyzed when the gas concentrations are known which are available to the photosynthetic tissue in the aqueous phase at varying temperatures.

In $C_3$ plants at low temperature, the increased solubility of $CO_2$ and the decreased percentage inhibition of photosynthesis by $O_2$ increases the efficiency of photosynthesis. $C_3$ plants appear to have higher levels of RBP carboxylase per unit leaf Chl than $C_4$ plants. At low temperatures where $CO_2$ is less limiting, the high levels of RBP carboxylase in $C_3$ species may give them an equal or higher photosynthetic capacity than $C_4$ plants. In $C_4$ plants at high temperature, the enzyme activity/unit RBP carboxylase protein at saturating $CO_2$ would be high so that less enzyme would be needed to support a high rate. At high temperature the $CO_2$ concentrating mechanism in $C_4$ prevents atmospheric levels of $CO_2$ from becoming rate limiting.

The $Q_{10}$ for photosynthesis (i.e., the extent to which the rate increases as the result of a 10°C rise in temperature) may not be constant, even in saturating $CO_2$. Studies with algae and isolated $C_3$ chloroplasts have indicated $Q_{10}$ values as high as

9–14 at temperatures below 15°C, values of 2 in the 15–25°C range, and values of less than 2 at higher temperatures. This has been attributed to autocatalysis through the RPP pathway (Sections 6.3, 6.14). Thus, a rise in temperature of 10°C would double the rate of each step in the RPP pathway if each enzyme had a $Q_{10}$ of 2 but it would also simultaneously increase the potential rate of feedback so that at the higher temperature, the enzymes would also tend to have more substrate. At still higher temperatures, this tendency would become limited by the rate of supply of assimilatory power (Chapter 5). The overall consequence of these effects is that the photosynthetic carbon assimilation by $C_3$ chloroplasts may not respond in accordance with the Arrhenius law where a linear relationship exists between the logarithm of the maximum velocity of photosynthesis and the reciprocal of absolute temperature on the absolute scale.

In $C_4$ species at optimum temperature, light is the major rate-limiting factor as photosynthesis generally does not saturate up to full sunlight (Fig. 14.11). Light is important for the activation of certain photosynthetic enzymes, for stomatal opening, and for the generation of assimilatory power. For example, the degree of light activation of pyruvate,Pi dikinase and NADP-malate dehydrogenase of the $C_4$ pathway is dependent on light intensity. $C_4$ plants such as maize require relatively high light to achieve maximum stomatal opening in comparison to $C_3$ plants such as wheat. As light intensity increases, the generation of energy would increase providing energy both for the $C_4$ pathway ($CO_2$ pump) and the RPP pathway. Thus,

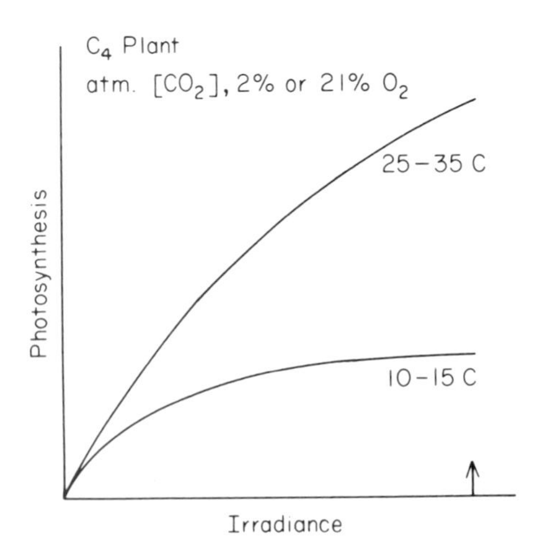


**Fig. 14.11.** Typical photosynthetic light response curve for a single leaf of a $C_4$ plant at high (25–35°C) versus low temperature (10–15°C) and atmospheric levels of $O_2$ and $CO_2$. Arrow on x-axis indicates full sunlight. (see Akita *et al.* 1969; Singh *et al.* 1974).

photosynthesis would continue to respond to light where $CO_2$ is no longer rate limiting.

At low temperatures, photosynthesis of $C_4$ plants is generally saturated well below full sunlight (Fig. 14.11). Maximum rates of leaf photosynthesis of $C_4$ plants may fall below that of $C_3$ at low temperatures. This is not due to limiting light or limiting [$CO_2$]. Rather, it may be due to limitations of metabolite transport, to cold inactivation of certain enzymes, and the reduction of activity of enzymes. For example, pyruvate transport by $C_4$ mesophyll chloroplasts of crabgrass is very sensitive to low temperatures. Pyruvate,Pi dikinase is a cold labile enzyme in many $C_4$ species and dissociates from a tetramer to dimers or monomers with cold treatment. The quantity of fraction 1 protein (RBP carboxylase) in some $C_4$ plants is only about 1/4 of that in $C_3$ plants (Section 14.4). Thus, at low temperatures, the catalytic capacity of this enzyme may be much lower in $C_4$ than in $C_3$ species.

In $C_3$ plants under atmospheric levels of $CO_2$ and $O_2$, photosynthesis is often saturated or near saturated at relatively low light over a wide range of temperatures (Fig. 14.12).

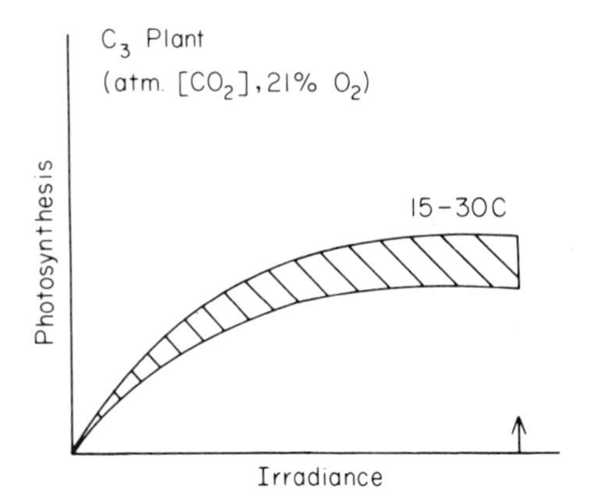


**Fig. 14.12.** Typical photosynthetic light response curve for a single leaf of a $C_3$ plant at atmospheric levels of $O_2$ and $CO_2$. Arrow on x-axis indicates full sunlight. (see Akita *et al.* 1969; Bohlar–Nordenkampf, 1976; Gaastra, 1959; Ku *et al.* 1977; Singh *et al.* 1974).

Above the light saturation point in $C_3$ plants, photosynthesis is limited by [$CO_2$] and by $O_2$ inhibition of photosynthesis. At atmospheric levels of $CO_2$, the percentage inhibition of photosynthesis by $O_2$ (21% vs 2%) is similar or decreases slightly with increasing light intensities. Therefore, the light saturation curve is similar at either 21% or 2% $O_2$ (illustrated in Fig. 14.13).

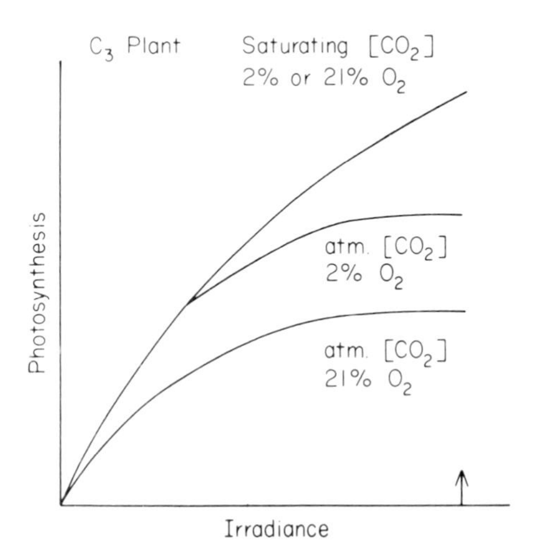


**Fig. 14.13.** Suggested photosynthetic light response curve for a single leaf of a $C_3$ plant at optimum temperature and at varying levels of $O_2$ and $CO_2$. Arrow on x-axis indicates full sunlight. (See Bohlar–Nordenkampf, 1976; Gaastra, 1959; Ku *et al.* 1977).

In $C_3$ plants at low $O_2$ and high light, $CO_2$ may still be rate limiting for photosynthesis. This is particularly true at relatively high temperatures (limiting $CO_2$ solubility) or at relatively high stomatal resistance (limiting $CO_2$ diffusion into the leaf). $C_3$ plants are thus able to benefit most from $CO_2$ enrichment at relatively high temperatures. When the $[CO_2]$ is saturating for $C_3$ plants, they respond much like $C_4$ species to increasing light intensity up to full sunlight (illustrated in Fig. 14.13). The $CO_2$ concentrating mechanism in $C_4$ plants has the dual effect of overcoming $O_2$ inhibition of photosynthesis and providing saturating levels of $CO_2$.

Also, exceptions to generalizations are often of particular interest. Some $C_3$ plants, such as sunflower, for unknown reasons have a relatively high light saturation curve for photosynthesis under atmospheric levels of $O_2$ and $CO_2$ (leaf thickness or Chl/leaf area may be important factors). On the other hand, *Panicum capillare*, a $C_4$ species, has a relatively low light saturation curve for photosynthesis. Within $C_3$ species and within $C_4$ species there is a broad range of adaption to the environment particularly with respect to temperature. *Tidestroma oblongifolia*, a $C_4$ species which grows in Death Valley, California, tolerates very high temperatures (up to 50°C). Most $C_4$ species examined are sensitive to low temperatures of around 10–15°C. However, certain species such as *Atriplex sabulosa*, *Atriplex confertifolia*, *Panicum virgatum* and *Spartina townsendii* are adapted to low temperatures (5–20°C). For such reasons, comparative studies on the regulation of photosynthesis

within $C_3$ and $C_4$ species from a variety of taxonomic groups and habitats are important.
[Refs. 1, 2, **7, 8, 10–14, 17, 19, 23, 24, 27, 30, 32, 33–35, 37, 39, 42, 43, 45, 46, 49, 52, 53**]

## 14.6 Water use efficiency

Transpiration and photosynthesis can be expressed in terms of Ohm's law where:

$$\text{rate} = \frac{\text{potential difference}}{\text{resistance}}$$

For transpiration: $T = \dfrac{W_L - W_a}{r'_a + r'_s}$ Eqn. 14.5

as illustrated in Fig. 14.14 where

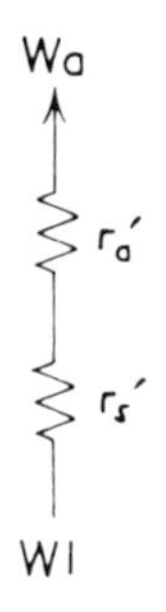


**Fig. 14.14.** Illustration of path of resistance to water loss from the leaf.

$T$ = rate of transpiration measured as water loss from the leaf, e.g., mg water cm$^{-2}$ leaf area s$^{-1}$. It is determined from the increase in relative humidity as an air stream of known humidity is passed over the plant in a leaf chamber.

$W_L$ = water vapour concentration in the stomatal cavity of the leaf, e.g. mg water cm$^{-3}$. This is assumed to be the saturation water vapour concentration in the stomatal cavity which is dependent on leaf temperature.

$W_a$ = water vapour concentration in the air, e.g., mg water cm$^{-3}$.

$r'_a$ = boundary layer resistance to diffusion of water. This resistance is due to an undisturbed layer of air surrounding the leaf and depends on the shape of the leaf and the wind speed. A narrow thin leaf and high wind speed minimize the boundary layer resistance. When making gas exchange measurements in a leaf chamber, it is important to have sufficient air flow to minimize $r'_a$. A high $r'_a$ leads to increased leaf temperature which can be detrimental to measurements of photosynthesis. With adequate air flow, the boundary layer resistance is normally a minor component (e.g., 0.5 s cm$^{-1}$ or less) of the total resistance to gaseous diffusion. $r'_a$ can be determined with a replicate of the leaf made of blotting paper, which is saturated with water. $r'_a$ is represented by its resistance to water loss (evaporation) in the leaf chamber. An estimate of $r'_a$ can also be obtained from the following equation.

$$r'_a = K' \sqrt{\frac{\text{leaf width (cm)}}{\text{width speed (cm s}^{-1})}} \qquad \text{Eqn. 14.6}$$

$K'$ is a constant and depends on the units used, e.g., 0.96 s$^{1/2}$ cm$^{-1}$ for diffusion of water. Thus, a leaf having a width of 2 cm and an air flow across the leaf of 100 cm s$^{-1}$ would have a boundary layer resistance to loss of water of 0.13 s cm$^{-1}$.

$r'_s$ = stomatal resistance to water loss. This will depend on the stomatal frequency and degree of stomatal opening. $r'_s$ can be calculated for a given leaf (s cm$^{-1}$) from determination of all other components in Equation 14.5. Alternatively, diffusion porometers are used with a sensor attached to the leaf which measures loss of water which is a function of stomatal opening.

The use of concentration difference, instead of difference in partial pressure, to express the potential gradient (Eqn. 14.5) is generally sufficient, particularly when the temperature between the leaf and air is similar and when a substantial gradient exists. However, strictly speaking, the gradient depends on the partial pressure difference of the water vapour where the vapour pressure is a function of the absolute temperature and concentration.

For photosynthesis: $$\mathrm{P} = \frac{[CO_2]_{air} - [CO_2]_{CP}}{r_a + r_s + r_m} \qquad \text{Eqn. 14.7}$$

as illustrated in Fig. 14.15 where

P = rate of photosynthesis, e.g., ng $CO_2$ fixed cm$^{-2}$ leaf s$^{-1}$.
$[CO_2]_{air}$ = $CO_2$ concentration in the air, e.g., ng cm$^{-3}$.
$[CO_2]_{CP}$ = $CO_2$ concentration in the chloroplast, e.g., ng cm$^{-3}$.
$r_s$ = stomatal resistance to diffusion of $CO_2$, e.g., s cm$^{-1}$. It can be calculated from $r'_s$ where $(1.56)\ r'_s = r_s$.

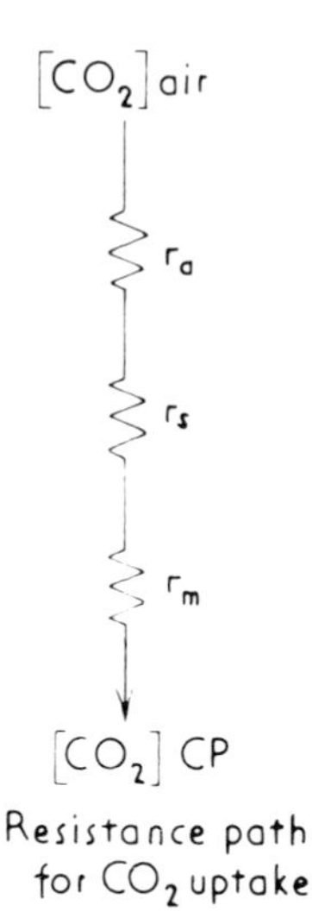


**Fig. 14.15.** Illustration of path of resistance in the leaf to $CO_2$ uptake.

[Water having a lower molecular weight than $CO_2$ has a higher diffusion coefficient (D) where:

$$\frac{D_{H_2O}}{D_{CO_2}} = \left[\frac{1/\text{mol. wt. } H_2O}{1/\text{mol. wt. } CO_2}\right]^{1/2} = \left[\frac{1/18}{1/44}\right]^{1/2} = 1.56$$

Thus, other conditions being equal, water diffuses 1.56 times faster than $CO_2$.]

$r_a$ = boundary layer resistance of leaf to diffusion of $CO_2$, e.g., s $cm^{-1}$. It can be calculated from $r'_a$ as follows:

$$r_a = (1.35)(r'_a)$$

[The ratio $D_{H_2O}/D_{CO_2}$ is raised to the 2/3 power. $(D_{H_2O}/D_{CO_2})^{0.66} = (1.56)^{0.66} = 1.35$), in order to account for that part of transfer in the boundary layer due to turbulence rather than diffusion. This correction is somewhat arbitrary since the degree of turbulence in the boundary layer will depend on the leaf geometry and air flow through the chamber.]

$r_m$ = the mesophyll resistance to $CO_2$, e.g., s $cm^{-1}$. This is a residual resistance (sometimes called ignorance resistance because of the difficulty in measuring its components) which includes diffusion of $CO_2$ in the aqueous part of the mesophyll cell (transfer resistance), carboxylation resistance (resistance due to limiting carboxylation potential), and photochemical resistance. With limiting $[CO_2]$ and high light intensity, the photochemical resistance would be eliminated and then $r_m = 1/CE$ where CE is the carboxylation efficiency (Section 13.10). An estimation of $r_m$ can be obtained from Equation 14.7. All values in the equation can be determined other than $[CO_2]_{CP}$ which is usually assigned a value of $\Gamma$, the $CO_2$ compensation point. It is clear that the actual value of $r_m$ will be overestimated if the value assigned $[CO_2]_{CP}$ is too low.

$C_4$ species are characterized by a lower $r_m$ than $C_3$ species (Table 14.2). This may be due to the fact that the PEP carboxylase of $C_4$ species has a high carboxylation capacity, a high affinity for carbon dioxide, and a lack of $O_2$ inhibition (Sections 14.4, 14.5). As the $O_2$ levels in the atmosphere are reduced, the $r_m$ values in $C_3$ plants decrease to those of $C_4$ species. Since $O_2$ inhibits RBP carboxylase, it follows that increasing $O_2$ will increase the mesophyll resistance to $CO_2$ uptake.

Plants with a low $r_m$ have an advantage for efficiency in water use (grams of water transpired/gram of $CO_2$ fixed), since $r_m$ is a component of photosynthesis but not of

**Table 14.2.** Some comparisons of leaf resistance to $CO_2$ flux in $C_3$ and $C_4$ species. (After Gifford, 1974 and S. B. Ku, personal communication.)

| Species | (reported minimal values) $r_m$* | $r_a + r_s$ |
|---|---|---|
| | s cm$^{-1}$ | |
| $C_3$ | | |
| *Glycine max* | 2–3 | 1.3 |
| *Atriplex hastata* | 2.6 | 1.1 |
| *Phaseolus* spp. | 2.6 | 1.1 |
| *Triticum aestivum* | 2.8 | 1.1 |
| *Solanum tuberosum* | 5.4 | 2.5 |
| *Medicago sativa* | 2.8 | 1.0 |
| $C_4$ | | |
| *Amaranthus* spp. | 0.3 | |
| *Atriplex spongiosa* | 0.4–0.6 | 1.9 |
| *Zea mays* | 0.7–0.9 | 1.5 |
| *Saccharum officinarum* | 0.3 | 1.5 |
| *Sorghum bicolor* | 1.8 | 1.8 |
| *Panicum virgatum* | 2.0 | 2.9 |

* These values of $r_m$ are not equivalent to 1/CE since measurements were made under atmospheric $[CO_2]$ and give the residual resistance including any photochemical limitation (see Section 13.10 for conditions where $r_m = 1/CE$).

transpiration. A decrease of $r_m$ with given values of $r_a + r_s$ will increase photosynthesis while transpiration remains constant. $C_4$ plants tend to have a lower mesophyll resistance and a higher water use efficiency than $C_3$ plants. Data collected by Shantz and Piemeisel more than 50 years ago illustrates the greater water use efficiency of $C_4$ species as compared to $C_3$ species (Table 14.3)

**Table 14.3.** Water use efficiency of different plants. (After Shantz & Piemeisel, 1927; Black *et al.* 1969).

| Species and subsequent classification | g $H_2O$ used/g dry matter produced |
|---|---|
| $C_3$ | |
| *Triticum aestivum* | 557 |
| *Oryza sativa* | 682 |
| *Chenopodium album* | 658 |
| *Citrullus vulgaris* | 577 |
| *Cucumis sativus* | 686 |
| *Phaseolus vulgaris* | 700 |
| *Medicago sativa* | 844 |
| $C_4$ | |
| *Setaria italica* | 285 |
| *Zea mays* | 349 |
| *Amaranthus graecizans* | 260 |
| *Portulaca oleracea* | 281 |

Comparisons of minimum resistance values of some $C_3$ and $C_4$ species indicate that while $r_m$ is lower in $C_4$ than in $C_3$, $r_a + r_s$ tends to be slightly higher in $C_4$ (Table 14.2). Whether the higher diffusive resistance in $C_4$ is due to a lower stomatal frequency (Section 15.14) or a difference in stomatal apertures is not clear. A low $r_m$ and relatively high diffusive resistance in $C_4$ provides conditions which favour a high water-use efficiency (transpiration per unit photosynthesis). For comparisons of species, a high ratio of the total resistance to water loss to the total resistance to $CO_2$ uptake will indicate a high potential water use efficiency.

$$\frac{r'_a + r'_s}{r_a + r_s + r_m} \qquad \text{Eqn. 14.8}$$

In practice, an increased stomatal resistance will also tend to increase the leaf temperature, since transpiration removes heat from the leaf as water evaporates. This, in turn, will increase the water vapour concentration gradient from the leaf to the atmosphere by increasing $W_L$ and will have a positive influence on transpiration not reflected in Equation 14.8. Thus, with increased stomatal resistance, the final rate of transpiration will be dependent on changes in both $r_s'$ and $W_L$. Changes in leaf temperature caused by changes in $r_s$ will also influence the rate of photosynthesis. Models to account for these interactions have been presented. Actual water use efficiency (transpiration per unit dry matter produced) will depend on other factors including respiratory losses and relative humidity. In climates with low relative humidity the water vapour concentration gradient between the leaf and atmosphere will be higher and transpiration efficiency lower.

Although the minimal diffusive resistance ($r_a + r_s$) may be slightly higher in $C_4$ than in $C_3$ plants, the total resistance to $CO_2$ influx is lower at reasonably high temperatures and hence, photosynthetic rates will be higher.

Two general examples can be given to indicate why water-use efficiency could be higher in $C_4$ than $C_3$ plants.

Example I

| Species | Resistance (s cm$^{-1}$) | | | Total |
|---|---|---|---|---|
| | $r_a$ | $r_s$ | $r_m$ | |
| $C_3$ | 1 | 1 | 3 | 5 |
| $C_4$ | 1 | 3 | 1 | 5 |

In this example the total resistance to $CO_2$ uptake is the same in both species which means they will have similar rates of photosynthesis. However, the resistance to water loss is twice as high in the $C_4$ as in the $C_3$ plant; thus, the water use efficiency would be 2-fold higher in the $C_4$, other conditions being equal [i.e. at the same water vapour concentration gradient ($W_L - W_a$)]. This example also indicates that the $C_4$ plant can afford a higher stomatal resistance (as may occur with some water stress) and yet maintain a rate of photosynthesis equivalent to that of a non-stressed $C_3$ plant.

In the second example the boundary layer and stomatal resistances are the same in both species which means that they both would have the same rates of transpiration for a given water vapour concentration gradient.

Example II

| Species | Resistance (s cm$^{-1}$) | | | Total |
|---|---|---|---|---|
| | $r_a$ | $r_s$ | $r_m$ | |
| $C_3$ | 1 | 1 | 3 | 5 |
| $C_4$ | 1 | 1 | 1 | 3 |

However, the total resistance is lower in $C_4$ plants due to the lower $r_m$ which will give the $C_4$ leaf an almost two-fold higher rate of photosynthesis. In this example the water use efficiency will be higher in the $C_4$ leaf because of a higher rate of photosynthesis at a given rate of transpiration. Under natural conditions at reasonably high temperatures (25–35°C) the actual situation may fall somewhere between these two examples if the $C_4$ leaf has a slightly higher stomatal resistance.

Temperature will influence photosynthesis in both $C_3$ and $C_4$ species by its effect on mesophyll and stomatal resistance. As the temperature increases, the $r_m$ values for $C_3$ plants may increase. In part this could be due to a decreased carboxylation potential caused by a decreased solubility of $CO_2$, an increased $O_2/CO_2$ solubility ratio (allowing greater $O_2$ inhibition) or perhaps a change in the kinetic properties of RBP carboxylase-oxygenase which would favour the oxygenase reaction (Section 14.5). This effect of temperature may be partially offset by a temperature-dependent increase in the carboxylation capacity (i.e., increased *V*max).

In contrast, $C_4$ plants have a relatively low $r_m$ at high temperatures. At low temperatures (10–20°C), $r_m$ may be high in $C_4$ due to limiting *V*max of enzymes (perhaps RBP carboxylase) or cold inactivation of certain enzymes. Increasing temperature causes a decrease in the ratio of $r_m(C_4)/r_m(C_3)$ which favours $C_4$ photosynthesis. Temperature-dependent changes in $r_s$ in $C_3$ and $C_4$ species will affect both photosynthesis and transpiration.
[Refs. **9, 16, 25, 41, 44, 54**]

## 14.7 $C_3$, $C_4$ and crop yield

There has been much interest in the possible agronomic benefit of $C_4$ photosynthesis (or agronomic detriment with serious $C_4$ weeds). In selecting for agronomically favourable characteristics among crops, a plant should have a high carboxylation capacity (which would be reflected in a low $r_m$). This would increase its water-use efficiency and, other conditions being favourable, give high rates of photosynthesis.

At their respective temperature optima, $C_4$ plants tend to have higher rates of photosynthesis than $C_3$ plants per unit leaf area, leaf Chl, or leaf nitrogen. Alternatively, $C_4$ plants can afford a higher stomatal resistance, conserve water and yet carry an equivalent rate of photosynthesis as $C_3$ (Section 14.6). In part, $C_4$ plants may have a higher rate of photosynthesis per unit nitrogen due to lower amounts of fraction 1 protein in comparison to $C_3$ species.

Comparisons of growth rates have been made of $C_3$ and $C_4$ crops growing in different environments, but interpretation is complicated due to the number of factors controlling growth. Temperature, light intensity, canopy structure, respiratory losses and harvest index (economic weight/biological weight) are some of the factors which need to be considered when comparisons in productivity among species are made. Comparisons of productivity need to distinguish between short term growth versus growth over the entire life of the plant.

$C_4$ plants are particularly suited to some environments; and under optimum conditions, including high temperature and high light, species like maize and sugarcane possess high growth rates. Likewise, some $C_3$ crops like potatoes and sugarbeets are very productive under temperate conditions (also see Section 3.11). [Refs. **15–18, 21, 38**]

## General Reading


1 BJÖRKMAN O. (1973) Comparative studies on photosynthesis in higher plants. *Photophysiology*. (ed. A. Giese.) Vol. 8, pp. 1–63, New York: Academic Press.

2 BJÖRKMAN O. & BERRY J. (1973) High efficiency photosynthesis. *Sci. Amer.*, **229**, 80–93.

3 BLACK C. C. (1973) Photosynthetic carbon fixation in relation to net $CO_2$ uptake. *Ann. Rev. Plant Physiol.*, **24**, 253–86.

4 HATCH M. D. & OSMOND C. B. (1976) Compartmentation and transport in $C_4$ photosynthesis. In *Encyclopedia of Plant Physiol.* vol. 3. Transport in Plants. New series. (eds. C. R. Stocking and U. Heber) pp. 144–84. New York: Springer-Verlag.

5 JENSEN R. G. & BAHR J. T. (1977) Ribulose 1, 5–bisphosphate carboxylase-oxygenase. *Ann. Rev. Plant Physiol.*, **28**, 379–400.

6 JOLIVET E. (1976) Les différentes modalités de carboxylation photosynthétique chez les espèces de type $C_3$ et de type $C_4$. *Physiol. Veg.*, **14**, 563–94.


## Specific Citations


7 AKITA S., MIYASAKA A. & MURATA Y. (1969) Studies on the differences of photosynthesis among species. I. Differences in the response of photosynthesis among species in normal oxygen concentration as influenced by some environmental factors. *Proc. Crop Sci. Soc., Japan.* **38**, 509–23.

8 AKITA S. & MOSS D. N. (1972) Differential stomatal response between $C_3$ and $C_4$ species to atmospheric $CO_2$ concentration and light. *Crop Sci.*, **12**, 789–93.

9 AKITA S. & TANAKA I. (1973) Studies on the mechanisms of differences in photosynthesis among species. IV. The differential response in dry matter production between $C_3$ and $C_4$ species to atmospheric carbon dioxide enrichment. *Proc. Crop. Sci. Soc., Japan.* **42**, 288–95.

10 BADGER M. R. & ANDREWS T. J. (1974) Effects of $CO_2$, $O_2$, and temperature on a high affinity form of ribulose diphosphate carboxylase-oxygenase from spinach. *Biochem. Biophys. Res. Commun.*, **60**, 204–10.

11 BADGER M. R. & COLLATZ G. J. (1977) Studies on the kinetic mechanism of ribulose 1,5-bisphosphate carboxylase and oxygenase reactions, with particular reference to the effect of temperature on kinetic parameters. *Carnegie Institute Yearbook 1976*. pp. 355–61.

12 BALDRY C. W. BUCKE C. & WALKER D. A. (1966) Temperature and photosynthesis. I. Some effects of temperature on carbon dioxide fixation by isolated chloroplasts. *Biochim. Biophys. Acta*, **126**, 207–13.

13 BIRD I. F., CORNELIUS M. J. & KEYS A. J. (1977) Effects of temperature on photosynthesis by maize and wheat. *J. Exp. Bot.*, **28**, 519–24.

14 BJÖRKMAN O., PEARCY R. W., HARRISON A. T. & MOONEY H. (1972) Photosynthetic adaptation to high temperatures. A field study in Death Valley, California. *Science*, **175**, 786–789.

15 BLACK C. C. (1971) Ecological implications of dividing plants into groups with distinct photosynthetic production capacities. *Advan. Ecol. Res.*, **7**, 87–114.

16 BLACK C. C., CHEN T. M. & BROWN R. H. (1969) Biochemical basis for plant competition. *Weed Sci.*, **17**, 338–44.

17 BOLHAR-NORDENKAMPF H. R. (1976) Measurements and mathematical models for the $CO_2$ gas exchange of *Phaseolus vulgaris* var. nanus L. with special consideration of photorespiration and atrazine effect. *Biochem. Physiol. Pflanzen.* **169**, 121–61.

18 BROWN R. H. (1978) A difference in N use efficiency in $C_3$ and $C_4$ plants and its implications in adaption and evolution. *Crop Sci.*, **18**, 93–8.

19 CALDWELL M. M., OSMOND C. B. & NOTT D. L. (1977) $C_4$ pathway photosynthesis at low temperatures in cold-tolerant *Atriplex* species. *Plant Physiol.*, **60**, 157–64.

20 GAASTRA P. (1959) Photosynthesis and crop plants as influenced by light, carbon dioxide, temperature, and stomatal diffusive resistance. *Meded Land Hogesch Wageningen*, **59**, 1–68.

21 GIFFORD R. M. (1974) A comparison of potential photosynthesis, productivity, and yield of plant species with differing photosynthetic metabolism. *Aust. J. Plant Physiol.*, **1**, 107–17.

22 HOGETSU D. & MIYACHI S. (1979) Role of carbonic anhydrase in photosynthetic $CO_2$ fixation in *Chlorella. Plant Cell Physiol.* **20**, 747–56.

23 HUBER S. C. & EDWARDS G. E. (1977) Transport in $C_4$ mesophyll chloroplasts. Characterization of the pyruvate carrier. *Biochim. Biophys. Acta*, **462**, 583–602.

24 JOHNSON H. & HATCH M. D. (1970) Properties and regulation of leaf NADP-malate dehydrogenase and malic enzyme in plants with the $C_4$-dicarboxylic acid pathway of photosynthesis. *Biochem. J.*, **119**, 273–80.

25 JONES H. G. (1976) Crop characteristics and the ratio between assimilation and transpiration. *J. Appl. Ecol.*, **13**, 605–22.

26 JONES H. & SLATYER R. (1972) Estimation of the transport and carboxylation components of the intracellular limitation to leaf photosynthesis. *Plant Physiol.*, **50**, 283–8.

27 KAWANABE S. (1968) Temperature responses and systematics of the Gramineae. *Proc. Japan. Soc. Plant Taxon.*, **2**, 17–20.

28 KU S. B. & EDWARDS G. E. (1975) Photosynthesis in mesophyll protoplasts and bundle sheath cells of various types of $C_4$ plants. IV. Enzymes of respiratory metabolism and energy utilizing enzymes of carbon assimilation. *Z. Pflanzenphysiol.*, **77**, 16–32.

29 KU S. B. & EDWARDS G. E. (1977a) Oxygen inhibition of photosynthesis. I. Temperature dependence and relation to $O_2/CO_2$ solubility ratio. *Plant Physiol.*, **59**, 986–90.

30 KU S. B. & EDWARDS G. E. (1977b) Oxygen inhibition of photosynthesis. II. Kinetic characteristics as affected by temperature. *Plant Physiol.*, **59**, 991–9.

31 KU S. B. & EDWARDS G. E. (1978) Photosynthetic efficiency of *Panicum hians* and *Panicum milioides* in relation to $C_3$ and $C_4$ plants. *Plant & Cell Physiol.*, **19**, 665–75.

32 KU S. B., EDWARDS G. E. & SMITH D. (1978) Photosynthesis and non-structural carbohydrate concentration in leaf blades of *Panicum virgatum* L. as affected by night temperature. *Can. J. Bot.*, **56**, 63–8.

33 KU S. B., EDWARDS G. E. & TANNER C. B. (1977) Effects of light, carbon dioxide, and temperature on photosynthesis, oxygen inhibition of photosynthesis, and transpiration in *Solanum tuberosum. Plant Physiol.*, **59**, 868–72.

34 KU S. B., SCHMITT M. R. & EDWARDS G. E. (1979) Quantitative determination of RuBP carboxylase-oxygenase protein in leaves of several $C_3$ and $C_4$ plants. *J. Exp. Bot.*, **30**, 89–98.

35 LAING W. A., OGREN W. L. & HAGEMAN R. H. (1974) Regulation of soybean net photosynthetic $CO_2$ fixation by the interaction of $CO_2$, $O_2$, and ribulose-1, 5-diphosphate carboxylase. *Plant Physiol.*, **54**, 678–85.

36 LILLEY R. McC. & WALKER D. A. (1975) Assimilation by leaves, isolated chloroplasts and ribulose bisphosphate carboxylase from spinach. *Plant Physiol.*, **55**, 1087–92.

37 LONG S. P., INCOLL L. D. & WOOLHOUSE H. W. (1975) $C_4$ photosynthesis in plants from cool temperature regions, with particular reference to *Spartina-townsendii. Nature*, **257**, 622–4.

38 MONTEITH J. L. (1978) Reassessment of maximum growth rates for $C_3$ and $C_4$ crops. *Expl. Agric.*, **14**, 1–5.
39 MURATA Y. & IYAMA J. (1963) Studies on the photosynthesis of forage crops. II. Influence of air-temperature upon the photosynthesis of some forage and grain crops. *Proc. Crop Sci. Soc., Japan.* **31**, 315–22.
40 RABINOWITCH E. I. (1945) *Photosynthesis and Related Processes.* Vol. 1. pp. 326–51 New York: Wiley Interscience.
41 RAWSON H. M. & BEGG J. E. (1977) The effect of atmospheric humidity on photosynthesis, transpiration and water use efficiency of leaves of several plant species. *Planta*, **134**, 5–10.
42 SELWYN M. J. (1966) Temperature and photosynthesis. II. A mechanism for the effects of temperature on carbon dioxide fixation. *Biochim. Biophys. Acta*, **126**, 214–24.
43 SERVAITES J. C. & OGREN W. L. (1977) Oxygen inhibition of photosynthesis and stimulation of photorespiration in soybean leaf cells. *Plant Physiol.*, **61**, 62–7.
44 SHANTZ H. L. & PIEMEISEL L. N. (1927) The water requirements of plants at Akron, Colorado. *J. Agric. Res.*, **34**, 1093–189.
45 SHIRAHASHI K., HAYAKAWA S. & SUGIYAMA T. (1978) Cold lability of pyruvate, orthophosphate dikinase in the maize leaf. *Plant Physiol.*, **62**, 826–30.
46 SINGH M., OGREN W. L. & WIDHOLM J. M. (1974) Photosynthetic characteristics of several $C_3$ and $C_4$ plant species grown under different light intensities. *Crop Sci.*, **14**, 563–6.
47 STRANG R. H. C. (1981) Estimation of *K*m values of enzymes requiring molecular $O_2$ as a substrate. *Biochem. J. Letters.* **193**, 1033–4.
48 TENHUNEN J. D., WEBER J. A., YOCUM C. S. & GATES D. M. (1979) Solubility of gases and the temperature dependency of whole leaf affinities for carbon dioxide and oxygen. An alternative perspective. *Plant Physiol.*, **63**, 916–23.
49 TERRI J. A. & STOWE L. G. (1976) Climatic patterns and the distribution of $C_4$ grasses in North America. *Oecologia*, **23**, 1–12.
50 UMBREIT W. W., BURRIS R. H. & STAUFFER J. F. (1972) *Manometric techniques and related methods for the study of tissue metabolism.* Minneapolis, Minnesota: Burgess Publishing Co.
51 WALKER D. A. & LILLEY R. MCM (1975) Ribulose bisphosphate carboxylase—an enigma resolved? In: *Perspectives in Experimental Biology.* Vol. 2. (ed. N. Sunderland.) pp. 189–198, Oxford: Pergamon Press.
52 WILLIAMS G. J. (1974) Photosynthetic adaptation to temperature in $C_3$ and $C_4$ grasses. *Plant Physiol.*, **54**, 709–11.
53 YAMAMOTO E., SUGIYAMA T. & MIYACHI S. (1974) Action spectrum for light activation of pyruvate phosphate dikinase in maize leaves. *Plant Cell Physiol.*, **15**, 987–92.
54 ZELITCH I. (1971) *Photosynthesis, Photorespiration, and Plant Productivity.* pp. 347, New York: Academic Press.

# Chapter 15
# Crassulacean Acid Metabolism



## SUMMARY

The history of studies on plants with Crassulacean Acid Metabolism (CAM) is longer than either that of photorespiration or $C_4$ photosynthesis. The earliest

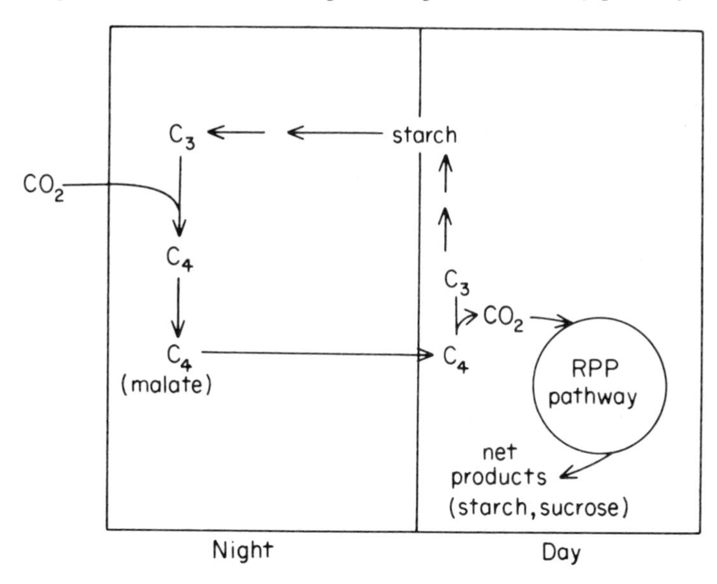


**Fig. 15.1** Crassulacean acid metabolism.
Plants which have Crassulacean Acid Metabolism (named after the family Crassulaceae in which it was first discovered) are very conservative in water usage due to a metabolic adaption in photosynthesis. $CO_2$ is fixed at night into malate. During the day malate is decarboxylated, and the released $CO_2$ assimilated through the RPP pathway. This allows the stomata to be closed during the day when the potential for water loss is high due to high day temperatures. Photosynthesis in CAM and $C_4$ plants have a common feature in that carbon is donated from $C_4$ acids to the RPP pathway.

observations showed dark acidification, and a dark respiratory quotient indicative of both $CO_2$ and $O_2$ uptake in the dark. However, it was many years before the link was made between dark fixation of $CO_2$ and dark acidification.

Dark acidification is now known to be accompanied by starch degradation, glycolytic formation of phosphoenolpyruvate (PEP), and carboxylation through PEP carboxylase utilizing atmospheric $CO_2$. The oxaloacetate formed is reduced to malate and stored in the vacuole. In the light, malate is decarboxylated through 'malic enzyme' or converted to oxaloacetate and decarboxylated through PEP carboxykinase. The $CO_2$ is then donated to the RPP pathway.

Some CAM plants can switch to a $C_3$-like photosynthesis under certain environmental conditions. In the 'CAM mode', the stomata are open by night and closed by day, while the reverse is true in the $C_3$ mode. CAM species can often show patterns of gas exchange involving both day and night uptake of $CO_2$. Water stress and high day—low night temperatures favour the CAM mode. The isotope composition ($^{13}C/^{12}C$) of a plant is useful to distinguish between these modes of $CO_2$ assimilation. PEP carboxylase shows little discrimination against fixing $^{13}CO_2$ whilst discrimination is more pronounced with RBP carboxylase.

Ecologically, CAM is best described as an adaptation to water stress and water conservation. CAM plants have a higher water-use efficiency than $C_3$ or $C_4$ plants but a lower productivity. Water is conserved during night fixation because the water vapour concentration gradient between the leaf and the atmosphere is low. The relatively high stomatal resistance and low mesophyll resistance (the latter due to $CO_2$ fixation through PEP carboxylase) provide favourable conditions for minimizing the transpiration/photosynthesis ratio.

## 15.1 Introduction

We are all accustomed to the proposition that green plants take up $CO_2$ and evolve $O_2$ by day (in photosynthesis) and take up $O_2$ and give out $CO_2$ by night (in respiration). At one time it was even the custom to remove flowers from hospital wards at night because of their supposed ill-effect on the air although this failed to take into account either the fact that non-green tissues evolve $CO_2$ continuously or that the contribution made by a bunch of flowers is likely to be substantially less than that of one person. In fact, green leaves probably also respire continuously, but under normal conditions, in bright light, $CO_2$ fixation greatly exceeds, and therefore masks, $CO_2$ evolution. There is still some controversy centred round the proposal that 'dark' respiration in green plants may not continue in the light. This is too large a question to enter into here, but while it seems certain that 'dark' respiration is unlikely to continue unchanged in a tissue which is simultaneously carrying out photosynthesis (and probably also photorespiration, Chapter 14) there is no good evidence that the tricarboxylic acid cycle ceases to function and reasonably definitive evidence to the contrary. Moreover, it is beyond dispute that most plants will grow in continuous light and that, in order to do so, they must draw on TCA cycle intermediates (such as succinate which is a precursor of $\delta$-aminolevulinic acid in chlorophyll synthesis).

There is also little doubt that most green tissues fix some $CO_2$ in the dark. Certainly incorporation of $^{14}CO_2$ into organic acids has been encountered wherever it has been sought. Usually, however, it is less than 1 % of full photosynthetic $CO_2$ fixation and is entirely masked by respiratory $CO_2$ evolution. In some plants, however, there is a massive fixation of $CO_2$ into malate (and some other acids) by night and a process of 'deacidification' by day in which decarboxylation is followed by photosynthetic refixation of $CO_2$. This diurnal fluctuation is well marked and was most studied in species belonging to the Crassulaceae. In the papers of Meirion Thomas, therefore, the term 'acid metabolism', which had been previously used by Bennet-Clark became 'Crassulacean Acid Metabolism' and this, or its abbreviation 'CAM', is now often applied to all those plants which exhibit similar behaviour. It should be noted, however, that this metabolic variant is not restricted to the Crassulaceae. Although it is normally associated with succulence, there are many succulents (e.g. halophytes) which do not exhibit CAM and there are CAM plants which can, at best, be described as semi-succulent.

CAM plants often have large mesophyll cells in the leaf ($\sim$ 100–200 microns in diameter) several times the size of those in $C_3$ or $C_4$ plants and the relatively large vacuole (the site of malate storage) of the cell is surrounded by a thin layer of cytoplasm. There is considerable anatomical variation in leaf anatomy among CAM species. Anatomical studies alone are not a means of distinguishing between $C_3$ and CAM species (whereas, for example, species of the family Gramineae can be classified as $C_3$ or $C_4$ from anatomical studies since no CAM plants are found within this family). Although pineapple, an important commercial plant having CAM, is succulent it has relatively small mesophyll cells (Fig. 15.2) without any anatomical features which readily distinguish it from $C_3$ species (Fig. 10.7b,d).

CAM is known to occur among species of the Agavaceae, Aizoaceae, Asclepiadaceae, Bromeliaceae, Cactaceae, Compositae, Crassulaceae, Cucurbitaceae Dideraceae, Euphorbiaceae, Liliaceae, Orchidaceae, Polypodiaceae, Portulacaceae, Vitaceae, and Welwitschiaceae. It is suspected to occur in some species of the families Geraniaceae, Labiatae, Oxalidaceae and Piperaceae. Without conclusive evidence one or more species of the families Butaceae, Caryophyllaceae, Chenopodiaceae, Convolvulaceae and Plantaginaceae were suggested to be CAM plants. Sometimes only a few species of a family are CAM plants. Many CAM plants exhibit either $C_3$

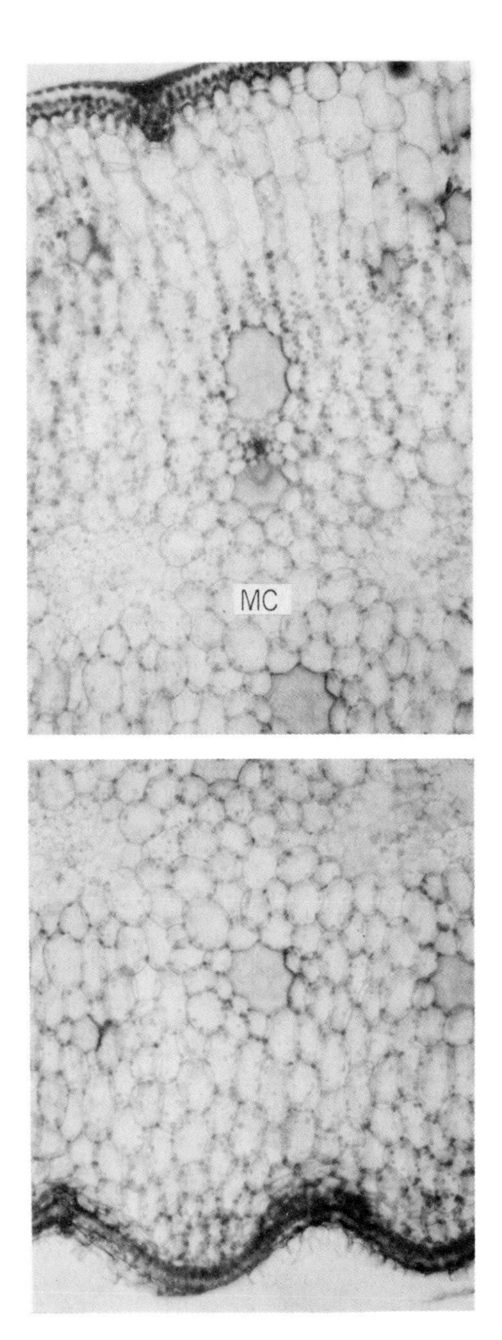


**Fig. 15.2.** Light micrographs of cross sections of leaf of a CAM plant (*Ananas comosus*, pineapple) × 288. Top, section towards upper epidermis; bottom, section towards lower epidermis. The diameter of mesophyll cells (MC) is approximately 30 microns. (Kindly provided by S. B. Ku).

photosynthesis or CAM depending on plant age or environmental conditions (e.g. CAM may be faculative, Section 15.11). Obviously, this flexibility in metabolism can make the identification of species which have the *capacity* for CAM more difficult than the identification of $C_3$ and $C_4$ plants. Some plants are considered to possess CAM at all times (obligate). In this case, e.g. some species of Cactaceae and Crassulaceae, only the magnitude of CAM may change with a change in environmental conditions.
[Ref. 7, **14, 15, 16, 54, 59**]

## 15.2 Discovery

CAM is an old subject and provides a remarkable example of the way in which a number of investigators were unable or unwilling to fit known facts into a feasible hypothesis. It is easy, of course, to be wise after the event and it is the general experience of most research workers that they are unable to see what is staring them in the face until it leaps from the bench and beats them about the head.

Carbon dioxide uptake in the dark was first reported as long ago as 1804 by de Saussure who carried out extensive measurements of gaseous exchange in plant organs, building upon the earlier work of Lavoisier with animals. For many years the term Respiratory Quotient (RQ) was used to describe the volumetric ratio of $CO_2/O_2$ in animal and plant respiration. RQ is now often restricted to the net gaseous exchange associated with such processes at the molecular level, e.g. the oxidation of sugar or lipid and the term 'apparent RQ' or 'gaseous exchange quotient (GEQ)' is applied to the gross gaseous exchange exhibited by whole organs or organisms which may reflect several simultaneous processes. De Saussure was correctly able to interpret $O_2$ uptake and concomitant $CO_2$ release by plants in the dark as an aspect of what is now referred to as 'respiration' or 'dark respiration' (i.e. an oxidative consumption of organic reserves which provides the 'energy' and metabolites for growth and other processes). He was unable, however, to explain the *simultaneous uptake* of $CO_2$ and $O_2$ (the so-called *negative* RQ) which he observed when he placed the cactus *Opuntia* in $CO_2$-enriched atmosphere in the dark. Similarly, Heyne, who discovered dark/light acidification/deacidification in 1813 by resorting to the mildly eccentric expedient of tasting *Bryophyllum* leaves, looked for an explanation in terms of $O_2$ uptake and release rather than incorporation of $CO_2$. The following is a letter, sent by Heyne to A. B. Lambert, Esq., Vice-President of the Linnean Society, and read by the Secretary on April 20th, 1813.

Dear Sir,

I had an opportunity some time ago of mentioning to you a remarkable, deoxidation of the leaves of a plant in day-light. As the circumstance is in itself curious, and throws great light on the opinion of those celebrated philosophers who have written on the subject, I will state it shortly in this letter, which if you please, you may in extract, or in any other way you think proper, lay before the Society.

The leaves of the Cotyledon calycina, the plant called by Mr. Salisbury *Bryophyllum calycinum*, which on the whole have an herbaceous taste, are in the morning as acid as sorrel, if not more so. As the day advances, they lose their acidity, and are tasteless about noon; and become almost bitterish towards evening. This is the case in India, where this plant is pretty generally cultivated in our gardens and it remains to be seen if the same takes place in the hot-houses in England, where it has been lately introduced.

I have seen this plant but once in this country, and that was at Mr. Loddiges', at Hackney, about twelve o'clock in the day-time, when I found it quite tasteless. The distance of that place from my habitation has hitherto prevented me from attending to it at an earlier hour in the morning, I have, however, but little doubt it will be found as acid as I have described it to be in India.

I need scarcely observe, that the acidity which these leaves possess in the morning cannot be ascribed to anything else than to the oxygen which the plant has absorbed during the darkness of the night, or which has been transferred from other constituent principles of the plant during that period. I think it has been absorbed, as it is so loosely united at its base, that even the light of the day has an immediate effect of disengaging it again.

Both Priestly and Ingenhousz have concluded, from numerous experiments, that all plants exhale vital air in the day-time, and fix air or carbonic acid gas during the night; but these conclusions have been called in question by some, from the various results of experiments since made on this subject. What I have now related is therefore not destitute of interest, as it seems incontrovertibly to establish the theory of these celebrated philosophers.

I was in hopes of learning something new or pertinent on this interesting subject in Sprengel's work on the Structure and Nature of Plants: but, to my great disappointment, there is nothing to be found but what has been advanced by the two philosophers mentioned, and by Saussure and Sennebier in later times.

Sprengel expatiates much on the exhalation and absorption of carbonic gas, and only once mentions oxygen, when he notices Sennebier's observations; according to which, more carbonic gas is exhaled by plants during the night in close vessels, than there is oxygen disengaged in sunshine.

I beg leave further to observe, that the plant above treated of is, in my opinion, truly a species of Cotyledon, with which it perfectly agrees, in habit and generic characters; the only difference being in the number of the parts of fructification,

which in Cotyledon calycina are one-fifth less than in the other species of the genus, a difference, however, that according to the principles of the Linnean System, does not form a sufficient grounds for separation.
I have the honour to be, etc.
20th April, 1813.

Benjamin Heyne

Heynes letter was then overlooked for half a century and de Saussure's observations were attributed to carbon dioxide entering into solution in the (acid!) sap of the Cactus. Nevertheless, by the late 1890s both Aubert and Kraus had established a correlation between the extent of dark acidification and the degree of light assimilation during the previous day. Both authors demonstrated that such plants could accumulate starch when illuminated in a $CO_2$-free atmosphere and both suggested that deacidification involved the oxidation of organic acids and the release of $CO_2$ which was then re-fixed in photosynthesis. Despite the fact that Warburg (1886) had also shown that acidification could be correlated with $CO_2$ uptake, and deacidification with $CO_2$ release, neither these workers nor Bennet-Clark, Vickery and Wolf (who worked in this field in the 1930s and 1940s) were able to make the final connection as it is now seen. It was left to Thomas and his colleagues in the 1940s to rediscover the 'de Saussure effect' when they observed a simultaneous uptake of $CO_2$ and $O_2$ by *Bryophyllum* in the dark. This plant was known to accumulate acid and Thomas was led to the suggestion that malate was formed from internal and external $CO_2$ by carboxylation of an acceptor formed in the breakdown of starch and sugars (Fig. 15.3). The newly discovered Wood and Werkman reaction (Eqn. 15.3) was invoked to illustrate the biochemical feasibility of such a reaction.
[Refs. 1, 8, **14, 41, 59–62, 65**]

## 15.3 The Wood and Werkman Reaction

In 1936 Wood and Werkman found that $CO_2$ was utilized by propionic acid bacteria during their fermentation of glycerol. Subsequently work with the heavy carbon isotope $^{13}C$ showed that $CO_2$ was incorporated into propionic acid and propyl alcohol and into the carboxyl groups of succinate. Carbon dioxide was also fixed during the aerobic dissimilation of pyruvate by pigeon breast muscle. These observations led to the proposal that the initial reaction involved the carboxylation of *pyruvate or a derivative of pyruvate.* Kalnitsky and Werkman also proposed that the energy required might be derived from phosphoenolpyruvate. When proteins

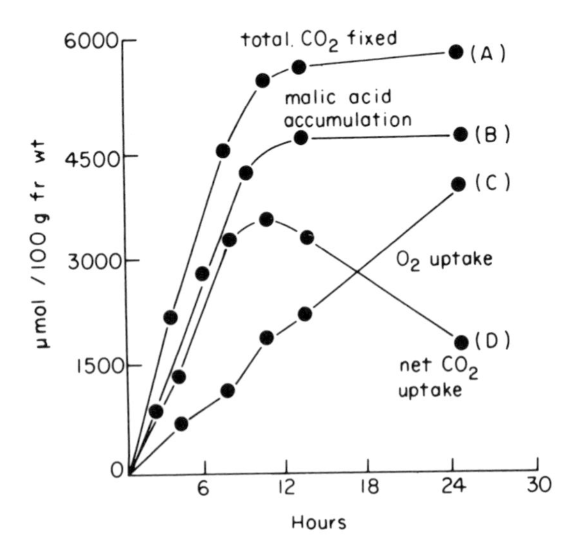


**Fig. 15.3.** Demonstration by Ranson that malate accumulation in *Bryophyllum* leaves can be accounted for in terms of $CO_2$ fixed ● (Adapted from Fig. 19.3, Thomas *et al.*, 1973)

Measurements were in the dark at 12°C. (A) Total $CO_2$ fixed equals the net uptake of $CO_2$ (curve D) + respiratory loss of $CO_2$ (calculated from curve C). (B) Malic acid accumulation measured from titratable acidity. (C) Oxygen uptake from respiration (considered due to normal oxidation of carbohydrate). (D) Net $CO_2$ uptake from an initial atmospheric level of 5%. From many such experiments with species from several genera the Newcastle group obtained ratios of about 2.6 mg malate formed per mg $CO_2$ fixed (approaching 1/1 on a molarity basis) and the quantitative conclusion was drawn that malate accumulated *could be* accounted for by $CO_2$ fixation.

which displayed oxaloacetic decarboxylase activity were isolated from bacteria and other organisms (Eqn. 15.1),

$$\text{oxaloacetate} \rightarrow \text{pyruvate} + CO_2 \qquad \text{Eqn. 15.1}$$

it was wrongly assumed, in some quarters, that the reverse of this reaction might account for the above results. Such proposals disregarded the very unfavourable equilibrium position of pyruvate carboxylation, as such ($\Delta F' = +6$kcal approx.). In theory this could be overcome by linking it to a very favourable reaction such as that catalysed by malate dehydrogenase

$$\text{pyruvate} + CO_2 \rightarrow \text{oxaloacetate} \qquad \text{Eqn. 15.1}$$
$$\text{oxaloacetate} + \text{NADH} + \text{H}^+ \rightarrow \text{malate} + \text{NAD}^+ \qquad \text{Eqn. 15.2}$$

and, in fact, Herbert was able to demonstrate the incorporation of small quantities of labelled $CO_2$ using appropriate enzyme mixtures. The rates, however, were entirely inadequate, reflecting as they did, the fact that such linked systems must satisfy kinetic as well as thermodynamic criteria in order to function well.

Thus, if an unfavourable reaction $A \rightleftharpoons B$ is to be displaced to the right by linking it to $B \rightleftharpoons C$ and the overall reaction sequence is to be catalysed by two enzymes, it is essential that the second has a very high affinity for the end-product (B) of the first reaction. If this is not the case, the catalysis of the second stage will be so slow (in the presence of the vanishingly small quantities of substrate produced by the first stage) that the overall equilibrium position will not be approached within a finite time. An initial failure to appreciate these aspects earned pyruvate carboxylation the dubious distinction of being called 'the wouldn't work man' reaction, a phrase which did scant credit to its originators who were never themselves in any doubt that the real mechanism was yet to be characterized. In higher plants the problem was resolved when Bandurski and Greiner isolated PEP carboxylase from spinach. For the first time two key criteria had been satisfied, i.e. a reaction with a favourable equilibrium position catalysed by an enzyme with a high affinity for $CO_2$. The Wood and Werkman reaction, as first conceived, had become a reality.
[Refs. 8, **11**, **27**, **30**, **78**]

## 15.4 The path of carbon in CAM

Present evidence suggests that CAM plants degrade starch and the resulting sugars and sugar derivatives to triose phosphates by conventional routes including the EMP (or glycolytic) pathway and the oxidative pentose phosphate pathway. The oxidative pentose phosphate pathway could contribute to oxidation of carbohydrate without the double carboxylation originally proposed (Section 15.8). A substantial fraction of the phosphoenolpyruvate so generated is converted to malate by the combined action of PEP carboxylase and malate dehydrogenase.

$$\text{PEP} + CO_2 + H_2O \rightarrow \text{oxaloacetate} + \text{Pi} \qquad \text{Eqn. 15.3}$$
$$\text{oxaloacetate} + \text{NADH} + H^+ \rightarrow \text{malate} + \text{NAD}^+ \qquad \text{Eqn. 15.2}$$

These events occur in the dark and account for $CO_2$ fixation, acidification and the consequent change in taste observed by Heyne at the beginning of the last century. During the following day malate is decarboxylated and the $CO_2$ which is released is refixed in photosynthesis. Decarboxylation may be effected by Ochoa's NADP-malic enzyme,

$$\text{malate} + \text{NADP}^+ \rightarrow \text{pyruvate} + \text{NADPH} + H^+ + CO_2 \qquad \text{Eqn. 15.4}$$

NAD-specific malic enzyme,

$$\text{malate} + \text{NAD}^+ \rightarrow \text{pyruvate} + \text{NADH} + H^+ + CO_2 \qquad \text{Eqn. 15.5}$$

or by a combination of malate dehydrogenase and PEP carboxykinase.

malate + $NAD^+$ → oxaloacetate + NADH + $H^+$ Eqn. 15.2

oxaloacetate + ATP → PEP + ADP + $CO_2$ Eqn. 15.6

In the latter sequence, the decarboxylation product is PEP and this could be reassimilated into carbohydrates by a reversal of glycolysis or be metabolized through the TCA cycle. If the decarboxylation product is pyruvate, as in the reaction catalysed by malic enzyme, then there is an added barrier to gluconeogenesis but this may be overcome by the following reaction

pyruvate + ATP + Pi → PEP + AMP + PPi Eqn. 15.7

and the pyruvate,phosphate dikinase required to catalyse this reaction has been reported in several malic enzyme type CAM plants. If PEP is produced by either of the above methods then there is evidently a need for compartmentation or control to prevent its immediate consumption by the PEP carboxylase reaction in what would then constitute a futile cycle. It is believed that the 3-carbon product of decarboxylation is largely metabolized to starch. To the extent the 3-carbon product is oxidized through the TCA cycle the $CO_2$ released may be reassimilated.
[Refs. 4, **28, 41**]

## 15.5 Carbon dioxide as a metabolite in the dark

Radioactive tracers have been used so extensively in the elucidation of metabolic sequences in the immediate past that it sometimes comes as a surprise to learn that the formulation of the glycolytic pathway and the tricarboxylic acid cycle owed little to their employment. The same can be said of the dark fixation of $CO_2$ in CAM where it is clear that the essential evidence was obtained before tracers were employed. Thus, if malate is formed by a process of dark fixation, there must obviously be an equivalence between $CO_2$ fixed and malate formed. This was demonstrated in Thomas's laboratory by Beevers and by Ranson in the period 1945–1947. Although Thomas mentioned these results (and his conclusion that $CO_2$ was a metabolite) in the third edition of his textbook which appeared in 1947, he was characteristically slow to publish in the more usual way and the papers with his co-workers did not appear until 1949 and 1954 respectively. Working independently in the United States, Bonner and Bonner arrived at similar conclusions (1948) when they demonstrated that malate accumulation varied as a function of external $CO_2$ concentration in the range of 0–0.1%. Finally radioactive tracers came into their own with the demonstration that $^{14}CO_2$ was taken up by *Bryophyllum* in the dark and incorporated into malate. Later, Saltman and his colleagues, using the same methods employed in tracing the path of carbon in photosynthesis found label

restricted almost entirely to malate and aspartate after exposure of only 6 seconds and concluded that these compounds were derived by reduction and transamination of a common precursor—a conclusion consistent with the prior formation of oxaloacetate by the carboxylation of PEP.
[Refs. **17, 46, 59–64, 69**]

## 15.6 The inverse relationship between substrate and product

Malate is believed to be formed in CAM plants by dark carboxylation of a substrate derived from carbohydrate catabolism. Extensive work by Wolf, Bennet-Clark and Vickery *et al* has shown that starch breakdown occurs during dark acidification and that, when the latter is at its maximum, $CO_2$ release is at its minimum. This would be consistent with utilization of $CO_2$ in acid formation (See Fig. 15.4).
[Refs. 9, **14, 45, 65**]

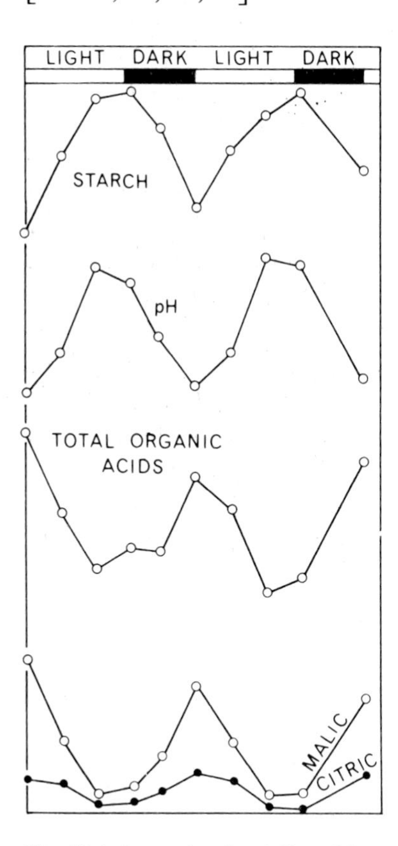


**Fig. 15.4.** Inversely related diurnal fluctuations in *Bryophyllum*. Starch and pH increase by day and fall by night. Total organic acid (and particularly malate) falls by day and increases by night. (After Thomas, 1947)

## 15.7 The enzymes concerned in acidification and deacidification

As indicated above there is evidence that one of the enzymes responsible for malate synthesis in CAM plants should be able to carboxylate a carbohydrate cleavage product and yield a product which could be as readily converted into aspartate as malate. The role for PEP carboxylase was proposed by Walker in 1956 following its isolation from *Kalanchoe* and *Sedum.* (The enzyme was first studied in spinach extracts by Bandurski and Greiner in 1953). Similarly, the role of malic enzyme in decarboxylation was proposed by Walker in 1960.

Work at Newcastle had also suggested that the carboxylase concerned must have a high affinity for $CO_2$ (CAM plants in the dark can remove all the $CO_2$ from small enclosed atmospheres) and that one or more steps in the entire sequence must be susceptible to inhibition by higher $CO_2$ concentrations (since dark acidification is similarly sensitive—Fig. 15.5). PEP carboxylase displays both of these characteristics.

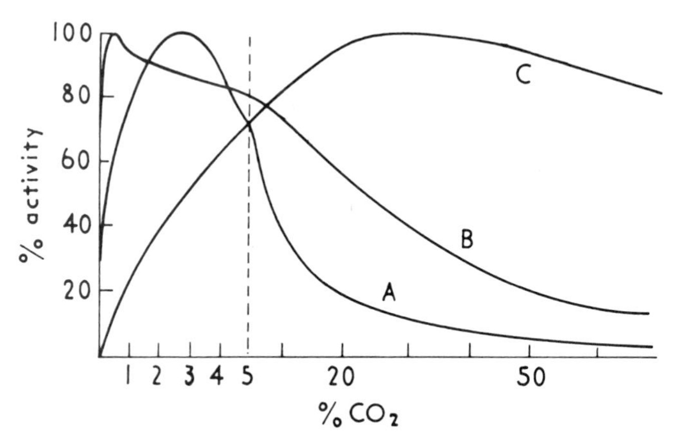


**Fig. 15.5.** Dark acidification in leaves (Curve A) is stimulated by $[CO_2]$ up to about 3% and is then progressively inhibited. PEP carboxylase (Curve B) is also inhibited by high $[CO_2]$ whereas carboxylation of pyruvate by Ochoa's malic enzyme (Curve C) is poor at low $[CO_2]$ and reaches a maximum as dark acidification approaches zero. All data for *Kalanchoe* (or *Kalanchoe* extracts) from Thomas' laboratory at Newcastle. (After Wolf, 1960)

Although Bandurski and Greiner and Tchen and Vennesland both commented on the high affinity of PEP carboxylase for $CO_2$ (displayed respectively by preparations from spinach and wheat germ), the first actual measurements were made by Walker and Brown (1957) who put the apparent *K*m (bicarbonate) of *Kalanchoe* extracts at 0.22 mM (pH 7.4), see also Section 14.4. More recent

measurements have yielded a variety of values, some as low as 0.02 mM and it also seems that the enzyme may exist in two forms with equally disparate affinities for PEP. The possibility that these differences reflect active and inactive forms of the enzyme is worthy of consideration. Inhibition by malate might in itself result in a decrease in activity by day which would diminish the possibility of a futile cycle in which immediate recarboxylation by PEP carboxylase would defeat the purpose of decarboxylation of malate which is an integral feature of CAM. On the other hand, this end might be equally well achieved by simultaneous increases in the $K$m values for both bicarbonate and PEP. To date, however, the question of activation of PEP carboxylase, comparable to activation of RBP carboxylase (Section 6.11), has been neglected, presumably because the highest $K$m's yet recorded are still indicative of a very high affinity for bicarbonate. As mentioned elsewhere (Section 15.4), decarboxylation may proceed via malic enzyme or PEP carboxykinase. The presence of pyruvate, phosphate dikinase might also account in some part for the results of Wolf who found that when pyruvate was fed to leaf segments of *Bryophyllum calycinum* (either during acidification or deacidification) oxygen-uptake increased and the GEQ (Gas Exchange Quotient) decreased. Since pyruvate oxidation would itself enhance $CO_2$ production more than $O_2$ uptake, it was argued that pyruvate must affect some other process and that its utilization in malate formation was the most likely possibility. Pyruvate utilization in the pyruvate, phosphate dikinase reaction is regarded as a first step in gluconeogenesis but if normal compartmentation or regulation is bypassed by external feeding through cut surfaces it seems possible that PEP formed from pyruvate might be immediately consumed in malate formation. It should be noted however that the dikinase is a light-activated enzyme and is unlikely to function effectively in the dark.
[Refs. 7, 9, **11, 55, 62, 66–70**]

## 15.8 The double carboxylation hypothesis

Incorporation of $^{14}CO_2$ via the PEP carboxylase reaction leads to the formation of malate and aspartate specifically labelled in carbon 4 (the $\beta$ carboxyl group). PEP carboxylase is sometimes referred to as a $\beta$-carboxylase because $CO_2$ entering the product does so at the $\beta$ position, i.e. at C-3 where numbering of the carbon atoms begins at the existing carboxyl group of the substrate; labelling of carbon atoms with Greek letters starts with C-2 and designate position of carbon atoms relative to C-1.

$$\underset{1}{\text{HOOC}}\cdot\underset{2}{\overset{\alpha}{\text{C}}}\text{O}\cdot\underset{3}{\overset{\beta}{\text{C}}}\text{H}_2\cdot\underset{4}{\text{C}}\text{OOH}$$

In the plant, however, malate and aspartate are found to be labelled in both carboxyl groups and an approximately 1/3–2/3 distribution of label between carbons 1 and 4 was originally found under a variety of conditions and after both long and short exposures to $^{14}C$. There are various ways in which carbon, initially incorporated into carbon 4 could be subsequently transferred to carbon 1. For example fumarate is symmetrical and interconversion of malate and fumarate promoted by fumarase could lead to randomization.

This possibility was always recognized but originally discounted because of the constancy of the $C_1/C_4$ ratio. More recent work has shown that the use of *Lactobacillus plantarum*, in the determination of the intra-molecular distribution of radioactivity in malate, is questionable. However, in many experiments, the preponderance of label in carbon 4, which would be predicted on the basis of $\beta$-carboxylation, has actually been observed. However, during the period that the ratio was regarded as more or less constant it was necessary to suggest a feasible explanation and Ranson, Bradbeer, and Stiller were led to propose a double carboxylation. According to their scheme a modified pentose phosphate pathway would lead to the formation and subsequent carboxylation of ribulose bisphosphate in the dark. In the presence of $^{14}CO_2$ this scheme would give two molecules of PGA, one labelled and the other not (Eqn. 6.15). After conversion to PEP these would be again carboxylated so that one would contain an atom of $^{14}C$ in both carboxyl groups and, in the other, only one of the two carboxyls (the $\beta$ carboxyl) would be labelled. In short, the observed ratios would result from the presence of some molecules of malate labelled in *both* carboxyls and not from a mixture of molecules labelled *only* in carbons 1 *or* 4.

This proposal received no support from Sutton and Osmond or from Kluge *et al.* who stimulated $^{14}C$ incorporation into malate by feeding PEP and PGA to leaf slices of *Bryophyllum* but found no change in the distribution of label within the malate molecule. Finally Cockburn and McAuley carried out an elegant and definitive experiment using a mass spectrograph and $^{13}C$. In principle, this allowed them to weigh the molecules concerned and showed conclusively that none had the extra mass that would have been derived from the presence of two atoms of $^{13}C$ in the same molecule of malate. The double carboxylation hypothesis has therefore been discounted, although it remains a mystery why about 2/3 of the label is consistently found in carbon 4 of malate.
[Refs. **18, 20, 29, 31, 53, 76**].

## 15.9 Light acidification

Normally malic acid is consumed in the light. In attempts to understand the regulation of acidification-deacidification, Thomas and Beevers found that enriching the atmosphere with $CO_2$ retarded the deacidification process. Since the [$CO_2$] has recently been shown to increase in the leaf during deacidification (Section 15.14), it is possible that $CO_2$ exerts some control on the rate of deacidification *in vivo*.

Further, Ranson showed when deacidified tissue is illuminated and given enriched atmospheres of $CO_2$, acidification occurs. A maximum is reached at about 5 % $CO_2$. Values near this maximum are maintained even in atmospheres of 10–20 % $CO_2$ despite the fact that normal photosynthetic uptake of $CO_2$ (and its associated $O_2$ evolution) are inhibited and dark fixation falls to between 18 and 38 % of its maximum in similar atmospheres. Thomas and his co-workers suggested that the suppression of photosynthesis in high [$CO_2$] might be explained in terms of inhibited formation of ATP and NADPH. This would favour the conversion of PGA to PEP rather than its reduction to triose phosphate. Why $CO_2$ should inhibit the generation of assimilatory power is not discussed but, conversely, it has been reported that high $CO_2$ may inhibit photosynthetic carbon assimilation by lowering the stromal pH and it may also be inferred from work on the reconstituted chloroplast system that RBP carboxylase (like PEP carboxylase) is sensitive to high $CO_2$. The notion that PEP might undergo carboxylation in the light to any large extent would also be difficult to equate with its proposed role in gluconeogenesis from malate. An alternative explanation is that such high levels of $CO_2$ would displace the equilibrium of the decarboxylating reactions. It has been shown for example that in 10 % $CO_2$ the formation of malate from pyruvate, catalysed by the malic enzyme, may exceed 80 % of its maximum value.
[Refs. 7, **61, 62, 68**]

## 15.10 Energy requirements in CAM

The energy requirements per $CO_2$ fixed in CAM can be considered in four phases: dark acidification, $C_4$ acid decarboxylation, $CO_2$ fixation in RPP pathway, and conversion of $C_3$ precursors of decarboxylation to starch.

### (a) Dark acidification

During starch breakdown and malate synthesis at night (Fig. 15.6) there is no net change in reductive power. The NADH formed during glycolysis through NAD-triose phosphate dehydrogenase is used for reducing oxaloacetate to malate. If starch breakdown occurs through phosphorylase there would be net synthesis of 0.5ATP/PEP formed. If starch hydrolysis occurs through amylases then no ATP would be synthesized. In either case there is little or no net change in assimilatory power during dark acidification. The mechanism of malate accumulation in the vacuole during acidification is uncertain although it may require energy.

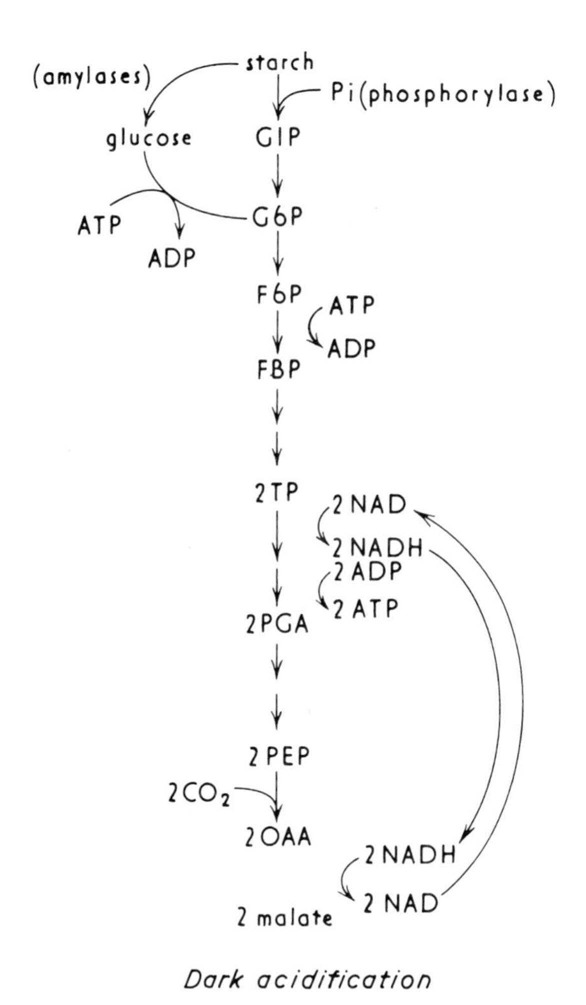


**Fig. 15.6.** Dark acidification, showing energy requirements.

(b) $C_4$ ACID DECARBOXYLATION

With decarboxylation through NADP-malic enzyme one NADPH is formed per $CO_2$ released (Eqn. 15.4). Malate conversion to PEP through NAD-malate dehydrogenase and PEP carboxykinase requires one ATP and generates one NADH (Eqns. 15.2 and 15.6).

### (c) RPP PATHWAY

Energy required for $CO_2$ fixation through the RPP pathway is 3 ATP, 2 $NADPH_2$ per $CO_2$ (Chapter 6). If there is $O_2$ inhibition of photosynthesis, then energy requirements would be higher (Section 13.6). But CAM plants, like $C_4$ plants may provide sufficient $CO_2$ during $C_4$ acid decarboxylation to repress $O_2$ inhibition of photosynthesis (Section 15.14).

### (d) METABOLISM OF THE $C_3$ PRODUCT OF DECARBOXYLATION TO STARCH

Where NADP-malic enzyme is the decarboxylase, metabolism of pyruvate to starch would require 3.5 ATP, 1 NADH per pyruvate. Three of these ATP are required in the conversion of pyruvate to triose phosphate (sum of Eqns. 15.7–15.13).

| | | |
|---|---|---|
| pyruvate + ATP + Pi → PEP + AMP + PPi | (pyruvate, Pi dikinase) | Eqn. 15.7 |
| PPi → 2 Pi | (pyrophosphatase) | Eqn. 15.8 |
| AMP + ATP → 2 ADP | (adenylate kinase) | Eqn. 15.9 |
| PEP → 2-PGA | (enolase) | Eqn. 15.10 |
| 2-PGA → 3-PGA | (phosphoglyceromutase) | Eqn. 15.11 |
| 3-PGA + ATP → DPGA + ADP | (PGA kinase) | Eqn. 15.12 |
| DPGA + NADH + $H^+$ → TP + $NAD^+$ + Pi | (NAD-triose phosphate dehydrogenase) | Eqn. 15.13 |

**Sum:** pyruvate + 3 ATP + NADH + $H^+$ → TP + $NAD^+$ + 3 ADP + 2 Pi  Eqn. 15.14

The other 0.5 ATP is required per $C_3$ equivalent of triose phosphate converted to starch, since the synthesis of ADPG, a precursor for starch synthesis, requires one ATP per glucose phosphate.

G1P + ATP → ADPG + PPi  (ADPG pyrophosphorylase)

Where PEP carboxykinase is the decarboxylase, PEP would be metabolized to starch requiring 1.5 ATP, 1 NADH per PEP. Overall, the energy requirements would then be 6.5 ATP, 2 NADPH per $CO_2$ fixed in CAM species using NADP-malic enzyme and 5.5 ATP, 2 NADPH in CAM species using PEP carboxykinase. Energy requirements per net $CO_2$ fixed will increase at very high temperatures where respiratory losses increase and under extreme drought conditions where stomata remain closed and CAM functions to recycle respiratory carbon.

Among CAM plants studied, a given species has NADP-malic enzyme or PEP-carboxykinase as the major decarboxylase. Thus, they can be classified into subgroups, i.e., NADP-ME and PEP-CK types analogous to $C_4$ plants (Chapter 11). There is also some evidence for NAD-malic enzyme in some CAM plants. However, the extent to which decarboxylation occurs through this enzyme is uncertain. Presently CAM species comparable to NAD-ME type $C_4$ plants have not been identified (i.e., not shown to be the major decarboxylase). The energy requirements for decarboxylation through NAD-malic enzyme, where it occurs in CAM, may be similar to that through NADP-malic enzyme.

The intracellular localization of enzymes of photosynthesis in CAM plants is not well established. In particular, the location of PEP carboxylase and the decarboxylases will determine the requirements for metabolite transport between the chloroplasts and cytosol.

Malate is stored in the vacuoles of the cells during dark acidification. Recent studies on intracellular localization of enzymes suggest PEP carboxylase, NADP-malic enzyme and PEP carboxykinase are in the cytoplasm and NAD-malic enzyme in the mitochondria of CAM species. Then the terminal reactions leading to acidification and the initial reaction of deacidification occur outside the chloroplasts (Fig. 15.7).

It appears that the energy requirements per $CO_2$ fixed in CAM plants are very similar to or slightly higher than in $C_4$ plants (Section 11.6). Also in $C_3$ plants and $C_4$ plants, the energy requirements per $CO_2$ fixed under atmospheric conditions are similar due to the $O_2$-dependent decrease in quantum yield in $C_3$ (Section 13.12). Therefore, all three photosynthetic groups may have similar energy requirements for photosynthesis under atmospheric conditions.

In spite of considerable knowledge of metabolism in CAM species its regulation is not understood. This includes regulation of enzyme activity in the dark versus the light, the mechanism of malic acid storage in and release from the vacuole, metabolite transport by the chloroplast, and control of sucrose versus starch synthesis. For instance PEP carboxylase and the decarboxylases, NADP-malic enzyme and PEP carboxykinase, function in the cytoplasm. Carboxylation and decarboxylation through these enzymes are regulated so they generally function at different times; otherwise there would be futile cycles of carboxylation-decarboxylation. Recent studies by Winter indicate that PEP carboxylase changes its properties between light and dark in a CAM species, such that during the day the

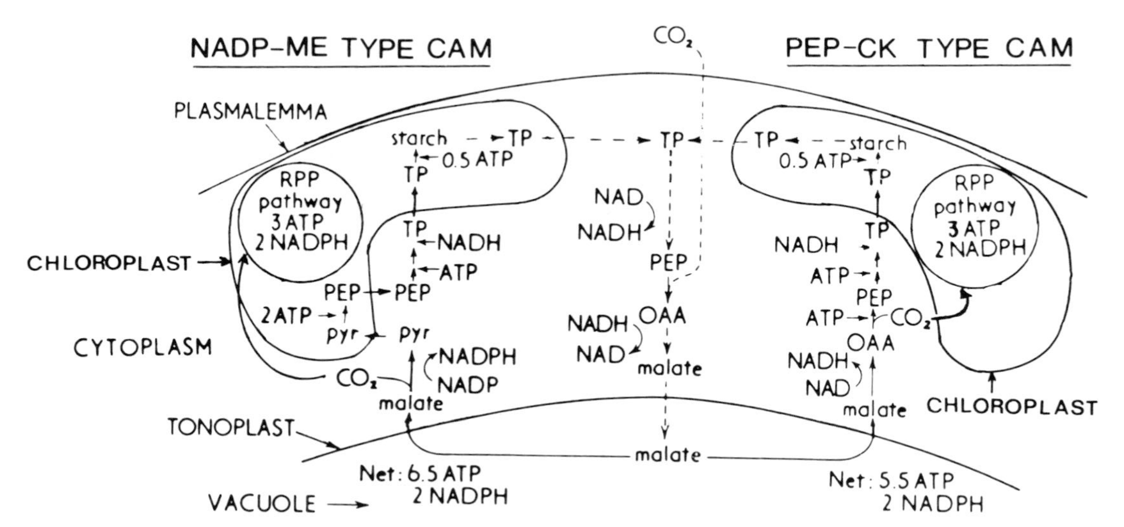


**Fig. 15.7.** Models for intracellular transport and photosynthetic pathways in CAM. (-------) dark metabolism, (——) light metabolism. Note that the net energy requirement per $CO_2$ fixed is that which must be generated photochemically in the light to drive carbon assimilation. As illustrated the $C_3$ product originating from malate decarboxylation is directly metabolized to starch. However, the triose phosphate formed from the $C_3$ product of decarboxylation is likely mixed with a common pool of triose phosphate in the chloroplast. See text for discussion. [Adapted from results with *Sedum praealtum* (NADP-ME type) and several PEP-CK type species. (Spalding *et al*, 1979a; Ku *et al*, 1980).

enzyme is particularly sensitive to inhibition by malate. Such findings are important towards understanding the control of CAM (see Fig. 15.8).
[Refs. **15, 24, 25, 32, 50, 52, 74**]

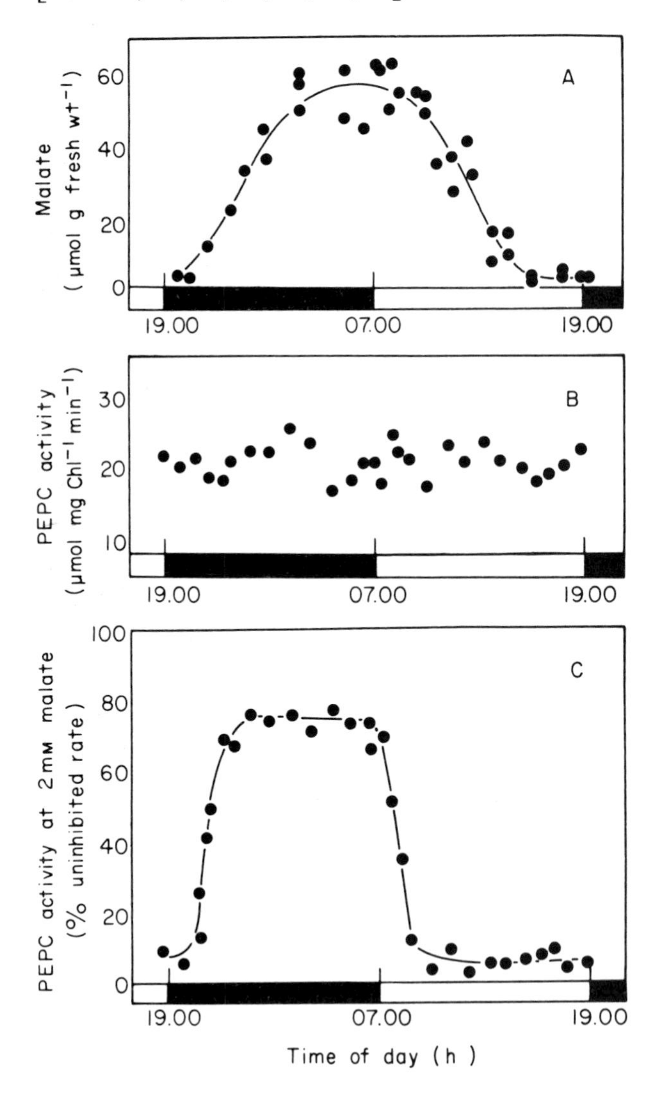


**Fig. 15.8.** Changes in the sensitivity of PEP carboxylase to inhibition by malate with the inducible CAM plant *Mesembryanthemum crystallinum*. The enzyme was extracted from fully expanded leaves during a 12 hour dark/12 hour light cycle. Malate content (A), extractable PEP carboxylase activity at pH 8.0 and 2 mM PEP in the absence (B) and presence of 2mM malate (C). (unpublished data kindly provided by Dr. Klaus Winter, also see Winter, 1980).

## 15.11 CAM mode versus $C_3$ mode

Some CAM plants have a plasticity in photosynthesis which is not seen in $C_3$ or $C_4$ species. Depending on the environmental conditions, certain CAM species may alter their degree of day versus night fixation of atmospheric $CO_2$. When CAM plants fix atmospheric $CO_2$ during the day, it is believed to be fixed directly through RBP carboxylase analogous to $C_3$ photosynthesis.

Temperature, water stress, and photoperiod are environmental factors found to shift the mode of photosynthesis in some CAM species. High day temperature, low night temperature, and water stress favour dark assimilation of $CO_2$. In some species of the genus *Kalanchoe* short days and long nights favour the CAM mode and induce flowering. In some *Sedum* species, long days and short nights favour flowering and appear to favour the CAM mode of photosynthesis.

The induction of dark fixation of $CO_2$ in some CAM species by saline conditions may be mimicking water stress. For example, either water stress or watering plants of *Mesembryanthemum crystallinum* and *Portulacaria afra* with a high salt media shifts them from fixing atmospheric $CO_2$ in the day to a CAM mode of photosynthesis. These treatments induced both diurnal fluctuation in leaf acidity and night fixation of $CO_2$.

The influence of temperature on photosynthesis in CAM is illustrated from studies by Neales on pineapple. A maximum proportion of the atmospheric fixation of $CO_2$ in the dark occurs under low night temperatures (Fig. 15.9).

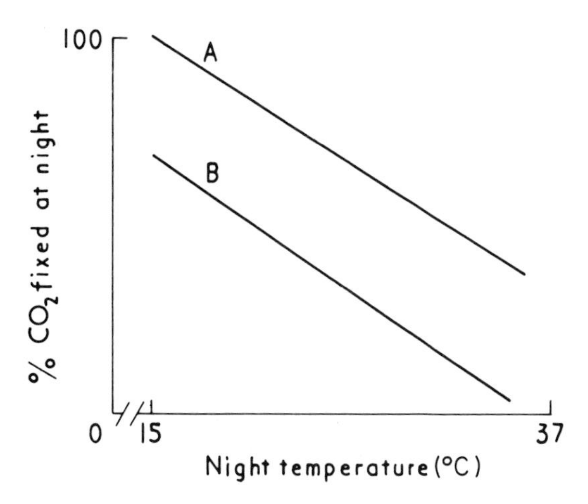


**Fig. 15.9.** $CO_2$ fixation by pineapple. Day temperature was 25 °C, day length was (A) 14 hours dark/10 hours light; (B) 8 hours dark 16 hours light. (Adapted from Neales 1973b).

The low rates of net $CO_2$ fixation in the dark at high temperature could be due to increased respiration rather than reduction in the rate of CAM. The average minimum night temperature during the growing season for pineapple in Hawaii is 10° C; in Australia, 12° C; and in South Africa, 5° C. This suggests that in these areas, pineapple would be growing primarily in the CAM mode.

Osmond (1978) notes that many CAM species, when functioning in the CAM mode, exhibit four phases of metabolism during a 24-hour cycle. Phase I is $CO_2$ fixation and acid synthesis in the dark. Phase II occurs at the beginning of the light period and consists of an initial burst of fixation of atmospheric $CO_2$. Both PEP and RBP carboxylase may be active at this time. Phase III (deacidification phase) occurs in the light and consists of $C_4$ acid decarboxylation and refixation of $CO_2$ with the stomata closed. Phase IV occurs during the latter part of the light period after deacidification is completed, and consists of fixation of atmospheric $CO_2$ predominantly via RBP carboxylase, as the stomatal conductance increases.
[Refs. 2–5, **19, 22, 26, 35, 38–40, 43, 44, 56, 72**]

## 15.12 Isotope fractionation in CAM relative to $C_3$ and $C_4$

Atmospheric $CO_2$ contains small quantities of the heavy carbon isotope $^{13}C$ (about 1% as compared to $^{12}C$). Measurements of $^{13}C/^{12}C$ in various material and in the atmosphere are related to a cretaceous limestone from South Carolina (fossil carbonate skeleton of *Belemnitella americana*).

The atmosphere contains less $^{13}CO_2$ relative to $^{12}CO_2$ than does the standard. Because the differences are small, the relative amounts of $^{13}C$ as compared to $^{12}C$ in a sample is expressed as $\delta^{13}C$ value in units per mil‰ (i.e. units per 1000). Against the standard, atmospheric $CO_2$ has a $\delta^{13}C$ value of about $-7‰$. There are slight variations in the value of $^{13}C/^{12}C$ in the atmosphere, but on average this means the atmosphere has 7 parts per 1000 less $^{13}CO_2$ than the standard. Even though the difference is small the amount of $^{13}C$ in a sample relative to $^{12}C$ can be very accurately determined using a mass spectrometer.

$$\delta^{13}C = \left\{\frac{^{13}C/^{12}C \text{ sample}}{^{13}C/^{12}C \text{ standard}} - 1\right\} \times 10^3$$

Where the standard has 1.116% $^{13}C$ and the atmosphere has on average 1.108% $^{13}C$ then:

$$\delta^{13}C = \left\{\frac{^{11.08}/1000}{^{11.16}/1000} - 1\right\} \times 10^3 = -7‰$$

[i.e. 7/1000 less $^{13}CO_2$ in atmosphere than in sample

$7/1000\ (1.116\%) = 0.0078\%$
$1.116\% - 0.008\% = 1.108\%$ $^{13}C$ in atmosphere]

While the $\delta^{13}C$ value is an expression of the isotope composition relative to the standard, isotope fractionation (or discrimination) is the difference in $\delta$ value between source and product of a particular reaction or process.

$$\text{Fractionation} = \frac{\delta(\text{source}) - \delta(\text{product})}{1 + [\delta(\text{source}) \div 1000]}$$

If the $\delta$ value of source is $-7‰$ and the $\delta$ value of the product is $-27‰$ then the fractionation is $+20‰$. Then the $\delta$ value of the product is dependent on the degree of fractionation and the amount of the minor isotope in the source where the latter is corrected for by the denominator in the equation. This correction becomes significant when the source has a relatively large amount of the minor isotope.

During carbon assimilation all plants discriminate against $^{13}CO_2$ but $C_3$ plants discriminate more than $C_4$ or CAM. Thus the average values for tissue of $C_3$ plants are $-25$ to $-35‰$ whereas $C_4$ values are $-10$ to $-17‰$ and CAM values (during active acidification) are near to those of $C_4$. Table 15.1 shows steps between uptake of atmospheric $CO_2$ and carbon assimilation where fractionation may occur. Since the isotope fractionation for RBP carboxylase ($+34‰$) is much greater than that for PEP carboxylase ($+2‰$) the difference in isotope composition in $C_3$ versus $C_4$

**Table 15.1.** Factors in the initial phase of $CO_2$ uptake which may influence the carbon isotope composition of plants. (also see O'Leary 1981).

| Factors | Fractionation values ‰ |
|---|---|
| Diffusion of $CO_2$ in air ($CO_{2\,ext}$ to $CO_{2\,int}$) | +4 |
| Equilibration of $CO_2$ to $HCO_3^-$ [a] ($CO_2 + H_2O \rightleftharpoons HCO_3^- + H^+$) | −8 |
| PEP carboxylase [b] | +2 |
| RBP carboxylase [c] | +34 |

[a] The heavier isotope general concentrates where bonds are strongest, in this case in $HCO_3^-$ rather than $CO_2$.
[b] The fractionation value of $+2‰$ is correct provided $HCO_3^-$ is used as the reference.
[c] Value given for RBP carboxylase is representative of results from several studies with a range of fractionation values of about 28–38‰. There is no good evidence that temperature influences the degree of fractionation of the carboxylase.

and CAM plants is thought to reflect this difference in fractionation by the primary carboxylases. But this difference does not totally account for the difference in isotope composition of the tissue. Where there is resistance for $CO_2$ uptake by the leaf (diffusion limited), fractionation of 4‰ may occur. In a $C_3$ plant the substrate for RBP carboxylase is $CO_2$ and not $HCO_3^-$. If carbon assimilation was completely limited by RBP carboxylase and not by diffusion then $\delta^{13}C$ values of −41‰ $[-7‰_{\text{(atmosphere)}} - 34‰_{\text{(RBP carboxylase)}}]$ would be expected. The predicted $\delta^{13}C$ value is obtained by subtracting fractionation from the $\delta^{13}C$ of the atmosphere

$$\delta^{13}C \text{ of product} = \delta^{13}C_{\text{(atmosphere)}} - \text{fractionation}.$$

If carbon assimilation in $C_3$ photosynthesis were totally limited by diffusion, $\delta$ values as low as −11‰ (−7‰−4‰) would be expected. Thus values may be greater than −41‰ to the extent there is a diffusional limitation. In some seagrasses where $CO_2$ is fixed by the $C_3$ pathway, the isotope composition of the tissue (−11‰) is close to that of the free $CO_2$ in seawater (Section 10.7). In this case the major rate limiting step for carbon assimilation is considered to be the rate of diffusion of $CO_2$ into the tissue such that nearly all the $CO_2$ taken up ($^{13}C + ^{12}C$) is fixed by RBP carboxylase.

In $C_4$ and CAM plants, little fractionation occurs via PEP carboxylase. $\delta^{13}C$ values of the tissue of −10‰ to −17‰ may be influenced by diffusional fractionation and fractionation in equilibration of $CO_2$ to $HCO_3^-$. PEP carboxylase utilizes $HCO_3^-$, and conversion of $CO_2$ to $HCO_3^-$ leads to an *enrichment* in $^{13}C$ with a fractionation value of −8‰. The minimum value expected if the system is diffusion limited, is −11‰(−7‰− 4‰); whereas, if the process is limited by carboxylation the maximum theoretical value would be −1‰

$$[-7‰ - (-8‰_{(CO_2 \text{ to } HCO_3^-)} + 2‰_{\text{(PEP carboxylase)}})].$$

$^{13}C$, the heavier isotope, diffuses slower through air than $^{12}C$ such that diffusional fractionation will occur as long as the intercellular $[CO_2]$ is lower than the external $[CO_2]$. Diffusional fractionation may be relatively greater in $C_4$ than in $C_3$ photosynthesis since under atmospheric conditions the intercellular $CO_2$ concentration in $C_4$ plants (about 100 ppm) is substantially lower than that of $C_3$ plants (about 220–250 ppm).

A proper analysis of isotope fractionation up to the point of carboxylation requires that the isotope composition of the carbon atom of $CO_2$ be measured after the initial steps of assimilation. However, generally only the isotope composition of the total tissue or of certain components have been reported. In one study, $\delta^{13}C$ values of C-4 of malate in *Kalanchoe daigremontiana* a CAM plant was found to be about −8‰, which is within the limits of the expected values (M. O'Leary, personal communication). Further fractionation, resulting in $\delta^{13}C$ values of about $-13‰$, may occur when the carbon is processed into biomass.

In $C_3$ plants RBP carboxylase brings about an enrichment of $^{12}C$ in its products (compared with the atmosphere) because it fixes $^{12}C$ faster than it fixes $^{13}C$ per unit of isotope available. However, in $C_4$ plants, in the enclosed space of the bundle sheath compartment, virtually *all* of the $CO_2$ which is offered as a result of decarboxylation is fixed. For this reason it can no longer discriminate effectively and the degree of enrichment in its products is close to that in the oxaloacetate which arises as the result of PEP carboxylase activity in the mesophyll.

CAM in some species has been shown to be facultative, i.e. the diurnal fluctuation in acid is most marked when the plants are not well watered. The $\delta^{13}C$ values vary in a similar fashion being least negative (most like $C_4$) when the diurnal fluctuation is most marked (Table 15.2). Sucrose synthesized by a $C_3$ plant (e.g. sugar beet) can be distinguished from sucrose synthesized by a $C_4$ plant (e.g. sugar-cane) due to differences in $\delta$ values. The differences in isotope fractionation among plants can be used in part to classify species into photosynthetic groups. One advantage of using this method of classification is that large surveys can be made using herbarium species since live tissue is not required. For example, in the Gramineae family Smith and Brown classified many species as either $C_3$ or $C_4$ based on $\delta^{13}C$ values (as no species of the Gramineae are CAM). This method correlates well with other methods used to distinguish between $C_3$ and $C_4$ species, i.e. $CO_2$ compensation point, anatomical differences, etc.

CAM plants have a range of $\delta\ ^{13}C$ values between those typical of $C_3$ and $C_4$ depending on the relative amount of $CO_2$ fixation at night (CAM mode) versus direct fixation of atmospheric $CO_2$ during the day through RBP carboxylase ($C_3$

**Table 15.2.** Changes in malate levels during the night ($\Delta$ malate) and $\delta^{13}C$ values at various dates during 1977 in fully developed leaves of the annual *Mesembryanthemum crystallinum* growing at the Mediterranean Sea shore near Caesarea, Israel. ND = not determined. (Modified after Winter *et al.*, 1978, Winter, 1979).

| Season | Date | $\Delta$ malate ($\mu$eq g fresh weight$^{-1}$) | $\delta^{13}C$ (‰) |
|---|---|---|---|
| wet | 26/27 January | −3 | −26.1 |
| wet | 24/25 February | −4 | −26.7 |
| wet | 17/18 March | −1 | −25.4 |
| dry | 26/27 April | +113 | −22.7 |
| dry | 25/26 May | +113 | −19.1 |
| dry | 26 June | ND | −15.7 |

mode). Thus the $\delta^{13}C$ of CAM has been used as a relative measure of night fixation through PEP carboxylase versus day fixation of atmospheric $CO_2$ through RBP carboxylase. Initially it was thought that the high $\delta\ ^{13}C$ values in $C_4$ species was in support of a transcarboxylation mechanism (see Section 10.2 for discussion of transcarboxylation). With $C_4$ acid decarboxylation and refixation the RBP carboxylase as a secondary carboxylase might still discriminate against fixing $^{13}CO_2$. It is now believed that in $C_4$ or CAM plants $C_4$ acid decarboxylation occurs in a confined space (bundle sheath cells in $C_4$ or CAM leaf with closed stomata). This may result in in nearly complete fixation of the $CO_2$ released ($^{13}CO_2 + {}^{12}CO_2$) although the RBP carboxylase tends to discriminate slightly against fixation of $^{13}CO_2$. (see Fig. 15.10). CAM plants often have $\delta^{13}C$ values somewhat intermediate to those of $C_3$ and $C_4$ plants. Evidence that night fixation results in high values and day fixation of

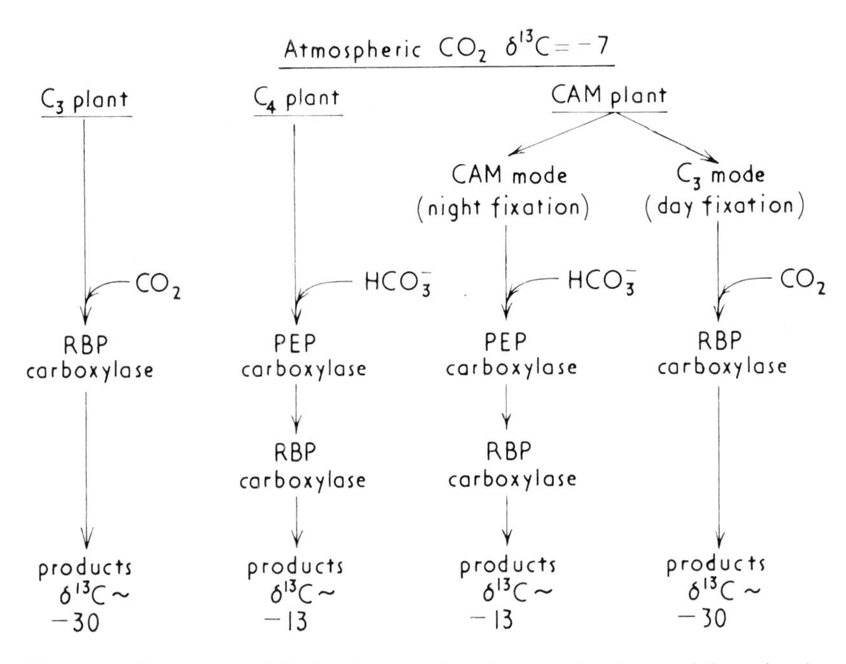


**Fig. 15.10.** Illustration of $CO_2$ fixation through carboxylases in plants and the carbon isotope composition of plant tissue.

atmospheric $CO_2$ results in low values in a CAM plant was elegantly demonstrated in a simple experiment by Nalborezyk, La Croix and Hill. They supplied $CO_2$ to *Kalanchoe daigremontia* either in the dark, the light, or both dark + light and obtained the following isotope composition upon subsequent analysis of the leaves.

| $CO_2$ supplied | $\delta^{13}C$ |
|---|---|
| Dark | −10.6 |
| Light | −25.9 |
| Light + Dark | −15.0 |

[Refs. **12, 13, 23, 34, 35, 37, 42, 44, 48, 49, 71**]

## 15.13 The relationship between CAM and $C_4$

The obvious feature shared by $C_4$ and CAM plants is that both groups initially incorporate $CO_2$ through PEP carboxylase, with the formation of oxaloacetate, and subsequently refix this (following some form of decarboxylation) in the reactions of the RPP pathway. In $C_4$, the initial fixation, the decarboxylation and the refixation all occur simultaneously but the first fixation is separated in space from the other two processes. In CAM the three processes all occur within the same cells but the initial fixation occurs in the dark and the other two processes occur in the light. In $C_4$ the product of $C_4$ acid decarboxylation is metabolized to PEP which is directly used by PEP carboxylase. In CAM starch breakdown through glycolysis leads to PEP formation. Then the PEP formed following decarboxylation of $C_4$ acids in the light is metabolized to starch and conserved until the following dark period. In $C_4$ the pool of $C_4$ acids is 1–2 equivalents per gram fresh weight with a turnover half time ($t_{\frac{1}{2}}$) of 5 s. In CAM the pool of $C_4$ acids is 100–200 equivalents per gram fresh weight with a $t_{\frac{1}{2}}$ up to 10 000 s.
[Refs. 5, **33**]

## 15.14 Ecological significance of CAM

CAM plants are particularly adapted to semi-arid environments. Their metabolic pattern is such that under drought conditions, they conserve respiratory $CO_2$ and assimilate external $CO_2$ for little water loss in the dark. Some succulent plants carry on $C_3$ photosynthesis as an additional means of growth when water is plentiful. The transpiration/photosynthesis ratios (grams water transpired/gram $CO_2$ fixed) are about 50–100 in CAM plants (in CAM mode), 250–300 in $C_4$ plants, and 400–500 in $C_3$ plants. These are typical values found in the literature for species representing these groups. They do not consider differences in leaf temperature and water vapour concentration in the atmosphere at the time of measurement. Stricter comparisons could be made by comparing the values of $r_{s,H_2O}/r_{s,CO_2} + r_m$ where $r_{s,H_2O}$ represents the stomatal resistance to water loss, $r_{s,CO_2}$ is the stomatal resistance to $CO_2$ uptake and

$r_m$ is the mesophyll resistance to $CO_2$ fixation. Thus the higher stomatal resistance and the lower $r_m$ will favour a low ratio of transpiration/photosynthesis (Section 14.6).

At a given stomatal resistance, a CAM plant can conserve more water by fixing $CO_2$ at night than during the day. The potential for water loss from a leaf depends on the difference between the water vapour concentration in the leaf versus the water vapour concentration in the atmosphere. Thus, in a hot arid environment with a high leaf temperature and low humidity in the atmosphere, the potential for water loss is very high. With high irradiance and low transpiration, the leaf temperature can easily exceed the air temperature. [Transpiration removes heat from the leaf as heat is absorbed when water is converted from the liquid to vapour.]

Consider a cactus plant whose leaf tissue is 40° C during the day versus 20° C during the night. At 40° C, the water vapour concentration in the intercellular space of the leaf at saturation will be $51 g/m^3$ compared to a value of $17 g/m^3$ at the 20° C leaf temperature. If the water vapour concentration in the atmosphere, for example, is $10 g/m^3$, then the gradient for water loss from the leaf would be $51 - 10 = 41 g/m^3$ during the day and $17 - 10 = 7 g/m^3$ during the night. In this case, the potential for water loss during the day is about six times that of the night.

The actual rate of transpiration can be expressed by the following equation (see Section 14.6)

$$T = \frac{W_L - W_a}{r'_a + r'_s}$$

where:

$T$ = transpiration
$W_L$ = water vapour concentration in the leaf
$W_a$ = water vapour concentration in the air
$r'_a$ = boundary layer resistance to water loss
$r'_s$ = stomatal resistance to water loss

As can be seen from the equation, the higher the boundary layer and stomatal resistance, the lower the transpiration. CAM plants have a lower stomatal frequency (Table 15.3) than either $C_3$ or $C_4$ species. Minimal values of leaf resistance in CAM under water stress of 100 s/cm are common although values as low as 2 s/cm may be reached under optimum conditions.

Besides being adapted to conserve water, CAM plants conserve carbon and minimize respiratory losses of $CO_2$. When photosynthesizing in the CAM mode, they may go through a complete 24-hour cycle without a net efflux of $CO_2$ at any time (Fig. 15.11). At night, there is a net uptake of $CO_2$ and any respired $CO_2$ may be

**Table 15.3.** Examples of stomatal frequency and stomata location on leaves of some CAM, $C_3$ and $C_4$ species. (After Meidner & Mansfield, 1968; Ting *et al*, 1972; also see Kluge & Ting, 1978)

| species | number of stomata/$mm^2$ upper epidermis | lower epidermis |
|---|---|---|
| CAM | | |
| *Agave americana* | 21 | 21 |
| *Crassula argenta* | 24 | 33 |
| *Portulacaria afra* | 45 | 33 |
| $C_4$ | | |
| *Zea mays* | 98 | 108 |
| $C_3$ | | |
| *Helianthus annuus* | 150 | 230 |
| *Quercus robur* | 0 | 340 |
| *Vicia faba* | 153 | 170 |

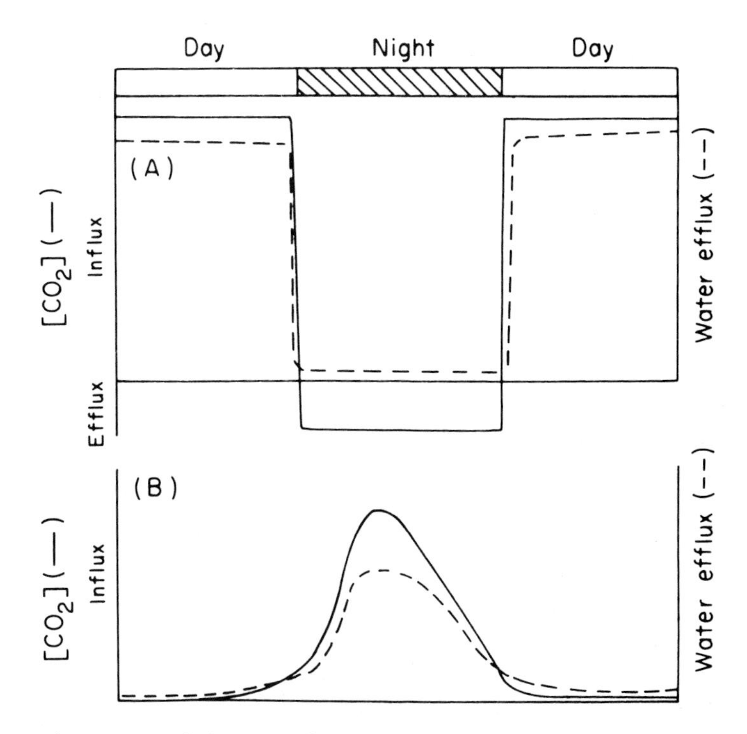


**Fig. 15.11.** Typical patterns of gas exchange for $C_3$ and $C_4$ plants (A) and CAM plants in a strict CAM mode (B).

refixed through PEP carboxylase. During the day the stomata are closed and the $CO_2$ released by $C_4$ acid decarboxylation or respiration is assimilated through the RPP pathway.

Also, $C_4$ acid decarboxylation behind closed stomata during CAM may generate saturating levels of $CO_2$ for RBP carboxylase. This would be analogous to the proposed $CO_2$-concentrating mechanism in $C_4$ (Sections 12.4, 14.4), which would minimize $O_2$ inhibition of photosynthesis. In fact, recently Cockburn *et al* took gas samples from a variety of CAM plants growing under natural environmental conditions. Analysis of the samples by gas chromatography revealed that $[CO_2]$ in the CAM leaves were 0.5–2% during the day and less than atmospheric levels at night. This high $[CO_2]$ in the leaves during the day was associated with a high stomatal resistance and occurred when the acidity (malate content) was high. Limiting the $CO_2$ fixation at night reduced the tissue acidity and the $[CO_2]$ in the leaf the following day. Spalding *et al.* measured both $[O_2]$ and $[CO_2]$ in samples from leaves of several CAM species during deacidification and found $O_2$ levels up to 42% and $CO_2$ levels up to 0.35%. On average the $CO_2/O_2$ ratio in the leaf was about four-fold higher than that in the atmosphere. The higher internal $[CO_2]$ and the higher $CO_2/O_2$ ratio in the CAM leaf during malate decarboxylation provides conditions favourable for photosynthesis and unfavourable for $O_2$ inhibition of photosynthesis (Sections 13.3, 14.3). During photosynthesis in the CAM mode, the $[CO_2]$ in the leaf derived from malate decarboxylation may directly control the stomatal resistance. The high $[CO_2]$ in leaves of the CAM species during the day is correlated with a high stomatal resistance. Exposing *Opuntia ficus-indica* to 2% $CO_2$ in air during the dark period causes stomatal closure as indicated by a striking increase in stomatal resistance from about 8 s/cm to 150 s/cm. Under severe water stress, some CAM plants maintain a high stomatal resistance throughout the 24-hour cycle but exhibit CAM-like day/night fluctuations in tissue acidity. Apparently, the respired $CO_2$ during the night is conserved through malate formation with subsequent donation of the carbon to the RPP pathway during the day. 'Idling' pads of *Opuntia basilaris* were also found to have a high internal $[CO_2]$ in the leaf during the day and a low concentration at night. In this case, some additional factor(s) may regulate the stomatal function since the stomata remain closed at night even though the internal $[CO_2]$ is low.

[Refs. 2–4, **21, 22, 26, 38–40, 43, 51, 56–58**]

### General reading

1 BRUINSMA J. (1958) Studies on the Crassulacean Acid Metabolism. *Acta Botanica Neerlandica* **7**, 531–90 and also North Holland Pub. Co., Amsterdam. [This proefschrift contains many references to the literature of the nineteenth century (Aubert, Kraus, de Saussure, de Vries, Warburg etc.)]

2 BURRIS R. H. & BLACK C. C. (1976) *$CO_2$ Metabolism and Plant Productivity*. Baltimore, Maryland: University Park Press.

3 KLUGE M. & TING I. P. (1978) *Crassulacean Acid Metabolism. Analysis of an ecological adaptation*. pp. 209 Berlin: Springer-Verlag.

4 MARCELLE R. (1975) *Environmental and Biological Control of Photosynthesis*. The Hague: Junk.

5 OSMOND G. B. (1978) Crassulacean Acid Metabolism: A curiosity in context. *Ann. Rev. Plant Physiol.*, **29,** 379–414.

6 RANSON S. L. & THOMAS M. (1960) Crassulacean Acid Metabolism. *Ann. Rev. Plant Physiol.*, **11,** 81–110.

7 THOMAS M., RANSON S. L. & RICHARDSON J. A. (1973) *Plant Physiology*. Fifth Edition. London: Longman.

8 WALKER D. A. (1962) Pyruvate carboxylation and plant metabolism. *Biol. Rev.*, **37,** 215–256.

9 WOLF J. (1960) Der Diurnale Saurerhythmus. *Encyclopedia of Plant Physiology* (ed. von W. Ruhland Vol. XII/2) pp. 809–89 Berlin: Springer-Verlag. [This review also contains a great many references to the old literature cf. Bruisma]

10 WOOD H. G. & UTTER M. F. (1965) The role of $CO_2$ fixation in metabolism. *Essays in Biochem.* **1,** 1–27. London: Academic Press.

## Specific citations

11 BANDURSKI R. S. & GREINER C. M. (1953) The enzymatic synthesis of oxalacetate from phosphoryl-enolpyruvate and carbon dioxide. *J. Biol. Chem.*, **204,** 781–6.

12 BENDER M. M. (1971) Variations in the $^{13}C/^{12}C$ ratios of plants in relation to the pathway of photosynthetic carbon dioxide fixation. *Phytochemistry*, **10,** 1239–44.

13 BENDER M. M., ROUHANI I., VINES H. M. & BLACK C. C. (1973) $^{13}C/^{12}C$ ratio changes in Crassulacean acid metabolism plants. *Plant Physiol.*, **52,** 427–30.

14 BENNET-CLARK T. A. (1933) The role of organic acids in plant metabolism. *New Phytol.*, **32,** 37–71, 128–61, 197–230.

15 BLACK C. C. (1973) Photosynthetic carbon fixation in relation to net $CO_2$ uptake. *Ann. Rev. Plant Physiol.*, **24,** 253–86.

16 BLACK C. C. & WILLIAMS S. (1976) Plants exhibiting characteristics common to Crassulacean acid metabolism. In *$CO_2$ Metabolism and Plant Productivity* (eds. R. H. Burris and C. C. Black). University Park Press, Baltimore, 407–24.

17 BONNER W. & BONNER J. (1948) The role of carbon dioxide in acid formation by succulent plants. *Amer. J. Bot.*, **35,** 113–117.

18 BRADBEER J. W., RANSON S. L. & STILLER M. (1958) Malate synthesis in Crassulacean leaves. I. The distribution of $^{14}C$ in the malate of leaves exposed to $^{14}CO_2$ in the dark. *Plant Physiol.*, **33,** 66–70.

19 BRULFERT J., GUERRIER D. & QUEIROZ O. (1975) Photoperiodism and enzyme rhythms: kinetic characteristics of the photoperiodic induction of Crassulacean acid metabolism. *Planta*, **125,** 33–44.

20 COCKBURN W. & MCAULEY A. (1975) The pathway of carbon dioxide fixation in Crassulacean plants. *Plant Physiol.*, 35, 67–89.

21 COCKBURN W., TING I. P. & STERNBERG L. O. (1979) Relationships between stomatal behaviour and internal carbon dioxide concentration in Crassulacean acid metabolism plants. *Plant Physiol.*, **63,** 1029–32.

22 CREWS C. E., WILLIAMS S. L., VINES H. M. & BLACK C. C. (1976) Changes in the metabolism and physiology of Crassulacean acid metabolism plants grown in controlled environments. In *$CO_2$ Metabolism and Plant Productivity* (eds. R. H. Burris and C. C. Black). pp. 235–50 Baltimore, Maryland: University Park Press.

23 DELEENS E. (1976) La discrimination du $^{13}C$ et les trois types de métabolisme des plantes. *Physiol. Veg.*, **14,** 641–55.

24 DITTRICH P. (1976) Nicotinamide adenine dinucleotide-specific 'malic enzyme' in *Kalanchoe*

*daigremontiana* and other plants exhibiting Crassulacean Acid Metabolism. *Plant Physiol.*, **57,** 310–4.
25 DITTRICH P., CAMPBELL W. H. & BLACK C. C. (1973) Phosphoenolpyruvate carboxykinase in plants exhibiting Crassulacean acid metabolism. *Plant Physiol.*, **52,** 357–61.
26. HARTSOCK T. L. & NOBEL P. S. (1976) Watering converts a CAM plant to daytime $CO_2$ uptake. *Nature*, **262,** 574–6.
27 HERBERT D. (1951) Oxalacetic decarboxylase and carbon dioxide assimilation in bacteria. *Symp. Soc. exp. Biol.*, **5,** 52–71.
28 HEYNE B. (1815) On the deoxidation of the leaves of *Cotyledon calycina. Trans. Linn. Soc. London*, **213,** 815–6.
29 JOLCHINE G. (1959) Sur la distribution du $^{14}C$ dans les molecules d'acide malique synthetisées par fixation de $^{14}CO_2$ dans les feuilles de *Bryophyllum daigremontianum* Berger bull. *Soc. Chim. Biol.*, **41,** 227–234.
30 KALNITSKY G. & WERKMAN C. H. (1944) Enzymatic decarboxylation of oxaloacetate and carboxylation of pyruvate. *Arch. Biochem.*, **4,** 25–40.
31 KLUGE M., KRIEBITZSCH C. H. & WILLERT D. J. V (1974) Dark fixation of $CO_2$ on Crassulacean acid metabolism: are two carboxylation steps involved? *Z. Pflanzenphysiol.*, **72,** 460–5.
32 KU M. S. B., SPALDING M. H., & EDWARDS G. E. (1980) Intracellular localization of phosphoenolpyruvate carboxykinase in leaves of $C_4$ and CAM plants. *Plant Sci. Letts.*, **19,** 1–8.
33 LAETSCH W. M. (1971) Chloroplast structural relationships in leaves of $C_4$ plants. In *Photosynthesis and Photorespiration*, (eds. M. D. Hatch, C. B. Osmond and R. C. Slatyer) pp. 323–49 New York: Wiley-Interscience.
34 LERMAN J. C. (1975) How to interpret variations in the carbon isotope ratio of plants: biologic and environmental affects. In *Environmental and Biological Control of Photosynthesis* (ed. R. Marcelle) pp. 323–35 The Hague: Junk.
35 LERMAN J. C. & QUEIROZ O. (1974) Carbon fixation and isotope discrimination by a crassulacean plant: dependence on the photoperiod. *Science*, **183,** 1207–9.
36 MEIDNER H. & MANSFIELD T. A. (1968) *Physiology of Stomata.* p. 178 London: McGraw-Hill.
37 NALBORCZYK E., LACROIX L. J. & HILL R. D. (1975) Environmental influences on light and dark $CO_2$ fixation by *Kalanchoe daigremontiana. Canad. J. Bot.*, **53,** 1132–8.
38 NEALES T. F. (1973a) The effect of night temperature on $CO_2$ assimilation, transpiration and water use efficiency in *Agave americana L. Aust. J. Biol. Sci.*, **26,** 705–14.
39 NEALES T. F. (1973b) Effect of night temperature on the assimilation of carbon dioxide by mature pineapple plants, *Ananas comosus* (L) Merr. *Aust. J. Biol. Sci.*, **26,** 539–46.
40 NEALES T. F., PATTERSON A. A. & HARTNEY V. J. (1968) Physiological adaption to drought in the carbon assimilation and water loss of xerophytes. *Nature*, **219,** 469–72.
41 OCHOA S., MEHLER A. & KORNBERG A. (1947) Reversible oxidative decarboxylation of malic acid. *J. Biol. Chem.*, **167,** 871–2.
42 O'LEARY M. H. (1981) Carbon isotope fractionation in plants. *Phytochemistry*, **20,** 553–67.
43 OSMOND C. B. (1976) $CO_2$ assimilation and dissimilation in the light and dark in CAM plants. In *$CO_2$ Metabolism and Plant Productivity*. pp. 217–34 Baltimore, Maryland: University Park Press.
44 OSMOND C. B., BENDER M. M. & BURRIS R. H. (1976) Pathways of $CO_2$ fixation in the CAM plant *Kalanchoe daigremontiana.* III. Correlation with $^{13}C$ value during growth and water stress. *Aust. J. Plant Physiol.*, **3,** 787–99.
45 PUCHER G. W., LEAVENWORTH C. H. S., GINTER W. D. & VICKERY H. B. (1947) The diurnal variation in organic acid and starch content of *Bryophyllum calycinum. Plant Physiol,*, **22,** 360–76.
46 SALTMAN P., KUNITAKE G., SPOLTER H. & STITTE C. (1956) The dark fixation of $CO_2$ by succulent leaves: the first products. *Plant Physiol.*, **31,** 464–8.
47 DE SAUSSURE T. H. (1804) *Recherches chimiques sur la végétation* Paris: Nyon.
48 SMITH B. N. (1972) Natural abundance of the stable isotopes of carbon in biological systems. *Bioscience*, **22,** 226–31.

49 Smith B. N. & Brown W. V. (1973) The Kranz syndrome in the Gramineae as indicated by carbon isotopic ratios. *Amer. J. Bot.*, **60**, 505–13.
50 Spalding M. H., Schmitt M. R., Ku S. B. & Edwards G. E. (1979a) Intracellular localization of some key enzymes of Crassulacean Acid Metabolism in *Sedum praealtum. Plant Physiol.*, **63**, 738–43.
51 Spalding M. H., Stumpf D. K., Ku M. S. B., Burris R. H. & Edwards G. E. (1979b) Crassulacean acid metabolism and diurnal variations of internal $CO_2$ and $O_2$ concentrations in *Sedum praealtum* DC. *Aust. J. Plant Physiol.*, **6**, 557–67.
52 Sutton B. G. (1975) Glycolysis in CAM plants. *Aust. J. Plant Physiol.*, **2**, 489–92.
53 Sutton B. G. & Osmond C. B. (1972) Dark fixation of $CO_2$ by Crassulacean plants—evidence for a simple carboxylation step. *Plant Physiol.*, **50**, 360–5.
54 Szarek S. R. & Ting I. P. (1977) The occurrence of Crassulacean acid metabolism among plants. *Photosynthetica*, **11**, 330–42.
55 Tchen T. T. & Vennesland B. (1955) Enzymatic carbon dioxide fixation into oxaloacetate in wheat germ. *J. Biol. Chem.*, **213**, 533–46.
56 Ting I. P. & Hanscom Z. (1977) Induction of acid metabolism in *Portulacaria afra. Plant Physiol.*, **59**, 511–4.
57 Ting I. P., Johnson H. B. & Szarek S. R. (1972) Net $CO_2$ fixation in Crassulacean acid metabolism plants. In *Net Carbon Dioxide Assimilation in Higher Plants* (ed. C. C. Black) pp. 26–53 Raleigh, North Carolina: Cotton Inc.
58 Ting I. P., Thompson M. & Dugger W. (1967) Leaf resistance to water vapor transfer in succulent plants: Effect of thermoperiod. *Amer. J. Bot.*, **54**, 245–51.
59 Thomas M. (1947) *Plant Physiology*. 3e. pp. 320–40 London: J. and A. Churchill.
60 Thomas M. (1951) Carbon dioxide fixation and acid synthesis in Crassulacean acid metabolism. *Symp. Soc. Exp. Biol.*, **5**, 72–93.
61 Thomas M. & Beevers H. (1949) Physiological studies on acid metabolism in green plants. II. Evidence of $CO_2$ fixation in *Bryophyllum* and the study of diurnal fluctuation of acidity in this genus. *New Phytol.*, **48**, 421–7.
62 Thomas M. & Ranson S. L. (1954) Physiological studies on acid metabolism in green plants. III. Further evidence of $CO_2$ fixation during dark acidification of plants showing Crassulacean acid metabolism. *New Phytol.*, **53**, 1–30.
63 Thurlow J. & Bonner J. (1948) Fixation of atmospheric $CO_2$ in the dark by leaves of *Bryophyllum. Arch. Biochem.*, **19**, 500–11.
64 Varner J. E. & Burrell R. C. (1950) Use of $C^{14}$ in the study of the acid metabolism of *Bryophyllum calycinum. Arch. Biochem.*, **25**, 280–7.
65 Vickery H. B. (1952) The behaviour of isocitric acid in excised leaves of *Bryophyllum calycinum* during culture in alternating light and darkness. *Plant Physiol.*, **27**, 9–11.
66 Walker D. A. (1956) Malate synthesis in a cell-free extract from a Crassulacean plant. *Nature*, **178**, 593–4.
67 Walker D. A. (1957) Physiological studies on acid metabolism. 4. Phosphoenolpyruvic carboxylase activity in extracts of Crassulacean plants. *Biochem. J.*, **67**, 73–9.
68 Walker D. A. (1960) Physiological studies on acid metabolism. 7. Malic enzyme from *Kalanchoe crenata*: effects of carbon dioxide concentration. *Biochem. J.*, **74**, 216–23.
69 Walker D. A. & Brown J. M. A. (1957) Physiological studies on acid metabolism. 5. Effects of carbon dioxide concentration on phosphoenolpyruvic carboxylase activity. *Biochem. J.*, **67**, 79–83.
70 Walker D. A. & Ranson S. L. (1958) Physiological studies in acid metabolism in green plants. Transaminases in cell-free extracts from Kalanchoe leaves. *Plant Physiol.*, **33**, 226–30.
71 Whelan T., Sackett W. M. & Benedict C. R. (1973) Enzymatic fractionation of carbon isotopes by phosphoenolpyruvate carboxylase from $C_4$ plants. *Plant Physiol.*, **51**, 1051–4.
72 Winter K. (1973) $CO_2$-fixation metabolism in the halophytic species *Mesembryanthemum crystallinum* grown under different environmental conditions. *Planta*, **114**, 75–85.

73 WINTER K. (1979) Photosynthetic and water relationships of higher plants in a saline environment. In *Ecological Processes in Coastal Environments.* (eds. R. L. Jefferies, A. J. Dary) pp. 297–320. Oxford: Blackwell Scientific Publications.
74 WINTER K. (1980) Malate sensitivity of PEP carboxylase during CAM. *Plant Physiol.*, **65**, 792–6.
75 WINTER K., FOSTER J. G., EDWARDS G. E. & HOLTUM J. A. M. (1982) Intracellular localization of enzymes of carbon metabolism in *Mesembryanthemum crystallinum* exihibiting $C_3$ photosynthetic characteristics or performing crassulacean acid metabolism. *Plant Physiol.* **69**, 300–7.
76 WINTER K. & KANDLER O. (1976) Misleading data on isotope distribution in malate-$^{14}C$ from CAM plants caused by fumarase activity of *Lactobacillus plantarum. Z. Pflanzenphysiol.*, **78**, 103–12.
77 WINTER K., LUTTGE U., WINTER E. & TROUGHTON J. M. (1978) Seasonal shift from $C_3$ photosynthesis to Crassulacean acid metabolism in *Mesembryanthemum crystallinum* growing in its natural environment. *Oecologia*, **34**, 225–37.
78 WOOD H. G. & WERKMAN G. H. (1936) The utilization of $CO_2$ in the dissimilation of glycerol by the propionic acid bacteria. *Biochem. J.*, **30**, 48–53.

# Chapter 16
# Comparative Studies of $C_3$, $C_4$ Metabolism in Other Plant Tissue



## SUMMARY

Ribulose-bisphosphate and RBP carboxylase have a unique function in plants in the photosynthetic assimilation of $CO_2$. Conversely, the involvement of phosphoenolpruvate (PEP), PEP carboxylase, and $C_4$ acids in plant metabolism is quite

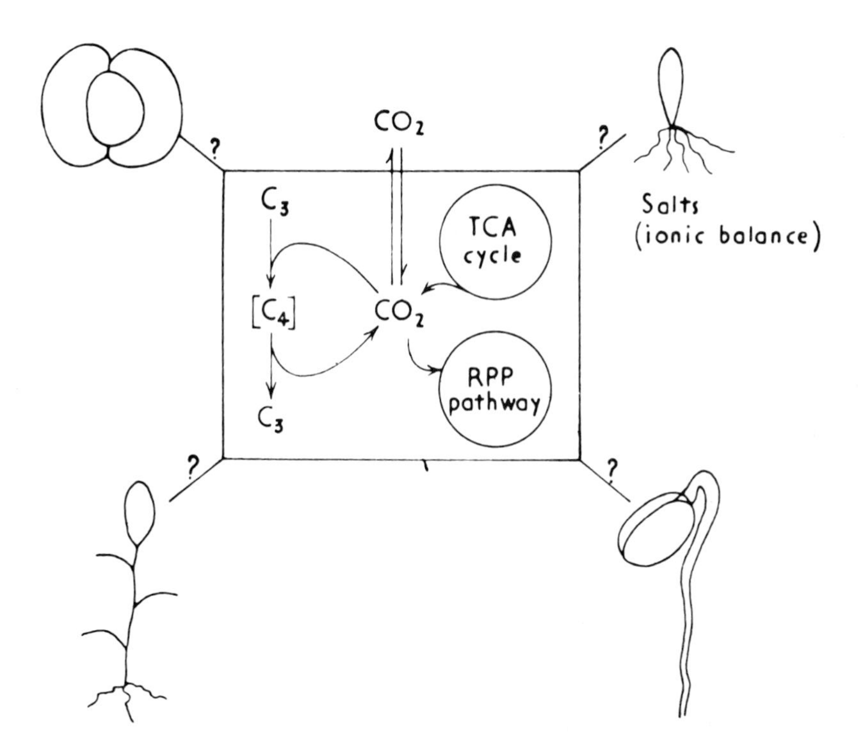


**Fig. 16.1.**
Among higher plants, metabolism involving carbon dioxide is not limited to leaves. Many chloroplast containing tissues may assimilate $CO_2$ through the RPP pathway. $CO_2$ can also be fixed into $C_4$ acids and it can be released through their decarboxylation and by metabolism in the TCA cycle.

diverse, extending beyond CAM and $C_4$ photosynthesis, as shown by a number of examples.

During stomatal opening, malate is synthesized via PEP carboxylase in some species and the malate concentration increases in the guard cells in the light. Along with $K^+$, this increase in malate contributes to the increased osmotic potential in the guard cells, which is needed to bring about stomatal opening.

Fruits, for example the pericarp of small grains, green tomatoes, and pods of legumes, have substantial levels of PEP carboxylase, with malate as a major product of $^{14}CO_2$ assimilation. A primary function of this $C_4$ metabolism may be to minimize respiratory losses of carbon.

Malate synthesis plays an important role in plant nutrition. Generally plants contain an excess of inorganic cations over inorganic anions, and the surplus of positive charges is balanced by organic anions (often malate). In addition, conversion of nitrate to amino nitrogen causes alkalization which can be neutralized by synthesis of malic acid.

In the glyoxylate cycle of germinating seedlings, malate is an intermediate in lipid degradation leading to sucrose synthesis. In this metabolism, PEP carboxykinase is responsible for conversion from oxaloacetate to PEP in order to provide the $C_3$ precursor for gluconeogenesis.

Wherever it has been sought in different plant tissue, some evidence for PEP carboxylase and metabolism of malate has always been found. RBP carboylase also occurs in tissues other than the leaves, although not as frequently as PEP carboxylase. In addition to the unique role of dicarboxylic acids in $C_4$ plant photosynthesis (Section 10.7) and CAM (Section 15.4), metabolism of $C_4$ acids may serve special functions in other tissues.
[Ref. **43**]

## 16.1 Stomata

Malate is synthesized in the guard cells of epidermal tissue and contributes to the increased turgor required for stomatal opening. Guard cells generally have chloroplasts but guard cells appear to be largely heterotrophic because their primary supply of reduced carbon comes from photosynthesis within the leaf. Thus, unlike $C_4$ photosynthesis or CAM, the metabolism of $C_4$ acids in guard cells is not directly linked to the RPP pathway.

There is now substantial evidence that the increase in solutes in the guard cells during stomatal opening is due to increases in the concentration of potassium, chloride, and malate. The relative proportion of malic acid and chloride may vary with species, e.g. in *Vicia faba* the anion is largely malate, while maize has substantial amounts of chloride as the counter ion to potassium.

Stomatal opening is dependent upon osmotic changes within the guard cells. An increase in the concentration of solutes within guard cells results in uptake of water from adjacent cells. Thus, the uptake of water which accompanies an increased solute concentration causes increased turgor in the guard cells (hydrostatic pressure in the guard cells increases relative to that in adjacent cells) and opening of the stomata. In dicots the outer walls of a pair of guard cells are thinner than the inner walls (Fig. 16.2). The change in the shape of the guard cell of dicots during stomatal opening is somewhat analogous to changes in the form of a balloon which is blown up and has a restriction along one side.

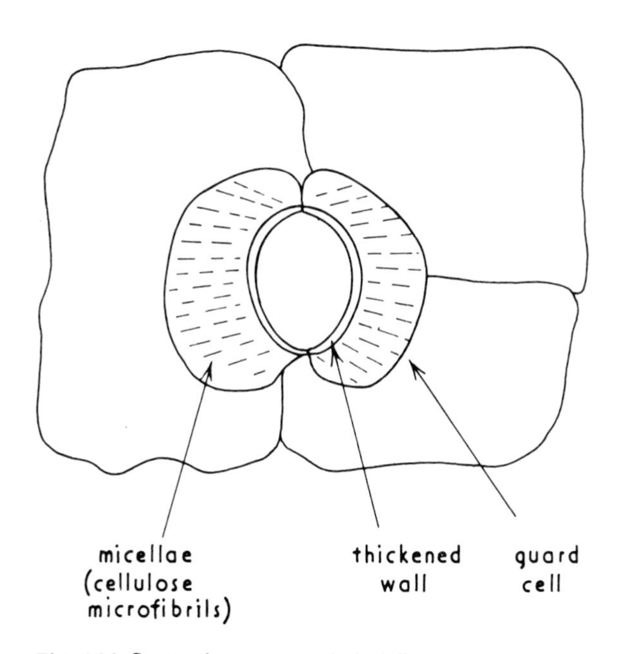


**Fig. 16.2.** Stomatal structure typical of dicots.

In grasses, the walls of the guard cells are thick in the middle and thin at the ends. With increasing osmotic pressure, the ends of cells expand which results in stomatal opening (Fig. 16.3).

Guard cell chloroplasts generally contain starch which in $C_3$ and $C_4$ plants is partially degraded by day and synthesized at night. This is the reverse of the pattern

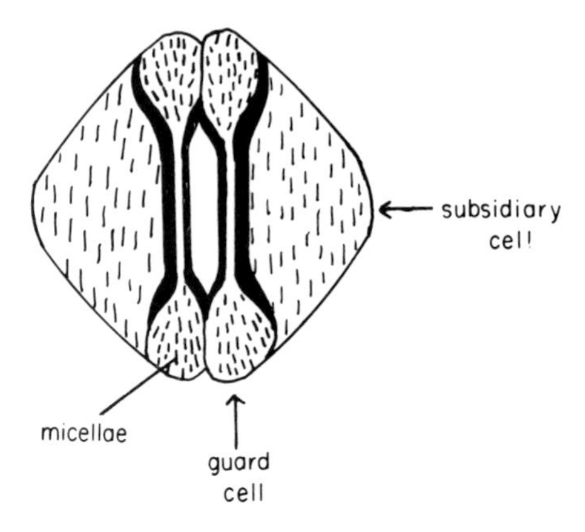


Fig. 16.3. Stomatal structure typical of grasses.

of starch metabolism seen in plastids of mesophyll ($C_3$, $C_4$ and CAM) and bundle sheath cells of the leaf.

The source of malate during stomatal opening is uncertain. One suggestion is that during stomatal opening, starch may be broken down in the guard cells, providing PEP as a precursor for malate formation through PEP carboxylase and malate dehydrogenase. This would be analogous to malate metabolism during the dark phase of CAM (Sections 15.5, 15.6). There is evidence that PEP carboxylase is located in guard cells of epidermal tissue. The malic acid synthesized in the guard cells would provide protons for counter exchange with the $K^+$ of adjacent subsidiary cells (Fig. 16.4).

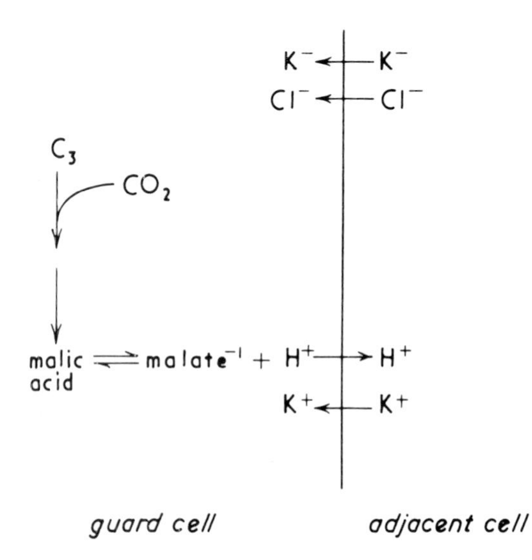


Fig. 16.4. Malate synthesis and $H^+/K^+$ exchange during stomatal opening.

During the night, the $C_4$ acid may be decarboxylated and the $C_3$ precursors metabolized back to starch. The $CO_2$ released would be lost rather than reassimilated in the RPP pathway as it is in CAM or $C_4$ plants. The conversion of pyruvate to starch would require input of energy which might be provided by mitochondrial respiration.

It appears that guard cells (e.g. of *Tulipa* and *Commelina* spp.) have little capacity for net assimilation of $CO_2$ through the RPP pathway. Malate and aspartate are the primary initial products of $^{14}CO_2$ fixation by either the attached or detached epidermal tissue. With epidermal strips removed from the parent tissue, rates of $^{14}CO_2$ fixation are low and are only about two-fold higher in the light than in the dark. Assimilated $^{14}CO_2$ appears in *attached* epidermal tissue at rates 100–200 times higher in the light than the dark. In the light the rate of appearance of labelled metabolites in the attached epidermal tissue was similar to that of the mesophyll tissue on a per g nitrogen basis (*Commelina* sp., Thorpe & Milthorpe).

Apparently, some of the carbon assimilated in mesophyll tissue of the leaf is rapidly transported to the epidermal tissue *in vivo*. Outlaw *et al.* showed that guard cell chloroplasts of *Vicia faba* have little or no RBP carboxylase, phosphoribulokinase, and NADP-triose phosphate dehydrogenase. Guard cell chloroplasts may serve primarily as a source of energy during the day rather than catalyzing net assimilation of carbon through the RPP pathway. Speculation exists for an ATP dependent $H^+ - K^+$ exchange at the plasma membrane between the guard and subsidiary cells.

There have been few studies of the stomatal physiology of CAM plants. During CAM, the stomata are open by night and closed by day. Thus, the starch in the guard cells is degraded in the dark, providing a precursor for malate synthesis. This is followed by starch synthesis during the day. In this case, the pattern of starch metabolism would parallel that of the mesophyll tissue of the CAM leaf. Whether guard cells of these plants have the capacity for CAM is, therefore, of interest. It may be noted that stomata of $C_3$ and $C_4$ plants can often be induced to close in the light in high external $[CO_2]$ and open in the dark in low external $[CO_2]$. It is conceivable, therefore, that it is the $[CO_2]$ in the substomatal space which is the overriding factor in stomatal control in CAM plants and that stomatal behaviour, therefore, follows the internal $[CO_2]$ during dark acidification and light deacidification (Section 15.14).

A number of factors are involved in the regulation of stomatal function, e.g. $CO_2$-dependent closing of stomata, relatively greater effectiveness of blue light over red light for stomatal opening, and the effect of growth regulators such as abscisic acid on stomatal closing induced by water stress (see refs. at end of this section). The underlying mechanism(s) of these effects and the role of carbon metabolism in stomatal physiology are not fully understood and there may be species-dependent variations.

The degree of development of chloroplasts in guard cells varies among different species, for example, chloroplasts of *Vicia faba* are much more obvious than those of maize. Photosynthetic metabolism and provision of energy by the plastids of guard cells is not necessarily the same in different species.
[Refs. **1, 11, 17, 27, 28, 30, 31, 32, 39, 46, 47**]

## 16.2 $C_3$, $C_4$ photosynthesis in reproductive tissue

### (a) Panicle photosynthesis

Considerable research has been made into the photosynthesis of the panicle of small grains such as barley, wheat, and rice. The spikelet consists of palea, fruit, lemma, awns, and glumes; the latter three are probably the major contributors to the assimilation of atmospheric $CO_2$ by the spikelet (Fig. 16.5).

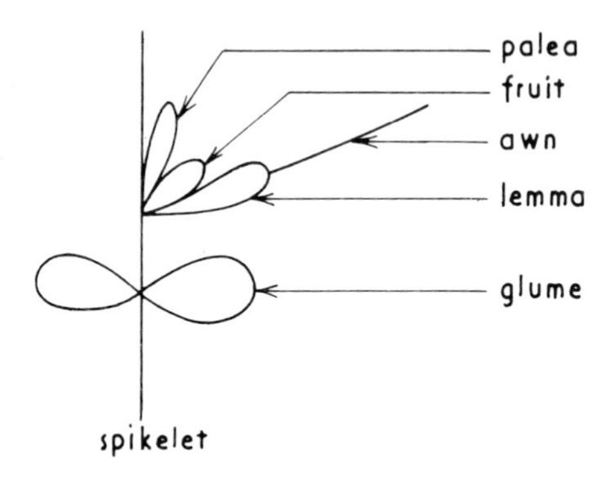


**Fig. 16.5.** General illustration of a spikelet in the panicle of small grains.

Panicle photosynthesis offers certain advantages. The photosynthetic product is formed close to the sink (fruit) which minimizes translocation; the panicle is exposed to the highest light intensity in the canopy and generally, senesces later than the leaves. In some varieties of cereals, up to half of the photosynthate of the grain may come from photosynthesis in the panicle. Varieties with awns tend to have the highest photosynthetic capacity in the panicle.

The chloroplast-containing tissue of the panicle appears to have the capacity for $CO_2$ fixation by the RPP pathway. Wirth *et al.* found RBP carboxylase and products of $^{14}CO_2$ fixation of the RPP pathway in the pericarp (outer layer of fruit which is derived from the ovary wall), palea, lemma, and glume of oat and wheat. PEP carboxylase was higher in the tissue of the panicle than in the leaf tissue, as was the ratio of PEP carboxylase/RBP carboxylase. Levels of $C_4$ acid decarboxylases were low. Palea, lemma, and pericarp may reassimilate some of the $CO_2$ respired during reproductive development. In the pericarp, unlike other photosynthetic tissue of the small grains, a large portion of the $CO_2$ is fixed into malic acid (see below). [Refs. **12, 18, 24, 38, 44, 48**]

### (b) FRUIT PHOTOSYNTHESIS

Developing fruits are largely heterotrophic and generally 95% or more of their carbon is imported from the leaves. Respiration of *imported* carbon provides precursors for the synthesis of organic acids including malate (unlike the origin of malate in $C_4$ or CAM). Some $CO_2$ fixation in the RPP pathway will reduce carbon losses from respiration.

In the developing fruit of barley, Nutbeam and Duffus suggested that pericarp tissue has some characteristics of $C_4$ photosynthesis. With isolated pericarp tissue taken from the panicles 30 days after anthesis (grain matured 60 days after anthesis) after a 1 min. exposure to $^{14}CO_2$ in the light, 84% of the labelled products was malate. Feeding $^{14}CO_2$ for longer periods resulted in label appearing in sucrose. Since the outer layers of the developing fruit are without stomata, PEP carboxylase may trap some of the respired $CO_2$ minimizing respiratory losses of carbon. Elucidation of labelling pattern of intermediates will be required to establish the pathway. Donation of carbon from C-4 of malate to C-1 of PGA would be consistent with carbon donation as occurs in $C_4$ photosynthesis (Sections 10.2, 12.3). However, labelling from $^{14}CO_2$ to malate to sucrose can occur without carbon donation from malate to the RPP pathway (e.g. as occurs with $^{14}CO_2$ metabolism in castorbean endosperm, Section 16.6).

In tomato, a $C_3$ plant, the green fruit has high rates of dark respiration which are reduced in the light due to light-dependent $CO_2$ fixation. With detached outer layers of the green fruit, rates of $^{14}CO_2$ fixation in the light are about 3-fold higher than in the dark. In both cases, malate is the major labelled product. There is also substantial labelling of citrate (metabolite of the TCA cycle) and metabolites of the RPP pathway. The relative levels of PEP and RBP carboxylase and NADP-malic enzyme in the outer layer of the green fruit are similar to those found in $C_4$ and CAM leaves. Labelled malate fed to the fruit is metabolized in the TCA cycle and possibly decarboxylated through NADP-malic enzyme. Part of the $CO_2$ released is reassimilated by the RPP pathway. Since stomata are not present on the tomato fruit, the $CO_2$ fixation may again represent a mechanism for trapping respiratory $CO_2$ and minimizing loss of carbon. Malate, as a product of this $CO_2$ fixation, may either be stored in the vacuole, decarboxylated with $CO_2$ donation to the RPP pathway, or fed into the TCA cycle to maintain its pool of metabolites.

Fruit of legumes such as peas and soybeans fix $^{14}CO_2$ in the light and dark. In the light, there may be net uptake of $CO_2$ during the early stages of fruit development while in the latter stages, light represses respiratory losses (as measured by infra-red gas analysis). Similar to fruits of tomato and barley, the pods refix part of the carbon lost by respiration. The pod wall, testa (seed coat), and embryo all have substantial levels of PEP and RBP carboxylase, with generally a higher level of PEP carboxylase. The pod wall probably has the highest capacity for $CO_2$ assimilation. Developing seeds of soybean may have a higher capacity for $CO_2$ fixation than seeds of peas. Flinn, *et al.* suggested that in illuminated peas, about 20% of the carbon of the fruit could be accounted for by pod photosynthesis. There are two photosynthetically active layers in the pod wall of peas, an outer layer composed of the chlorenchyma of the mesocarp and an inner layer composed of the inner epidermis of the endocarp. Atmospheric $CO_2$ is assimilated by chloroplasts of the mesocarp while $CO_2$ respired from the seed is assimilated by chloroplasts in the inner epidermis. Sambo, *et al.* estimated that the $CO_2$ fixed by the soybean pod was only about 4% of the total carbon imported from the leaf, although 50–70% of the carbon respired by the pods may be reassimilated. The stomatal diffusive resistance and boundary layer resistance (Section 14.6) of the pods were about twice that of leaves.

In grape berries, organic acids (particularly malic acid) accumulate during early growth. This is followed by breakdown of organic acids and synthesis of sugars during ripening. The grapes have PEP carboxylase, PEP carboxykinase, and NADP-malic enzyme. Accumulation of organic acids may be linked to $CO_2$ fixation through PEP carboxylase (although PEP carboxykinase could be involved) and malate dehydrogenase. During deacidification, $C_4$ acids could be decarboxylated through PEP carboxykinase and/or NADP-malic enzyme, and the $C_3$ products, PEP and pyruvate, respectively, converted to sugars via gluconeogenesis. In order to convert pyruvate, a product of decarboxylation through NADP-malic enzyme, to sugar, pyruvate,Pi dikinase is required for the synthesis of PEP (Sections 15.4, 15.10). Whether pyruvate,Pi dikinase is present in grapes is uncertain. It has been found in the pericarp of barley fruit. Acidification and deacidification in berries may be analogous to acidification and deacidification in $C_4$ and CAM plants in terms of the enzymes utilized, but the fixation of $CO_2$ through the RPP pathway following $C_4$ acid decarboxylation is likely less efficient than in $C_4$ and CAM plants.

Many fruits have a relatively high rate of respiration. Imported carbon, in the form of sucrose and amino acids, may be assimilated primarily into lipids, proteins, and oligosaccharides utilizing energy provided by respiration. In order for the TCA cycle to function, it must continue to form oxaloacetate to react with acetyl CoA. With each turn of the cycle, oxaloacetate is regenerated if there is no metabolite loss.

$$\text{acetyl CoA} + \text{oxaloacetate} \rightarrow 2\ CO_2 + \text{oxaloacetate} \qquad \text{Eqn. 16.1}$$

However metabolites of the cycle will be used in biosynthesis such that an external source of $C_4$ acids is required for the cycle to continue to function. This could be provided by malate synthesized from a $C_3$ precursor. Also, some fruits are acidic by nature, so synthesis and storage of malate and other acids during some phase of their development is to be expected. Fruits which store substantial citric acid (e.g. lemon, tomato) may also have a strong requirement for synthesis of the $C_4$ acids, oxaloacetate or malate. Net synthesis of citrate through the TCA cycle requires an equivalent input of pyruvate and a $C_4$ acid. During glycolysis metabolism of PEP to pyruvate through $C_4$ acid synthesis and decarboxylation may function to convert NADH to NADPH (Section 16.4). In green fruit, photosynthesis is a secondary process where some respired $CO_2$, or $CO_2$ from malate decarboxylation, may be reassimilated through the RPP pathway (Fig. 16.6).
[Refs. **2, 3, 9, 13–16, 20, 21, 29, 35, 36, 45**]

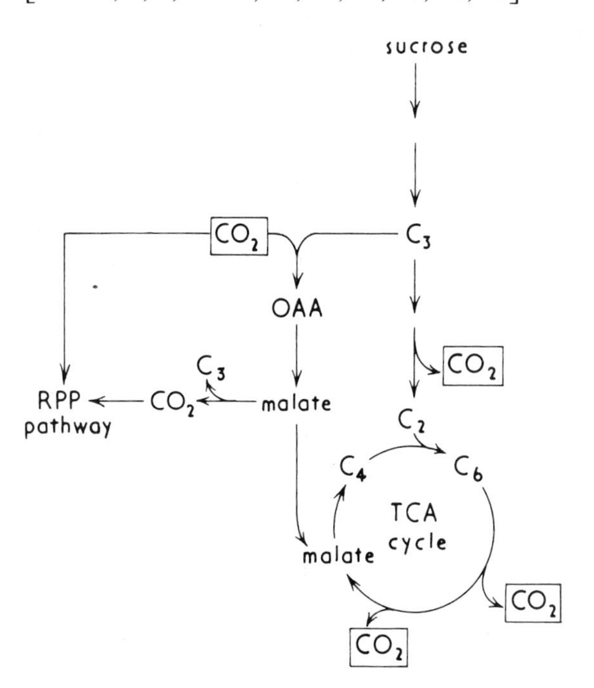


**Fig. 16.6.** Some ways in which malate may be synthesized and metabolized in fruits.

## 16.3 Greening leaves

Green leaves in the dark and etiolated leaves in the light and dark fix $^{14}CO_2$ primarily into organic acids with substantial label in malate. For dark-grown barley leaves ($C_3$ plant), Tamas *et al.* found that the main products of $^{14}CO_2$ assimilation were malate, aspartate, glutamate, and citrate in the light or dark. After illumination of leaf tissue for two to four hours, malate, aspartate, and glutamate syntheses were greatly accelerated in the light with malate as the primary product. Synthesis of these organic acids may be important in respiratory metabolism and for formation of proteins and lipids during chloroplast development. After six hours of pre-illumination of dark-grown tissue, sucrose was the major product of $^{14}CO_2$ fixation as in mature $C_3$ leaves.
[Ref. **37**]

## 16.4 Roots

In roots as in other tissues, malate is a metabolite of the TCA cycle. In addition to pyruvate uptake by mitochondria, malate may enter the TCA cycle following its synthesis from $C_3$ precursors.

Ting and Dugger suggested that malate synthesis and decarboxylation in root tissue could effectively serve as a transhydrogenase, providing NADPH where needed for biosynthetic processes. NADH generated during glycolysis would be converted to NADPH through a coupling of NAD-malate dehydrogenase and NADP-malic enzyme. This would occur at the expense of 1 ATP which is generated during glycolysis through pyruvate kinase (Fig. 16.7).

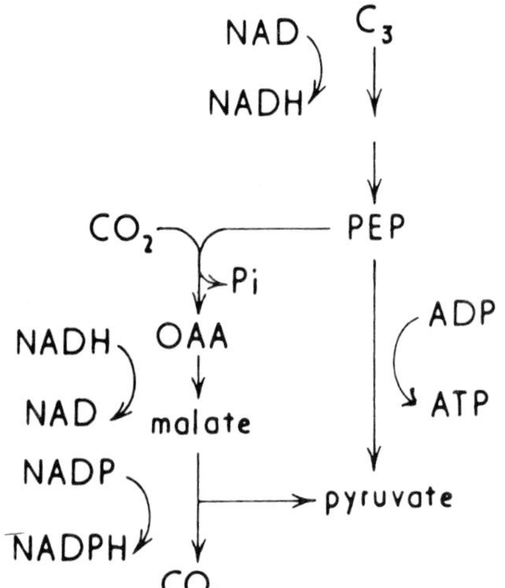

**Fig. 16.7.** Malate synthesis and decarboxylation during glycolysis may function like a transhydrogenase converting NADH to NADPH (Ting & Dugger, 1965; 1967).

One hypothesis for flooding tolerance (tolerance to temporary anaerobiosis) is linked to PEP carboxylase and malate synthesis. During glycolysis under anaerobic conditions ethanol is synthesized and its accumulation becomes toxic to the tissue. Ethanol synthesis through alcohol dehydrogenase utilizes excess reductive power which allows glycolysis to continue to provide ATP. It is proposed that in flooding-tolerant species, PEP formed during glycolysis would be metabolized by PEP carboxylase and the resultant oxaloacetate converted to malate via malate dehydrogenase. This would affect recycling of the pyridine nucleotide as an alternative to ethanol synthesis (Fig. 16.8). A low level of malic enzyme in flooding-tolerant species would allow malate to accumulate.

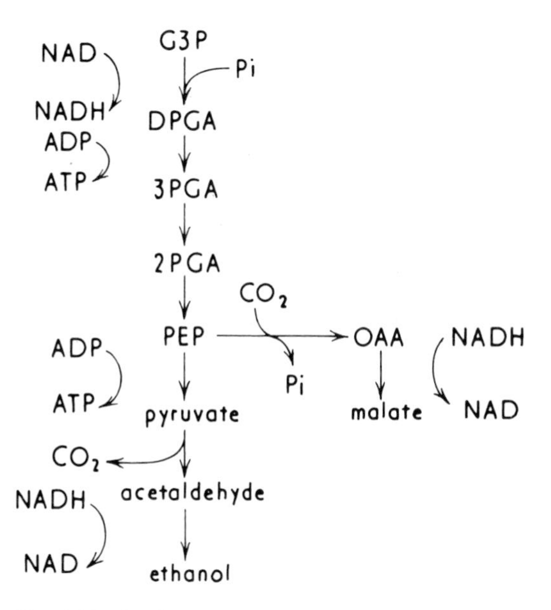


**Fig. 16.8.** One hypothesis for flood tolerance is linked to malate synthesis.

Prolonged flooding does cause some increase in malate concentration in roots of flooding-tolerant plants. However, Avadhani *et al.* found no support for the hypothesis in studies with flooding-tolerant rice seedlings. Exposure of the seedlings to low [$O_2$] resulted in only a small increase in malate whereas the concentration of ethanol in the tissue after one day at low $O_2$ was 20–30-fold higher than the levels of malate.

[Refs. **4, 5, 23, 25, 40, 41**]

## 16.5 Ionic balance and pH-stat

Plants commonly have an excess of inorganic cations over inorganic anions. The electrochemical balance is made up by organic anions, in many cases the malate anion. This relationship can be expressed as $C - A = Y$ where

$C$ = mequivalents of cations (e.g. $K^+$, $Na^+$, $Ca^{2+}$, $Mg^{2+}$)
$A$ = mequivalents of inorganic anions (e.g. $NO_3^-$, $HPO_4^{-2}$, $SO_4^{-2}$ and $Cl^-$)
$Y$ = mequivalents of organic anions.

Y is fairly constant for a given species. Representative values of Y in meq/kg dry matter are about 1000 for grasses, 2000 for unnodulated legumes, and 4000 for sugarbeet. Loomis and Gerakis note that plants may carry with them a massive amount of their own sewage as a result of this need to synthesize organic anions. For example, potassium uptake by roots in excess of inorganic anions can result in a $K^+ - H^+$ exchange. The synthesis of malic acid can provide protons to exchange out of roots for potassium, resulting in the formation of potassium malate in the tissue. In such cases, a stoichiometric synthesis of malate can offset the deficiency of anion uptake.

It has been proposed that in both roots and leaves, a balance between malate synthesis and degradation serves to control the pH in the cells. Metabolism, or metabolite uptake which leads to alkalization, could be offset by malate synthesis through PEP carboxylase and NAD-malate dehydrogenase. Acidification in cells could be balanced by malate decarboxylation through NAD- or NADP-malic enzyme (Fig. 16.9). In this hypothesis the compartmentation of enzymes and the extent to which pH may control enzyme reactions leading to acidification versus deacidification must be considered.

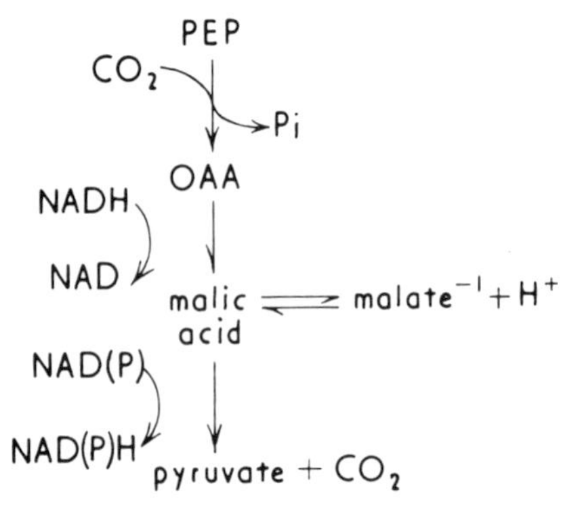


**Fig. 16.9.** A potential for malate synthesis and its decarboxylation may help to control pH in cells.

A large part of organic acid synthesis in plants is associated with reduction of nitrate. When nitrate is metabolized to protein, a hydroxyl ion is generated which can be balanced by malate synthesis.

$$NO_3^- + NADH + H^+ \rightarrow NO_2^- + NAD^+ + H_2O$$
$$NO_2^- + 6H^+ + 6e^- \rightarrow NH_3 + H_2O + OH^- \qquad \text{Eqn. 16.2}$$

If $KNO_3$ is taken up by roots, transported to leaves and the nitrate converted to amino nitrogen, KOH is generated. Synthesis of malic acid then results in formation of potassium malate, which may be stored in the vacuole. The synthesis of malate would then be equivalent to the amount of nitrate assimilated to amino nitrogen. In some plants, other organic anions (e.g. oxalate) may have a primary role in compensating for alkalization during nitrate reduction (see Fig. 16.10).

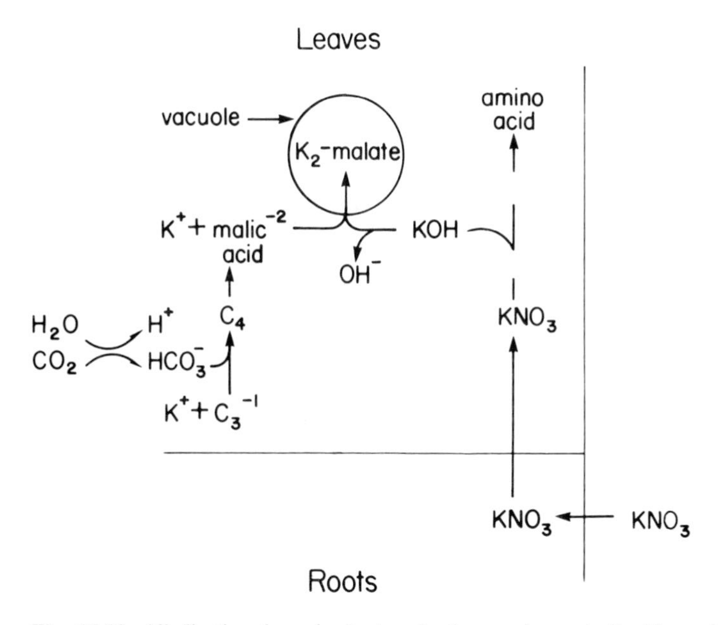


**Fig. 16.10.** Alkalization through nitrate reduction can be neutralized by malic acid synthesis.

An alternative to storage of potassium malate in the leaf tissue is its transport to roots and decarboxylation through malic enzyme. This would provide $HCO_3^-$ to exchange out of the roots for $NO_3^-$ uptake. Formation of free $CO_2$ from $HCO_3^-$ then results in generation of hydroxyl ions. In this case, the net effect is transfer of hydroxyl ions from the leaf to the soil (Fig. 16.11).
[Refs. **8, 10, 19, 22, 26, 33, 34, 42**]

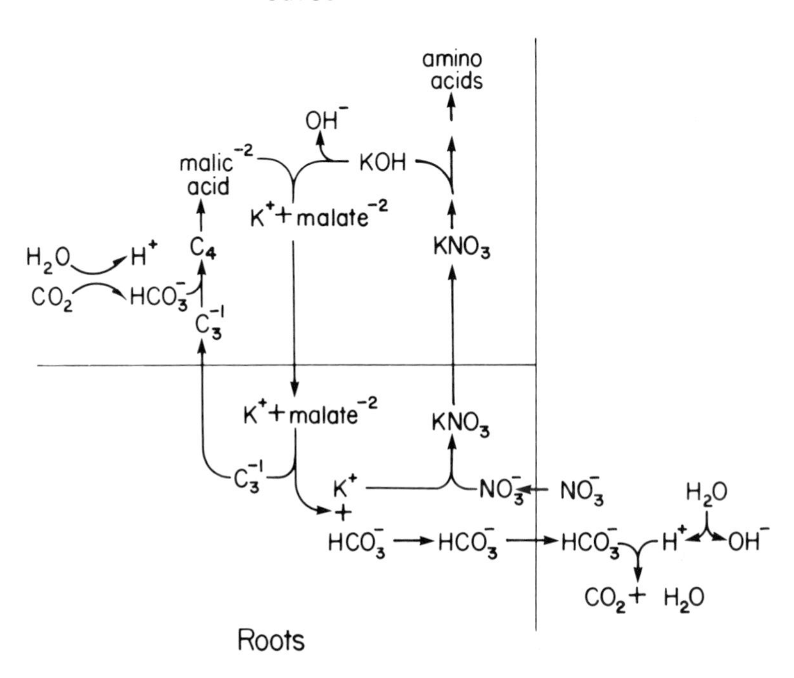


**Fig. 16.11.** A $C_4$ cycle between leaves and roots linked to nitrate uptake.

## 16.6 Malate metabolism in the glyoxylate cycle of germinating seedlings

Malate is an intermediate in the glyoxylate cycle during the metabolic conversion of lipids to sucrose in the endosperm tissue (e.g. in castorbean endosperm) or cotyledons of germinating seedlings. It is synthesized through malate synthetase of the glyoxylate cycle.

Acetyl CoA + glyoxylate → malate + CoA (malate synthetase) Eqn. 16.3

Malate then serves as a precursor to sucrose synthesis through malate dehydrogenase and PEP-carboxykinase, which results in PEP formation.

Castorbean endosperm tissue has PEP carboxykinase, PEP carboxylase and RBP carboxylase. This is the only nonchlorophyllous tissue in plants which has been found to have RBP carboxylase. RBP carboxylase activity increases during germination and reaches a maximum after four days, similar to that of malate synthetase and isocitrate lyase of the glyoxylate cycle. However, it appears that the RPP pathway is of minor importance in the endosperm tissue.

When $^{14}CO_2$ is fed to endosperm tissue, malate becomes labelled and labelling of sucrose follows. Malate could be labelled via PEP carboxykinase (either carboxylation or exchange reaction, PEP + $^{14}CO_2$ + ADP ⇄ $^{14}$C-oxaloacetate + ATP) or via PEP carboxylase coupled to malate dehydrogenase. The hexose moiety of sucrose contains most of the label in carbon atoms 3 and 4. One explanation is that labelling in malate becomes randomized between carbons 1 and 4 through fumarase (Section 15.8). Subsequently decarboxylation through carboxykinase results in label in the carboxyl carbon of PEP which then appears in carbons 3 and 4 of hexoses following gluconeogenesis. If substantial $CO_2$ fixation occurred through the RPP pathway, either directly or following $C_4$ acid decarboxylation, metabolites of the cycle would become uniformly labelled leading to randomization of label in sucrose. This, however, has not been observed which suggests that sucrose is synthesized from the 3 carbon product of malate decarboxylation rather than through the RPP pathway. [Refs. **5, 6, 7, 26**].

## Specific Citations


1 ALLAWAY W. G. (1973) Accumulation of malate in guard cells of *Vicia faba* during stomatal opening. *Planta*, **110**, 63–70.

2 ANDREWS A. K. & SVEC L. V. (1975) Photosynthetic activity of soybean pods at different growth stages compared to leaves. *Canad. J. Plant Sci.*, **55**, 501–5.

3 ATKINS C. A., KUO J., PATE J. S., FLINN A. M. & STEELE T. W. (1977) The photosynthetic pod wall of peas (*Pisum sativum* L.): Distribution of $CO_2$ fixing enzymes in relation to pod structure. *Plant Physiol.*, **60**, 779–86.

4 AVADHANI P. N., GREENWAY H., LEFROY R. & PRIOR L. (1978) Alcoholic fermentation and malate metabolism in rice germination at low oxygen concentrations. *Aust. J. Plant Physiol.*, **5**, 15–25.

5 BEEVERS H. (1961) *Respiratory Metabolism in Plants.* 232 pp. White Plains, New York: Row, Peterson & Co.

6 BENEDICT C. R. (1973) The presence of ribulose 1,5-bisphosphate carboxylase in the non-photosynthetic endosperm of germinating castor bean. *Plant Physiol.*, **51**, 755–9.

7 BENEDICT C. R. & BEEVERS H. (1961) Formation of sucrose from malate to phosphoenolpyruvate. *Plant Physiol.*, **36**, 540–4.

8 BEN-ZIONI A., LIPS S. H. & VAADIA Y. (1970) Correlations between nitrate reduction, protein synthesis and malate accumulation. *Plant Physiol.*, **23**, 1039–47.

9 CROOKSTON R. K., O'TOOLE J. & OZBUN J. L. (1974) Characterization of the bean pod as a photosynthetic organ. *Crop Sci.*, **14**, 708–12.

10 DAVIES D. D. (1973) Control of and by pH. In *Symp. Soc. Exp. Biol.*, **27**, 513–29.

11 DITTRICH P. & RASCHKE K. (1977) Malate metabolism in isolated epidermis of *Commelina communis* L. in relation to stomatal functioning. *Planta*, **134**, 77–81.

12 ENYL B. A. C. (1962) The contribution of different organs to grain weight in upland swamp rice. *Ann. Bot.*, **26**, 529–31.

13 FARINEAU J. & LAVAL-MARTIN D. (1977) Light versus dark carbon metabolism in cherry tomato fruits. II. Relationship between malate metabolism and photosynthetic activity. *Plant Physiol.*, **60**, 877–80.

14 Flinn A. M., Atkins C. A. & Pate J. S. (1977) The significance of photosynthetic and respiratory exchanges in the carbon economy of the developing pea fruit. *Plant Physiol.*, **60**, 412–8.

15 Harvey D. M., Hedley C. L. & Keely R. J. (1976) Photosynthetic and respiratory studies during pod and seed development in *Pisum sativum* L. *Ann. Bot.*, **40**, 993–1001.

16 Hedley C. L., Harvey D. M. & Keely R. J. (1975) Role of PEP carboxylase during seed development in *Pisum sativum. Nature*, **258**, 352–4.

17 Humble G. D. & Raschke K. (1971) Stomatal opening quantitatively related to potassium transport. Evidence from electron probe analysis. *Plant Physiol.*, **48**, 447–53.

18 Jennings V. M. & Shibles R. M. (1968) Genotypic differences in photosynthetic contributions of plant parts in grain yield in oats. *Crop Sci.*, **8**, 173–5.

19 Kirkby E. A. & Mengel K. (1967) Ionic balance in different tissues of tomato plant in relation to nitrate, urea, or ammonium nutrition. *Plant Physiol.*, **42**, 6–14.

20 Lakso A. N. & Kliewar W. M. (1975) The influence of temperature on malic acid metabolism in grape berries. *Plant Physiol.*, **56**, 370–2.

21 Laval-Martin D., Farineau J. & Diamond J. (1977) Light versus dark carbon metabolism in cherry tomato fruit. I. Occurrence of photosynthesis. Study of the intermediates. *Plant Physiol.*, **60**, 872–6.

22 Loomis R. S. & Gerakis P. A. (1975) Productivity of agricultural ecosystems. In *Photosynthesis and Productivity in Different Environments.* (ed. J. P. Cooper) pp. 145–72. Cambridge: Cambridge University Press.

23 McManmon M. & Crawford R. M. M. (1971) A metabolic theory of flooding tolerance: the significance of enzyme distribution and behaviour. *New Phytol.*, **70**, 299–306.

24 Nutbeam A. R. & Duffus C. M. (1976) Evidence for $C_4$ photosynthesis in barley pericarp tissue. *Biochem. Biophys. Res. Commun.*, **70**, 1198–203.

25 Osmond C. B. (1976) Ion absorption and carbon metabolism. In *Encyclopedia of Plant Physiology.* Vol. II. New series. Transport in Plants. (eds. U. Luttge and M. G. Pitman) pp. 347–72. New York: Springer-Verlag.

26 Osmond C. B., Akazawa T. & Beevers H. (1975) Localization and properties of ribulose diphosphate carboxylase from castor bean endosperm. *Plant Physiol.*, **55**, 226–30.

27 Outlaw W. H., Manchester J., DiCamellia C. A., Randall D. D., Rapp B. and Veith G. M. (1979) Photosynthetic carbon reduction pathway is absent in chloroplasts of *Vicia faba* guard cells. *Proc. Natl. Acad. Sci., USA* **76**, 6371–5.

28 Pearson C. J. (1973) Daily changes in stomatal aperture and in carbohydrates and malate within epidermis and mesophyll of leaves of *Commelina cyanea* and *Vicia faba. Aust. J. Biol. Sci.*, **26**, 1035–44.

29 Quebedeaux B. & Chollet R. (1975) Growth and development of soybean (*Glycine max* (L.) Merr.) pods. $CO_2$ exchange and enzyme studies. *Plant Physiol.*, **55**, 745–8.

30 Raghavendra A. S. & Das V. S. R. (1972) Control of stomatal opening by cyclic photophosphorylation. *Curr. Sci.*, **41**, 150–1.

31 Raschke K. (1975) Stomatal action. *Ann. Rev. Plant Physiol.*, **26**, 309–40.

32 Raschke K. & Dittrich R. (1977) [$^{14}$C] carbon-dioxide fixation by isolated leaf epidermis with stomata closed or open. *Planta*, **134**, 69–75.

33 Raven J. A. & Smith F. A. (1976) Nitrogen assimilation and transport in vascular land plants in relation to intracellular pH regulation. *New Phytol.*, **76**, 415–31.

34 Raven J. A. & Smith F. A. (1976) Cytoplasmic pH regulation and electrogenic $H^+$ extrusion. *Curr. Adv. Plant Sci.*, **24**, 649–60.

35 Ruffner H. P. & Kliewer W. M. (1975) Phosphoenolpyruvate carboxykinase activity in grape berries. *Plant Physiol.*, **56**, 67–71.

36 Sambo E. Y., Moorby J. & Milthorpe F. L. (1977) Photosynthesis and respiration of developing soybean pods. *Aust. J. Plant Physiol.*, **44**, 713–21.

37 TAMAS I. A., YEMM E. W. & BIDWELL R. G. S. (1970) The development of photosynthesis in dark-grown barley leaves upon illumination. *Canad. J. Bot.*, **48**, 2313–7.
38 THORNE G. N. (1965) Photosynthesis in ear and flag leaf in wheat and barley. *Ann. Bot.*, **29**, 317–29.
39 THORPE N. & MILTHROPE F. L. (1977) Stomatal metabolism. $CO_2$ fixation and respiration. *Aust. J. Plant Physiol.*, **4**, 611–21.
40 TING I. P. & DUGGER W. M. (1965) Transhydrogenation in root tissue: mediation by carbon dioxide. *Science*, **150**, 1727–8.
41 TING I. P. & DUGGER W. M. Jr. (1967) $CO_2$ metabolism in corn roots. I. Kinetics of carboxylation and decarboxylation. *Plant Physiol.*, **42**, 712–8.
42 ULRICH A. (1941) Metabolism of non-volatile organic acids in excised barley roots as related to cation-anion balance during salt accumulation. *Amer. J. Bot.*, **28**, 526–37.
43 WALKER D. A. (1962) Pyruvate carboxylation and plant metabolism. *Biol. Rev.*, **37**, 215–56.
44 WALPOLE P. R. & MORGAN D. G. (1972) Physiology of grain filling in barley. *Nature*, **240**, 416–7.
45 WILLMER C. M. & JOHNSTON W. R. (1976) Carbon dioxide assimilation in some aerial plant organs and tissues. *Planta*, **130**, 33–7.
46 WILLMER C. M., KANAI R., PALLAS J. E. Jr. & BLACK C. C. (1973) Detection of high levels of phosphoenolpyruvate carboxylase in leaf epidermal tissue and its significance in stomatal movements. *Life Sci.*, **12**, 151–5.
47 WILLMER C. M., PALLAS J. E. & BLACK C. C. (1973) Carbon dioxide metabolism in leaf epidermal tissue. *Plant Physiol.*, **52**, 448–52.
48 WIRTH E., KELLY G. J., FISCHBECK G. & LATZKO E. (1977) Enzyme activities and products of $CO_2$ fixation in various photosynthetic organs of wheat and oat. *Z. Pflanzenphysiol.*, **82**, 78–87.

# Appendix A Chloroplast Isolation and Criteria for Intactness



## A.1 Introduction

In order to study carbon metabolism, transport properties of chloroplasts and compartmentation of enzymes (the subject of much of the text), it is necessary to isolate chloroplasts which are intact, relatively pure, and biochemically functional. In some cases it is necessary to achieve a very high degree of purity (e.g. in localizing enzymes which are not present in large quantities) and in these instances, the absence of contamination may be more important than the chloroplasts' ability to support photosynthesis at high rates. In other circumstances, the highest level of purity may be sacrificed in the interests of maximal function, but contamination by cytoplasmic enzymes could complicate the interpretation of results (see Chapter 8). It is also important to note two facts which are often overlooked even though they are self-evident. The first is that it is not possible to prepare 'good' chloroplasts from 'poor' material (see A.3). The second is that 'good' chloroplasts will only perform well if they are properly treated. In retrospect, it is clear that, apart from preliminary isolation of protoplasts (A.5), there have been few real advances in the actual isolation of chloroplasts since the return, in the early 1960s, to Hills' practice of using a sugar as an osmoticum. Subsequent increases in activity owed more to improvements in assay and particularly to the introduction of inorganic pyrophosphate to the assay (Section 8.14).

The technique of isolating chloroplasts is therefore critical to the validity of the results obtained. Presently, researchers have developed relatively effective methods with only a few $C_3$ and $C_4$ species. There has been little success in isolating functional chloroplasts from CAM plants. This may be related to the acidity of the tissue and the presence of tannins or other compounds in the vacuole which have an immediate detrimental effect during chloroplast isolation.

It is now accepted that the chloroplast is the site of $CO_2$ fixation through the RPP pathway (Chapter 8). This applies to chloroplasts of $C_3$ and CAM plants, and bundle sheath chloroplasts of $C_4$ plants. The $C_4$ mesophyll chloroplasts lack the RPP pathway. These chloroplasts, in the presence of pyruvate and PEP carboxylase of the cytosol, drive the carboxylation phase of the $C_4$ pathway in the light (Chapter 12).

A reasonable measure of the photosynthetic performance of the isolated chloroplast is to make comparisons with the parent tissue. This would be most meaningful if measurements were made on the parent tissue ($CO_2$ exchange with an infrared gas analyzer) and a chloroplast preparation from the same tissue ($^{14}CO_2$ fixation or $O_2$ evolution with an oxygen electrode) at the same time under similar conditions of light, temperature, and $[CO_2]$. It is impractical to make such comparisons on a routine or daily basis. As a general assessment of the performance of isolated chloroplasts, average rates of photosynthesis of leaf tissue can be used (Table A.1).

**Table A.1.** Representative rates of whole leaf photosynthesis of the three photosynthetic groups under optimal conditions.

| Photosynthesis group | $\mu$mol $CO_2$ fixed $mg^{-1}$ Chl $h^{-1}$ | Temperature |
|---|---|---|
| $C_3$ | 100–200<br>(200–400)* | 15–25°C |
| $C_4$ | 200–400 | 20–35°C |
| CAM | 30–80 | 15–35°C |

* $CO_2$ saturated rates in $C_3$ will be higher than with atmospheric $CO_2$. Thus rates of photosynthesis by chloroplasts under saturating $CO_2$ should be compared with the parent tissue under $CO_2$-saturating conditions.

Thus, rates of $CO_2$ fixation with isolated chloroplasts of CAM plants of 1–5 and of $C_3$ and $C_4$ plants of 10–20 $\mu$mol $mg^{-1}$ Chl $h^{-1}$ which have been reported in some studies in the past are obviously unsatisfactory. Unless sufficiently high rates of photosynthesis are obtained with the isolated chloroplasts, their in-vitro metabolism may not be representative of the parent tissue. This is illustrated with the requirement for Pi (Chapter 8). $C_3$ chloroplasts, in the absence of Pi, have low rates of photosynthesis. The discovery of a requirement for Pi contributed greatly towards understanding the in-vivo mechanism of photosynthesis.

## A.2 Methods

In general, methods for the isolation of chloroplasts involve either mechanical disruption of leaf tissue or gentle disruption of the protoplasts. In the latter method which was developed only recently, enzymes (mixture of cellulase + pectinase) digest the cell wall of mesophyll cells, releasing protoplasts. The protoplasts are then gently broken to obtain intact organelles.

## A.3 Growth of Plants

This is the first important factor in determining the potential for isolating functional chloroplasts. Naturally, different species will have differences in their requirements for optimum growth. Much of the variability in rates of photosynthesis (and contradictions in conclusions made) among researchers using a given species (e.g. spinach) may be attributed to variability in the plant material. For example, in order to obtain good spinach chloroplasts, special attention must be given to the length of the photoperiod and to the nutrient conditions employed in plant growth. There is an additional concern when protoplasts are used as a means for isolating chloroplasts. Under some growth conditions, leaf tissue may become resistant to enzymatic digestion. Considerable attention has been given to the influence of growth conditions including light, humidity, and nutrition on isolation of protoplasts from tobacco and tomato leaves. With grasses, most success in isolating protoplasts has been through the use of relatively young leaf tissue (one to three weeks of age depending on the species).

### Material

Spinach is still preferred for many purposes because of its lack of contaminating phenolics and the fact that there is already an enormous literature relating to chloroplasts from this species. Until recently it was the only plant (with the possible exception of young pea leaves) from which really intact and active chloroplasts could be isolated. For some purposes peas are a convenient alternative, although it is becoming increasingly clear that chloroplasts isolated from young pea shoots have different permeability characteristics to those from mature spinach. Lettuce has been used extensively by Avron and others but mainly for work on electron transport and photophosphorylation. The protoplast procedures (A.5) have, however, made it possible to isolate 'good' chloroplasts from many more species.

### Culture

Although true spinach (*Spinacia oleracea*) is an easy plant to use, it is a difficult plant to grow. In some parts of North America, Australia, India, and southern Europe it may be successfully cultivated in the field even in the height of summer, but in more northern latitudes (as in the United Kingdom) it produces virtually no leaf before flowering in long days and is grown, for experimental purposes, under conditions in which the photoperiod is usually limited to $8\frac{1}{2}$ hours. In the absence of devices for shortening the day length, some extension of the effective growing season may follow the utilization of long-day varieties. It is doubtful whether any particular variety can be recommended for all environments. U.S. Hybrid 424 (Ferry-Morse Seed Co. P.O. Box 100, Mountain View, California 94042) which is also sold as Hybrid 102 Yates (244–254 Horsley Road, Milperra, N.S.W. 2214, Australia or The Seed Centre, Withyfold Drive, Macclesfield, Cheshire, United Kingdom) is evidently very useful in many circumstances. For example, the highest rates of carbon assimilation ever recorded (350 $\mu mol\ mg^{-1}\ Chl\ h^{-1}$ at 20°C and better) were obtained by Ulrich Heber using 'Yates 102' grown in water culture in Canberra. The same variety grown in the same medium at Sheffield (UK) led to substantial improvements. Rates in excess of 100 $\mu mol\ mg^{-1}\ Chl\ h^{-1}$ became commonplace, and rates in excess of 200 were occasionally recorded. These may of course may be equalled by field grown spinach at its best, but for many purposes there is no doubt that water culture is worth the extra effort involved. At Sheffield the standard nutrient solution (see Table A.2) now employed is based on that used in Canberra. The nutrient solution is usually delivered to plants in one of two ways. Either the roots are directly immersed in shallow trays through which aerated nutrient flows at a low rate or plants are grown in plastic pots in vermiculite (heat expanded mica) or a mixture of vermiculite and peat and periodically irrigated from below with nutrient solution. Seeds are germinated on vermiculite moistened with nutrient solution and

**Table A.2.** Nutrient Solution for Spinach Water Culture

| NO | Solution | Stock concentration | Volume (ml) per 20 litres nutrient solution made up with distilled water |
|---|---|---|---|
| 1 | $KNO_3$ | 1 M | 120 |
| 2 | $Ca(NO_3)_2$ | 1 M | 80 |
| 3 | $MgSO_4$ | 1 M | 40 |
| 4 | $KH_2PO_4$ | 1 M | 20 |
| 5 | $MgCI_2$ | 1 M | 80 |
| 6 | Trace elements | (B, Mn, Zn, Cu, Mo) | 20 |
| 7 | NaFe-EDTA | (3.86 gm/250 ml) | 20 |

| Trace elements | Quantity (mg) in 250 ml $H_2O$ |
|---|---|
| $H_3BO_3$ | 715 |
| $MnCI_2 \cdot 4H_2O$ | 452 |
| $ZnSO_4 \cdot 7H_2O$ | 55 |
| $CuSO_4 \cdot 5H_2O$ | 20 |
| $NaMoO_4 \cdot 2H_2O$ | 7.25 |

when the cotyledons are fully expanded are held in polyurethane foam plugs fitting into 3.5 cm apertures cut into the lid of aerated aquarium tanks. Each tank accommodates 12–20 plants, and when these have developed leaves of 2–3 cm, they are transferred to larger, shallower 'tanks' or pots. In the summer months, day-length is shortened to $8\frac{1}{2}$ hours by an automatic blind which prevents illumination during the early morning and early evening. In the winter, daylight is augmented by 400-W Wotan lamps (metal halide with dysprosium) at 1 m centres delivering approximately 50 W/m$^2$ of photosynthetically active light. (To date we have not been able to grow spinach which would yield really active Type A chloroplasts under fluorescent light.) (Section A.7)

In full daylight, high temperatures appear to be unimportant, and, indeed, some of the best material has been grown under glass at summer temperatures reaching 50°C (in the sun). Conversely, in poor winter light, 25°C should not be exceeded.

### PEAS

Seeds are soaked overnight in aerated water and germinated, in vermiculite, in flat trays at a density of 125 g (dry weight) per tray (38 × 23 × 8 cm). Again the combination of low light and high temperature is to be avoided but in relatively strong light (> 50 W/m$^2$), temperatures of 20–25°C and photoperiods of 9–12 h, good material can be grown in 9–11 days, by which time the shoots of a variety such as Progress No. 9 (Suttons Ltd., London Road, Earley, Reading, United Kingdom) are 3–4 cm high.

## A.4 Mechanical procedures

These involve disrupting the parent tissue (usually with a mechanical blender) while trying to avoid undue damage to the chloroplasts. Homogenization is therefore usually brief (5 s or less) and deliberately inefficient so that only about 1% of the chloroplasts are recovered. Rapid separation from the brei appears to be advantageous and the chloroplasts are usually spun-out of the grinding medium (by centrifugation) within 2–3 min of homogenization.

Grinding media are many and varied but they always contain an osmoticum (see Section 8.10) which is usually a sugar (e.g. sucrose or glucose) or a sugar alcohol (such as sorbitol or mannitol) at about 0.33 M. In addition, they usually contain a buffer to maintain the hydrogen ion concentration (often but not always at a slightly acid pH) and an anti-oxidant such as ascorbate. For many purposes a chelating agent also seems to be essential and if inorganic pyrophosphate is used as the buffer, this may also fill the requirement for chelation. Other protective agents such as BSA (bovine serum albumin) may also be added but with 'good' spinach, excellent chloroplasts can be prepared in solutions containing nothing more than 0.33 M sorbitol and 10 mM $Na_4P_2O_7$ adjusted to pH 6.5 with HCl.

Mechanical procedures have not been very successful for isolation of intact chloroplasts from $C_4$ plants. Mesophyll chloroplasts generally have a low percentage intactness (based on enzyme retention or ferricyanide test for intactness, Section A.7). Bundle sheath cells have a thick cell wall and are therefore more resistant to breakage than mesophyll cells and thus generally yield poor chloroplasts. For $C_4$ plants, an additional problem can occur with mechanical procedures due to cross-contamination of mesophyll and bundle sheath preparations (Chapter 10).

## A.5 Chloroplasts from protoplasts

This method of obtaining functional chloroplasts is still in its infancy. It has proved particularly useful with $C_4$ plants for obtaining, in a pure form, mesophyll protoplasts from which intact chloroplasts can be isolated. It has also been used successfully with the $C_3$ monocots (wheat and barley), to a limited extent with $C_3$ dicots (spinach, sunflower, tobacco and peas), and with the CAM plants, e.g. *Sedum praealtum* and *Mesembryanthemum crystallinum*.

### (a) Preparation of leaf tissue

The method used for preparing leaf tissue for enzymatic digestion is dependent on the species. In monocots, leaf segments can be cut manually with a razor blade (transverse sections about 0.7–0.8 mm). Alternatively, use of a mechanical cutter provides uniform segments, eliminating experimental variation in preparations. It seems a simple task, but there is variation from person to person in the manual sectioning of leaf tissue. In dicots, the epidermal layer can occasionally be removed from the lower surface of the leaf with a pair of forceps (e.g. with tobacco and *Kalanchoë daigremontiana*). Otherwise, in dicots the surface of the leaf can be rubbed with carborundum or brushed gently (e.g. with a tooth brush) in order to break through the epidermal tissue. The leaf tissue is then incubated for a maximum of 2.5–4 h at 25–30°C under low light. Longer incubations are undesirable as they may lead to loss of photosynthetic functions. Vacuum infiltration of the leaf segments with the enzyme medium or shaking during incubation are treatments to be considered with different species. Neither of these methods seem of benefit when small leaf sections of monocots are used.

An example of the isolation medium:

2% cellulase (source-*Trichoderma viride*)
pectinase (source-*Rhizopus* sp.), 0.1% for $C_4$, 0.3% for $C_3$ and CAM
0.5 M sorbitol
1 mM $CaCl_2$
20 mM MES buffer, pH 5.5
0.05% bovine serum albumin (defatted)

The enzymes from the above fungi, which are commercially available, have been found particularly useful with a number of species including most monocots. In some cases desalting or further purification of the enzymes may provide for greater protoplast stability and more consistent yields. For species resistant to digestion, either some variation in growth conditions or alternative sources of digestive enzymes may prove useful.

### (b) Isolation and purification

Following incubation of the leaf tissue with digestive enzymes, the isolation medium can be gently removed and discarded. Successive washes with an osmoticum (e.g. in 0.5 M sorbitol, 1 mM $CaCl_2$) releases a mixture of protoplasts, chloroplasts, and vascular tissue. Undigested material including epidermal and vascular tissue can be removed by filtration through nylon sieves ($C_3$ and CAM material through nets having 1 mm and 200 $\mu$m openings and $C_4$ material through nets having 1 mm, 500 and 80 $\mu$m openings). Bundle sheath strands from $C_4$ plants are usually resistant to digestion and can be collected on 80 $\mu$m nylon mesh. Low speed centrifugation of the extract, at 100g for 5 min, gives a pellet containing protoplasts and chloroplasts. There are two means of purifying protoplasts. With some species mesophyll protoplasts will float when resuspended in a solution of sufficiently high density. For example, if mesophyll protoplasts of wheat, barley, or sunflower are suspended in a solution containing 0.5 M sucrose and centrifuged at low speeds (200 g for 5 min), they float to the top of the medium. If necessary, the density of the medium can be increased by adding, for example, 2–10% (w/v) Dextran (e.g. $T_{20}$ or $T_{40}$) or Ficoll. Layering a lower density osmoticum (e.g. sorbitol) on top of the sucrose medium prior to centrifugation results in protoplast partitioning at the *interphase*. They are then easily collected with a Pasteur pipette.

Protoplasts which do not float readily can be purified in an aqueous two-phase system formed by a mixture of two polymers: dextran and polyethylene glycol. The mixture is composed of 5.5% (w/v) polyethylene glycol 6000, 10% (w/v) dextran $T_{20}$, 10 mM sodium phosphate (pH 7.5), and 0.46 mM sorbitol. If 0.6 ml of the unpurified protoplast preparation is thoroughly mixed with 5.4 ml of the two-phase solution and centrifuged at 300 g for 5 min at 4°C, the protoplasts partition at the interphase and chloroplasts in the lower phase. Protoplasts collected from the interphase can be resuspended and washed in an appropriate storage medium.

### (c) Storage

The mesophyll protoplasts are most stable if stored at a relatively low pH and with some divalent cations (e.g. 0.4 M sorbitol, 1 mM $CaCl_2$, and 20 mM MES, pH 6.0). Studies with barley, wheat, and tobacco indicate that sorbitol is an adequate osmoticum for at least 8–10 h if the protoplasts are stored on ice. If stored at 25°C, the protoplasts or chloroplasts from protoplasts lose their photosynthetic capacity when a sorbitol medium is used. However, the protoplasts are photosynthetically stable for at least 10–20 h at 25°C with sucrose as the osmoticum.

### (d) Chloroplast isolation from protoplasts

The mesophyll protoplasts of $C_3$ and $C_4$ plants have an average diameter of 30–40 $\mu$m whilst those of CAM plants are at least 60–100 $\mu$m in diameter. A quick, effective procedure for isolating chloroplasts is to pass the protoplasts several times through a nylon mesh (20 $\mu$m for $C_3$ and $C_4$, 44 $\mu$m for CAM). This can be achieved by fitting nylon net to the end of a 1ml disposable syringe, which has had the tip excized to provide a pore of about 3 mm. An aliquot of the protoplasts will be ruptured when taken up and ejected two or three times from the syringe. Intact chloroplasts (which have a diameter of 3–5 $\mu$m) will then be released. Although the above method is adequate and simple, other methods may be employed. For example, passing protoplasts through a Yeda Press at 75 psi will yield largely intact chloroplasts. Also, CAM protoplasts may be broken by expelling them from a syringe fitted with a 25 gauge needle. The protoplast extract can be centrifuged at low speed, i.e. 250 g for 1 min, to obtain a chloroplast pellet largely free of the cytosol fraction. This is sufficient for many studies on photosynthesis with chloroplasts from various species. For studies on the intracellular localization of enzymes, the total protoplast extract can be layered on a sucrose density gradient and then centrifuged and fractionated by standard procedures (Section 12.1).

### (e) Material

Cell walls are composed mainly of cellulose, hemicellulose and protein while pectin and calcium pectate are normally considered as the cementing material between cells.

The following are examples of sources of some of the special materials which may be used for cell or protoplast isolation. Commercial sources of digestive enzymes used for cell or protoplast isolation are generally a composite of several enzymes. Depolymerization of pectate and pectin can occur through several types of depolymerases including polygalacturonase, polymethyl galacturonase, pectate lyase and pectin lyase. Apparently, there is sufficient difference among species in the composition of the pectin-pectate which cements cells together such that certain depolymerases are more effective in some species than in others. Interestingly there has been little or no success in enzymatically isolating cells from the grasses.

A number of enzymes are required for degrading cellulose and various hemicelluloses. Some enzymes concerned with cellulose degradation are endo-1,4-$\beta$-glucanase, cellobiohydrolase, cellobiase and possibly an enzyme which converts crystalline cellulose into an amorphous form prior to the action of other enzymes. Hemicellulose is a complicated, heterogeneous group of polysaccharides which may contain xylan, arabinans, galactans, mannans, and glucuronic and galacturonic acids. Differences in the composition of hemicellulose among species may well influence the effectiveness of commercial sources of digestive enzymes in degrading the cell wall.

i. Enzymes

Cellulase from *Trichoderma viride* (contains cellulose and hemicellulose degrading enzymes)

| | | |
|---|---|---|
| As Onozuka R10 and RS | Yakult Biochemical Co., Ltd.<br>Enzyme Products<br>8–21, Shingikancho<br>Nishinomiya JAPAN | |
| As Cellulysin | Calbiochem-Behring Corp.<br>Hoechst U.K. Ltd.<br>Hoechst House<br>Salisbury Road<br>Hounslow, Middx.<br>TW4 6JH<br>ENGLAND | Calbiochem-Behring Corp.<br>P. O. Box 12087<br>San Diego, CA 92112<br>U.S.A. |
| As Onozuka R10 | Unwin and Co., Ltd.<br>Prospect Place<br>Welwyn, Bedfordshire<br>ENGLAND | |

| | | |
|---|---|---|
| As Meicelase | Meiji Seika Kaisha Ltd.<br>8–2 Chome Kyobashi,<br>Chuo-Ku<br>Tokyo, JAPAN | |
| Pectinase from *Rhizopus* sp. (contains polygalacturonase) | | |
| As Macerozyme R10 | Yakult Biochemical Co., Ltd.<br>(see above) | Unwin and Co., Ltd.<br>(see above) |
| As Macerase | Calbiochem-Behring Corp.<br>(see above) | |
| Pectinase from *Aspergillus* sp. | | |
| As Extractase PC | Fermco Biochemics, Inc.<br>2638 Delta Lane<br>Elk Grove Village, IL 60007<br>U.S.A. | |
| As Rohament P | Rohm G. m. b. H.<br>D-6100 Darmstadt<br>Kirschenallee<br>Postfach 4242<br>WEST GERMANY | |
| As pectolyase Y23-Pectinase from *Aspergillus japonicus* (active components polygalacturonase, pectin lyase and an unidentified protein factor). | Seishin Pharmaceutical Co., Ltd.<br>Noda, Chiba<br>JAPAN | |
| ii. Nylon mesh | Henry Simon, Ltd.<br>P. O. Box 31<br>Stockport, Cheshire<br>SK3 ORT<br>ENGLAND | Tetko, Inc.<br>Precision Woven Screening Media<br>420 Saw Mill River Road, Elmsford,<br>NY 10523<br>U.S.A. |

| | | |
|---|---|---|
| iii. Dextran<br>($T_{20}$-Avg. MW 20,000)<br>($T_{40}$-Avg. MW 40,000) | U.S. Biochemicals Corp.<br>21000 Miles Parkway<br>Cleveland, OH 44128<br>U.S.A.<br>(Dextran $T_{20}$ and $T_{40}$) | Pharmacia Fine Chemicals<br>Uppsala, Sweden<br>(Dextran $T_{40}$) |

## A.6 Advantages and disadvantages of mechanical versus enzymatic procedures

The effectiveness of the two methods for obtaining photosynthetically functional chloroplasts currently depends very much on the species used. Chloroplasts showing very high rates of photosynthesis can be isolated mechanically from $C_3$ species such as spinach and peas.

Recent reports also indicate that functional chloroplasts can be obtained from mesophyll protoplasts of these plants. As a routine procedure for studying $CO_2$ assimilation by chloroplasts of these species, the protoplast method currently offers no advantages. However, it is useful for determining the intracellular localization of enzymes since a high percentage of the organelles can be isolated intact. Spinach seems to present some of the difficulties found with tobacco and tomatoes in that variation in plant material due to growth conditions may result in some tissue being resistant to enzymatic digestion.

Relatively intact and active chloroplasts have been isolated from the $C_3$ mesophyll protoplasts of wheat, barley, tobacco, and sunflower. Also, intact chloroplasts isolated from $C_4$ mesophyll protoplasts of *Digitaria sanguinalis* (crabgrass), *Eleusine indica* (goosegrass) and *Urochloa panicoides* effectively catalyze light-dependent photosynthetic metabolism of the $C_4$ cycle. The usefulness of protoplasts for studying $C_4$ photosynthesis is covered in Chapters 10 and 12. Generally neither enzymatic nor mechanical procedures have yielded chloroplasts from CAM plants able to catalyze rates of $CO_2$ assimilation greater than a few $\mu$mol mg$^{-1}$ Chl h$^{-1}$. Chloroplasts isolated from protoplasts of the CAM plant *Sedum praealtum* have average rates of $CO_2$ fixation of 20–30 $\mu$mol mg$^{-1}$ Chl h$^{-1}$.

For studying the intracellular localization of enzymes in the mesophyll cell, the enzymatic procedure has several advantages. Firstly, the vascular and epidermal tissues are eliminated as a potential source of enzymes, mitochondria and microbodies. Secondly, a high percentage of intact organelles can be obtained. This is naturally an advantage for studying the intracellular compartmentation of enzymes. Aggregation of organelles, nonspecific association of soluble enzymes to organelles, and formation of protein-tannin complexes may be more likely with mechanical than with protoplast procedures.

Protoplasts may offer a useful system of studying the reconstitution of metabolism between the cytoplasm and organelles. In $C_4$ plants, mesophyll protoplast extracts have a functional carboxylation phase of the $C_4$ pathway dependent on the chloroplast and cytosol (Section 12.3). In $C_3$ plants photosynthetic metabolism requiring links between the cytosol fraction, mitochondria or peroxisomes (e.g. sucrose synthesis and photorespiration) might be studied through protoplast extracts or reconstitution of the subcellular fractions.

Once a preparation of protoplasts is obtained, multiple isolations of chloroplasts with consistent activity can be made from a homogeneous protoplast stock. This minimizes experimental variability which may occur with mechanical preparations of different leaf tissue and maximizes the possibility of observing differences between actual treatments (conditions of isolation, resuspension, and assay). In some cases, results obtained with protoplasts may be used to develop effective mechanical procedures with a given species.

The disadvantages of the protoplast method are the failure for some species to provide adequate enzymatic digestion, the additional time required for protoplast isolation (not a limitation if the results warrant the effort) and possibly limited yields (1–5 mg Chl/10 g leaf tissue). It is also imperative that the digestion time is minimized and an adequate protoplast purification procedure is used. The possibility of inhibitors in digestive enzymes, or of side effects from enzymatic digestion, should not be overlooked.

This discussion has concentrated on methods available for isolating chloroplasts. Similar considerations are necessary in the isolation of functional mitochondria and peroxisomes from plant leaves.

## A.7 Criteria for determining intactness of chloroplasts

Isolated chloroplasts can exist in a number of states, both in terms of morphology and function. David Hall provided a useful nomenclature for $C_3$ chloroplasts which correlates microscopic appearance or structure with function. Type A, 'complete chloroplasts,' are prepared in sugar solutions and retain intact envelopes with rates of $CO_2$ fixation of 50–250 $\mu mol\,mg^{-1}\,Chl\,h^{-1}$. Exogenous $NADP^+$ and ferricyanide, a Hill oxidant, do not penetrate. Type B, 'unbroken chloroplasts' prepared in sugars or salts, have low rates of $CO_2$ fixation. The chloroplasts retain soluble proteins but ferricyanide and ADP penetrate, indicating that the envelope has lost some of its semipermeable properties. Types C to F are variations on types of broken chloroplasts (all having lost their outer envelopes) which result from different methods of preparation.

There is a useful method for distinguishing between Type A chloroplasts and the other types. Since ferricyanide does not permeate intact chloroplast envelopes, Type A chloroplasts do not show ferricyanide-dependent $O_2$ evolution. If a preparation of chloroplasts is subjected to osmotic shock (rapidly lowering the osmoticum, inducing exosmosis), the outer envelope will rupture and maximum activity of ferricyanide-dependent $O_2$ evolution can be measured. By measuring the rate of ferricyanide-dependent $O_2$ evolution before (B) and after (A) osmotic shock, the percentage intactness can be calculated.

$$\% \text{ intactness} = \frac{A - B}{A}(100)$$

Glyceraldehyde is added to prevent any $CO_2$-dependent $O_2$ evolution and $NH_4Cl$ to uncouple electron flow from phosphorylation and allow maximum rates. This method of determining percentage intactness of chloroplasts obtained from wheat protoplasts is illustrated in Figure A.1.

The percentage intactness as determined by the ferricyanide method suggests that the envelope of the chloroplast has retained its semipermeable properties, but this may not directly correlate with its $CO_2$ fixation potential. There is evidence that during chloroplast isolation, a few of them rupture and then reseal with a loss of stromal enzymes. These would show a high degree of intactness via ferricyanide test, but a low capacity for $CO_2$-dependent $O_2$ evolution. This indicates that the ferricyanide test may overestimate the percentage of functional chloroplasts. From a plot of soluble protein/mg Chl versus the percentage intactness as determined with ferricyanide, a correction can be made for this overestimation of intactness.

For other reasons, the percentage intactness by the ferricyanide test may not be indicative of the capacity for photosynthetic assimilation of $CO_2$. For example, sunflower chloroplasts isolated mechanically have a high degree of intactness but low rates of $CO_2$ fixation. Thus, there are additional requirements, perhaps involving the proper function of translocators in the chloroplast envelope (Chapter 8), which, in some cases, may be impaired during chloroplast isolation. Different subgroups may then exist within Type A chloroplasts. (An analogous method has been used for determining percentage intactness of isolated mitochondria by Douce *et al.* The activity of succinate-cytochrome c oxidoreductase is measured by following reduction of exogenous cytochrome c, before and after osmotic shock. With intact mitochondria, cytochrome c will not penetrate the outer membrane. The cytochrome c must have access to the inner membrane in order to be reduced.)

When chloroplasts are isolated from protoplasts, an additional method can be used to determine the percentage intactness. Protoplast extracts are centrifuged at low speeds to separate the chloroplasts from the extrachloroplastic enzymes. An enzyme considered to be exclusively localized in chloroplasts *in vivo*, e.g. NADP-glyceraldehyde phosphate dehydrogenase, is assayed in extracts of the chloroplast pellet (CP) versus that in the supernatant (S).

$$\% \text{ intactness} = \frac{\text{CP}}{\text{CP} + \text{S}}(100)$$

The accuracy of this method depends on complete recovery of the enzyme activity in both the chloroplast and supernatant fraction. Some enzymes such as RBP carboxylase may become labile or inhibited when exposed to the cell extract. It is thus important that the activity of the enzyme in the total cell extract (isolated

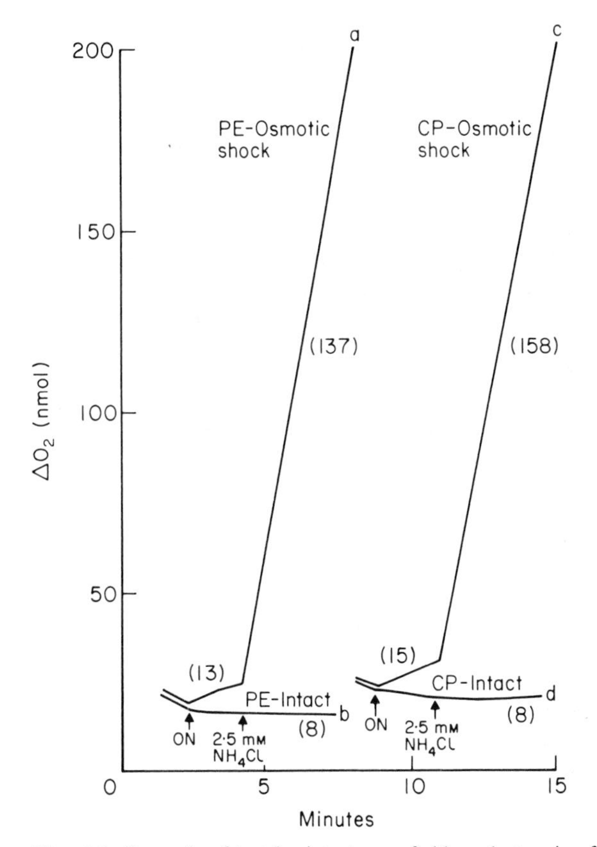


**Fig. A.1.** Example of test for intactness of chloroplasts using ferricyanide as a Hill oxidant. Chloroplasts were isolated from protoplasts of wheat. Protoplast extract (PE), chloroplasts from protoplasts, once washed (CP). Percentage intactness was calculated at 94% (PE) and 95% (CP). Numbers in parenthesis are $\mu$mol $O_2$ evolved $mg^{-1}$ Chl $h^{-1}$ and are corrected for uptake of $O_2$ by the $O_2$ electrode.

without osmoticum) equals the activity in the chloroplast plus supernatant fraction.

There has been considerable variation in media used to study photosynthetic metabolism by isolated chloroplasts. Important requirements in the assay medium with $C_3$ (Chapter 8) and $C_4$ chloroplasts (Chapter 12) have been discussed.

## A.8 Purification

Whether chloroplasts are isolated mechanically or from protoplasts, it is expedient and often sufficient to separate the intact chloroplasts by centrifugation. The extrachloroplastic fraction (cytoplasm, peroxisomes and mitochondria) can be largely retained in the supernatant when the intact chloroplasts have sedimented by centrifugation for 1–2 min (about 300–2000 g, depending on the amount of medium). Many of the broken chloroplasts will also be retained in the supernatant fraction and the resulting chloroplast pellet under optimal conditions will contain from 70 to nearly 100% intact chloroplasts.

When necessary, further removal of thylakoids is possible by centrifugation in a salt-free medium (0.33 M sorbitol brought to pH 7.5 with Tris base) which helps to retain the broken chloroplasts in the supernatant fraction. However, the resulting preparation of intact chloroplasts should be stored in a salt-containing medium to avoid substantial loss of activity during storage.

Sucrose gradients have been used for purifying chloroplasts and other organelles. However, the chloroplasts are subjected to abnormally high concentrations of osmoticum which results in low photosynthetic capacity. In order to obtain a high density but maintain a relatively low osmotic concentration in gradients, large molecular weight polymers can be used. For example, pure and functional chloroplasts can be obtained with peas and spinach after purification using a colloidal silica sol coated with polyvinylpyrrolidone (Percoll). The crude chloroplast preparation can be layered on 40% (v/v) Percoll containing 0.33 M sorbitol and a buffer. After centrifugation for 1 min at 2500 g, extrachloroplastic material is in the supernatant and the pellet contains intact chloroplasts. Alternatively, a linear gradient containing Percoll (approximately 10–90%) can be used. The crude chloroplast preparation is layered on the gradient and centrifuged for 15 min at approximately 10 000 g. The intact chloroplasts, which will form a band in the gradient according to their density, can be collected by fractionation and washed to remove the Percoll. One of the most recent advances in isolation techniques is a method of separating chloroplasts from extrachloroplastic material within a few seconds of protoplast lysis (Section 8.3). This is particularly useful for studying the compartmentation of metabolites and photosynthetic products within the cell during the course of carbon assimilation.

## References


1 DOUCE R. E., CHRISTENSEN L. & BONNER W. D. (1972) Preparation of intact plant mitochondria. *Biochem. Biophys. Acta*, **275**, 148–60.

2 EDWARDS G. E., HUBER S. C. & GUTIERREZ M. (1976) Photosynthetic properties of plant protoplasts. In *Microbial and Plant Protoplasts*. (eds. J. F. Peberdy, A. H. Rose, H. J. Rogers and E. C. Cocking). pp. 299–322 New York: Academic Press.

3 EDWARDS G. E., ROBINSON S. P., TYLER N. J. C., & WALKER D. A. (1978) Photosynthesis by isolated protoplasts, protoplast extracts and chloroplasts of wheat. Influence of orthophosphate, pyrophosphate and adenylates. *Plant Physiol.*, **62**, 313–7.

4 EDWARDS G. E., LILLEY R. McC. & HATCH M. D. (1979) Isolation of intact and functional chloroplasts from mesophyll and bundle sheath protoplasts of the $C_4$ plant *Panicum miliaceum*. *Plant Physiol.*, **63**, 821–7.

5 HALL D. O., (1972) Nomenclature for isolated chloroplasts. *Nature New Biol.*, **235**, 125–6.

6 HUBER S. C. & EDWARDS G. E. (1975) An evaluation of some parameters required for the enzymatic isolation of cells and protoplasts with $CO_2$ fixation capacity from $C_3$ and $C_4$ grasses. *Physiol. Plant.*, **35**, 203–9.

7 HUBER S. C., HALL T. C. & EDWARDS G. E. (1977) Light dependent incorporation of $^{14}CO_2$ into protein by mesophyll protoplasts and chloroplasts isolated from *Pisum sativum*. *Z. Pflanzenphysiol.*, **85**, 153–63.

8 ISHII, S. (1975) Enzymatic maceration of plant tissues by endopectin lyase and endo-polygalacturonase from *Aspergillus japonicus*. *Phytopathology*, **66**, 281–289.

9 KANAI R. & EDWARDS G. (1973) Purification of enzymatically isolated mesophyll protoplasts from $C_3$, $C_4$ and Crassulacean Acid Metabolism plants using an aqueous dextran polyethylene glycol two-phase system. *Plant Physiol.*, **52**, 484–90.

10 LILLEY R. McC., FITZGERALD M. P., RIENITS K. G. & WALKER D. A. (1975) Criteria of intactness and the photosynthetic activity of spinach chloroplast preparations. *New Phytol.*, **75**, 1–10.

11 MILLS W. R. & JOY K. W. (1980) A rapid method for isolation of purified, physiologically active chloroplasts, used to study the intracellular distribution of amino acids in pea leaves. *Planta*, **148**, 75–83.

12 NAGATA, I. and S. ISHII. (1979) A rapid method for isolation of mesophyll protoplasts. *Can. J. Bot.* **75**, 1820–1823.

13 NAKATANI H. Y. & BARBER J. (1977) An improved method for isolating chloroplasts retaining their outer membranes. *Biochim. Biophys. Acta*, **461**, 510–2.

14 SLABAS A. R., POWELL A. J. & LLOYD C. W. (1980) An improved procedure for the isolation and purification of protoplasts from carrot suspension culture. *Planta* **147**, 283–6.

15 SPALDING M. H. & EDWARDS G. E. (1980) Photosynthesis in isolated chloroplasts of the Crassulacean acid metabolism plant *Sedum praealtum*. *Plant Physiol.*, **65**, 1044–8.

16 SPALDING M. H., SCHMITT M. R., KU S. B. & EDWARDS G. E. (1979) Intracellular localization of some key enzymes of Crassulacean Acid Metabolism in *Sedum praealtum*. *Plant Physiol.*, **63**, 738–43.

17 TAKABE T., NISHIMURA M. & AKAZAWA T. (1979) Isolation of intact chloroplasts from spinach leaf by centrifugation in gradients of the modified silica 'Percoll'. *Agric. Biol. Chem.*, **43**, 2137–42.

18 WALKER D. A. (1980) Preparation of higher plant chloroplasts. In: *Methods in Enzymology*, Vol. 69. pp. 94–104 New York: Academic Press.

# Appendix B. Enzyme Nomenclature



From Enzyme Nomenclature Recommendations (1978) of the International Union of Biochemistry, Academic Press, New York.
Sequence of listings: *Recommended name*, E.C. No., (other names), systematic name. [Reaction]

## B.1 Reductive pentose phosphate pathway.

*ribulosebisphosphate carboxylase*, E.C. No. 4.1.1.39., (carboxydismutase), 3-phospho-D-glycerate carboxy-lyase (dimerizing). [D-ribulose 1,5-bisphosphate + $CO_2$ = 2,3-phospho-D-glycerate. The enzyme also catalyzes D-ribulose 1,5-bisphosphate + $O_2$ = 3-phospho-D-glycerate + phosphoglycollate]

*phosphoglycerate kinase*, E.C. No. 2.7.2.3., (1,3 Diphosphoglycerate-ADP transphosphatase) ATP:3-phospho-D-glycerate 1-phosphotransferase.
[ATP + 3-phospho-D-glycerate = ADP + 3-phospho-D-glyceroyl phosphate]

*glyceraldehyde-phosphate dehydrogenase* ($NADP^+$) (phosphorylating), E.C. No. 1.2.1.13., (triosephosphate dehydrogenase ($NADP^+$), D-glyceraldehyde-3-phosphate:$NADP^+$ oxidoreductase (phosphorylating). [3-phospho-D-glyceroyl phosphate + NADPH + $H^+$ = D-glyceraldehyde 3-phosphate + orthophosphate + $NADP^+$]

*triosephosphate isomerase*, E.C. No. 5.3.1.1., (phosphotriose isomerase), D-glyceraldehyde-3-phosphate ketol-isomerase.
[D-glyceraldehyde 3-phosphate = dihydroxyacetone phosphate]

*fructose-bisphosphate aldolase*, E.C. No. 4.1.2.13., (zymohexase), D-fructose-1,6-bisphosphate D-glyceraldehyde-3-phosphate-lyase.
[3-phospho-D-glyceroyl phosphate + NADPH + $H^+$ = glyceraldehyde 3-phosphate + orthophosphate + $NADP^+$] The second reaction catalysed by this enzyme in the RPP pathway is D-erythrose 4-

phosphate + dihydroxyacetone phosphate = sedoheptulose 1,7-bisphosphate

*fructose bisphosphatase*, E.C. No. 3.1.3.11., (hexosediphosphatase) D-fructose-1,6-bisphosphate 1-phosphohydrolase.
[D-fructose 1,6-bisphosphate + $H_2O$ = D-fructose 6-phosphate + orthophosphate]

*transketolase*, E.C. No. 2.2.1.1., (glycolaldehydetransferase), sedoheptulose 7-phosphate:D-glyceraldehyde 3-phosphate glycolaldehyde-transferase.
[sedoheptulose 7-phosphate + D-glyceraldehyde 3-phosphate = D-ribose 5-phosphate + D-xylulose 5-phosphate]. This enzyme also catalyzes D-fructose 6-phosphate + D-glyceraldehyde 3-phosphate = D-xylulose 5-phosphate + D-erythrose-4-phosphate in the RPP pathway.

*sedoheptulose-bisphosphatase*, E.C. 3.1.3.37., sedoheptulose 1,7-bisphosphate 1-phosphohydrolase.
[sedoheptulose 1,7-bisphosphate + $H_2O$ = sedoheptulose 7-phosphate + orthophosphate]

*ribosephosphate isomerase*, E.C. No. 5.3.1.6., (phosphopentosisomerase), D-ribose 5-phosphate ketol-isomerase.
[D-ribose 5-phosphate = D-ribulose 5-phosphate]

*ribulosephosphate 3-epimerase*, E.C. No. 5.1.3.1., (phosphoribulose epimerase) D-ribulose 5-phosphate 3-epimerase.
[D-ribulose 5-phosphate = D-xylulose 5-phosphate]

*phosphoribulokinase*, E.C. No. 2.7.1.19., (phosphopentokinase), ATP:D-ribulose-5-phosphate 1-phosphotransferase.
[ATP + D-ribulose 5-phosphate = ADP + D-ribulose 1,5-bisphosphate]

## B.2 The glycolate pathway

*phosphoglycollate phosphatase*, E.C. No. 3.1.3.18., 2-phosphoglycollate phosphohydrolase.
[2-phosphoglycollate + $H_2O$ = glycollate + orthophosphate]

*glycollate oxidase*, E.C. No. 1.1.3.1., (glycollate − $O_2$ transhydrogenase), glycollate:oxygen oxidoreductase.
[glycollate + $O_2$ = glyoxylate + $H_2O_2$]

*catalase*, E.C. No. 1.11.1.6., hydrogen-peroxide:hydrogen-peroxide oxido-reductase.
[$H_2O_2 + H_2O_2 = O_2 + 2\,H_2O$]

*glycine aminotransferase*, E.C. No. 2.6.1.4., glycine:2-oxoglutarate aminotransferase.
[glyoxylate + L-glutamate = glycine + 2-oxoglutarate]
*serine – glyoxylate aminotransferase*, E.C. No. 2.6.1.45., L-serine:glyoxylate aminotransferase.
[L-serine + glyoxylate = 3 hydroxypyruvate + glycine]

*glycine synthase*, E.C. No. 2.1.2.10., 5,10-methylenetetrahydrofolate: ammonia hydroxymethyltransferase (carboxylating, reducing).
[tetrahydrofolate + glycine + oxidized hydrogen-carrier protein = 5,10-methylenetetrahydrofolate + $CO_2$ + $NH_3$ + reduced hydrogen-carrier protein]

*serine hydroxymethyltransferase*, E.C. No. 2.1.2.1., (serine aldolase), 5,10-methylenetetrahydrofolate: glycine hydroxymethyltransferase.
[5,10-methylenetetrahydrofolate + glycine + $H_2O$ = tetrahydrofolate + L-serine]

*glyoxylate reductase*, E.C. No. 1.1.1.26., (NADH – glyoxylate transhydrogenase) glycollate:$NAD^+$ oxidoreductase.
[hydroxypyruvate + NADH + $H^+$ = D-glycerate + $NAD^+$]
[glyoxylate + NADH + $H^+$ = glycollate + $NAD^+$]
Comment: reduces hydroxypyruvate to D-glycerate or glyoxylate to glycollate.

*glycerate kinase*, E.C. No. 2.7.1.31., ATP:D-glycerate 3-phosphotransferase.
[ATP + D-glycerate = ADP + 3-phospho-D-glycerate]

Associated metabolism in the glycolate pathway for shuttle of reductive power.
*malate dehydrogenase*, E.C. No. 1.1.1.37., (L-malate-NAD transhydrogenase), L-malate:$NAD^+$ oxidoreductase.
[L-malate + $NAD^+$ = oxaloacetate + NADH + $H^+$]

Enzymes for reassimilating ammonia are given in Section B.7.

## B.3. Metabolism of triose phosphate to sucrose.

*triosephosphate isomerase*, E.C. No. 5.3.1.1., (phosphotriose isomerase), D-glyceraldehyde 3-phosphate ketol-isomerase.
[dihydroxyacetone phosphate = D-glyceraldehyde 3-phosphate]

*fructose-bisphosphate aldolase*, E.C. No. 4.1.2.13., (zymohexase), D-fructose 1,6-bisphosphate D-glyceraldehyde 3-phosphate-lyase.
[dihydroxyacetone phosphate + D-glyceraldehyde 3-phosphate = D-fructose 1,6-bisphosphate]

*fructose bisphosphatase*, E.C. No. 3.1.3.11., (hexose diphosphatase) D-fructose 1,6-bisphosphate 1-phosphohydrolase.
[D-fructose 1,6-bisphosphate + $H_2O$ = D-fructose 6-phosphate + orthophosphate]

*glucosephosphate isomerase*, E.C. No. 5.3.1.9., (phosphohexose isomerase), D-glucose-6-phosphate ketol-isomerase.
[D-fructose 6-phosphate = D-glucose 6-phosphate]

*phosphoglucomutase*, E.C. No. 2.7.5.1., (glucose(1→6)-phosphomutase), α-D-glucose 1,6-bisphosphate:α-D-glucose 1-phosphate phosphotransferase.
[α-D-glucose 6-phosphate + α-D-glucose 1,6-bisphosphate = α-D-glucose 1,6-biphosphate + α-D-glucose 1-phosphate]

*glucose-l-phosphate uridylyltransferase*, E.C. No. 2.7.7.9., (UDPglucose pyrophosphorylase), UTP:α-D-glucose-1-phosphate uridylyltransferase.
[UTP + α-D-glucose 1-phosphate = pyrophosphate + UDPglucose]

*nucleosidediphosphate kinase*, E.C. No. 2.7.4.6., (ATP →nucleosidediphosphate transphosphatase), ATP:nucleosidediphosphate phosphotransferase.
[ATP + nucleoside diphosphate = ADP + nucleoside triphosphate]

*inorganic pyrophosphatase*, E.C. No. 3.6.1.1., pyrophosphate phosphohydrolase.
[pyrophosphate + $H_2O$ = 2 orthophosphate]

*sucrose-phosphate synthase*, E.C. No. 2.4.1.14., (UDPglucose-fructosephosphate glucosyltransferase), UDPglucose:D-fructose 6-phosphate 2-α-D-glucosyltransferase.
[UDPglucose + D-fructose 6-phosphate = UDP + sucrose 6-phosphate]

*sucrose-phosphatase*, E.C. No. 3.1.3.24., sucrose-$6^F$-phosphate phosphohydrolase.
[sucrose $6^F$-phosphate + $H_2O$ = sucrose + orthophosphate]

Alternatively:

*sugar-phosphatase*, E.C. No. 3.1.3.23., sugar-phosphate phosphohydrolase.
[sugar phosphate + $H_2O$ = sugar + orthophosphate]
Comment: In this context for converting fructose-6-phosphate →fructose + Pi.

*sucrose synthase*, E.C. No. 2.4.1.13., (UDPglucose-fructose glucosyltransferase)
[UDPglucose + D-fructose = UDP + sucrose]
Comment: In leaves may be involved in sucrose degradation.

## B.4. Starch synthesis from triose phosphate.

*triosephosphate isomerase*, E.C. No. 5.3.1.1., (phosphotriose isomerase), D-glyceraldehyde-3-phosphate ketol-isomerase.
[D-glyceraldehyde 3-phosphate = dihydroxyacetone phosphate]

*fructose-bisphosphate aldolase*, E.C. No. 4.1.2.13., (zymohexase), D-fructose 1,6-bisphosphate D-glyceraldehyde-3-phosphate-lyase.
[dihydroxyacetone phosphate + D-glyceraldehyde 3-phosphate = D-fructose 1,6 bisphosphate]

*fructose-bisphosphatase*, E.C. No. 3.1.3.11., (hexosediphosphatase) D-fructose 1,6-bisphosphate 1-phosphohydrolase.
[D-fructose 1,6-bisphosphate + $H_2O$ = D-fructose 6-phosphate + orthophosphate]

*glucosephosphate isomerase*, E.C. No. 5.3.1.9., (phosphohexose isomerase), D-glucose 6-phosphate ketol-isomerase.
[D-fructose 6-phosphate = D-glucose 6-phosphate]

*phosphoglucomutase*, E.C. No. 2.7.5.1., (glucose(1→6)-phosphomutase), α-D-glucose 1,6-bisphosphate:α-D-glucose-1-phosphate phosphotransferase.
[α-D-glucose 6-phosphate + α-D-glucose 1,6-bisphosphate = α-D-glucose 1,6-bisphosphate + α-D-glucose 1-phosphate]

*glucose-1-phosphate adenylytransferase*, E.C. No. 2.7.7.27., (ADPglucose pyrophosphorylase), ATP:α-D-glucose-1-phosphate adenylyltransferase.
[ATP + α-D-glucose 1-phosphate = pyrophosphate + ADPglucose]

*inorganic pyrophosphatase*, E.C. No. 3.6.1.1., pyrophosphate phosphohydrolase.
[pyrophosphate + $H_2O$ = 2 orthophosphate]

*starch (bacterial glycogen) synthase*, E.C. No. 2.4.1.21., (ADPglucose −starch glucosyltransferase), ADPglucose:1,4-α-D-glucan 4-α-glucosyltransferase.
[ADPglucose + (1,4-α-D-glucosyl)$_n$ = ADP + (1,4-α-D-glucosyl)$_{n+1}$]

The enzymes for conversion of triose to hexose phosphates in starch and sucrose synthesis are catalytically the same. However, those for starch synthesis are located in the chloroplasts and those for sucrose synthesis in the cytosol. Thus isozymes may exist with different regulatory properties.

## B.5. $C_4$ pathway

### (a) CARBOXYLATION PHASE

*pyruvate, orthophosphate dikinase*, E.C. No. 2.7.9.1., ATP:pyruvate, orthophosphate phosphotransferase.
[ATP + pyruvate + orthophosphate = AMP + phosphoenolpyruvate + pyrophosphate]

*inorganic pyrophosphatase*, E.C. No. 3.6.1.1., pyrophosphate phosphohydrolase.
[pyrophosphate + $H_2O$ = 2 orthophosphate]

*adenylate kinase*, E.C. No. 2.7.4.3., (myokinase), ATP:AMP phosphotransferase
[ATP + AMP = ADP + ADP]

*phosphoenolpyruvate carboxylase*, E.C. No. 4.1.1.31., orthophosphate: oxaloacetate carboxy-lyase (phosphorylating).
[phosphoenolpyruvate + $CO_2$ + $H_2O$ = oxaloacetate + orthophosphate]

*malate dehydrogenase* ($NADP^+$), E.C. No. 1.1.1.82., L-malate:$NADP^+$ oxidoreductase
[oxaloacetate + NADPH + $H^+$ = L-malate + $NADP^+$]

*alanine aminotransferase*, E.C. No. 2.6.1.2., (glutamic-pyruvic transminase) L-alanine:2-oxoglutarate aminotransferase.
[L-alanine + 2-oxoglutarate = pyruvate + L-glutamate]

*aspartate aminotransferase*, E.C. No. 2.6.1.1., (glutamic-oxaloacetic transaminase), L-aspartate:2-oxoglutarate aminotransferase.
[oxaloacetate + L-glutamate = L-aspartate + 2-oxoglutarate]

(b) DECARBOXYLATION PHASE (NADP-MALIC ENZYME)

*malate dehydrogenase (decarboxylating) (NADP$^+$)*, E.C. No. 1.1.1.40., ('malic' enzyme), L-malate:NADP$^+$ oxidoreductase (oxaloacetate-decarboxylating)
[L-malate + NADP$^+$ = pyruvate + $CO_2$ + NADPH + $H^+$]

(c) DECARBOXYLATING PHASE (NAD-MALIC ENZYME)

*aspartate aminotransferase*, E.C. No., 2.6.1.1., (glutamic-oxaloacetic transaminase), L-aspartate:2-oxoglutarate aminotransferase.
[L-aspartate + 2-oxoglutarate = oxaloacetate + L-glutamate]

*malate dehydrogenase*, E.C. No. 1.1.1.37., (L-malate-NAD transhydrogenase) L-malate:NAD$^+$ oxidoreductase.
[oxaloacetate + NADH + $H^+$ = L-malate + NAD$^+$]

*malate dehydrogenase (decarboxylating)*, E.C. No. 1.1.1.39., ('malic' enzyme) L-malate:NAD$^+$ oxidoreductase (decarboxylating).
[L-malate + NAD$^+$ = pyruvate + $CO_2$ + NADH + $H^+$]

*alanine aminotransferase*, E.C. No. 2.6.1.2., (glutamic-pyruvic transaminase), L-alanine:2-oxoglutarate aminotransferase.
[pyruvate + L-glutamate = L-alanine + 2-oxoglutarate]

(d) DECARBOXYLATING PHASE (PEP-CARBOXYKINASE)

*aspartate aminotransferase*, E.C. No. 2.6.1.1., (glutamic-oxaloacetic transaminase), L-aspartate:2-oxoglutarate aminotransferase.
[L-aspartate + 2-oxoglutarate = oxaloacetate + L-glutamate]

*phosphoenolpyruvate carboxykinase* (*ATP*), E.C. No. 4.1.1.49., (phosphopyruvate carboxylase ATP), ATP:oxaloacetate carboxy-lyase (transphosphorylating)
[ATP + oxaloacetate = ADP + phosphoenolpyruvate + $CO_2$]

## B.6. Crassulacean acid metabolism

### (a) ACIDIFICATION

*α-amylase*, E.C. No. 3.2.1.1., (diastase), 1,4-α-D-glucan glucanohydrolase [Endohydrolysis of 1,4-α-glycosidic linkages in polysaccharides containing three or more 1,4-α-linked D-glucose units].

*β-amylase*, E.C. No. 3.2.1.2., (diastase), 1,4-α-glucan maltohydrolase. [Hydrolysis of 1,4-α-glycosidic linkages in polysaccharides so as to remove successive maltose units from the non-reducing ends of the chains.]

α-D-*glycosidase*, E.C. No. 3.2.1.20, (maltase), α-D-glucoside glucohydrolyase. [Hydrolysis of terminal non reducing 1,4-linked α-D-glucose residues with release of α-glucose.]

*hexokinase*, E.C. No. 2.7.1.1., (ATP→glucose transphosphatase), ATP D-hexose 6-phosphotransferase.
[ATP + D-hexose = ADP + D-hexose 6-phosphate]

*phosphorylase*, E.C. No. 2.4.1.1., (P-enzyme (only for the plant enzyme)), 1,4-α-D-glucan:orthophosphate α-D-glucosyltransferase.
[$(1,4\text{-}\alpha\text{-D-glucosyl})_n$ + orthophosphate = $(1,4\text{-}\alpha\text{-D-glucosyl})_{n-1}$ + α-D-glucose 1-phosphate]

*phosphoglucomutase*, E.C. No. 2.7.5.1., (glucose-phosphomutase), α-D-glucose-1,6-bisphosphate:α-D-glucose 1-phosphate phosphotransferase.
[α-D-glucose 1,6-bisphosphate + α-D-glucose 1-phosphate = α-D-glucose 6-phosphate + α-D-glucose 1,6-bisphosphate]

*glucosephosphate isomerase*, E.C. No. 5.3.1.9., (phosphohexose isomerase), D-glucose 6-phosphate ketolisomerase. [D-glucose 6-phosphate = D-fructose 6-phosphate]

6-*phosphofructokinase*, E.C. No. 2.7.1.11., (phosphohexokinase), ATP: D-fructose-6-phosphate 1-phosphotransferase.
[ATP + D-fructose 6-phosphate = ADP + D-fructose 1,6-bisphosphate]

*fructose-bisphosphate aldolase*, E.C. No. 4.1.2.13., (zymohexase), D-fructose 1,6-bisphosphate D-glyceraldehyde-3-phosphate-lyase.
[D-fructose 1,6-bisphosphate = dihydroxyacetone phosphate + D-glyceraldehyde 3-phosphate]

*triosephosphate isomerase*, E.C. No. 5.3.1.1., (phosphotriose isomerase), D-glyceraldehyde 3-phosphate ketol-isomerase.
[D-glyceraldehyde 3-phosphate = dihydroxyacetone phosphate]

*glyceraldehyde-phosphate dehydrogenase*, E.C. No. 1.2.1.12., (triose phosphate dehydrogenase), D-glyceraldehyde 3-phosphate:$NAD^+$ oxidoreductase (phosphorylating).
[D-glyceraldehyde 3-phosphate + orthophosphate + $NAD^+$ = 3-phospho-D-glyceroyl phosphate + NADH + $H^+$]

*phosphoglycerate kinase*, E.C. No. 2.7.2.3., (1,3-diphosphoglycerate →ADP transphosphatase), ATP:3-phospho-D-glycerate 1-phosphotransferase.
[3-phospho-D-glyceroyl phosphate + ADP = ATP + 3 phospho-D-glycerate]

*phosphoglyceromutase*, E.C. No. 2.7.5.3., (glycerate (3→2)-phosphomutase), 2,3-bisphospho-D-glycerate:2-phospho-D-glycerate phosphotransferase.
[2,3-bisphospho-D-glycerate + 3-phospho-D-glycerate = 2-phospho-D-glycerate + 2,3-bisphospho-D-glycerate]

*enolase*, E.C. No. 4.2.1.11., (phosphopyruvate hydratase), 2-phospho-D-glycerate hydro-lyase.
[2-phospho-D-glycerate = phosphoenolpyruvate + $H_2O$]

*phosphoenolpyruvate carboxylase*, E.C. No. 4.1.1.31., orthophosphate: oxaloacetate carboxy-lyase (phosphorylating).
[phosphoenolpyruvate + $CO_2$ + $H_2O$ = oxaloacetate + orthophosphate]

*malate dehydrogenase*, E.C. No. 1.1.1.37., (L-malate-NAD transhydrogenase) L-malate: $NAD^+$ oxidoreductase.
[oxaloacetate + NADH + $H^+$ = L − malate + $NAD^+$]

(b) DECARBOXYLATION PHASE (NADP-MALIC ENZYME)

*malate dehydrogenase (decarboxylating) (NADP$^+$)*, E.C. No. 1.1.1.40., ('malic' enzyme), L-malate:NADP$^+$ oxidoreductase (oxaloacetate-decarboxylating).
[L-malate + NADP$^+$ = pyruvate + $CO_2$ + NADPH + $H^+$]

(c) DECARBOXYLATION PHASE (NAD-MALIC ENZYME)

*malate dehydrogenase (decarboxylating)*, E.C. No. 1.1.1.39., ('malic' enzyme) L-malate:NAD$^+$ oxidoreductase (decarboxylating).
[L-malate + NAD$^+$ = pyruvate + $CO_2$ + NADH + $H^+$]

(d) DECARBOXYLATING PHASE (PEP-CARBOXYKINASE)

*aspartate aminotransferase*, E.C. No. 2.6.1.1., (glutamic-oxaloacetic transaminase), L-aspartate:2-oxoglutarate aminotransferase.
[L-aspartate + 2-oxoglutarate = oxaloacetate + L-glutamate]

*phosphoenolpyruvate carboxykinase (ATP)*, E.C. No. 4.1.1.49., (phosphopyruvate carboxylase ATP), ATP:oxaloacetate carboxy-lyase (transphosphorylating)
[ATP + oxaloacetate = ADP + phosphoenolpyruvate + $CO_2$]

(e) CONVERSION OF 3C PRODUCT OF DECARBOXYLATION TO STARCH

*pyruvate, orthophosphate dikinase*, E.C. No. 2.7.9.1., ATP:pyruvate, orthophosphate phosphotransferase.
[ATP + pyruvate + orthophosphate = AMP + phosphoenolpyruvate + pyrophosphate] Comment: Proposed enzyme for converting pyruvate → PEP following decarboxylation through NADP- or NAD-malic enzyme.

*enolase*, E.C. No. 4.2.1.11., (phosphopyruvate hydratase), 2-phospho-D-glycerate hydro-lyase.
[phosphoenolpyruvate + $H_2O$ = 2-phospho-D-glycerate]

*phosphoglyceromutase*, E.C. No. 2.7.5.3., (glycerate (3→2)-phosphomutase) 2,3-bisphospho-D-glycerate:2-phospho-D-glycerate phosphotransferase.
[2,3-bisphospho-D-glycerate + 2-phospho-D-glycerate = 3-phospho-D-glycerate + 2,3-bisphospho-D-glycerate]

*phosphoglycerate kinase*, E.C. No. 2.7.2.3., (1,3-diphosphoglycerate $\rightarrow$ADP transphosphatase), ATP: 3-phospho-D-glycerate 1-phosphotransferase
[ATP + 3-phospho-D-glycerate = ADP + 3-phospho-D-glyceroyl phosphate]

*glyceraldehyde-phosphate dehydrogenase*, E.C. No. 1.2.1.12., (triose phosphate dehydrogenase), D-glyceraldehyde-3-phosphate:$NAD^+$ oxidoreductase (phosphorylating).
[D-glyceraldehyde 3-phosphate + orthophosphate + $NAD^+$ = 3-phospho-D-glyceroyl phosphate + NADH + $H^+$]

See Section B.4 for further metabolism of triose phosphate to starch.

## B.7. Nitrate metabolism

*nitrate reductase (NADH)*, E.C. No. 1.6.6.1., (assimilatory nitrate reductase), NADH:nitrate oxidoreductase.
[NADH + nitrate + $H^+$ = $NAD^+$ + nitrite + $H_2O$]

*nitrite reductase (NAD(P)H)*, E.C. No. 1.6.6.4., (NADH$\rightarrow$nitrite transhydrogenase), NAD(P)H:nitrite oxidoreductase.
[3 NAD(P)H + 3$H^+$ + nitrite = 3 $NAD(P)^+$ + $NH_4OH$ + $H_2O$]

*glutamine synthetase*, E.C. No. 6.3.1.2., L-glutamate: ammonia ligase (ADP-forming).
[ATP + L-glutamate + $NH_3$ = ADP + orthophosphate + L-glutamine]

*glutamate synthase* (ferredoxin), E.C. No. 1.4.7.1., L-glutamate: ferredoxin oxidoreductase (transaminating)
[L-glutamine + 2-oxoglutarate + 2-reduced ferredoxin = 2 glutamate + oxidized ferredoxin]

*glutamate dehydrogenase ($NAD(P)^+$)*, E.C. No. 1.4.1.3., (glutamic dehydrogenase) L-glutamate:$NAD(P)^+$ oxidoreductase (deaminating)
[L-glutamate + $H_2O$ + $NAD(P)^+$ = 2-oxoglutarate + $NH_3$ + NAD(P)H + $H^+$]

# Index